D1602272

ASTRONOMY AND ASTROPHYSICS LIBRARY

Series Editors: I. Appenzeller, Heidelberg, Germany
G. Börner, Garching, Germany
M.A. Dopita, Canberra, ACT, Australia
M. Harwit, Washington, DC, USA
R. Kippenhahn, Göttingen, Germany
J. Lequeux, Paris, France
A. Maeder, Sauverny, Switzerland
P.A. Strittmatter, Tucson, AZ, USA
V. Trimble, College Park, MD, and Irvine, CA, USA

Springer
New York
Berlin
Heidelberg
Hong Kong
London
Milan
Paris
Tokyo

ASTRONOMY AND ASTROPHYSICS LIBRARY

Series Editors: I. Appenzeller • G. Börner • M.A. Dopita • M. Harwit •
R. Kippenhahn • J. Lequeux • A. Maeder •
P.A. Strittmatter • V. Trimble

The Design and Construction of Large Optical Telescopes
By P.Y. Bely

Stellar Physics (2 volumes)
Volume 1: Fundamental Concepts and Stellar Equilibrium
Volume 2: Stellar Evolution and Stability
By G.S. Bisnovatyi-Kogan

Theory of Orbits (2 volumes)
Volume 1: Integrable Systems and Non-perturbative Methods
Volume 2: Perturbative and Geometrical Methods
By D. Boccaletti and G. Pucacco

Galaxies and Cosmology
By F. Combes, P. Boissé, A. Mazure, and A. Blanchard

The Solar System 2nd Edition
By T. Encrenaz and J.-P. Bibring

Physics of Planetary Rings
Celestial Mechanics of Continuous Media
By A.M. Fridman and N.N. Gorkavyi

Compact Stars
Nuclear Physics, Particle Physics, and General Relativity
2nd Edition
By N.K. Glendenning

The Physics and Dynamics of Planetary Nebulae
By G.A. Gurzadyan

Asymptotic Giant Branch Stars
Editors: Harm J. Habing and Hans Olofsson

Stellar Interiors
Physical Principles, Structure, and Evolution
By C.J. Hansen and S.D. Kawaler

Astrophysical Concepts 3rd Edition
By M. Harwit

Physics and Chemistry of Comets
Editor: W.F. Huebner

Stellar Structure and Evolution
By R. Kippenhahn and A. Weigert

Continued after Index

Harm J. Habing
Hans Olofsson
Editors

Asymptotic Giant Branch Stars

With 176 Illustrations

Springer

Harm J. Habing, Leiden Observatory, University of Leiden, 2300 RA Leiden, The Netherlands. Email: habing@strw.leidenuniv.nl

Hans Olofsson, Stockholm Observatory, Stockholm University, Alba Nova, SE-10691 Stockholm, Sweden. Email: hans@astro.su.se

Series Editors

Immo Appenzeller
Landessternwarte, Königstuhl
D-69117 Heidelberg
Germany

Gerhard Börner
MPI für Physik und Astrophysik
Institut für Astrophysik
Karl-Schwarzschild-Str. 1
D-85748 Garching
Germany

Michael A. Dopita
The Australian National University
Institute of Advanced Studies
Research School of Astronomy
 and Astrophysics
Cotter Road, Weston Creek
Mount Stromlo Observatory
Canberra, ACT 2611
Australia

Martin Harwit
Department of Astronomy
Space Sciences Building
Cornell University
Ithaca, NY 14853-6801
USA

Rudolf Kippenhahn
Rautenbreite 2
D-37077 Göttingen
Germany

James Lequeux
Observatoire de Paris
61, Avenue de l'Observatoire
75014 Paris
France

André Maeder
Observatoire de Genève
CH-1290 Sauverny
Switzerland

Peter A. Strittmatter
Steward Observatory
The University of Arizona
Tuscon, AZ 85721
USA

Virginia Trimble
Astronomy Program
University of Maryland
College Park, MD 20742
and Department of Physics
University of California
Irvine, CA 92717
USA

Library of Congress Cataloging-in-Publication Data
Asymptotic giant branch stars / editors, Harm J. Habing, Hans Olofsson.
   p. cm. — (Astronomy and astrophysics library)
   Includes bibliographical references and index.
   ISBN 0-387-00880-2 (acid-free paper)
   1. Asymptotic giant branch stars. I. Habing, H.J. (Harm Jan), 1937– II. Olofsson, Hans. III. Series.
QB843.A89A78 2003
523.8′8—dc21   2003045456

ISBN 0-387-00880-2   Printed on acid-free paper.

The figure on the cover has been provided by the 2MASS Project (UMASS/IPAC).

© 2004 Springer-Verlag New York, Inc.
All rights reserved. This work may not be translated or copied in whole or in part without the written permission of the publisher (Springer-Verlag New York, Inc., 175 Fifth Avenue, New York, NY 10010, USA), except for brief excerpts in connection with reviews or scholarly analysis. Use in connection with any form of information storage and retrieval, electronic adaptation, computer software, or by similar or dissimilar methodology now known or hereafter developed is forbidden.
The use in this publication of trade names, trademarks, service marks, and similar terms, even if they are not identified as such, is not to be taken as an expression of opinion as to whether or not they are subject to proprietary rights.

Printed in the United States of America.

9 8 7 6 5 4 3 2 1   SPIN 10922026

www.springer-ny.com

Springer-Verlag   New York Berlin Heidelberg
A member of BertelsmannSpringer Science+Business Media GmbH

Preface

This book deals with stars during a short episode before they undergo a major, and fatal, transition. Soon the star will stop releasing nuclear energy, it will become a planetary nebula for a brief but poetic moment, and then it will turn into a white dwarf and slowly fade out of sight. Just before this dramatic change begins the star has reached the highest luminosity and the largest diameter in its existence, and while it is a star detectable in galaxies beyond the Local Group, its structure contains already the inconspicuous white dwarf it will become. It is called an "asymptotic giant branch star" or "AGB star".

Over the last 30 odd years AGB stars have become a topic of their own although individual members of this class had already been studied for centuries without realizing what they were. In the early evolution, so called "E-AGB"-phase, the stars are a bit bluer than, but otherwise very similar to, what are now called red giant branch stars (RGB stars). It is only in the second half of their anyhow brief existence that AGB stars differ fundamentally from RGB stars. In this phase, the "thermally pulsing phase", the TP-AGB star is more luminous than any RGB star, it very probably pulsates with a period of one year or more, and it ejects matter at a high rate − in fact it is this loss of matter which will cause the end of nuclear burning and the rapid transition to a planetary nebula. In this monograph it is the TP-AGB star which will receive most of the attention.

There are several reasons why TP-AGB stars have become such well-studied objects. The two most outstanding reasons are, first, the new possibilities of making extensive infrared observations of continuously increasing quality and, second, the availability of fast computers which allows more and more detailed, realistic simulations of these very complex objects. Because AGB stars are best detected in the infrared, scientific progress has been strongly stimulated by the measurements of space observatories such as IRAS, ISO, MSX, the infrared camera on the HST, and by new telescopes on the ground at high and dry sites. The large DENIS and 2MASS databases of $IJHK$-photometry, obtained by dedicated sky surveys from the ground, provide basic information which often turns out to be essential. Radio observations at (sub)-mm wavelengths give important additional, and sometimes decisive, information, and finally, the data obtained in recent (optical) monitoring surveys such as EROS, MACHO, and OGLE shine a totally new light on the properties of pulsating TP-AGB stars.

AGB stars are useful to other fields of astronomy because as evolved objects they inform us about the star formation history in other galaxies. They play also an important role in the production and ejection of new materials: new elements, such as the s-process nuclei, and dust particles – those solid state things which have evolved from an interstellar nuisance to a necessary agent in the formation of new stars: without dust the Earth and its mankind would probably never have existed.

Our knowledge of AGB stars has now sufficiently matured that a dedicated monograph is possible and then also necessary. This is the main reason for writing this book. Because AGB stars have many connections to other fields in astronomy, we must limit the subject. Therefore, we do not include planetary nebulae and white dwarfs for which excellent monographs already exist. Equally so, there are several good and up-to-date books on stellar evolution in which the red giant and horizontal branches are dealt with. We have also excluded red supergiants, such as NML Cygni and VY CMa, which in several aspects are similar to AGB stars; we leave them to comprehensive books on stars of high mass.

We hope that this book will be useful for graduate students who work in the field, for teachers who want to incorporate these stars in their courses, and for all aficionades of these stars and their astrophysics. Each chapter is written by (a) separate author(s), and it has been the intention that they should be readable as stand-alone entities, while still containing ample cross-references to other chapters. It has been impossible to reference all the interesting work on AGB stars and their environment, and the reference list at the end of the book should only serve as an introduction to the subject, and does not necessarily contain the original nor the most extensive work on a subject. In addition, the reference philosophy may vary somewhat between the chapters.

Leiden and Stockholm,   *Harm J. Habing*
June 2003   *Hans Olofsson*

Contents

1 AGB Stars: History, Structure, and Characteristics
Harm J. Habing, Hans Olofsson 1
1.1 Bits of History 1
1.2 The Structure of AGB-Stars 12
1.3 Observational Characteristics of AGB Stars 15
1.4 Distinctive Properties of AGB stars 17
References 19

2 Evolution, Nucleosynthesis, and Pulsation of AGB Stars
John C. Lattanzio, Peter R. Wood 23
2.1 Basic Observational Properties 23
2.2 Pre-AGB Evolution 27
2.3 Stellar Evolution on the AGB 31
2.4 Evolution Beyond the AGB: Planetary Nebula Nuclei and White Dwarfs 41
2.5 Nucleosynthesis in AGB Stars 43
2.6 Variability 80
2.7 Conclusions and Outlook 99
References 100

3 Synthetic AGB Evolution
Martin A.T. Groenewegen, Paula Marigo 105
3.1 The Role of Synthetic Evolutionary Models 105
3.2 A Historical Overview 106
3.3 The Main Ingredients of a Synthetic AGB Model 108
3.4 Stellar Yields 125
3.5 From One Star to Population Synthesis 129
3.6 Observational Constraints 132
3.7 Conclusions and Outlook 144
References 145

4 Atmospheres of AGB Stars
Bengt Gustafsson, Susanne Höfner 149
4.1 Introduction 149
4.2 Observations 152

4.3	Physics and Characteristic Conditions	159
4.4	Microscopic State of Matter	165
4.5	The Radiation Field	182
4.6	The Modelling of AGB Star Atmospheres	189
4.7	Dynamics	201
4.8	Mass Loss	219
4.9	Abundances and Other Fundamental Parameters	230
4.10	Conclusions and Recommendations	236
References		239

5 Molecule and Dust Grain Formation
Tom J. Millar 247

5.1	Introduction	247
5.2	Chemical Processes for Molecule and Dust Formation	250
5.3	Detailed Models: Carbon-Rich CSEs	270
5.4	Detailed Models: Oxygen-Rich CSEs	279
5.5	Complications	282
5.6	Conclusions and Outlook	286
References		287

6 Dynamics and Instabilities in Dusty Winds
Yvonne Simis, Peter Woitke 291

6.1	Introduction	291
6.2	Modeling the Dust-Driven AGB Wind	293
6.3	Instabilities and Structure in the Outflow	310
6.4	Conclusions and Outlook	319
References		321

7 Circumstellar Envelopes
Hans Olofsson 325

7.1	Mass Loss and Circumstellar Envelopes	325
7.2	A "Standard" Gaseous AGB CSE	326
7.3	Circumstellar Line Observations	348
7.4	A "Standard" Dusty AGB CSE	361
7.5	Circumstellar Dust Observations	367
7.6	Geometry and Kinematics of AGB CSEs	371
7.7	Mass-Loss-Rate Estimators	383
7.8	Mass-Loss Rate	391
7.9	Conclusions and Outlook	400
References		402

8 AGB Stars as Tracers of Stellar Populations
Harm J. Habing, Patricia A. Whitelock 411

8.1	Introduction	411
8.2	The Milky Way Galaxy (MWG)	415
8.3	The Magellanic Clouds	431

8.4	M 31, M 32, NGC 205, and M 33	439
8.5	The Remaining Galaxies of the Local Group (LG)	441
8.6	AGB Stars in Galaxies Outside of the Local Group	452
8.7	Conclusions and Outlook	453
References		454

9 AGB Stars in Binaries and Their Progeny
Alain Jorissen 461

9.1	The Binary–AGB Connection	461
9.2	AGB Stars in Binary Systems	462
9.3	Impact of Binarity on Intrinsic Properties of AGB Stars	475
9.4	The Progeny of AGB Stars in Binary Systems	482
9.5	Conclusions and Outlook	512
References		514

10 Post-AGB Stars
Christoffel Waelkens, Rens B.F.M. Waters 519

10.1	Introduction	519
10.2	Observational Definition of a Post-AGB Star	520
10.3	Observed Properties of PAGB Stars: The Central Star	522
10.4	Observed Properties of Post-AGB Stars: The Circumstellar Envelope	530
10.5	Binary Post-AGB Stars	539
10.6	Confrontation of Observations with Theory	543
10.7	Conclusions and Outlook	547
References		547

Index 555

1 AGB Stars: History, Structure, and Characteristics

Harm J. Habing[1] and Hans Olofsson[2]

[1] Leiden Observatory
[2] Stockholm Observatory

1.1 Bits of History

"Asymptotic giant branch stars" developed into a broad, coherent field of research in the last quarter of the preceding century, but its roots go much deeper into the past and involve several serendipitous discoveries. We will discuss briefly when and how its most significant concepts came to the surface. To that end we have arranged the historical developments into five broad areas. We do not intend to write a full history of astronomy, and we will skip over the fundamental developments in physics. When no specific reference is given, we have used North's History of Astronomy and Cosmology [56] and A Source Book in Astronomy and Astrophysics, edited by Lang and Gingerich [48]. We also limit most of our review to the developments before 1984 for two reasons: (i) In a long review published in 1983 Iben and Renzini [39] showed that the theoretical side of the AGB evolution was broadly and consistently understood; (ii) in 1984 the results of the IRAS full-sky survey at 12, 25, 60, and 100 μm became public; the catalogue contained thousands of AGB stars, and a new era of observation began.

1.1.1 First Discoveries Mostly by Serendipity

Red giants, dwarfs, and the HR-diagram: Different stars have different colors, and this must have been known from the earliest times of star-gazing. For example, the contrast between the red star Betelgeuse (α Ori, M2, $V=0.5$) and (within 10° on the sky) the blue star Alnitak (ζ Ori, O9.5, $V=2$) must have been noticed by the Arabic astronomers who gave these stars their names. The significance of the colors became clear in the last decades of the nineteenth century when the connection was made between thermodynamics and electromagnetic radiation and ultimately resulted in 1900 in the "Planck equation." Since then, we have understood that red stars are relatively cool.

The terms "giant" and "dwarf" date back to the epoch between 1905 and 1915. At Harvard Observatory a gigantic program was carried out to classify the spectra, obtained with objective prisms, of a few hundred thousand stars: the production of the Henry Draper (HD) catalogue. Maury, of the Harvard team, noticed a subclass of the A, F, G, and K stars with narrower and deeper absorption lines than most; she labeled the stars with the letter c. In 1905 Hertzsprung [34] showed that when you reduce all stars of the same spectral class to the same apparent magnitude, the c stars have much smaller proper motions. He drew the conclusion that the c stars are more distant and thus have higher luminosity. The terms "dwarf" and "giant" were introduced some time between this publication and the publication of the first "Hertzsprung-Russell" diagrams (1911: Hertzsprung [35]; 1914: Russell [65]). Russell's diagram is dominated by the main sequence, but there are some red giants and one white dwarf, the companion of o^1 Eri. Russell was aware of the peculiar location of the latter star, but in his presentation of the diagram he dismisses it and suggests, somewhat vaguely, that the assigned spectral type might be wrong (which it wasn't!). The discovery of dwarfs and giants led immediately to the question, what causes a star to be either a dwarf or a giant? The answer had to wait for fifty years.

Long-period variables, LPVs: Many AGB stars are long-period variables. The first "LPV" was discovered in two steps. In 1596 the East Frisian astronomer/clergyman Fabricius studied Jupiter and incidentally noticed the disappearance of a 3rd magnitude star in the constellation of the Whale (*Cetus*). He assumed that this was another "Stella Nova," similar to the one of 1572 described by Tycho Brahe ("Tycho's supernova"). Fabricius communicated his findings to Brahe, and the news was published later by Kepler, who was the inheritor of Brahe's scientific notes. In 1638 the Dutch astronomer Holwarda observed that the star had reappeared. By watching it regularly he noticed its weakening, disappearance, and reappearance, and he established that the dis- and reappearances recurred with a period of about one year; because of this behavior the star later received the name "Stella Mira", the "miraculous" or "amazing" star. In the following centuries many other LPVs were detected. In spite of the increase in observations a good physical understanding did not result. Even the classification into Miras, Semiregulars, and Irregulars is based only on the appearance of the light curves without an understanding of the physical processes at work. Hard information came in 1963 when Feast [20] showed that the statistical properties of the radial velocities of Mira variables are different for stars of different period, and he drew the conclusion that Miras with shorter periods are older and of less mass than the Miras with longer periods. A few years later, in 1981, came the discovery by Glass and Lloyd Evans [28] of a linear relation between the K-magnitude and the logarithm of the period in Mira variables.

Carbon stars: Soon after Kirchoff and Bunsen had published the correct interpretation of spectral lines in 1860, several programs were started to classify the stars according to their spectra. At first, the spectra were observed by eye, it was still too early to consider taking photographs. Around 1868 Father Secchi at the Vatican Observatory [70] classified some four thousand stars.

Among these he recognized a small group of very red stars with a spectrum totally different from all other red stars; he even noticed the similarity of their spectra to that of the light in carbon arcs. This discovery posed the question, why there are two very different spectral classes of red stars. A partial answer came in 1934 from Russell [64], who showed that the high binding energy of CO leads to M-type spectra when the abundance of oxygen is higher than that of carbon, and to carbon-type spectra otherwise. The answer to the question *why* some stars have more C-nuclei than O-nuclei had to wait until the 1970s.

1.1.2 Stellar Models: Main-Sequence, RGB, and AGB

What is a star, or phrasing it in today's terminology, what are suitable physical stellar models? The question is very old indeed, and finding the answer is one of the major accomplishments of astrophysical research in the previous century. We will briefly discuss the search for models of main-sequence stars and then those of red giants. In a separate item below, the history of the detection of white dwarfs, and what they are, will be told briefly.

Models of main-sequence stars: In the search for models of these stars one name stands out: Eddington. He was vital in showing that the Sun and other stars are spheres of gas in which gravity and pressure are everywhere in equilibrium. His book "The Internal Constitution of the Stars", published in 1926 [18], is a classic, and all present-day textbooks owe a lot to it. A major lack of essential knowledge concerned the production of the radiation energy. Eddington explicitly states that the conversion of hydrogen into helium is probably the source of the energy, but he does not know how the mechanism works and he cannot calculate how the energy conversion depends on the temperature and pressure of the gas at the stellar center. Furthermore, it was assumed that the atomic composition of the Sun was the same as that of the Earth, which is true, if you ignore hydrogen and helium. It is often stated that in her thesis in 1925 Payne-Gaposchkin had found the large relative abundance of H and He, but Gingerich [26] has pointed out that she rejected this result as probably wrong, and this she did under the influence of Russell. In 1928 Unsöld [78] drew the correct conclusion, but assumed that his result was true only for the solar atmosphere; the lightest elements might float on top of the more heavy material. In 1929 Russell [63] drew the correct conclusion. The next problem, finding how the fusion proceeds in detail, took one more decade. The last stone put into the foundation of the building was the demonstration in 1939 by Bethe (1939, [7]) that in main-sequence stars of about one solar mass the pp-reaction provides the energy, while in more massive stars the CNO-cycle dominates. This last cycle had been identified independently by von Weizsäcker (1938, [81]).

Though satisfying models had been found for main-sequence stars, the structure of the red giants remained uncertain, and it was still unknown where and how they fitted into the evolution of stars; as an example, in his paper just mentioned von Weizsäcker speculated that red giants are young and contracting, and this was clearly a defendable scientific position at that time.

Models of red giants: After World War II things had changed. The epicenter of scientific research had moved over the Atlantic from Europe to the US. It was clear that the search for the structure of the red giants was high on the list of important topics. Progress in the field profited from at least two developments: (*i*) Deeper and more accurate photometry of individual stars in globular clusters came with the completion of the 200-inch telescope on Mount Palomar, and with the introduction of "photoelectric" detection techniques, and (*ii*) the more and more powerful "electronic devices" made numerical calculations of stellar models less time-consuming, and the analytic approximations so bitterly needed for the early stellar models gradually lost their usefulness.

Inspection of the new, deeper, and more accurate color magnitude diagrams of globular clusters led to an unexpected discovery that at first was described as a "bifurcation of the red giant branch" (Arp, Baum, and Sandage, 1953, [2]). The choice of words shows that the observers had found the feature by looking downward from the brightest magnitudes. Only later was the phenomenon considered upward from the faintest stars and the same phenomenon became known as the "asymptotic giant branch."

Lang and Gingerich [48] point out the large consequences that "just" a detail may have on subsequent research: The fact that in the color-magnitude diagram of some stellar clusters the giant branch appeared to connect to the main sequence inspired Sandage and Schwarzschild (1952, [66]) to analyze red giant models starting with a main-sequence star so old that it had a helium core. A set of hybrid models, representing a sequence in time and each consisting of a helium core and a hydrogen envelope,[3] showed that the core contracted, the envelope expanded to a large size and the luminosity increased: the first correct explanation of the red giant branch (RGB). This was a breakthrough, and further developments were fast. A paper by Hoyle and Schwarzschild (1955, [36]) shows that these authors understood the evolution of stars through the red giant branch to a maximum luminosity and then down to the horizontal branch. This paper is also a beautiful illustration of how numerical models helped the authors in finding their way through the

---

[3] Here is a case of confusing terminology. Practically all theoreticians of stellar evolution call the matter between core and atmosphere the "envelope." "Envelope," however, has also been used in the relevant literature to indicate the ejected gas that surrounds the star. The first usage has first-born rights, although it remains less appropriate; "mantle" would have been a better word. We will follow the custom, and will assume that when the word "envelope" is used in this book, the context defines which of the two meanings is meant. In ambivalent cases we will use "stellar envelope" and "circumstellar envelope."

first part of the post-main-sequence evolution. Understanding the next part, evolution after the horizontal branch, came a decade later and after a major surprise.

Thermal pulses: After a horizontal branch star has burned the helium in its core, the nuclear fusion continues in a shell around the carbon/oxygen core. When the He-burning shell approaches the hydrogen envelope, an instability develops: a "He-shell flash" (later named a "thermal pulse"), serendipitously discovered by Schwarzschild and Härm (1965, [68]) and independently by Weigert (1966, [82]). At first, Schwarzschild and Härm suspected numerical errors, but once the reality of thermal pulses was established, they became a distinctive mark of the late evolution of AGB stars. In a subsequent paper Schwarschild and Härm [69] showed that after a number of such pulses a convection zone extending from the He-burning shell makes contact with the convective H-envelope, and the mixing that follows generates conditions favorable for nucleosynthesis. In addition, the new elements may be brought to the stellar surface by convection. In a short paper following [69], Sanders [67] argued that the s-process, see below, might work well under the conditions of the He-burning shell.

Nucleosynthesis: Studies of the details of nuclear reactions became possible when quantum physics reached maturity in the late 1920s. Soon these studies were extended by asking the question whether the abundances of all elements could be explained by nucleosynthesis under astrophysical conditions. The most important paper, studying a wide variety of nuclear reactions under astrophysical conditions, appeared in 1957 (Burbidge, Burbidge, Fowler and Hoyle = B^2FH, [11]). One chain of nuclear reactions, labeled by B^2FH the "s-process," is a sequence of absorption of free neutrons followed by β-decay; it appears to be efficient in thermally-pulsing (TP) stars with the short but repeating high-energy conditions in the He-burning layer.

Much numerical modeling was needed, however, before self-consistent models could be constructed. During a thermal pulse the critical parts of the stellar structure vary rapidly; a large number of models must be calculated to cover the pulse, and the steady increase in computing power was more than welcome. In 1975 Iben [38] and Sugimoto and Nomoto [73] presented models that showed not only dredge-up of s-process elements but also of carbon produced in the He-burning layer by the 3α-process. After several thermal pulses, enough new carbon may have been injected into the convective envelope to turn the star into a carbon star, and explain the existence of Father Secchi's peculiar red stars. The success of this explanation was, however, not complete, because first, the theoretical carbon stars found in this way were too luminous and too massive when compared with the observations of AGB carbon stars in the Large Magellanic Cloud, and second, many carbon stars have a luminosity too low to be on the AGB. To what extent the first problem has been resolved may be read in Chapters 2 and 3; the solution to the second problem may be found in Chapter 9. Luckily, one unexpected piece of evidence proves that newly created elements are dredged up: In 1952 Merrill [52] discovered lines of technetium in Mira variables; later, lines were seen in more Miras [49, 50]. The s-process element, Tc ($Z = 43$), has only radioactive isotopes. ^{99}Tc, which has a half-life time of about 200,000 yr and which is produced by the s-process, has been seen, and this proves that these stars have had a recent dredge-up and may thus be considered as true TP-AGB stars. A small segment of a spectrum containing Tc-lines is shown in Figure 9.10.

1.1.3 New Results from New Observing Techniques

Infrared astronomy, circumstellar envelopes: In the 1960s infrared astronomy took off, finally, and after a long incubation time. Most developments took place in the US, a fact that is partially, but not fully, explained by the military interest in infrared technology and the consequential "classification" of IR detector materials by the (US) Department of Defense. Slowly, but persistently, the wavelengths of the observations increased until around 1970, observations were made in all telluric "windows" from 1 to 20 µm.

A landmark is the 2.2 µm IRC-survey by Neugebauer and Leighton (1969, [54]), in which the detection of 5000 sources north of $\delta = -33°$ was reported. Subsequent studies showed that a majority of the sources were red giants with huge circumstellar dust envelopes. In some objects the surrounding dust has such a large optical depth that most of the stellar energy is emitted at wavelengths longer than 2.2 µm; examples are IRC+10011 (= WX Psc, $I-K = 8$, $K-N = 5$) and IRC+10216 (= CW Leo, $I-K = 8$, $K-N = 9$). Circumstellar envelopes (CSEs) became important objects for study. In particular, IRC+10216, which is probably the nearest carbon star, has become the archetype of AGB stars with high mass-loss rates and thick CSEs. Another result was the detection of red giant stars near the Galactic center and associated with the puzzling radio source Sgr A [6].

In 1976 appeared another catalogue of almost 2400 point sources, detected at four wavelengths between 4 and 28 µm during space flights primarily for US military objectives (Price and Walker [59]). It contained some new, very interesting, almost unique objects, which since then have been studied with great interest at many wavelengths. Several were later found to be post-AGB stars; prime examples are the "Egg-nebula" (AFGL 2688) and the "Red Rectangle" (AFGL 915). The next space telescope was IRAS (1983), which made a survey of the full sky (except for a few percent) at 12, 25, 60, and 100 µm. A few hundred thousand new sources were detected. The IRAS catalogue, and the products derived later, opened a large observational basis of AGB and post-AGB stars.

The photospheric radii of AGB stars, estimated from the stellar luminosity and color, are at least 100 $R_\odot$ or about 1 AU in diameter. At a distance of

100 pc the star has a diameter of 10 marcsec. To resolve the star this resolution should be achieved at $\lambda \approx 1\,\mu$m. The CSEs are bigger; the diameter of the warmest part is a few times the photospheric diameter, but the wavelengths where they are best seen is longer (5 to 10 µm). Special techniques must be applied to resolve the objects; the development of three such techniques started in the 1970s. Lunar occultation was used for the first time in 1972 (Toombs et al. [76]; $\lambda = 2.2$ and 4.8 µm). The resolution achieved by this method is limited by how accurate and how often one can measure the increase (decrease) of the flux during egress (ingress). The technique is still used successfully, but it has the disadvantage that the resolution is almost one-dimensional, and only a very small fraction of the sky is accessible. Speckle interferometry, invented in 1970 by Labeyrie [45], offers resolutions below the diffraction limit. The technique was applied, with some success [9, 46], by using the then largest optical telescope, the 200 inch, but the diameter of even this telescope limited application to only the nearest and largest AGB stars: the ratio λ/D corresponds to 40 marcsec ($\lambda = 1\,\mu$m, $D = 5$ m). Interferometry with movable mirrors and baselines of tens of meters appears to offer the best prospects; infrared astronomy does not differ from radio astronomy in this sense. Some interferometric measurements at wavelengths between 8 and 11 µm were published by McCarthy and Low (1975, [51]), but it seems that the technique of heterodyne interferometry developed by Townes and associates has become the most successful in the long run. Their first astronomical publication is from 1977 [74]. In the 1980s several other groups started to develop comparable interferometers, and major breakthroughs came in the 1990s, well beyond the time limit we set for this summary.

Polarimetry is another technique that gives unique information. If there is no magnetic field involved, polarized optical and near-IR radiation is produced by scattering. In spite of some promising early work (Dyck et al. 1971, [17]), relatively few studies have been devoted to AGB stars. Polarimetry is of particular importance for the study of post-AGB stars because their structure is definitely nonspherical, and light from the star often reaches us indirectly after scattering by dust particles.

Radio astronomy: The first detection of a circumstellar line at radio wavelengths was that of an OH line at 1612 MHz (18 cm) toward the supergiant NML Cyg by Wilson and Barrett (1968, [84]). The line was so strong that it had to be due to a natural maser. In several cases the line flux showed large-amplitude periodic variations with periods of 300 to over 1000 days (Harvey et al. 1974, [32]), and this confirmed the association of these masers with LPVs, even when the star proper could not be seen. Strong maser emissions from H_2O and SiO molecules were also found among AGB stars of spectral class M. The line profile turned out to be another characteristic feature of the masers, in particular, that of the OH maser (see Figure 7.3) which is explained by a model in which matter flows through a thin shell at a distance of a few hundred AU from the star (1977, Reid et al. [60], Olnon [58]). This model received a beautiful confirmation when interferometric techniques showed that the brightness of the 1612 MHz maser was distributed in a circular ring (1981, MERLIN array: Booth et al. [10], VLA: Baud [4]). The high intensity of the maser lines made it profitable to search for AGB stars at radio wavelengths, and several surveys were carried out during the 1970s; the detection of maser stars near the Galactic center showed that a large part of our Milky Way is accessible.

Further progress came from observations at millimeter wavelengths, made possible after considerable technical developments. Not only had new cryogenically cooled receivers to be developed, new antennae had to be built with higher surface precision and preferentially at a high and dry site. In the 1970s and 1980s a large number of telescopes operating at millimeter and submillimeter wavelengths were constructed, and many different molecules were detected in a short time. The first detection at millimeter wavelengths of a thermally excited circumstellar line was the CO ($v=0$, $J=1\rightarrow 0$) line at 2.6 mm toward IRC + 10216 (Solomon et al. 1971, [72]). The angular resolution of millimeter observations was greatly improved when in the early 1980s interferometers operating at millimeter wavelengths came on-line. The analysis of maps of line brightness distributions, at different radial velocities, has become a very important tool for the study of the mass loss process of AGB stars.

1.1.4 Mass Loss

Mass loss and circumstellar envelopes: In all calculations of stellar evolution before, say, 1980 the assumption was made that the mass of the star did not change; in the model calculations there was neither mass accretion nor mass loss. The first observational evidence of mass loss is from 1951, when Biermann [8] calculated that solar photons do not transfer enough momentum to line up the tail of a comet. He then proposed that the alignment was due to matter particles flowing out of the Sun, a proposition confirmed very early in the space age. The solar mass-loss rate is, however, very low ($3\times 10^{-14}\,M_\odot\,\mathrm{yr}^{-1}$), and at this rate the Sun will lose less than 4×10^{-4} of its mass during its total lifetime.

What about red giants? Circumstellar lines had been seen around M-giants, and this suggested mass loss, but it could not be proven that the matter would escape into space. This changed through the observations in 1956 of the binary system α Her by Deutsch [16]. He noticed that the circumstellar absorption lines in the MII component of the binary were seen also in the spectrum of the companion GII star. This implied that circumstellar gas extended to $2 \times 10^5\,R_\odot$ from the M-giant. With an outflow velocity of $10\,\mathrm{km\,s}^{-1}$ this gas escapes the system. Deutsch estimated a mass-loss rate of $3\times 10^{-8}\,M_\odot\,\mathrm{yr}^{-1}$. Collecting data on similar systems Reimers (1975, [61]) derived an empirical relation of the form $\dot{M}\propto LR/M$, where L, R, and M

are the stellar luminosity, radius, and mass, respectively. This relation was later named for Reimers.

Direct evidence of much stronger winds in late-type giants came to light at the end of the 1960s, but before that, several indirect arguments had been brought forward to prove that stars must lose a large fraction of their mass during their final evolution. One such argument is due to Auer and Woolf (1965, [3]): The Hyades cluster contains at least a dozen white dwarfs, and each will have a mass below the Chandrasekhar limit of $1.4\,M_\odot$. However, the Hyades is a young group, and stars with a mass of $2\,M_\odot$ are still on the main sequence. Thus, the stars that developed into the white dwarfs each lost at least $0.6\,M_\odot$. But why hadn't they been seen? What do stars look like when they lose matter at a high rate?

In the late 1960s the first infrared measurements made at $\lambda > 5\,\mu m$ showed that red giants with spectral type M5 or later frequently radiate much more strongly in the infrared than was expected, an excess attributed to circumstellar dust. In 1968/1969 Gillett et al. [24] and Woolf and Ney [85] together identified for the first time an emission band around $10\,\mu m$, in the spectra of M-type giants, as due to small particles of silicate. In 1971 Gehrz and Woolf [23] derived the mass of circumstellar dust around a number of M-type stars. Using the expansion velocity taken from Deutsch's result they found mass loss rates between 10^{-7} and $10^{-6}\,M_\odot\,yr^{-1}$ (and around $10^{-5}\,M_\odot\,yr^{-1}$ for a few supergiants). Carbon stars are also surrounded by dust particles, but not with the same infrared spectrum (Woolf and Ney 1969, [85]). A welldefined emission band at $11.5\,\mu m$ is seen in the spectra of carbon stars in a 1972 paper by Hackwell [31]; later, the band became attributed to SiC grains.

Apparently, there are two kinds of dust grains, one produced by M-type stars, the other by carbon stars. This dichotomy was explained by Gilman (1969, [25]) as the consequence of the high binding energy of CO, an explanation rather similar to that by Russell in 1934 for the appearance of the two types of red giants; Gilman seems to have been unaware of Russell's paper. These discoveries also made clear that late-type giants are the birthplaces of the interstellar dust grains, a question open since around 1930, when interstellar extinction was discovered. The discoveries gave birth to "astromineralogy," a subject that now also covers the formation of PAHs (an IRAS result) and of crystalline particles (an ISO result). It is interesting that a stimulating role in all these discoveries was played by a highly speculative article by Hoyle and Wickramasinghe (1962, [37]) on the possible formation of graphite dust grains around carbon stars.

In 1966 Wickramasinghe et al. [83] showed convincingly that light pressure on grains pushed not only graphite grains out of the stellar gravitational field, but also the surrounding gas, because of momentum exchange between dust and gas. In 1976 Goldreich and Scoville [29] developed a simple spherical model for a CSE with matter flowing through it. They also explained the circumstellar OH maser as due to infrared pumping of OH molecules in a thin layer formed by the dissociation of H_2O molecules into OH and H by the interstellar radiation field (and subsequently of the OH molecule into O and H atoms).

The existence of dense CSEs around AGB stars required mass-loss rates much higher than predicted by Reimers' equation, originally proposed for the low mass-loss rates of red giant stars. In 1981 Renzini and Voli [62] came back to the idea that RGB stars experience a very mild wind, but that in the last TP-AGB phase a very dense wind ("superwind") occurred. The issue of the existence of superwinds was closed with the identification of mass-loss rates up to 10^{-4} $M_\odot$ yr^{-1} using the OH 1612 MHz line emission (Engels et al. 1983, [19]; Baud and Habing 1983, [5]).

At this point in time the grand overview had been reached: The life of red giants is cut off by nonexplosive, gradual, but heavy mass loss on the AGB. This makes the star undetectable at optical wavelengths and sometimes even in the near-IR. AGB stars were understood *in essentia*, and the often quoted review paper by Iben and Renzini (1983, [39]) marked the completion of the theoretical groundwork. In 1984 the observational results of the IRAS survey opened a wide field for observers; they could begin to do their job.

1.1.5 Descendants: Post-AGB Stars, Planetary Nebulae, and White Dwarfs

When a star leaves the AGB it will disguise itself successively as a post-AGB star, a planetary nebula, and a white dwarf. The first examples of each group were detected, serendipitously, in another order.

Planetary nebulae: In the eighteenth century, comet hunting became a popular sport among amateur astronomers. While searching the skies the hunters were plagued by interlopers: cosmic nebulae. Messier's list of 109 "noncomet nuisances" from around 1770 already contains four planetary nebulae (PNe), the "Ring Nebula" (M57), the "Dumbbell" (M27), the "Owl" (M97), and M76 (also called "the Little Dumbbell"). The misleading term "planetary" nebula seems to have been invented by Herschel because their greenish color reminded him of the planet Uranus, whose discovery in 1781 had made him famous. It is one more example of the frequent difficulty in astronomy of having to give a name to something newly discovered when the nature of the object is still a mystery. The PNe became a fascinating subject for study with photographic spectroscopy early in the twentieth century. The nebulae consist of an expanding mass of gas, and their very short life (between 10^4 and 10^5 yr) emphasized the need for an explanation of their origin. In 1956 Shklovskii [71] proposed that PNe are blownup examples of red giants; the faint white star at the center of the nebula has been the core of the red giant and in the future it will become a white dwarf. Shklovskii's suggestion appealed to many, but it posed a hard problem: The total mass of the ionized nebular gas is much smaller than the matter in the stellar envelope of a red giant. A major step forward was taken in 1978 by Kwok et al. [44], who proposed that PNe consist of hot gas ejected by the central star inside a hollow sphere made in an earlier stage by the slow AGB wind.

White dwarfs: The first white dwarf may have been seen around 1840 by the Earl of Rosse with his 36inch reflector, then one of the largest telescopes in the world. He studied M57 (the "Ring Nebula") and discovered a faint star at the center of the nebula, but he was not aware (how could he have been?) of the fact that this (white dwarf) star was fundamentally different from all other stars known. In 1862 Clark saw for the first time Sirius B, the companion to Sirius whose existence had been predicted by Bessel when he analyzed (visual!) measurements of Sirius's variable proper motion. Though Sirius is the brightest star in the sky ($V = -1.4$), its companion turned out to be a faint star of the eighth magnitude, roughly 10^{-4} times as luminous as Sirius A, and yet Sirius B had half the mass of Sirius A. The first photographic spectrum of Sirius B without an annoying amount of stray light from Sirius A, taken by Adams in 1915 [1], showed beyond doubt that the two stars have about the same spectral class in spite of the enormous luminosity difference. This implied that the radiating surface of Sirius B had to be about 10^4 times smaller than that of Sirius A, that the volume of Sirius B was 10^6 times smaller and the density the same factor larger: one solar mass of matter in a sphere the size of the Earth. Such a density could not be explained within the existing physical laws, and the astrophysicists of the time were aware of the problem. In 1926 Fowler [22] explained that this density might be reached when one applied the new quantum statistics developed shortly before by Fermi. In 1931 Chandrasekhar [12] proved that white dwarfs must have a mass below $1.4\,M_\odot$.

Post-AGB: In the 1970s a few very unusual and intriguing objects were detected in the catalogue of the AFGL-survey, such as the "Egg nebula" [55] and the "Red Rectangle" [14]. Analysis of data obtained at all available wavelengths led to a model in which two blobs of scattered visual light are separated by a dark band behind which the strong IR source is located: a PN *in statu nascendi*. Ten years later, many more objects of this kind "came to light" through the IRAS survey. In spite of all efforts the census of post-AGB stars is still small, and may be incomplete in the sense that some types of post-AGB stars may exist but have not yet been recognized, see Chapter 10. Two examples of post-AGB objects not first detected in the infrared are "Minkowski's Footprint" (M1-92) [33] and the "Calabash" or "Rotten Egg nebula" (OH231.8+4.2) [53, 77].

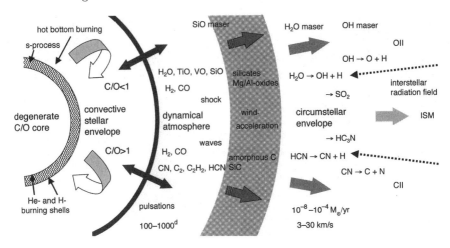

Fig. 1.1. Overview of an AGB star (not drawn to scale). The division into four major parts, and some important physical/chemical processes are indicated (adapted by J. Hron from an original idea by Th. Le Bertre)

1.2 The Structure of AGB-Stars

AGB stars have a complex structure; the reader who is not yet convinced should first browse through this book. They are stars of "moderate" mass that have come to the end of their evolution, where "moderate" means $0.8\,M_\odot < M < 8\,M_\odot$ – the precise boundaries are model-dependent (and stars down to about $0.5\,M_\odot$ may go through the AGB evolution, but they have not yet had time to do this in our universe).

An ABG star can, in principle, be described as consisting of four parts: (*i*) the small, very hot, and dense core, (*ii*) the large, hot, and less dense stellar envelope, (*iii*) the tenuous, warm atmosphere, and (*iv*) the very large, very diluted, and cool circumstellar envelope; Figure 1.1. Thus, an AGB star covers more than 10 orders of magnitude in size scale, 30 orders of magnitude in density scale, and 7 orders of magnitude in temperature scale. A full understanding of its evolution can be obtained only by studying the intricate interplay between various physical/chemical processes taking place in its different parts. We briefly describe here some of the more important aspects of the structure of an AGB star, and we introduce the different chapters of the book.

To begin to understand the basic properties of AGB stars, one must pay attention to their previous evolution. Let us pick up the evolution after the star has exhausted hydrogen at its core, has been a red giant, and has begun a new, but much shorter, life as an object that burns helium into carbon (3α-process) and some of the carbon into oxygen in its center. In the HR-diagram the star is located at the horizontal branch (if its initial mass was low and with a low metallicity), or it is a "clump" star.

The core and stellar envelope: When the helium in the center has been converted into carbon (and oxygen) the He-burning zone moves outward, and the core of carbon/oxygen contracts until it has reached the density of white dwarfs. It has now a size of about 10^8 cm and a temperature of about 10^8 K. The contraction of the core and the expansion of the envelope lead to a rapid increase in luminosity: the beginning of the E-AGB phase, when He-burning in a shell produces most of the energy. The size of the stellar envelope reaches about 10^{13} cm, and the surface temperature is about 3000 K. At some point during the early evolution the envelope becomes pulsationally unstable with a characteristic time scale on the order of hundred days.

Double-shell burning and thermal pulses: When a star has gained the luminosity of approximately that of the tip of the RGB, i.e., around $M_{bol} = -4$ or $3000\,L_\odot$, a new process starts: The star is able to burn both helium and hydrogen in shells, and on a regular basis the thin He-layer (around the core) burns rapidly into carbon, most of which is then added to the mass of the core. During this short intermission, which is called a "thermal pulse" or "He-shell flash," the star goes through a moderate luminosity modulation. Between thermal pulses the star again burns hydrogen. The phase of intermittent helium and hydrogen burning is called the "TP-AGB phase."

The dredge-up: A second important effect during a thermal pulse is "dredge-up." During a thermal pulse, convection reaches the layers where nuclear processing takes place and brings material to the stellar surface, which becomes enriched by the products of nuclear-burning, especially by "new" carbon. In some stars this continues sufficiently long for the abundance of carbon to exceed that of oxygen, and the atmosphere will change from "oxygen-rich" into "carbon-rich."[4] This is actually the second important dredge-up process for a low-mass star and the third for an intermediate-mass star. In other cases this conversion will not happen: In stars of sufficient mass the carbon will burn into nitrogen before the stellar surface is reached ("hot bottom burning"), and in low-mass stars all the matter between core and stellar surface will be ejected, and any further evolution will be terminated before the conversion can take place.

The structure of an AGB star, its nucleosynthesis, and its pulsational properties are covered in Chapter 2. A synthesis of this is presented in Chapter 3.

---

[4] The term "oxygen-rich" is somewhat inappropriate: In interstellar space there is more oxygen than carbon, and by default all newborn stars are oxygen-rich. Being carbon-rich is special. We will follow, however, the now well established practice of using the terms "oxygen-rich" and "carbon-rich" stars.

The atmosphere: The outer part of the stellar envelope is cool enough that formation of molecules will take place, and it is their presence that determines the spectral appearance of the stars (which can differ greatly depending on whether C/O < 1 or > 1), at least into the near-IR, and to some extent also the structure of their atmospheres. The pulsation of the star deposits mechanical energy into outer parts that are only weakly bound by gravity. As a consequence, the atmosphere becomes very extended compared to a hydrostatic one. Shocks will develop, and at high enough altitudes, grain condensation takes place in the postshock gas. It appears that gas particles reach escape velocity, under a broad range of conditions, due to the mechanical energy input augmented by radiation pressure on grains (and perhaps molecules). This is our current understanding of the mass-loss mechanism.

The physics and chemistry of the atmosphere are covered in Chapters 4 and 5, including the formation of the microscopic grain particles.

The circumstellar envelope: An AGB star will eventually lose mass at a considerable rate in the form of a slow ($\approx 15\,\mathrm{km\,s^{-1}}$) wind, and this forms a CSE of escaping gas and dust particles. The gas is to a large extent molecular with H_2 dominating. The molecules are located in envelopes of different sizes (the by far largest ones being those of CO and H_2), or shells of different radii and widths, depending on their resistance to UV photons and their formation routes. The dust grains survive much further from the star than do the molecules. The molecular setup, as well as the grain properties, are a strong function of the C/O ratio. At some point the CSE merges into the surrounding ISM. In this region the temperature is below 10 K, and the particle density less than $10\,\mathrm{cm^{-3}}$. This may occur at a distance of 10^{18} cm or more from the star, and it defines the outer limit of the AGB star.

The dynamics of a CSE, the circumstellar formation of molecules, and the observational information on CSEs are covered in Chapters 6, 5, and 7, respectively.

Termination of the AGB: Ultimately, the rate of ejection of matter is higher than the growth rate of the core, and from now on, the mass-loss process determines the evolution. Eventually, all the matter around the core is dispersed into space, and this marks the endpoint of AGB evolution, and the beginning of the post-AGB evolution. In practice, this starts when the stellar envelope mass becomes less than about $0.01\,M_\odot$. Since the mass loss at this phase may reach rates of $10^{-4}\,M_\odot\,\mathrm{yr^{-1}}$, the star must drastically "pull the brake." For a very short time (on the order of a few thousand years) the star becomes a "post-AGB star," and then, for a time of a few tens of thousands of years, it turns into a planetary nebula. The nebula is expanding rapidly; it fades, and only the stellar core remains as a white dwarf. However, during its short existence, a planetary nebula can present one of the most spectacular astronomical phenomena because of its complicated geometrical structures and because each structure has its own colors.

The early post-AGB evolution is presented in Chapter 10. Planetary nebulae are not covered and we refer here to the recent book by Kwok [43] and the proceedings of the series of IAU symposia on PNe, [30] being the latest.

AGB stars in binary systems: It is a complicating fact that AGB stars often have companions. If these are close enough, the evolution may follow completely different routes. On the positive side is that interesting and exotic objects may result from such a phenomenon. A more worrying fact is that interaction may form stars that disguise themselves as AGB stars whereas they are not.

The effects of binarity are covered in Chapter 9.

AGB stars as probes: The AGB stars are luminous and are therefore easily detected throughout our Galaxy, as well as in galaxies within about 10 Mpc. They are also old and define relaxed systems. These properties make them good probes of galactic structure and dynamics, as well as of star formation histories.

These aspects are covered in Chapter 8.

1.3 Observational Characteristics of AGB Stars

We summarize briefly the main observational characteristics of AGB stars in this section. They are described more thoroughly in the various chapters of this book.

1.3.1 Spectra

The most important spectral classes of AGB stars are M, S, and C. The M stars are characterized by TiO-bands, and in the late-M spectra also by the presence of VO. The S stars have characteristic ZrO bands. In C stars all molecular bands are from carbon compounds; metallic oxides are absent. In the case of the carbon stars, the alternative spectral classes N and R exist. This book deals with carbon stars of spectral class N. The R stars are clearly not on the AGB. For a detailed discussion of spectroscopic criteria we refer the reader to the book by Jaschek and Jaschek [41].

Spectra of M-giants between 380 and 900 nm have been recorded and discussed in quite some detail by [21]. For spectra from 500 nm to 2.40 μm [47] is a useful reference, and spectra between 2.36 and 4.1 μm obtained by ISO have been published by [79]. These three data sets have comparable wavelength resolutions (1100 to 1500). AGB stars with heavy mass loss are (almost) undetectable at optical wavelengths and have a maximum flux around 10 μm;

see [75] for detailed ISO spectra of O-rich objects between 2.4 and 197 μm. Illustrative spectra are shown in Chapters 2 and 4.

1.3.2 Variability

An important property of AGB stars is the long-period variation of their flux. The customary criteria for classifying the light curves of LPVs are those given in the *General Catalogue of Variable Stars* (GCVS) [42]. There are four classes of objects: *Mira-like* ("M"), regular variations with a large amplitude in the V-band ($\Delta V > 2.5$); *Semiregular of type a* ("SRa"), relatively regular with a smaller amplitude in the V-band ($\Delta V < 2.5$); *Semiregular of type b* ("SRb"), poor regularity with a small amplitude in the V-band ($\Delta V < 2.5$); *Irregular* ("L"), irregular variations of low amplitude.

The division between the various classes is based on the appearance of the light curves, often not well sampled, in the optical range, and it is far from sharp. There are astrophysical differences between the different classes, but there are many cases "in between." Over the last few years the surveys for microlensing events (EROS, MACHO, and OGLE) have produced thousands of well-sampled light curves with highly accurate photometry for thousands of red variables in the LMC; a very useful and unforeseen result. All kinds of variability are present, e.g., Cepheids, eclipsing binaries, and many LPVs. Equally unexpectedly, the light curves of the LPVs are often quite complex and contain multiple periods. These results seem to us to call for a new interpretation of the variability of AGB stars; see Chapter 2.

1.3.3 The Circumstellar Medium

The primary observational consequence of the mass-loss process is the formation of an expanding CSE. In a star with a low mass-loss rate the CSE manifests itself by a weak IR-excess above the expected stellar flux. At high mass-loss rates the star is totally obscured, and the object appears as a low-temperature, high-luminosity object, indeed, not too different from the appearance of a protostar. A more detailed investigation reveals broad features in the spectral energy distribution, SED, which can be attributed to solid particles, microscopic grains. These are specific to the chemistry, which is determined by the C/O ratio of the central star (except in rare cases); basically, we have silicate grains around M stars, and amorphous carbon grains around carbon stars. An even more detailed spectroscopic observation reveals the presence of spectral lines from various circumstellar atomic and molecular species. The molecular setup is also determined by the C/O ratio of the central star. The impressive chemical ability of carbon is apparent in the CSEs of carbon stars. In addition, the lines provide important kinematic information. The circumstellar medium has the advantage of carrying historical information on the evolution of the AGB star, and it plays an important role in the shaping of the descendants: the early post-AGB objects and the PNe.

1.4 Distinctive Properties of AGB stars

1.4.1 How Do We Recognize E- and TP-AGB Stars?

How do we recognize AGB stars? In particular, how do we distinguish AGB stars from RGB stars and supergiants? Even more problematic, how do we distinguish between E-AGB stars and TP-AGB stars? The first AGB stars were discovered in HR-diagrams of globular clusters: see Figure 2.1. While in this and in other clusters we can differentiate between AGB and RGB stars, the differences are slight, and it will be much more difficult to separate field AGB stars from field RGB stars in the same way, i.e., through absolute magnitudes and through colors.

The RGB has a maximum luminosity that is well explained theoretically. In addition, this luminosity is well determined observationally in the Magellanic Clouds (e.g., [13]): $M_{bol} = -3.9$, or $L = 2900 L_\odot$. Thus, stars more luminous than the tip of the RGB must be AGB stars (or supergiants, but these are rarer than the AGB stars). Below the tip, where photometry does not distinguish well enough between RGB and AGB, one must rely on other properties.

The first effect that comes to mind are the thermal pulses, because these are a unique property of AGB stars. To recognize a thermal pulse directly is very difficult, even in the unlikely event that one catches a star in the spike of high luminosity that occurs in the beginning of a pulse. It is more profitable to look for the effects of the dredge-up that follows the spike. The detection of s-type elements (e.g., Zr and V, both observed in some Mira variables) is one possibility; a surplus of carbon or nitrogen is another possibility (the latter points at hot bottom burning). If this surplus is mild, as it will be after the first thermal pulses, it may be difficult to detect. Carbon stars and S stars are probably TP-AGB stars, unless the elemental setup is the result of mass exchange in a binary. Another element that appears as a consequence of dredge-up is lithium. Last is the presence of Tc-lines, which implies a recent dredge-up because the half-life time of the relevant Tc-isotope is only about 2×10^5 yr.

Long-period pulsations is another frequently used criterion for recognizing TP-AGB stars. We lack good models for intrinsically pulsating stars with long periods, and what we know is predominantly based on empirical information. Observations of red giants in the Magellanic Clouds prove that Miras have on average $M_{bol} < -3.9$. The same is true for the Mira variables in globular clusters. Semiregulars appear to have luminosities similar to those of Miras: The recent discovery of several linear relations between the K-magnitude and $\log P$ suggests that these are also TP-AGB stars.

Mass loss is a third characteristic property. The stars that show a significant mass-loss rate, say above $10^{-7} M_\odot \text{ yr}^{-1}$, are AGB stars, supergiants such as NML Cyg and VY CMa, or luminous blue variables (LBV). The supergiants and the LBVs form such a small group that they are easily recognized

Table 1.1. Model predictions for the bolometric magnitudes in, and the duration of, various evolutionary phases according to Vassiliadis and Wood [80]

M_i	M_f	M_{bol}	t_{MS}	t_{EAGB}	$\frac{t_{EAGB}}{t_{MS}}$	t_{TPAGB}	$\frac{t_{TPAGB}}{t_{MS}}$
[M$_\odot$]	[M$_\odot$]		[Gyr]	[Myr]	[%]	[Myr]	[%]
			Z=0.016				
1.0	0.57	−4.0	11.3	12	0.16	0.50	0.004
1.5	0.60	−4.5	2.7	9.2	0.34	0.83	0.03
2.0	0.63	−4.9	1.2	7.9	0.66	1.20	0.10
2.5	0.67	−5.1	0.62	1.1	0.18	2.20	0.35
3.5	0.75	−5.7	0.23	2.8	1.2	0.43	0.19
5.0	0.89	−6.2	0.10	1.2	1.2	0.26	0.27
			Z=0.004				
1.0	0.59	−4.5	6.7	8.0	0.12	0.87	0.01
1.5	0.64	−4.9	2.1	6.3	0.30	0.97	0.05
2.0	0.67	−5.2	0.89	6.7	0.75	1.60	0.18
2.5	0.69	−5.5	0.46	5.2	1.1	1.30	0.27
3.5	0.85	−6.0	0.18	2.2	1.2	0.25	0.14
5.0	0.94	−6.5	0.08	0.6	7.3	0.31	0.39

and removed from any sample. It has often been assumed automatically that all AGB stars with a mass loss in excess of 10^{-7} M$_\odot$ yr^{-1} are in their TP phase. It is a fact that the large majority of such stars (not all!) are LPVs, so the conclusion has some support.

1.4.2 Bolometric Magnitudes and Time Scales

The evolution of AGB stars including large mass-loss rates has been calculated in several different studies. None of these may be called "ab initio:" not in the treatment of convection, nor in that of the pulsation and the mass loss. We use the calculations by Vassiliadis and Wood [80]. Their results may be compared to the more recent calculations of Girardi et al. [27], which are much more elaborate and cover a finer grid of free parameters. The two sets of models show modest, but significant, numerical differences; we consider these differences to be a measure of the minimum uncertainties of each model.

We consider now briefly the luminosities and the duration of the TP-AGB phase. In Table 1.1 we summarize numerical results for two different metallicities, $Z=0.016$ (representative for our Galaxy) and $Z=0.004$ (representative for the SMC). Metallicities in other galaxies will probably lie in between. Columns 1 and 2 contain the initial and the final mass of the star; column 3 M_{bol}(tAGB), the bolometric magnitude at the end of the TP-AGB phase, just before the star starts its subsequent transformation into a post-AGB star; column 4 gives the duration of the main sequence phase, $t_{\rm MS}$; column 5, the duration of the early AGB phase, $t_{\rm EAGB}$; column 7, the duration of the TP-AGB phase, $t_{\rm TPAGB}$. Columns 6 and 8 contain the ratios $t_{\rm EAGB}$ over $t_{\rm MS}$ and $t_{\rm TPAGB}$ over $t_{\rm MS}$, respectively, in percent.

The range of bolometric magnitudes in Table 1.1 has been measured for at least two sets of AGB stars with known distances: stars near the Galactic Center and stars in the Magellanic Clouds. Bolometric magnitudes of AGB stars near the former have been reported in several papers. In a recent analysis based on IRAS-detected AGB stars, Jackson et al. [40] obtained a narrow luminosity function with an average luminosity around $3000\,L_\odot$, which corresponds to the lowest value in Table 1.1, where we expect to find the most AGB stars. A luminosity distribution for RGB and AGB stars in the LMC and SMC is given in Figure 8.17, and it shows clearly the presence of AGB stars from the tip of the RGB to 2.5 magnitudes less, i.e., from $M_{bol} = -3.9$ to -6.4, i.e., the same range as given in Table 1.1.

A rather simple argument indicates that TP-AGB stars must be very rare. If one makes the naive assumption that the starforming rate in the Milky Way has been constant over the last 11 Gyr, then the ratio in the last column gives the ratio between the number of TP-AGB stars to that of the MS stars of the same mass (both numbers relating to the same volume of space). Olivier et al. [57] estimate the local column density of dust-enshrouded AGB stars to be $(15 \pm 4) \times 10^{-6}$ stars per pc^2. This may be compared with a local column density of 30 main-sequence stars per pc^2 and 3 white dwarfs [15]: AGB stars are rare!

References

1. Adams, W. S. *PASP*, 27, 236, 1915.
2. Arp, H. C., Baum, W. A., and Sandage, A. R. *AJ*, 58, 4, 1953.
3. Auer, L. H. and Woolf, N. J. *ApJ*, 142, 182, 1965.
4. Baud, B. *ApJ*, 250, L79, 1981.
5. Baud, B. and Habing, H. J. *A&A*, 127, 73, 1983.
6. Becklin, E. E. and Neugebauer, G. *ApJ*, 157, L31, 1969.
7. Bethe, H. A. *Physical Review*, 55, 434, 1939.
8. Biermann, L. *Zeitschrift für Astrophysik*, 29, 274, 1951.
9. Blazit, A., Bonneau, D., Koechlin, L., and Labeyrie, A. *ApJ*, 214, L79, 1977.
10. Booth, R. S., Norris, R. P., Porter, N. D., and Kus, A. J. *Nature*, 290, 382, 1981.
11. Burbidge, E. M., Burbidge, G. R., Fowler, W. A., and Hoyle, F. *Reviews of Modern Physics*, 29, 547, 1957.
12. Chandrasekhar, S. *ApJ*, 74, 81, 1931.
13. Cioni, M.-R. L., van der Marel, R. P., Loup, C., and Habing, H. J. *A&A*, 359, 601, 2000.
14. Cohen, M., Anderson, C. M., Cowley, A., et al. *ApJ*, 196, 179, 1975.
15. Cox, A. N., editor. *Allen's Astrophysical Quantities, 4th edition* . Springer Verlag: New York, 2000.

16. Deutsch, A. J. *ApJ*, 123, 210, 1956.
17. Dyck, H. M., Forrest, W. J., Gillett, F. C., et al. *ApJ*, 165, 57, 1971.
18. Eddington, A. *The Internal Constitution of the Stars*. Cambridge University Press: Cambridge, 1926.
19. Engels, D., Kreysa, E., Schultz, G. V., and Sherwood, W. A. *A&A*, 124, 123, 1983.
20. Feast, M. W. *MNRAS*, 125, 367, 1963.
21. Fluks, M. A., Plez, B., The, P. S., et al. *A&AS*, 105, 311, 1994.
22. Fowler, R. H. *MNRAS*, 87, 114, 1926.
23. Gehrz, R. D. and Woolf, N. J. *ApJ*, 165, 285, 1971.
24. Gillett, F. C., Low, F. J., and Stein, W. A. *ApJ*, 154, 677, 1968.
25. Gilman, R. C. *ApJ*, 155, L185, 1969.
26. Gingerich, O. *Ap&SS*, 267, 3, 1999.
27. Girardi, L., Bressan, A., Bertelli, G., and Chiosi, C. *A&AS*, 141, 371, 2000.
28. Glass, I. S. and Evans, T. L. *Nature*, 291, 303, 1981.
29. Goldreich, P. and Scoville, N. *ApJ*, 205, 144, 1976.
30. Habing, H. J. and Lamers, H. J. G. L. M., editors. *IAU Symp. 180: Planetary Nebulae*. Kluwer Academic Publishers: Dordrecht, 1997.
31. Hackwell, J. A. *A&A*, 21, 239, 1972.
32. Harvey, P. M., Bechis, K. P., Wilson, W. J., and Ball, J. A. *ApJS*, 27, 331, 1974.
33. Herbig, G. H. *ApJ*, 200, 1, 1975.
34. Hertzsprung, E. *Zeitschrift für Wissenschaftliche Photographie*, 3, 429, 1905.
35. Hertzsprung, E. *Publikationen des Astrophysikalischen Observatoriums zu Potsdam*, 22(63), 1911.
36. Hoyle, F. and Schwarzschild, M. *ApJS*, 2, 1, 1955.
37. Hoyle, F. and Wickramasinghe, N. C. *MNRAS*, 124, 417, 1962.
38. Iben, I. *ApJ*, 196, 525, 1975.
39. Iben, I. and Renzini, A. *ARA&A*, 21, 271, 1983.
40. Jackson, T., Ivezić, Ž., and Knapp, G. R. *MNRAS*, 337, 749, 2002.
41. Jaschek, C. and Jaschek, M. *The Classification of Stars*. Cambridge University Press: Cambridge, 1987.
42. Kholopov, P. N. *General Catalogue of Variable Stars, 5th edition*. Nakua Publishing House: Moscow, 1987.
43. Kwok, S. *Cosmic Butterflies: the Colorful Mysteries of Planetary Nebulae*. Cambridge University Press: Cambridge, 2001.
44. Kwok, S., Purton, C. R., and Fitzgerald, P. M. *ApJ*, 219, L125, 1978.
45. Labeyrie, A. *A&A*, 6, 85, 1970.
46. Labeyrie, A., Koechlin, L., Bonneau, D., Blazit, A., and Foy, R. *ApJ*, 218, L75, 1977.
47. Lançon, A. and Wood, P. R. *A&AS*, 146, 217, 2000.
48. Lang, K. R. and Gingerich, O. *A Source Book in Astronomy and Astrophysics, 1900-1975*. Harvard University Press: Cambridge, 1979.
49. Little, S. J., Little-Marenin, I. R., and Bauer, W. H. *AJ*, 94, 981, 1987.
50. Little-Marenin, I. R. and Little, S. J. *AJ*, 84, 1374, 1979.
51. McCarthy, D. W. and Low, F. J. *ApJ*, 202, L37, 1975.
52. Merrill, S. P. W. *ApJ*, 116, 21, 1952.
53. Morris, M. and Bowers, P. F. *AJ*, 85, 724, 1980.

54. Neugebauer, G. and Leighton, R. B. *Two-micron Sky Survey. A Preliminary Catalogue.* NASA SP, 1969.
55. Ney, E. P., Merrill, K. M., Becklin, E. E., Neugebauer, G., and Wynn-Williams, C. G. *ApJ*, 198, L129, 1975.
56. North, J. *The Fontana History of Astronomy and Cosmology.* Fontana Press: Hammersmith, London, 1994.
57. Olivier, E. A., Whitelock, P., and Marang, F. *MNRAS*, 326, 490, 2001.
58. Olnon, F. *Shells around Stars.* PhD thesis, Leiden University, 1977.
59. Price, S. D. and Walker, R. G. *The AFGL Four Color Infrared Sky Survey: Catalog of Observations at 4.2, 11.0, 19.8, and 27.4 Micrometer.* Air Force Geophysics Laboratory, Optical Physics Division, 1976.
60. Reid, M. J., Muhleman, D. O., Moran, J. M., Johnston, K. J., and Schwartz, P. R. *ApJ*, 214, 60, 1977.
61. Reimers, D. *Societé Royale des Sciences de Liège, Memoires*, 8, 369, 1975.
62. Renzini, A. and Voli, M. *A&A*, 94, 175, 1981.
63. Russell, H. N. *ApJ*, 70, 11, 1929.
64. Russell, H. N. *ApJ*, 79, 317, 1934.
65. Russell, H. *Popular Astronomy*, 22, 275, 1914.
66. Sandage, A. R. and Schwarzschild, M. *ApJ*, 116, 463, 1952.
67. Sanders, R. H. *ApJ*, 150, 971, 1967.
68. Schwarzschild, M. and Härm, R. *ApJ*, 142, 855, 1965.
69. Schwarzschild, M. and Härm, R. *ApJ*, 150, 961, 1967.
70. Secchi, A. *Mem. Soc. Ital. Scienze*, 2, 73, 1868.
71. Shklovskii, I. *Astron. Zh.*, 33, 315, 1956.
72. Solomon, P., Jefferts, K. B., Penzias, A. A., and Wilson, R. W. *ApJ*, 163, L53, 1971.
73. Sugimoto, D. and Nomoto, K. *PASJ*, 27, 197, 1975.
74. Sutton, E. C., Storey, J. W. V., Betz, A. L., Townes, C. H., and Spears, D. L. *ApJ*, 217, L97, 1977.
75. Sylvester, R. J., Kemper, F., Barlow, M. J., et al. *A&A*, 352, 587, 1999.
76. Toombs, R. I., Becklin, E. E., Frogel, J. A., et al. *ApJ*, 173, L71, 1972.
77. Turner, B. *Astrophysics Letters*, 8, 73, 1971.
78. Unsöld, A. *Zeitschrift für Physik*, 46, 765, 1928.
79. Vandenbussche, B., Beintema, D., de Graauw, T., et al. *A&A*, 390, 1033, 2002.
80. Vassiliadis, E. and Wood, P. R. *ApJ*, 413, 641, 1993.
81. von Weizsäcker, C. *Physikalische Zeitschrift*, 39, 633, 1938.
82. Weigert, A. *Zeitschrift für Astrophysik*, 64, 395, 1966.
83. Wickramasinghe, N. C., Donn, B. D., and Stecher, T. P. *ApJ*, 146, 590, 1966.
84. Wilson, W. J. and Barrett, A. H. *AJ*, 73, 209, 1968.
85. Woolf, N. J. and Ney, E. P. *ApJ*, 155, L181, 1969.

2 Evolution, Nucleosynthesis, and Pulsation of AGB Stars

John C. Lattanzio[1] and Peter R. Wood[2]

[1] School of Mathematical Sciences, Monash University
[2] Research School of Astronomy and Astrophysics, Australian National University

2.1 Basic Observational Properties

2.1.1 AGB Stars in the HR-Diagram

The Hertzsprung–Russell diagrams of low metallicity globular clusters provide excellent demonstrations of the asymptotic giant branch, or AGB. The HR-diagram of M 3, shown in Figure 2.1, is a good example. When HR-diagrams such as Figure 2.1 were first obtained in the 1950s, the asymptotic merging of the AGB and the red giant branch (RGB) was regarded as a "bifurcation" viewed looking down from the tip of the giant branch (e.g., Sandage [123]; Arp [7]). At this stage the direction of evolution was unknown. Probably the first authors to use the term "asymptotic" were Sandage and Walker [124]. This was just after Schwarzschild and Härm [129] had evolved a low mass star past the Horizontal Branch and *up* the AGB. Although the term asymptotic giant branch originated as a description of the sequence of stars in the HR diagrams of low-mass globular cluster stars, the term is now generally used to describe all stars with masses $\lesssim 8\,M_\odot$ that are on the second ascent into the red giant region of the HR-diagram.

2.1.2 The M, S, and C Stars

Luminous cool giants can be broadly divided into two distinct groups on the basis of low-resolution optical spectra: The M stars whose spectra are dominated by bands of the TiO molecule, and the C (carbon) stars whose spectra are dominated by bands of the C_2 and CN molecules (see Figure 2.2). It is now known that the reason for this division is that the surface matter in the M stars has a C/O number ratio less than unity, while the C stars have C/O > 1. Since CO is the most strongly bound molecule in the atmospheres of cool stars, if C/O < 1 there are surplus O atoms available to form oxygen-rich molecules such as TiO. On the other hand, if C/O > 1, all the O is bound up in CO, and surplus C atoms are available to form molecules such as CN and C_2; see Chapters 4 and 5. Since normal stars and the interstellar medium have C/O < 1, the existence of C stars indicates that the envelope material in these stars has had extra carbon added to it. This occurs during the *third dredge-up* which is explained in more detail in Section 2.3.3.

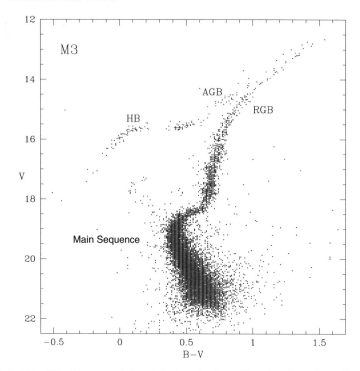

Fig. 2.1. The HR-diagram of the globular cluster M3, using data from Buonanno et al. [25]. The positions of the main-sequence, the red giant branch (RGB), the horizontal branch (HB), and the asymptotic giant branch (AGB) are indicated

There is another peculiarity of the spectra of some luminous cool giants that is readily seen in low-resolution spectra: Enhanced bands of molecules involving s-process elements. A prominent example is ZrO (see Figure 2.2). Since the s-process elements are produced by slow irradiation with neutrons in regions of the star where $T \approx 10^8$ K, the existence of s-process enhancements indicates that s-processed material from deep in the star has been brought to the stellar surface. The third dredge-up can readily explain the s-process enhanced stars.

Stars containing both TiO and ZrO molecular bands are known as MS stars. As a star evolves up the AGB, starting with an oxygen-rich atmosphere, the third dredge-up continues to enrich the envelope (througout this chapter, the term "envelope" refers to the "stellar envelope") in both C and s-process elements. Eventually, the bands of molecular oxides disappear when the increasing numbers of CO molecules soak up all the excess oxygen, leading to the pure S stars. Further enrichment causes molecular bands of carbon-rich molecules to appear, and the star becomes a C star. When the s-process elements are prominent in these stars, they are called SC stars.

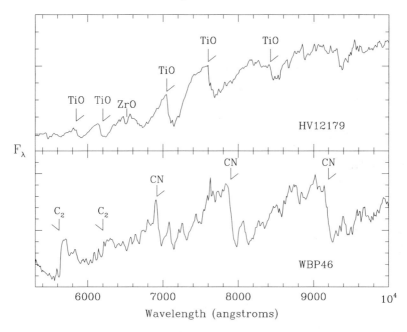

Fig. 2.2. Spectra of an MS star (**top** panel) and a C star (**bottom** panel) with some of the strongest molecular bands identified. The oxygen-rich M and MS stars have spectra dominated by TiO molecules (and VO for very cool stars), while the carbon-rich C stars have spectra dominated by C_2 and CN molecules. Both the stars shown are members of the Large Magellanic Cloud

A particularly interesting element found in the atmospheres of some AGB stars is technetium [101], which has no stable isotopes. The isotope ^{99}Tc is produced in the s-process and is presumed to be the one seen in AGB stars. Since ^{99}Tc has a half-life of $\approx 2 \times 10^5$ years, its presence means that it has been brought to the stellar surface of these stars in the last few times 10^5 years. This is direct observational evidence for the production of new elements inside stars.

2.1.3 Masses and Luminosities

As noted in Section 2.1.1, AGB stars occur in many globular clusters and they are identified by their position in the HR-diagram. These AGB stars must have had initial masses close to the turnoff masses in globular clusters, i.e., $\approx 0.85\,M_\odot$. Very few of the AGB stars in globular clusters are more luminous than the tip of the first giant branch, which occurs at $M_{\mathrm{bol}} \approx -3.6$ [153]. However, Mira variables occur in the more metal-rich globular clusters with [Fe/H] $\gtrsim -1$ [54], and they can be slightly more luminous than the RGB tip. The AGB stars in globular clusters are all observed to be oxygen-rich (K or M stars), so that carbon does not appear to have been dredged up into the envelopes during thermal pulses (globular clusters can contain carbon-rich CH stars, but the carbon overabundance in these stars is the product of mass transfer early in the life of the star from an intermediate-mass AGB companion (see [105] and Chapter 9).

In general, large-amplitude red variables are all AGB stars near the tip of the AGB, and their luminosities can be used to estimate AGB luminosities over a wide mass range. The Large Magellanic Cloud (LMC) provides a large sample of long-period variables (LPVs) at a known distance so that absolute luminosities can be derived. In addition, pulsation periods, combined with pulsation theory, can be used to derive the current stellar mass. All LPVs known in the LMC and SMC were plotted by [170] in the (M_{bol}, $\log P$) plane, and this showed that the LPVs occupy a luminosity range from that of globular cluster Miras ($M_{bol} \approx -3.8$) to the AGB limit luminosity $M_{bol} \approx -7.1$. This limit luminosity corresponds to the degenerate core mass approaching the Chandrasekhar limiting mass for white dwarfs of $\approx 1.4\,M_\odot$ (we now know that the luminosity limit is not a strict one for the more massive AGB stars [17], as discussed further in Sections 2.3.2 and 2.5.11). Using pulsation theory, approximate current masses could be assigned [170] and values were found from $<1\,M_\odot$ at $M_{bol} \approx -4$ to $\approx 8\,M_\odot$ at $M_{bol} \approx -7.1$. They also noted that many of their stars were MS stars or C stars, confirming that they belonged to the AGB. Thus, the LPVs show that AGB stars occupy a large mass and luminosity range.

Another way to observationally examine the masses and luminosities of AGB stars is to look at the most luminous stars in the rich clusters in the Magellanic Clouds by plotting the bolometric luminosities of luminous giant stars in Magellanic Cloud clusters against SWB type, the latter quantity being a measure of cluster age or, equivalently, initial red giant mass [55]. In Figure 2.3, we replot these data using cluster ages from the literature. Figure 2.3 clearly shows an increase in maximum AGB luminosity with increasing initial mass (decreasing age), although the maximum luminosity found in the clusters ($M_{bol} \approx -6$) is not as bright as found for the field LPVs [170]; this is probably because of the relatively small numbers of stars in the clusters compared to the field. The cluster data clearly show that C stars are particularly common in clusters of age 0.5–3 Gyr, corresponding to initial masses of ≈ 1.5–$3\,M_\odot$.

In the discussion above, it has been shown observationally that AGB stars range in luminosity from $M_{bol} \approx -3.6$ to -7.1 and that they have masses from ≈ 0.85 to $8\,M_\odot$. Some AGB stars are carbon stars, and some are S or MS stars. We now seek to explain these observational facts in terms of stellar evolution theory. Firstly, a brief review of the evolution of stars up to the beginning of the AGB is given.

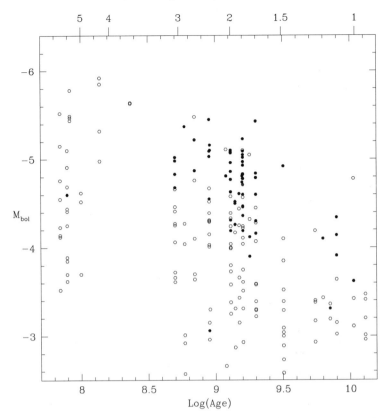

Fig. 2.3. AGB stars in Magellanic Cloud clusters. M_{bol} values are from [55], while cluster ages (yr) were obtained from the literature. Distance moduli of 18.50 and 18.90 have been assumed for the LMC and SMC, respectively. Carbon stars are shown as filled symbols. The top axis shows the mass (in $M_\odot$) at the beginning of the AGB for the corresponding cluster age

2.2 Pre-AGB Evolution

2.2.1 Evolution to the Early-AGB at $M \approx 1\,M_\odot$

The important features of the pre-AGB evolution of a $1\,M_\odot$ star are shown in Figure 2.4. On the zero-age main sequence, core H burning occurs radiatively. The central temperature and density grow in response to the increasing molecular weight caused by conversion of H to He until central H exhaustion (points 1–4 in Figure 2.4). Corresponding H abundance profiles are shown in inset (a). The star now leaves the main sequence and crosses the Hertzsprung Gap (points 5–7), while the central He core becomes electron degenerate, and nuclear burning of H is established in a shell surrounding this core. Inset (b) shows the advance of the H-burning shell during this evolution. Simultaneously, the star expands, and the outer layers become convective. As the star reaches the Hayashi limit (about point 7), convection extends deeply inward (in mass) from the surface, and the convective envelope penetrates into the region where partial H burning has occurred in earlier evolution, as shown in inset (c). This phenomenon is known as the "first dredge-up." As a result of first dredge-up, material enriched in ^{4}He and the products of CN cycling, primarily ^{14}N and ^{13}C, are mixed to the surface (point 8). Typical surface abundance changes are a decrease in the number ratio ^{12}C/^{13}C from ≈ 90 to ≈ 20 and a decrease in the ^{12}C/^{14}N ratio by a factor of ≈ 2.5 [35]. More details are given in Section 2.5.1.

As the star ascends the giant branch, the He core continues to contract and heat. Neutrino energy losses from the center cause the temperature maximum to move outward, as shown in inset (d). Eventually triple-alpha reactions are ignited at the point of maximum temperature. The resulting ignition is fierce, and is referred to as the "helium core flash" (point 9: [75, 141]) . After helium ignition, the star quickly moves to the Horizontal Branch (in low-metallicity stars with $M < 1\,M_\odot$) or the red giant clump (in solar metallicity stars). Here the star burns ^{4}He gently in a convective core and H in a shell (points 10–13). Helium burning increases the mass fraction of ^{12}C and ^{16}O [the latter through ^{12}C$(\alpha, \gamma)^{16}$O], and the outer regions of the convective core become stable according to the Schwarzschild convection criterion but remain unstable according to the Ledoux criterion, a situation referred to as "semiconvection" [32, 33]. The semiconvection causes the composition profile to adjust itself to produce convective neutrality, with the resulting profiles shown in inset (e).

Following He exhaustion (point 14) the star begins to ascend the giant branch for the second time. The core now becomes electron degenerate, and the star's energy output is provided by the He-burning shell (which lies immediately above the C-O core) and the H-burning shell. Above both is the deep convective envelope. This is the beginning of the Asymptotic Giant Branch phase.

2.2.2 Evolution to the Early-AGB at $M \approx 5\,M_\odot$

On the main sequence, the main difference between a $5\,M_\odot$ and a $1\,M_\odot$ star is the higher temperature in the core at $5\,M_\odot$. This means that the CNO cycle dominates H-burning, and the high temperature dependence of the CNO reactions causes a convective core to develop. As H is burned into ^{4}He, the opacity (due mainly to electron scattering, and hence proportional to the H content) decreases, and the extent of the convective core decreases with time. This corresponds to points 1–4 on the evolutionary track in Figure 2.5. Following core H exhaustion, there is a phase of shell burning as the star crosses the Hertzsprung Gap (points 5–7 and inset (b)), and then ascends the (first) giant branch. As with the $1\,M_\odot$ star, first dredge-up occurs and ^{13}C and ^{14}N are enhanced at the stellar surface (point 8 and inset (c)).

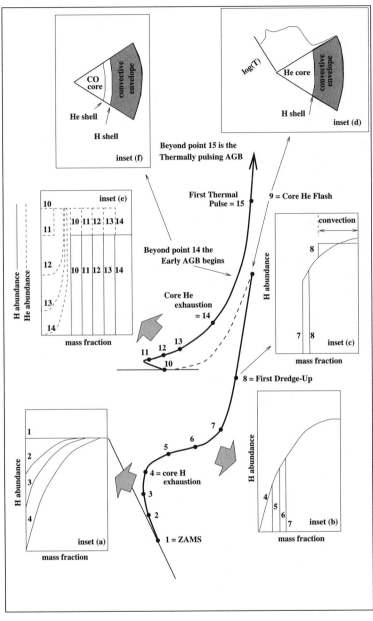

Fig. 2.4. Schematic evolution of a star of mass $M \approx 1\,M_\odot$

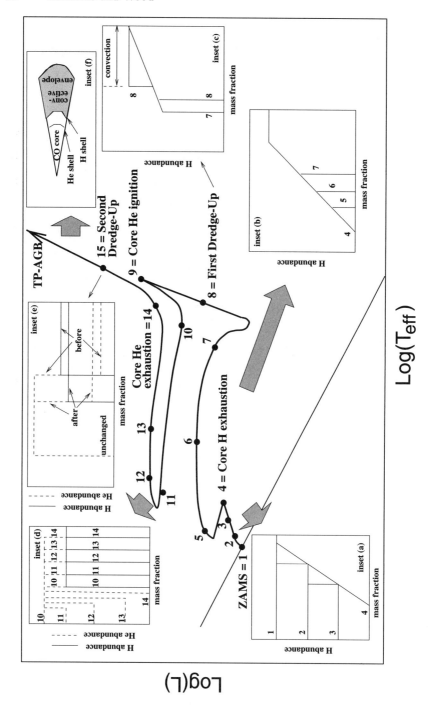

Fig. 2.5. Schematic evolution of a star of mass $M \approx 5\,M_\odot$

For these more massive stars, the ignition of ^{4}He occurs at the center and under nondegenerate conditions, and the star settles down to a period of quiescent He-burning in a convective core, together with H-burning in a shell (see inset (d)). The competition between these two energy sources determines the occurrence and extent of the blueward excursion in the HR-diagram [99]. At $5\,M_\odot$ the star crosses the Cepheid instability strip (points 10–14). Following core He exhaustion the structural readjustment to He shell burning results in a strong expansion, and the H shell is extinguished as the star begins its ascent of the AGB. At this time, the inner edge of the convective envelope penetrates the dormant H shell, and the products of complete H-burning are mixed to the surface in what is called the "second dredge-up" (point 15). This significantly alters the surface compositions of ^{4}He, ^{12}C, and ^{14}N, and reduces the mass of the H-exhausted core by mixing H inward (see inset (e)). Note that there is a critical mass (about $4\,M_\odot$, but dependent on composition) below which convection is not able to penetrate below the H shell, and second dredge-up does not occur. Following second dredge-up, the H shell is reignited, and the star continues to evolve up the AGB.

2.3 Stellar Evolution on the AGB

At the early-AGB stage where both H- and He-burning shells have been established, the structure of all AGB stars is qualitatively similar regardless of the stellar mass. As an illustration, we show the interior structure of a typical $1\,M_\odot$ AGB star in Figure 2.6. The star contains a carbon–oxygen core supported by the pressure of degenerate electrons and surrounded by He- and H-burning shells. The degeneracy of the C-O core ensures that no carbon burning will occur in the center of the star because of cooling from neutrino emission. The requirement that the C-O core be degenerate also sets the upper mass limit for AGB stars at about $8\,M_\odot$.

Figure 2.6 shows that although the region interior to the H-burning shell contains roughly half the mass of a $1\,M_\odot$ AGB star, the volume of this core is tiny and similar to that of a white dwarf. Almost the entire volume of the star is taken up by the convective envelope. The convective envelope is where stellar pulsation takes place, while the tiny core is where all the interesting stellar evolution and nucleosynthesis processes occur.

2.3.1 The Thermally Unstable He-Burning Shell

Schwarzschild and Härm [129], and independently Weigert [157], were the first to find that He-shell burning on the AGB does not proceed smoothly but is subject to thermal instabilities. We demonstrate the development of these instabilities in Figure 2.7, which shows the variation of total, H-burning and He-burning luminosities during the early AGB phase of a $1\,M_\odot$ star. At

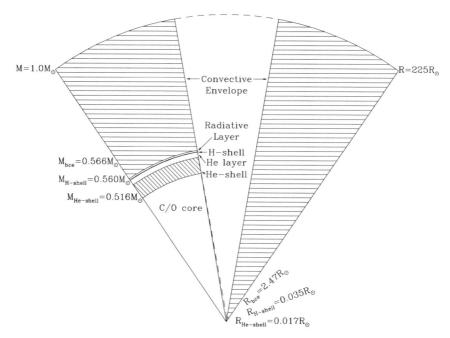

Fig. 2.6. A schematic view of a $1\,M_\odot$ AGB star interior. On the **left**, various regions in the star are plotted against mass fraction, while on the **right** the regions are plotted against radius. M_{bce} is the mass at the base of the convective envelope, while $M_{H-shell}$ and $M_{He-shell}$ are the masses at the middle of the hydrogen- and helium-burning shells, respectively

low luminosities, He-burning provides most of the luminosity as the He shell burns through the thick helium layer left behind by the H-burning shell during the core He-burning phase. As the intershell He layer becomes thinner, the H-burning shell becomes more prominent, and the He-burning shell declines in luminosity. Eventually, the He-burning luminosity starts to oscillate, with the increases in He-shell luminosity leading to a near-simultaneous decline in H-burning luminosity. These oscillations mark the beginning of the "thermally pulsing AGB," or TP-AGB, phase.

The basic physical explanation for AGB thermal pulses [129] is as follows. Suppose a positive temperature perturbation is introduced into the He-burning shell. This perturbation will cause (1) an increase in the rate of nuclear energy production in the shell, since the energy production rate per gram by a nuclear source is $\epsilon \propto \rho^\mu T^\nu$, with $\nu \approx 30\text{--}40$ and $\mu = 2$ for the triple-alpha reaction, and (2) an increase in the rate of energy *loss* from the shell because of the additional temperature gradient generated from shell center to edge. For a thermal instability to occur, the increased energy input should

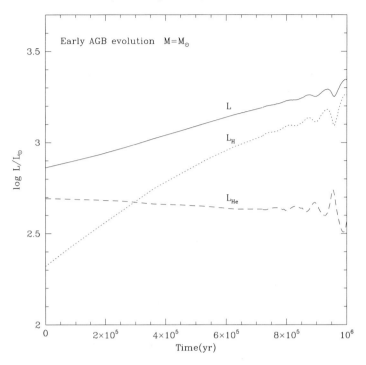

Fig. 2.7. The surface luminosity, H-burning luminosity and He-burning luminosity plotted against time on the early AGB for a $1\,M_\odot$ star of solar initial composition $Z=0.016$

be greater than the increased energy loss. The large value of ν for the triple-alpha reaction ensures a large increase in nuclear energy production rate for a modest temperature perturbation, while a shell that is not too thin ensures that the corresponding increases in energy loss are not large (the thinner the shell, the larger the *gradient* of perturbed temperature from shell center to edge, and the larger the outflow of energy). One further requirement is that when extra energy is deposited in the shell, it must lead to a further positive increase in shell temperature so that the initial temperature perturbation is enhanced and a thermal runaway will occur. One way to guarantee that adding energy leads to an increase in temperature is to add the heat at constant pressure (for a perfect gas at constant pressure $dQ = \frac{5}{2}\frac{P}{\rho}d\ln T$). Constant pressure can be maintained if the shell is thin (thickness/radius $\ll 1$), for the following reasons. The pressure in the shell is determined by the gravitational acceleration of the core acting on the matter above the shell. For a thin shell, any expansion of the shell as a result of a temperature perturbation will make an insignificant change in the radius of mass elements above the shell, so that the pressure in the shell will be effectively unaltered. This result holds even for a perfect gas. However, adding electron degeneracy reduces the requirement on shell thinness, since pressure becomes independent of temperature in a degenerate gas. A thicker shell retains energy deposited in it more efficiently than a thin shell, so it is more likely to become unstable. In summary, the thermal instability of the He shell arises because of the high temperature sensitivity of the triple-alpha reaction coupled with the thinness of the shell. Strong electron degeneracy is not required in the shell for the instability, although it enhances the instability mechanism.

2.3.2 AGB Evolution at Various Stellar Masses

Calculation of complete evolutionary sequences through the TP-AGB stage is very difficult and time-consuming and also involves many uncertainties such as convective energy transport and mixing, and the mass loss rate. This has led many authors to use "synthetic" AGB calculations, where the results of detailed calculations are parameterized in terms of a few quantities such as abundance, stellar mass, and core mass. Such synthetic calculations are discussed in Chapter 3. The most extensive existing sets of full AGB calculations are those of [15, 154]. Many other calculations have explored various aspects of AGB evolution; see [83] for a review of early works; more recent work is found in [18, 19, 20, 94, 95, 136, 154]; attempts to refine the convective mixing algorithm in AGB stars can be found in [64, 66]; the effect of rotation on mixing during AGB evolution is discussed in [93]. Here we outline the general properties of full TP-AGB evolution calculations and refer the reader to the above papers for more extensive results.

2.3.2.1 Surface Luminosity Variations

The variation in surface luminosity from the early AGB to the end of the TP-AGB phase is shown in Figure 2.8 for AGB stars of mass 1, 2.5, and $5\,M_\odot$. After the onset of thermal pulsing, the maximum luminosity of each pulse grows rapidly with pulse number until an approximate limiting amplitude is reached. TP-AGB evolution then continues through a variable number of thermal pulses until the hydrogen-rich convective envelope is dissipated by mass loss. A typical lifetime on the AGB is 10^6 yr, although this number clearly varies considerably and nonmonotonically with mass, as seen from the three examples in Figure 2.8.

The luminosity variations during three typical fully-developed thermal pulses are shown in detail in Figure 2.9. The broad behavior of surface luminosity (left panel) consists of a decline after each thermal pulse (shell flash) followed by a slow recovery to a maximum immediately before the next thermal pulse (this maximum is usually called the "quiescent" luminosity maximum, and the interval between thermal pulses is known as the quiescent evolutionary phase). During the surface luminosity decline, the He-burning

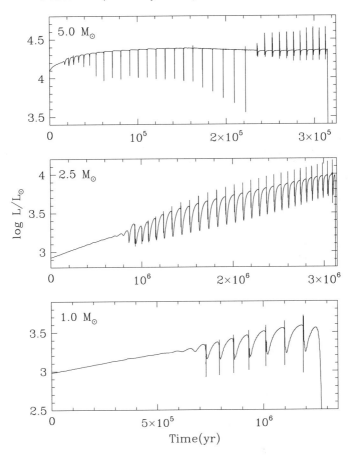

Fig. 2.8. The surface luminosity variation of thermally pulsing AGB stars of mass 1, 2.5, and 5 M$_\odot$ and initial composition $Y = 0.25$ and $Z = 0.016$. The plots start on the early AGB when the H- and He-burning luminosities are equal and end when the stars have left the AGB and moved to the planetary nebula part of the HR-diagram. From [154]

shell provides most of the star's luminosity, while during the recovery phase the H-burning shell is the main energy source. The right panel of Figure 2.9 shows a thermal pulse at high time resolution. When the He shell ignites, it burns at a very rapid rate, up to $\approx 10^6$ L$_\odot$ in the present case. The energy dumped into the shell region by this rapid burning causes a hydrostatic structural readjustment, which rapidly extinguishes the H-burning shell and leads to an initial drop in surface luminosity. When the deposited energy escapes from the stellar core, it leads to a peak in surface luminosity lasting several hundred years. Note that at about this time, the helium shell undergoes a secondary, weaker, flash.

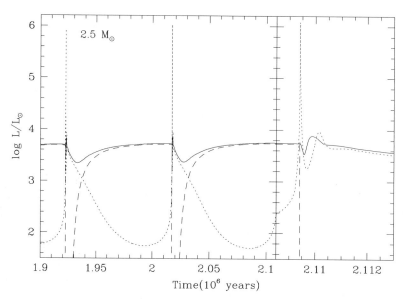

Fig. 2.9. The surface luminosity (solid line), H-burning luminosity (dashed line), and He-burning luminosity (dotted line) plotted against time for three consecutive thermal pulses in a 2.5 M$_\odot$ star. The **right** panel shows the third pulse with a highly expanded time axis. From [154]

The thermal pulse cycles described above are characteristic of the 1 and 2.5 M$_\odot$ star and the final evolution of the 5 M$_\odot$ star when the envelope mass has been reduced below ≈ 1.5 M$_\odot$ (time $> 2.2 \times 10^5$ yr in Figure 2.8). In all these cases, there is a radiative layer between the H-burning shell and the base of the envelope convection during quiescent evolution. However, when convection extends down to near the H-burning shell, the surface luminosity behaves quite differently, as seen in the 5 M$_\odot$ star over the time interval 0.5–2.2 $\times 10^5$ yr. With such a deep convective envelope, there is very little variation in surface luminosity over the thermal pulse cycle.

The existence of a radiative layer between the H-burning shell and the base of the envelope convection during quiescent evolution makes for a great simplification. In such a situation, it is easy to show theoretically that the evolution of the H-exhausted core is independent of the envelope mass [144], and so the core evolution should depend primarily on the core mass. This has led to the wide use of various relations between properties of TP-AGB stars and their core mass in synthetic AGB calculations. In 1970 it was shown that a linear relation exists between the mass of the H-exhausted core M_c (sometimes labeled M_H in the literature) and AGB luminosity L [116, 148]; Paczynski ([116]) gave this relation in the form

$$L = 59250\,(M_\text{c} - 0.522),\tag{2.1}$$

where L and M_c are in solar units. This equation now carries Paczynski's name. Very similar luminosity–core mass relations are still in use today [19]. The increase in average luminosity with time in Figure 2.8 is due to the existence of this relation. Another simple linear relation is the interpulse period–core mass relation (for fully developed thermal pulses) in the form [118]

$$\log \tau_\text{ip}\,(\text{yr}) = 3.05 - 4.5\,(M_\text{c} - 1.0).\tag{2.2}$$

The shortening of the interpulse period as the core mass grows is clearly seen in Figure 2.8 for the $2.5\,M_\odot$ star.

For stars of $M \gtrsim 3\,M_\odot$ and large envelope masses, there is no radiative buffer zone between the H-burning shell and the envelope convection zone, and the simple dependence of AGB evolution on core mass alone breaks down. This is shown in Figure 2.8, where it can be seen that the quiescent surface luminosity of the $5\,M_\odot$ star in the time interval $(0.5\text{–}2.2)\times 10^6$ yr (when the envelope mass is large and convection penetrates close to the H-burning shell) is larger than the quiescent surface luminosity at later times (when mass loss has reduced the envelope mass and a radiative buffer layer exists above the H-burning shell). This signifies that the luminosity–core mass relation given above has broken down and that stars with deep convective envelopes can have luminosities on the AGB significantly larger than that given by this relation [17, 21]. Similarly, the sudden shortening of the interpulse period for the $5\,M_\odot$ star when the pulse character changes (Figure 2.8) also shows that the interpulse period–core mass relation for stars with deep convective envelopes is different from that pertaining to low-mass AGB stars. For more information see Section 2.5.11.

2.3.2.2 Mass Loss

Mass loss is crucial to any study of AGB evolution since it is mass loss that leads to the termination of evolution on the AGB. At the present time, our knowledge of the dependence of AGB mass loss on stellar properties is rather poor, although advances are being made in the theory of AGB winds driven by the combination of pulsation and radiation pressure acting on dust grains (see Chapters 4 and 6). In the AGB evolutionary calculations of [15, 154] the mass-loss laws used were based on observational estimates and theoretical estimates, respectively. A feature of these laws is that the mass-loss rate increases very rapidly with increasing luminosity, so that the main mass losing interval is confined to the end of the AGB. This is demonstrated in Figure 2.10, where the mass-loss rate and the total stellar mass are plotted against time for a $1\,M_\odot$ AGB star. It is clear that the most extensive mass loss occurs during the high-luminosity part of quiescent evolution during the last few thermal pulse cycles. High mass-loss rates are generally confined to

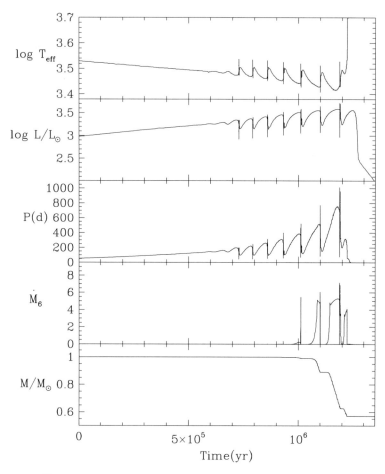

Fig. 2.10. Various properties plotted against time on the AGB for a $1\,M_\odot$ star of initial composition $(Y, Z) = (0.25, 0.016)$. **From the top**: $\log T_{\rm eff}$; $\log L/L_\odot$; pulsation period P in d; mass-loss rate in units of $10^{-6}\,M_\odot\,{\rm yr}^{-1}$; and total stellar mass. From [154]

the same time intervals, although the highest mass-loss rates actually occur during the luminosity peaks associated with the last few thermal pulse cycles. Because of the short duration of the luminosity peaks, only a small amount of mass is lost during them. The narrow circumstellar shells found around some AGB stars are probably produced by these brief increases in the mass-loss rate (see Chapter 7).

2.3.2.3 The Interaction of Evolution and Pulsation

The luminosity variations of AGB stars during the thermal pulse cycle lead to changes in the stellar radius and hence changes in the pulsation period.

Most AGB stars appear to be variable (see Section 2.6.2.2), so the period distribution of AGB stars will be affected in a complicated way by interior evolution. Since the pulsation period P is proportional to $R^\alpha M^{-\beta}$, where $\alpha \approx 1.5$–2.5 and $\beta \approx 0.5$–1.0 [53, 165], both radius and mass variations are important.

The variation with time of the fundamental mode period of a $1\,M_\odot$ star during AGB evolution is shown in Figure 2.10. Up to the beginning of the high-mass-loss "superwind" stage, which begins at a time of $\approx 10^6$ yr in the figure, the pulsation period varies regularly with the changes in $\log L$ and $\log T_{\rm eff}$. However, when the "superwind" starts and the stellar mass starts decreasing significantly, the period grows rapidly due to both (a) the mass dependence of the pulsation period and (b) the fact that $T_{\rm eff}$ for AGB stars decreases as the mass decreases. Stars of larger initial mass lose larger amounts of mass, and hence they undergo larger increases in period. The very long periods, often longer than 1000 d, exhibited by dust-enshrouded AGB stars (e.g., see [63, 174]) are a result of these effects.

2.3.3 The Third Dredge-Up: Making Carbon Stars

In Section 2.1.2, we noted that many AGB stars show evidence of the recent "dredge-up" to the stellar surface of the products of nuclear burning. In particular, carbon stars have enhanced surface abundances of ^{12}C, while MS, S, and SC stars show enhancements of s-process elements. We now describe how these surface enhancements are produced by the so-called third dredge-up. This process, which produces carbon stars, was discovered by Iben [77] and Sugimoto and Nomoto [138].

In Figure 2.11, we show in schematic form the variation with time of a number of quantities associated with the H- and He-burning shells. In particular, we show the shell luminosities and the position in mass of the convective regions. Following Iben [78] we identify four phases of a thermal pulse cycle in the description of the third dredge-up:

(a) The "off" phase is the time when the He-burning shell is essentially dormant, producing much less energy than the H-burning shell. It occupies the left and right portions of the top panel of Figure 2.11 and it takes up most of the thermal pulse cycle.

(b) The "on" phase begins when the He shell starts to burn very strongly, producing luminosities up to $\approx 10^8\,L_\odot$, depending on the core mass and whether limiting flash amplitude has been reached. The energy deposited by the He-burning reactions cannot be transported by radiation alone, and a convective shell develops immediately above the He-burning region. This convective shell is known variously as the *intershell convective zone* or sometimes the *pulse-driven convective zone*. This convection extends upward almost to the H shell at its maximum extent. The intershell convective zone is composed mostly of ^{4}He (about 75%) and ^{12}C (about 22%), and it lasts for a few hundred yr (its duration is highly dependent on the core mass).

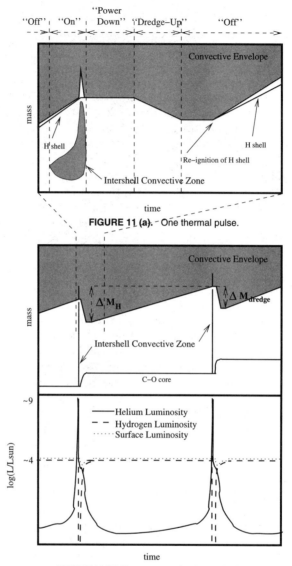

Fig. 2.11. Some details of interior evolution during one thermal pulse (**top**) and for two consecutive pulses (**bottom**). Convective regions are shaded. The line toward the bottom of the middle panel shows the position of the He-burning shell ($Y = 0.5$). The position of the H-burning shell is indistinguishable from the position of the base of the convective envelope in this panel. The "On," "Power Down," "Dredge-Up," and "Off" phases of the thermal pulse cycle are marked on the top panel

(c) The "power down" phase coincides with the decline of He shell burning and the disappearance of the intershell convection zone. The previously released energy drives a substantial expansion of the intershell region, pushing the H shell to such low temperatures and densities that it is extinguished (or very nearly so).

(d) The "dredge-up" phase corresponds to the escape from the stellar core into the convective envelope of the energy released by the shell flash. In response to the increasing luminosity coming out of the core, the convection extends inward in mass. If the envelope convection penetrates into the mass zone previously occupied by the intershell convection, it will dredge up some of the ^{12}C that was produced by the He-burning shell and mix it to the stellar surface. This is the "third dredge-up."

A measure of the efficiency of the third dredge-up is given by the so-called dredge-up parameter, λ. This is defined as $\lambda = \Delta M_{dredge}/\Delta M_c$, where, as shown in Figure 2.11, ΔM_{dredge} is the amount of mass dredged up by the convective envelope in one thermal pulse, and ΔM_c is the amount of mass through which the H shell has moved during the "off" phase. Evolutionary calculations show that $\lambda = 0$ for early flashes and for stars with low envelope mass, $\lambda \lesssim 0.3$ for lower mass stars ($\lesssim 2.5\,\mathrm{M_\odot}$) with well-developed thermal pulses, and $\lambda \lesssim 1.0$ for $M \gtrsim 5\,\mathrm{M_\odot}$. These estimates of λ are highly uncertain because of the difficulty of estimating the amount of convective overshoot that occurs at convective boundaries. Further discussion can be found in Chapter 3.

2.4 Evolution Beyond the AGB: Planetary Nebula Nuclei and White Dwarfs

For completeness, we include here a brief outline of the evolution of stars beyond the AGB (for further details see Chapter 10).

When superwind mass loss near the tip of the AGB reduces the hydrogen-rich envelope to small values ($< 10^{-3}\,\mathrm{M_\odot}$), the star is no longer able to maintain its large size, and it begins to shrink. At this stage, with a small envelope mass, there is always a radiative layer between the H-burning shell and the base of the convective envelope. As noted in Section 2.3.2.1, this means that the core luminosity is unaffected by the reduction in envelope mass. Hence, the star evolves at essentially constant luminosity to hotter temperatures in the HR-diagram, passing through the domain of the post-AGB stars or pre-planetary nebulae (pre-PNe) and passing on to the domain of the planetary nebula nuclei (PNNi) where $T_{\mathrm{eff}} \gtrsim 30,000$ K.

For a star of given core mass and luminosity, there is a well-defined relation between the effective temperature and the mass of the hydrogen-rich envelope (e.g., [127]) with T_{eff} increasing as the H-rich envelope mass decreases. There are two factors that lead to the reduction of the H-rich envelope mass: Mass loss and nuclear burning. Since the nuclear burning rate for a typical PNN

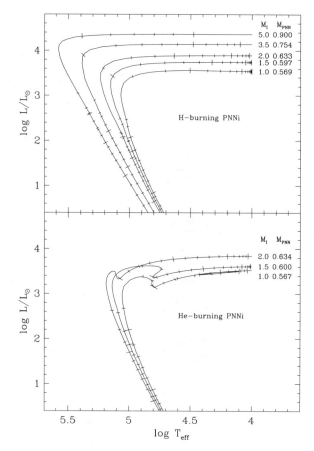

Fig. 2.12. Post-AGB evolutionary tracks for stars with initial composition $(Y, Z) = (0.25, 0.016)$. These tracks are shown starting at $\log T_{\text{eff}} = 4$, and they then pass through the region of the HR-diagram occupied by PN nuclei and by white dwarfs. The **top** panel shows the evolution of stars that are burning H, while the **bottom** panel shows the evolution of stars with He-burning shells. Tick marks on the tracks are at intervals of 0.2 in $\log t$, where $t = 0$ at $\log T_{\text{eff}} = 4$. The first tick is at $t = 10$ yr. Long tick marks are used for times that are integral powers of 10. From [155]

with core mass of 0.6 $M_\odot$ is $\approx 6 \times 10^{-8}$ $M_\odot$ yr^{-1} and the envelope mass when the star begins to cross the HR-diagram is $\approx 10^{-3}$ $M_\odot$, it will take $\approx 16,000$ yr for nuclear burning alone to move the star across the HR-diagram to the PN domain. This is the typical age of an old PN, whereas young PNe have ages $\lesssim 1000$ yr. Hence, mass loss is clearly important in the post-AGB star stage in order to hasten evolution across the HR-diagram. In practice, we have essentially no knowledge of the mass-loss processes in the immediate post-AGB stage, so the time scales for evolution there are essentially unknown. However, stellar evolution calculations through this region have used rough estimates of the mass-loss rates (e.g., [16, 126, 155]). Further across the HR-diagram, there have been many measurements of the mass-loss rates from PNNi, and these have been included in the calculations noted above. Eventually, consumption of the H-rich envelope means that there is not enough mass left above the H-burning shell to maintain the temperature required for burning, and the star declines rapidly in luminosity before spending the remainder of its life cooling off as a white dwarf.

Many computations of the evolution of PNNi have been made [16, 79, 80, 117, 126, 155, 171]. Typical evolutionary tracks for PNNi are shown in Figure 2.12. These tracks highlight the diversity of possible tracks brought about by the possibility of a star leaving the AGB during quiescent AGB evolution or at or near a thermal pulse. Estimates of the relative frequency of He-burning PNNi are given in [41, 127], although these two papers come to quite different conclusions.

2.5 Nucleosynthesis in AGB Stars

Having outlined the evolution and structure of AGB stars, we now discuss the nucleosynthesis that occurs on the AGB. Of course, the structure and nucleosynthesis are not totally independent, but many of the reactions we will discuss in this section are unimportant from a structural point of view (e.g., they produce negligible energy, or changes in the opacity). In contrast, some reactions (e.g., carbon production) are crucial to the structure, while the uncertainties concerning mixing affect both structure and nucleosynthesis.

It is not our aim to provide all the latest results, but rather to explain the principles and our understanding of the mechanisms. Pointers to the literature are provided for those who want more details.

2.5.1 Initial Composition

We start by describing the structure and composition of a star as it begins its evolution along the AGB. It is important to get a feel for the relative sizes of the different regions, and these are shown in Figure 2.13 in mass and radius, for both a $1\,M_\odot$ and a $5\,M_\odot$ model. The center consists of a small, dense, degenerate core of ^{12}C and ^{16}O, the exact proportions of which depend on the rate of the ^{12}C$(\alpha,\gamma)^{16}$O reaction. Sitting atop this is the He-burning shell. A very thin intershell zone then separates this shell from the H-burning shell. Between this and the convective envelope is a radiative buffer zone, which can actually disappear for the more massive models, resulting in "hot bottom burning."

The composition of the surface of the star reflects the initial composition with the addition of the effects of the first dredge-up (FDU), and for stars more massive than some (composition-dependent) limit of about $4\,M_\odot$, the second dredge-up (SDU).

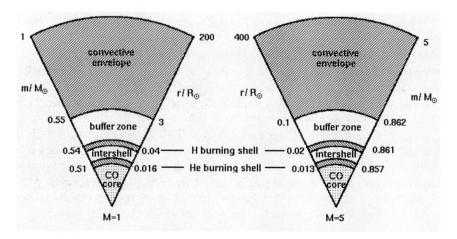

Fig. 2.13. Early AGB structure and dimensions for $1\,M_\odot$ (**left**) and $5\,M_\odot$ (**right**) stars. The coordinates of critical zones are given in both radius and mass

Qualitatively, both the FDU and SDU have the same effect: They mix to the surface regions that have undergone H-burning (primarily via the CNO cycles), and hence they increase the abundances of ^{13}C, ^{14}N, and some ^{4}He. There are associated decreases in ^{12}C, ^{16}O, and ^{18}O. Note that SDU, when it occurs, mixes to the surface a substantial amount of material that has burned all of its H, and hence it produces a relatively large change in the abundances. In contrast, the FDU mixes regions that have undergone only partial H-burning.

As the H-shell advances into the envelope it burns H into ^{4}He, and essentially all CNO elements end up as ^{14}N. This is the composition of the intershell region. It is this material that is processed by the He-shell, with most of the ^{14}N capturing two α-particles to form ^{22}Ne, and triple-alpha reactions burning the ^{4}He into ^{12}C.

2.5.1.1 First Dredge-Up

We digress briefly to discuss in more detail the abundance changes at the first and second dredge-up episodes.

As the star first ascends the giant branch, the inward penetration of the convective envelope means that the products of earlier partial H-burning are mixed to the surface. When this envelope retreats, it leaves behind an abundance discontinuity which is erased only when the H-burning shell reaches this position. This happens only for stars below about $2.5\,M_\odot$, since the more massive stars ignite their central helium supply before the shell can advance that far outward.

The changes produced by the FDU depend on the mass and composition of the star, but are roughly the following: ^{12}C decreases by $\approx 30\%$, ^{13}C increases by a factor of ≈ 2–3, ^{14}N increases by a factor of ≈ 3–4, ^{15}N decreases by a factor ≈ 2, and ^{18}O is reduced by $\approx 30\%$. There may also be a substantial increase in the ^{17}O content. Typical resulting isotopic ratios are ^{12}C/^{13}C ≈ 20 and ^{16}O/^{18}O ≈ 600. For more quantitative details see [40, 45, 46, 97].

From a quantitative viewpoint, the models are in reasonable agreement with observations for masses above about $2.5\,\mathrm{M}_\odot$, but there are substantial discrepancies below this mass. Specifically, the observed ^{12}C/^{13}C ratios decrease to as low as ≈ 10 (and sometimes lower) rather than the predicted value ≈ 20. An explanation seems to require some form of "extra-mixing" below the convective envelope to submit the material to further CN cycling (e.g., [36, 139]). Similarly, the observed oxygen isotope ratios seen in presolar meteorite grains also require some extra-mixing (e.g., [23, 24]). This mixing is believed to be due to rotation [139, 179] and is inhibited by steep composition gradients, so that the abundance discontinuity left by the retreating convective envelope may prevent any extra-mixing until the H-shell removes this discontinuity. This does not happen for $M \gtrsim 2.5\,\mathrm{M}_\odot$, so we may expect the extra-mixing to affect only stars below about $2.5\,\mathrm{M}_\odot$, which is consistent with the observed discrepancies. The need for, or effect of, such mixing on the AGB is yet to be determined (see [24, 114]).

2.5.1.2 Second Dredge-Up

For stars more massive than about $4\,\mathrm{M}_\odot$ (depending on the composition) there is a second dredge-up event, which occurs soon after the exhaustion of central helium, as the He-burning shell is being established [10, 11]. In this case, the convective envelope penetrates a region that has completely burned all hydrogen, and the resulting helium-rich material is added to the envelope. There are small changes in the oxygen isotope abundances, but because CNO cycling has burned most CNO nuclei into ^{14}N in this region, there are large changes in the ^{14}N/^{15}N ratio, by factors ≈ 6 (for details see [97]).

2.5.2 Sites of Nucleosynthesis in AGB Stars

The nucleosynthesis in AGB stars is mostly associated with H- and He-burning, complemented by neutron captures. Of particular importance is the action of repeated third dredge-up events that mix the products of He-burning to the stellar surface. The He-burning products in the deeper parts of the stellar envelope are subjected to proton captures in the H-burning shell during the next interpulse phase. Thus, we have to consider not only the thermal pulses themselves, but the effect of subsequent passages of the dredged-up material through the H-shell, giving rise to a combination of H and He processed material. These processes are basically responsible for the production of ^{4}He, ^{12}C, ^{14}N, possibly some ^{16}O, ^{19}F, ^{22}Ne, ^{23}Na, ^{25}Mg, ^{26}Mg, ^{26}Al, and ^{27}Al. Other species are made by hot bottom burning, and will be discussed in Sections 2.5.11–2.5.14. It is this combination of H and He processing, together with the conditions necessary for neutron capture, that makes AGB stars an essential ingredient in nucleosynthesis studies.

2.5.3 The Production of ^{12}C

During the thermal instability of the He-shell the main nuclear reaction to occur is the production of ^{12}C from ^{4}He via triple-alpha reactions. The growth of the intershell convection zone means that new ^{4}He is mixed down to the shell from further out in the star, and of course the freshly produced ^{12}C is simultaneously mixed outward by the convection, as shown in Figure 2.11. In most calculations there is negligible ^{16}O produced, and the dominant composition in the intershell convective zone (ISCZ) is about 20–25% ^{12}C, 70–75% ^{4}He, and a few percent ^{16}O (after the ISCZ has dissipated). There is also a relatively large amount of ^{22}Ne due to α-captures on the ^{14}N left over from the CNO cycle in the hydrogen shell. A different result is found if one includes some convective overshoot from the edges of the intershell convective zone. In this case, overshooting downwards into the CO core will penetrate regions that are richer in ^{16}O (from helium burning in the core earlier in the evolution), and hence the composition of ^{16}O in the intershell convective zone can be much higher [65].

The ISCZ homogenizes most of the intershell region, from the bottom of the He-shell almost to the H-shell. During the subsequent third dredge-up phase the outer layers of this region are mixed to the surface and result in an increased ^{12}C abundance in the envelope. Figure 2.14 shows the third dredge-up and resulting ^{12}C abundance during the AGB evolution of a 3.5 M$_\odot$ model with $Z = 0.02$. Eventually, repeated dredge-up may see the star become a C-star (i.e., one where C atoms outnumber O atoms in the envelope). A quantitative estimate of the total yield of ^{12}C from AGB stars is hampered by uncertainties in the depth and onset of the third dredge-up, but it appears that they produce roughly one third of the Galaxy's inventory of ^{12}C, providing roughly the same amount as supernovae and WR stars [42].

2.5.4 The Production of s-Process Elements

After ^{12}C production, perhaps the most important nucleosynthesis to occur in AGB stars is the slow neutron capture, which results in the production of the so-called s-process elements. The historical development of our understanding of this process is fascinating, but somewhat outside the thrust of this chapter; we refer the interested reader to [27, 109].

It was originally thought that the neutron source acting in AGB stars was the reaction ^{22}Ne$(\alpha, n)^{25}$Mg. The intershell region is rich in ^{14}N from CNO

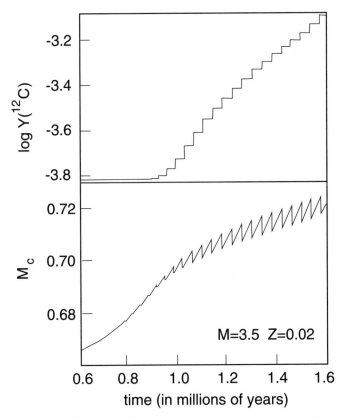

Fig. 2.14. Variation of the core-mass M_c (**bottom**) and the surface ^{12}C abundance (**top**) during the AGB evolution for a 3.5 M$_\odot$ model with $Z = 0.02$. Note that Y is the mole fraction, and is equal to the mass fraction divided by the mass number: $Y = X/A$. The time axis is offset by 276 million yr (since the ZAMS)

cycling, and during a pulse the reactions ^{14}N$(\alpha,\gamma)^{18}$F$(\beta^+\nu)^{18}$O$(\alpha,\gamma)^{22}$Ne occur, so that the intershell region should be quite rich in ^{22}Ne. If the temperature is high enough, say above 300 million K, then the reaction ^{22}Ne$(\alpha,n)^{25}$Mg follows. However, evidence began to mount that most s-element enhanced stars were from low-mass populations, and a search [133] for enhancements of ^{25}Mg, which would be dredged to the surface if ^{22}Ne were the neutron source, was negative. The alternative neutron source is ^{13}C$(\alpha,n)^{16}$O, which requires more modest temperatures, in the region of 100 million K. But the problem here is that the amount of ^{13}C present in the intershell, from CNO cycling, is insufficient to produce the required number of neutrons. The source of the required ^{13}C has been, and to some extent remains, a major uncertainty in the models.

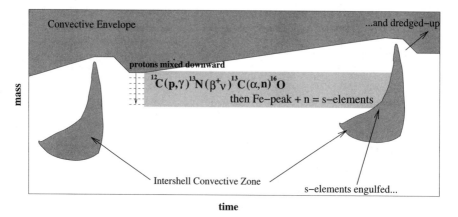

Fig. 2.15. Schematic showing internal evolution between two pulses. Partial mixing during the third dredge-up phase produces a ^{13}C pocket, which provides the neutrons for the s-processing. Note that this figure is not to scale: The protons are mixed downward a depth of only a few $\times 10^{-4}\,M_\odot$, whereas the intershell convective zone is up to a few $\times 10^{-2}\,M_\odot$ in extent

2.5.5 The ^{13}C Pocket

During a strong thermal pulse, the large expansion of the intershell region can push the outer edge of the previous ISCZ to sufficiently low temperatures (about 1 million K) that the carbon atoms begin to recombine, as discovered by Sackmann [120] and Iben and Renzini [81, 82]. This dramatically alters the opacity. A small semiconvective zone (a few $\times 10^{-4}\,M_\odot$) forms at the bottom of the H-envelope during the third dredge-up phase. This is shown schematically in Figure 2.15. This has the very important consequence of mixing some protons downward into the ^{12}C-rich region. During the next interpulse phase, as the star contracts and heats again, the protons are captured by ^{12}C to make ^{13}N which undergoes positive beta-decay to produce ^{13}C. It is important that there be not too many protons, because in that case the proton captures continue, and complete the CN cycle by producing mostly ^{14}N. But if there are fewer protons, the reactions can terminate with the production of a shell, or pocket, enriched in ^{13}C. During the next thermal pulse the ISCZ engulfs this ^{13}C, and neutrons are released in the ISCZ via the ^{13}C$(\alpha, n)^{16}$O reaction [88].

This picture was modified by Straniero et al. ([137], see also [136]) when it was realized that in fact, the ^{13}C burns in situ, under radiative conditions, during the interpulse phase rather than when engulfed by the ISCZ. Hence the neutrons are released in situ and the s-processing occurs between pulses in the same layers where the ^{13}C was deposited after the dredge-up phase. When the next ISCZ occurs it ingests this layer and hence the s-process elements previously produced by the ^{13}C source. In addition, at the peak of the pulse the temperature is often high enough for a brief burst of neutrons from the ^{22}Ne source mentioned above (e.g., [27]).

Unfortunately, the first semiconvection computations could not be satisfactorily repeated. Nevertheless, it illustrated the principle, and indeed other processes may produce a similar partial mixing at the border between the H-rich envelope and the C-rich intershell. Such mechanisms could be a partially mixed convective overshoot region (e.g., [65]), or some shear or other rotational instability [93]. In any event, the models do sometimes produce ^{13}C pockets and the nucleosynthesis resulting from a distribution in the sizes of such pockets seems to fit the observations well: There seems little doubt that nature also produces ^{13}C pockets, although we are still unsure about how exactly it achieves this.

2.5.6 Neutron Captures and the s-Process

A detailed discussion of neutron capture nucleosynthesis will not be given here. Rather, we will elucidate the main features and results. We will begin with the "classical" analysis, which tries to understand the process phenomenologically without reference to the details of the astrophysical site.

2.5.6.1 The Classical Approach

Two simple extremes are easily identified: Depending on the number of free neutrons available, either neutron captures dominate the (negative) β-decays or the β-decays dominate the neutron capture. The former means that neutron capture is *rapid* compared to β-decays and defines the r-process, with typical neutron densities $n_\mathrm{n} \gtrsim 10^{20}$ cm^{-3}; the latter is the s-process, where neutron capture is *slow* compared to β-decays, and $n_\mathrm{n} \lesssim 10^8$ cm^{-3}. It is the s-process that is relevant for AGB stars. Under this approximation and assuming no branchings (see the next subsection), the equation governing the abundance N_A of the (stable) isobar of mass A is

$$\frac{\mathrm{d}N_A}{\mathrm{d}t} = -n_\mathrm{n}\langle\sigma v\rangle_A N_A + n_\mathrm{n}\langle\sigma v\rangle_{A-1} N_{A-1}, \qquad (2.3)$$

where n_n is the neutron number density and $\langle\sigma v\rangle_A$ is the thermally averaged neutron-capture cross-section for the isobar of mass A. It is common to write $\langle\sigma v\rangle_A$ as $\sigma_A v_T$, where v_T is the thermal velocity of the neutrons and σ_A is an appropriate average cross-section. We define the neutron exposure τ by

$$\tau = \int n_\mathrm{n} v_T \, \mathrm{d}t, \qquad (2.4)$$

and thus we get

$$\frac{\mathrm{d}N_A}{\mathrm{d}\tau} = -\sigma_A N_A + \sigma_{A-1} N_{A-1}. \qquad (2.5)$$

Note that τ is a time-integrated neutron flux, with units of mbarn^{-1} (1 barn = 10^{-24} cm^2). If a steady state is attained, then $\mathrm{d}N_A/\mathrm{d}t$ goes to zero and hence $\sigma_A N_A$ is constant for all values of A, a fact noted in the foundation papers by Burbidge et al., Clayton et al., and Seeger et al. [26, 37, 130]. Indeed, the solar system distribution does show $\sigma_A N_A$ roughly constant (for s-only isotopes) when away from the isobars of magic neutron number ($N = 50, 82, 126$). The magic neutron numbers behave as bottlenecks in the buildup of s-process elements due to the low cross-section of such isobars. To create significant numbers of nuclei on the high-mass side of a magic neutron number, a large neutron exposure is needed.

Clayton et al. [37] showed that a single neutron exposure τ could not reproduce the solar system distribution, but this could be achieved by multiple exposures with an exponential[3] distribution of the neutron exposure [130]. Such a distribution of neutron exposures allows neutron addition to bypass the bottlenecks at the magic neutron numbers. Within this framework, if we let $\rho(\tau)\mathrm{d}\tau$ be the number of iron nuclei in the solar system that have experienced a neutron exposure between τ and $\tau + \mathrm{d}\tau$, then an exponential distribution may be written as

$$\rho(\tau) \propto \frac{N_{56}}{\tau_0} e^{-\tau/\tau_0}, \qquad (2.6)$$

where N_{56} is the initial abundance of ^{56}Fe, and τ_0 is the mean neutron exposure, an adjustable parameter.

As noted above, a single exposure cannot match the solar system distribution of s-process elements. It was soon recognized that three distinct components were required.

(1) *The Weak Component:* This produces most of the s-isotopes with $A \lesssim 90$, from Fe to Sr, and can be described by $\tau_0 \approx 0.06$ mbarn^{-1},
(2) *The Main Component:* This is responsible for the s-isotopes from $90 \lesssim A \lesssim 204$, from Sr to Pb, and requires $\tau_0 \approx 0.30$ mbarn^{-1},
(3) *The Strong Component:* This was devised primarily to produce the solar system's ^{208}Pb, and would require $\tau_0 \approx 7.0$ mbarn^{-1}.

To reproduce the solar system distribution we would need to add the three components together in a suitable mixture. The weak component is believed to come from central helium burning in the cores of massive stars, where the neutron source is ^{22}Ne$(\alpha,\mathrm{n})^{25}$Mg. The main component is believed to be associated with thermal pulses in AGB stars. When we consider real stars we will see that under some circumstances (e.g., very low metallicity), AGB stars can produce substantial amounts of ^{208}Pb. This removes the need for a separate strong component.

---

[3] Note that the exponential distribution is a convenient mathematical formulation and *not* a physical requirement!

When Ulrich [147] showed that the overlapping convective shells in AGB stars (see Figure 2.11) would lead to an exponential distribution of neutron exposures, the idea of such a distribution became synonymous with s-processing in AGB stars, although as we shall see below, this is no longer believed to be either necessary or correct. Indeed, the classical analysis, which discusses neutron capture within a purely mathematical context with no reference to astrophysical sites, may obscure the fact that the solar system s-process abundance distribution itself, even in the case of the main component, is not the result of a unique process, but rather the cumulative outcome of all previous generations of AGB stars.

2.5.6.2 Branchings in the Neutron Capture Pathways

A complication to the simple classical model, described above, is that there are times when the β-decay rate is not substantially higher than the neutron capture rate. This can lead to branchings at some isobars, i.e., to the transmutation of a nucleus into either of two possible daughter nuclei depending on whether a β-decay or a neutron addition occurs. Branching is particularly important for certain crucial unstable isotopes such as ^{85}Kr and ^{134}Cs. The branching around ^{134}Cs is shown in Figure 2.16. Neutron captures on ^{133}Cs produce ^{134}Cs, which can either decay to ^{134}Ba, if the neutron fluence is low, or can capture another neutron to produce ^{135}Cs. The path depends on the flux of neutrons and the β-decay lifetime of ^{134}Cs, which varies from around a year to about 30 d over the temperature range from 100 to 300 million K [140]. Further neutron captures along either path will produce ^{136}Ba (note that ^{135}Cs has a relatively long half-life). So the amount of ^{134}Ba produced (or the ratio ^{134}Ba/^{136}Ba) will depend critically on the branching, and hence can be used as a tracer of the neutron flux (and hence neutron density) and temperature of the site where the nucleosynthesis occurred. Note that ^{134}Ba and ^{136}Ba can be made *only* by the s-process.

The exact distribution of nuclei resulting from the neutron captures depends on the temperature and density of the site of the processing, and will be discussed further below. But it is worthwhile including a summary here. The ^{22}Ne source requires a temperature of some 300 million K or more to be activated, and this occurs in the ISCZ of AGB stars with masses above about $3\,\mathrm{M}_\odot$. This produces neutron densities n_n in excess of $10^{11}\mathrm{cm}^{-3}$, and a nonsolar distribution that is intermediate between the s- and r-processes, with many branches favoring the neutron-rich nuclides such as ^{86}Kr, ^{87}Rb, and ^{96}Zr. On the other hand, the ^{13}C source is activated at temperatures as low as 100 million K, and the resulting neutron density is $\log n_\mathrm{n} \approx 6\text{--}7$.

2.5.6.3 The Neutron Source in AGB Stars

For the more massive AGB stars it is believed that the dominant neutron source is indeed ^{22}Ne$(\alpha,\mathrm{n})^{25}$Mg, which requires a temperature in excess of

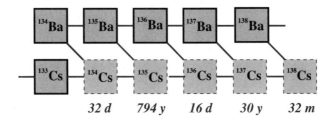

Fig. 2.16. Neutron capture path near ^{134}Cs showing the branching to ^{134}Ba and ^{136}Ba. Neutron number increases to the right, and proton number increases upward. Hence, a neutron capture moves a nucleus one box to the right, while a negative β-decay moves a nucleus one box up and one box to the left. Unstable isotopes are shown in dashed boxes, and average half-lives in the ISCZ of a typical low-mass AGB star are shown below the box. After [27]

300 million K. However, there is strong evidence that most giant stars enriched in s-process elements have masses below this, at around 1.5–2 M$_\odot$. In these stars the neutron source is ^{13}C$(\alpha,n)^{16}$O.

We have seen how partial mixing during the third dredge-up can produce a shell or pocket that is rich in ^{13}C. As the star contracts again, the H shell reignites, and the temperature in the ^{13}C pocket approaches 100 million K, where the time scale for α-capture decreases below the time between pulses [137]. Hence neutrons are released within the pocket, at quite low neutron densities ($n_n \lesssim 10^7$ cm^{-3}), but with a fairly large neutron exposure $\tau \approx 0.4$ mbarn^{-1} within the shell due to the long duration of the exposure.

The details of the exposure depend on the details of the ^{13}C pocket. Next to this ^{13}C pocket is a ^{14}N pocket, which forms in the region where the proton abundance was higher, and the CN cycle proceeded further. However, ^{14}N is a strong neutron absorber, or "poison," so that the details of the neutron flux depend on the initial distribution of the protons, and the resulting CN nuclei, as emphasized and explored by the Torino group (Gallino and Busso and coworkers, e.g., [5, 58]).

Figure 2.17 shows the ^{13}C pocket (the site of the s-processing) following the 27th pulse in a 2.5 M$_\odot$ model with solar metallicity. At the 28th pulse the s-elements are ingested by the ISCZ. Note that the outer edge of the ISCZ reaches into regions filled with the ashes of H-burning, where there is some ^{13}C that has not yet experienced α-captures, and hence has not yet released neutrons. Thus this ^{13}C is indeed mixed into the ISCZ, where it is burned at somewhat higher temperatures than seen by the ^{13}C pocket, but because the amount of ^{13}C involved is small, so is the contribution to the overall neutron capture nucleosynthesis.

The ISCZ mixes up material left over from the previous pulse (see Figure 2.17). Hence, already exposed nuclei will be exposed again to any neutron

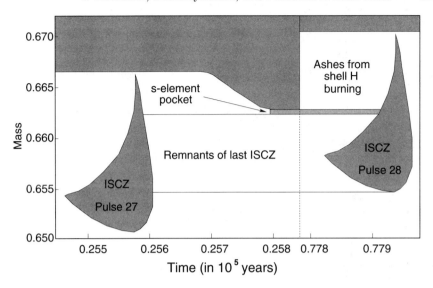

Fig. 2.17. The overlap between the 27th and 28th pulses in the evolution of a 2.5 M$_\odot$ model with $Z = 0.02$. The time axis is offset by 733.55 million yr (since the ZAMS)

flux that is created during the helium shell flash, and some of the already exposed nuclei will become part of the next ^{13}C pocket and suffer neutron exposure in the next interpulse interval. This is how AGB stars produce multiple neutron exposures, which explain the observed solar system s-process abundance pattern.

Even for lower-mass AGB stars, the peak temperature at the base of the ISCZ can *very briefly* ignite the ^{22}Ne source. A brief (few yr) burst of neutrons reaching $n_\mathrm{n} \approx$ few $\times\, 10^{10}$ cm^{-3} has been found in a 3 M$_\odot$ model of solar composition [58]. This burst is important because it affects, for example, the branchings near ^{86}Kr, ^{87}Rb, ^{96}Zr, ^{134}Ba, ^{137}Ba, ^{152}Gd, as well as producing ^{164}Er, which is made only by the s-process.

In summary, neutron capture nucleosynthesis in AGB stars is a multistep process:

1. proton penetration into the ^{12}C-rich zone during third dredge-up forms a "proton pocket";
2. proton captures on the abundant ^{12}C form the "^{13}C pocket";
3. ^{13}C consumption by α-capture, lasting about 10,000 yr, and the consequent release of neutrons forms an "s-process pocket";
4. ingestion of the s-process pocket by the ISCZ occurs at the next pulse;

5. continued growth of the ISCZ leads to ingestion of fresh ^{13}C into the convective zone and then to another burst of neutrons, and neutron captures, within the ISCZ from the ^{13}C source;
6. at the maximum temperature of the ISCZ there may be a brief burst of neutrons from the ^{22}Ne source.

Note that a very different picture arises if rotationally induced mixing occurs. This smears out both the ^{13}C and and the ^{14}N pockets, before the ^{13}C is consumed (Herwig et al., in press). Since ^{14}N is an efficient neutron absorber, this substantially alters the predicted neutron flux and abundance distribution.

2.5.6.4 The s-Element Distribution

Firstly, we briefly examine what can be learned from the observations. It is instructive to broadly group the s-elements into the lighter elements, designated "ls," and the heavier elements "hs" [103]. For the former, it is usual to define

$$[\text{ls/Fe}] = \frac{[\text{Sr/Fe}] + [\text{Y/Fe}] + [\text{Zr/Fe}]}{3},$$

and for the latter

$$[\text{hs/Fe}] = \frac{[\text{Ba/Fe}] + [\text{La/Fe}] + [\text{Nd/Fe}] + [\text{Sm/Fe}]}{4}.$$

Often, the ratios are taken with respect to H rather than Fe.[4] The ratio hs/ls is thus an indicator of the neutron exposure. Figure 2.18 shows the observed [ls/M] vs [hs/ls] for many AGB stars of various spectral classes (as noted in the figure). Here M is an average metal abundance, based on Ca, Ti, V, Fe, and Co (see [2]). The solid lines give the ratio $N(^{13}\text{C})/N(^{56}\text{Fe})$ normalized to the value that fits the solar system main component. Because this corresponds to a $\tau_0 \approx 0.3$ mbarn^{-1}, one can estimate the τ_0 values by multiplying the $N(^{13}\text{C})/N(^{56}\text{Fe})$ value by 0.3. Hence we see that the mean neutron exposure varies from about 0.1 to 0.6 mbarn^{-1} with an average near 0.3, just as found for the ^{13}C source in the previous section.

Neutron capture nucleosynthesis is mostly a *secondary* process, because it is dependent on the initial seed nuclei such as ^{56}Fe. Given what we think we know about the possible formation mechanisms for the ^{13}C pocket, this pocket is not in principle dependent on the initial metallicity of the star. Hence the neutron source is probably largely independent of the metallicity of the star. Although it may depend on the initial rotation rate or other

---

[4] Sometimes the individual element abundances [el/Fe], or [el/H], are weighted by the number of spectroscopic lines used to determine the abundance. Sometimes Sr is not included in ls due to more uncertainties in the abundance determination than for either Y or Zr; indeed, the elements used for hs and ls can vary between authors.

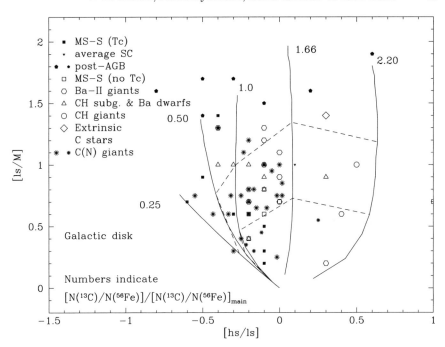

Fig. 2.18. Observed [ls/M] vs. [hs/ls] for various AGB stars. Each curve represents the evolution of the envelope composition as material is continually added via dredge-up. M is an average metal, based on Ca, Ti, V, Fe, and Co content. The solid lines give the ratio $N(^{13}C)/N(^{56}Fe)$ normalized to the value that fits the solar system main component, with $\tau_0 \approx 0.3$ mbarn^{-1}. The dashed lines connect the points corresponding to the fourth and eighth dredge-up episodes. Figure from [3] supplied by M. Busso

parameters, which vary from star to star, it is not expected to vary in a systematic way with the stellar metallicity. Thus we expect, and find, that the resulting distribution of elements depends strongly on the metallicity: A given flux of neutrons is absorbed by a certain number of seed nuclei, the latter being proportional to the metallicity. Thus seed nuclei in low-metallicity stars experience more neutron captures per seed nucleus than in higher metallicity stars. Thus the distribution moves from ls to hs as the initial [Fe/H] of the host star decreases. This is illustrated in Figures 2.19 and 2.20. Note in particular that the shift to heavier species at lower metallicity is sufficient to explain all of the galactic inventory of ^{208}Pb, without the need for an ad hoc "strong" component [58]. Also, as shown in Figure 2.20, the flat distribution for a typical AGB star with [Fe/H]=−0.3 indicates that the main component of the solar system distribution is easily reproduced by such AGB stars.

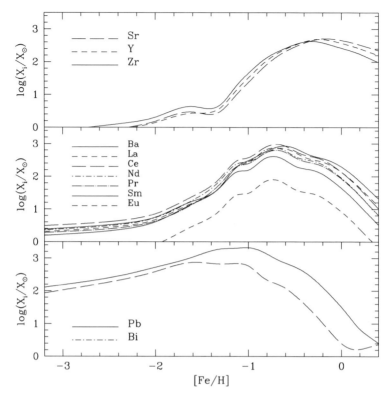

Fig. 2.19. Enhancement factors (relative to solar distribution) for a $2\,M_\odot$ model "standard" ^{13}C pocket, burned radiatively between pulses [58]. Note that lower-metallicity stars produce relatively more heavy nuclides than the more metal-rich stars. From [27]

Before ending our discussion of the nucleosynthesis of heavy elements, we note that even a true Population III star, with $Z = 0$, can produce s-process elements [60]. This surprising result is due to the fact that Ne and Mg, produced by the thermal pulses (see Section 2.5.9) can form the basis for neutron captures producing elements all the way through to Pb and Bi.

Finally, estimates of the contribution of AGB stars to the galactic content of the s-process elements are limited by our understanding of dredge-up; for a discussion see [142].

2.5.7 The Production of Fluorine

It was Jorissen, Smith and Lambert [85] who showed that [F/O] correlates with C/O in AGB stars, with C stars (of N spectral-type) showing up to 30

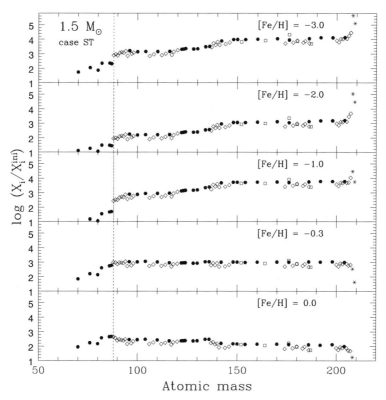

Fig. 2.20. Enhancement factors for neutron-rich nuclides in a 1.5 M$_\odot$ model of various metallicities using the "standard" ^{13}C pocket [58]. *Heavy dots* represent nuclei made only by the s-process; *open squares* show nuclei whose production is at least 80% from the s-process; *open diamonds* mark nuclei with a 60–80% s-process origin. Figure supplied by R. Gallino and S. Masera

times the solar fluorine abundance.[5] The increase in C abundance is clearly due to the third dredge-up and thermal pulses, and hence these are also implicated in the production of ^{19}F. These authors considered many possible routes to ^{19}F and concluded that the most likely chain is

$$^{14}\text{N}(\alpha,\gamma)^{18}\text{F}(\beta^+\nu)^{18}\text{O}(p,\alpha)^{15}\text{N}(\alpha,\gamma)^{19}\text{F}.$$

Quantitative studies of this sequence [52, 111] showed that the situation is quite complicated, involving neutron captures (and hence the ^{13}C pocket again [112]) to produce protons via (n, p) reactions on various species such as ^{14}N and ^{26}Al. The required ^{15}N can be produced by different nuclear paths,

---

[5] We note the seemingly anomalous result that the derived F abundances for the SC stars are greater than for the C stars. This is hard to understand, and it was pointed out in [85] that the SC stars have n(C)=n(O) and hence opacity and thermal atmospheric effects can lead to large and systematic errors [84].

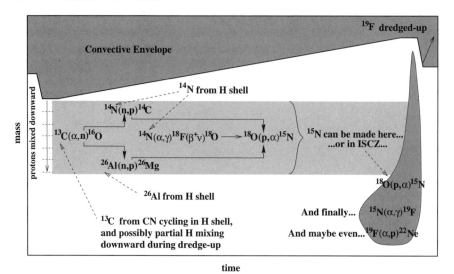

Fig. 2.21. Schematic (not to scale) showing the production of ^{19}F beginning with neutrons from ^{13}C and then (n, p) reactions on ^{14}N and ^{26}Al. The ^{14}N also produces ^{18}O, which then captures a proton to make ^{15}N, and finally α-captures on this nuclide produce ^{19}F. The ^{15}N production can occur either in the erstwhile ^{13}C pocket or the ISCZ itself, depending on the mass and composition of the star. For more massive stars the ^{19}F can also be destroyed by (p, α) in the envelope

and these can occur either in the intershell region between pulses or in the ISCZ during pulses, depending on the maximum temperatures reached, which in turn depend on the core (and hence initial) mass. To complicate things further, ^{19}F can be destroyed again at high enough temperatures. We try to clarify the situation below.

Figure 2.21 shows the main steps involved in the production of ^{19}F. We start with some ^{13}C, which is left over from CN burning in the H shell and is probably supplemented by the ^{13}C pocket. As the star contracts this ^{13}C captures α-particles and releases neutrons. These neutrons must serve for the production of the s-process elements, as we saw above, but they can also be captured by other species, and we note that ^{14}N and ^{26}Al are both strong neutron absorbers. Both of these species are produced by the H shell as it burns outward between pulses. Subsequent (n, p) reactions on these nuclides produce protons in a region otherwise devoid of H: ^{14}N(n, p)^{14}C and ^{26}Al(n, p)^{26}Mg. Simultaneously, the ^{14}N produces ^{18}O via ^{14}N(α, γ)^{18}F($\beta^+\nu$)^{18}O. So this region contains both ^{18}O and protons. Hence, if the temperature is high enough, we can produce ^{15}N by ^{18}O(p, α)^{15}N. Note that for stars of lower mass, the intershell may not reach high enough temperatures for this reaction to occur, in which case it can occur in the ISCZ during the next pulse. In any event,

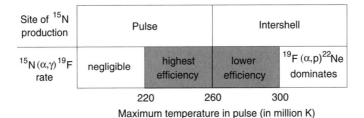

Fig. 2.22. Summary of ^{19}F production as a function of the temperature at the base of the ISCZ, which correlates with stellar mass

we succeed in producing some ^{15}N in the hot ISCZ, and this enables the production of ^{19}F through ^{15}N$(\alpha,p)^{19}$F.

With so many steps involved in the production of ^{19}F, it is not surprising that the details are quite complicated. For quantitative studies one must include the contribution from material left over from previous pulses, and other possible reactions such as ^{12}C$(n,\gamma)^{13}$C, proton capture on species other than ^{18}O, as well as reactions that destroy ^{19}F, such as ^{19}F$(n,\gamma)^{20}$F and ^{19}F$(\alpha,p)^{22}$Ne. The former is unimportant unless the neutron flux is high, and the latter is important only in the more massive (hotter) AGB stars. We summarize our understanding in Figure 2.22, but note that the situation regarding ^{19}F is particularly complex, and many of the reaction rates involved are quite uncertain and undergoing revision at the time of writing. Although AGB stars are likely to be the main source of ^{19}F in the Galaxy, a reliable quantitative estimate is not yet possible, since it requires accurate estimates of the size of the ^{13}C pocket as well as dredge-up, and detailed model calculations across a range of masses and compositions to allow for the sensitivity of the ^{19}F production mechanisms to the temperature of the ISCZ and the intershell region between pulses.

2.5.8 The Production of ^{22}Ne and ^{23}Na

The Ne–Na chain of reactions is shown in Figure 2.23. At quite low temperatures, about 15 million K, we find ^{22}Ne begins to burn into ^{23}Na. For temperatures above about 35 million K there is an important contribution to ^{23}Na from the burning of ^{20}Ne. Note that the decrease in ^{20}Ne is small, but this still produces significant ^{23}Na. At still higher temperatures, in excess of about 55 million K, ^{23}Na begins to burn by proton captures, causing a leakage into the Mg–Al chain. Indeed, whether the Ne–Na reactions are a *cycle* or a *chain* depends on the ratio of the (p,γ) to (p,α) rates on the ^{23}Na nucleus. Current estimates of the rates indicate that below about 55 million K the reactions act as a cycle, with significant leakage occurring only at higher temperatures [6].

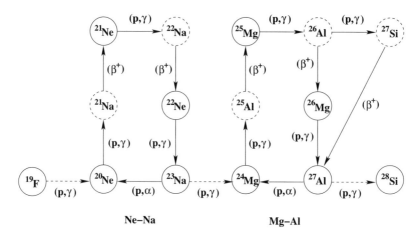

Fig. 2.23. Summary of Ne–Na and Mg–Al chains. Unstable isotopes are shown in dashed circles, and dashed lines show leakage into and out of the chains. After [6]

Prior to the AGB phase the surface of the star may be enriched in ^{23}Na by the first dredge-up, and also the second dredge-up if it occurs. The increase can be up to 100% (0.30 dex) depending on the mass and composition [46, 110]. Larger enrichments require thermal pulses.

The ISCZ is composed of the products of H-burning, as deposited by the H-burning shell. Thus this region is mostly ^{4}He and ^{14}N (from CNO cycling). During the thermal pulse the dominant reaction is the production of ^{12}C, but the abundant ^{14}N also captures α-particles via

$$^{14}\text{N}(\alpha,\gamma)^{18}\text{F}(\beta^+\nu)^{18}\text{O}(\alpha,\gamma)^{22}\text{Ne},$$

and thus primary[6] ^{22}Ne is produced. This will be dredged to the surface during the next third dredge-up episode. Note that the ^{22}Ne itself is subject to further α-captures if the temperature is sufficiently high (exceeding about 300 million K), so that the fate of the ^{22}Ne is sensitive to the stellar mass.

The H-burning shell is primarily powered by CNO cycling, of course. But other proton capture reactions occur there, such as the Ne–Na and and Mg–Al chains (Figure 2.23). We have just seen that thermal pulses can enrich the envelope in ^{22}Ne. The Ne–Na reactions then burn this into ^{23}Na. Note that this is primary ^{23}Na.

The ^{23}Na contribution from the original ^{20}Ne content is negligible compared to that produced as a result of thermal pulsing [110]. It is worthwhile examining Figure 2.24 to see the steps involved. Initially, primary ^{12}C is produced by the thermal pulse in ISCZ #1. This ^{12}C is dredged to the surface,

---

[6] In this context, *primary* indicates that the amount produced does not depend on the presence of any seed nuclei, but proceeds from the initial hydrogen content.

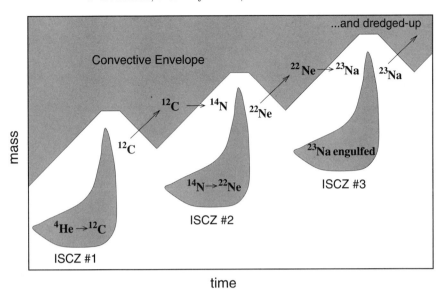

Fig. 2.24. The various stages involved in the production of ^{23}Na by thermal pulses. Short arrows indicate processing by α-captures in the ISCZ, intermediate arrows indicate processing via protons captures in the H-burning shell, and the long arrows show material being dredged to the surface

and then the advance of the H-burning shell through this material produces primary ^{14}N. When the next pulse occurs, this ^{14}N is engulfed by the ISCZ (#2 in the figure) and suffers α-captures to produce ^{22}Ne via the reactions shown above. During the subsequent dredge-up phase this (primary) ^{22}Ne is mixed to the surface. During the interpulse phase the H-burning shell processes this ^{22}Ne via the Ne–Na chain and produces primary ^{23}Na which is engulfed by ISCZ #3 and then mixed to the surface by the next dredge-up event. Thus is the surface of the star enriched in substantial amounts of primary ^{23}Na [110].

Of course, each thermal pulse and associated ISCZ plays the role of all three ISCZs shown in Figure 2.24, so that each pulse produces primary ^{12}C at the same time as it is producing primary ^{22}Ne as well as engulfing the primary ^{23}Na produced by the H-shell during the previous interpulse phase!

Since the ^{23}Na is fed from the primary ^{12}C made in the star, we expect lower-metallicity stars to be efficient producers of ^{23}Na. There is also a mass dependence through various factors, such as the dilution and overlap of the shells [110] as well as the activation of ^{22}Ne destruction in the ISCZ through α-captures. This occurs for temperatures above about 300 million K, present in the intermediate mass AGB stars. As a guide, for $M \gtrsim 4\,M_\odot$ models we find $\log Y(^{23}\text{Na})$ as high as ≈ -5. (An example may be seen in Figure 2.41, to be

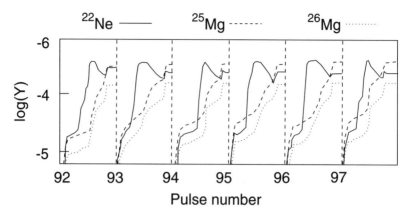

Fig. 2.25. Abundances in the ISCZ for pulses 92 to 97, in the evolution of a $6\,M_\odot$ model with $Z = 0.004$. Time within the pulse is scaled from start to end of pulse, and values are plotted only while the ISCZ exists. The solid line is for ^{22}Ne, the dashed line is for ^{25}Mg, and the dotted line is for ^{26}Mg

discussed below.) In summary, ^{23}Na production is expected to be important for more massive stars (say $M \gtrsim 4\,M_\odot$) and for metallicities below solar. It may also be important in the later pulses even for stars of lower mass.

2.5.9 The Production of ^{25}Mg and ^{26}Mg

We have seen how ^{22}Ne is produced in the ISCZ of the thermal pulse, via α-captures on ^{14}N. If the maximum temperature in the ISCZ exceeds about 300 million K, then further α-captures can make ^{25}Mg and ^{26}Mg via ^{22}Ne$(\alpha,n)^{25}$Mg and ^{22}Ne$(\alpha,\gamma)^{26}$Mg, respectively. The production of these heavy Mg isotopes can be substantial, especially at higher masses and/or lower metallicities. During the following dredge-up phase these heavy magnesium isotopes are then mixed to the surface.

Figure 2.25 shows the abundances of ^{22}Ne, ^{25}Mg, and ^{26}Mg within the ISCZ for pulses 92–97 in the evolution of a $6\,M_\odot$ model with $Z = 0.004$. The abundances are plotted only when there is an ISCZ present, and the time is scaled to the (normalized) pulse duration. Note that ^{22}Ne grows initially as the α captures on the abundant ^{14}N proceed. But ^{22}Ne soon reaches a peak and then decreases, as ^{22}Ne$(\alpha,n)^{25}$Mg begins producing ^{25}Mg and ^{22}Ne$(\alpha,\gamma)^{26}$Mg begins producing ^{26}Mg. The last point for each pulse shows the final composition of the ISCZ when the convection disappears, and hence this is the composition of the ISCZ material that is then dredged to the surface during the following dredge-up episode. Figure 2.26 shows the surface composition of the Mg isotopes for the same model (over the whole AGB lifetime). This clearly illustrates the increase in the surface abundances of

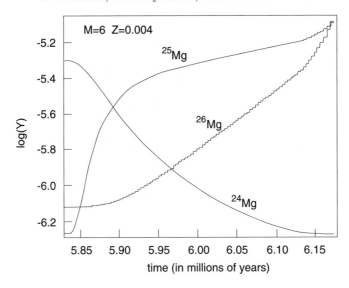

Fig. 2.26. Surface Mg abundances during the AGB evolution of the model in Figure 2.25. The time axis is offset by 60 million yr (since the ZAMS)

the heavy isotopes of Mg. There is also a decrease in ^{24}Mg in this star, due to processing via hot bottom burning (see Section 2.5.11).

2.5.10 The Aluminium Isotopes

Following dredge-up, the Mg isotopes produced within the ISCZ are subjected to further processing by the H-burning shell via the Mg–Al chain, which we now summarize (see Figure 2.23). Firstly, we note that ^{26}Al has both a ground state ^{26}Al^{g} with a half-life of $t_{1/2} = 740$ million yr as well as an isomeric state ^{26}Al^{iso} with a half-life of $t_{1/2} = 6.35$ s. These are well out of equilibrium at the temperatures relevant to AGB stars, and hence must be considered as separate species. Wherever we refer to ^{26}Al without specifying the state, it is understood that we mean ^{26}Al^{g}.

The Mg–Al chain begins when temperatures exceed about 35 million K with proton captures on ^{25}Mg producing ^{26}Al. When the temperature exceeds about 55 million K the next step begins, being proton captures on ^{26}Al to produce ^{27}Al via ^{26}Al$(p,\gamma)^{27}$Si$(\beta^{+}\nu)^{27}$Al. It is not until 65 million K that the abundant ^{24}Mg begins to suffer destruction via proton capture, overcoming the leakage into ^{24}Mg from the Ne–Na chain at these temperatures. ^{26}Mg plays only a minor role for two reasons: Firstly, the ^{26}Mg(p,γ) rate is quite slow, and secondly, it is largely bypassed by (p,γ) reactions on ^{26}Al. Note that the Mg–Al reactions are a chain, rather than a cycle, because the ^{28}Si(p,γ) rate greatly exceeds the ^{28}Si(p,α) rate for $T \gtrsim 60$ million K [6].

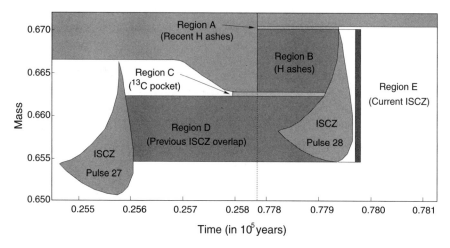

Fig. 2.27. The various regions involved in the production of ^{26}Al by thermal pulses

The Mg–Al chain produces ^{26}Al and ^{27}Al via proton captures on ^{25}Mg and ^{26}Mg, respectively. These Al isotopes are in turn dredged to the surface and then expelled into the interstellar medium by the stellar wind. The ^{26}Al decays with a $t_{1/2}$ =740 million yr into ^{26}Mg, and its presence can be detected in presolar meteorite grains (e.g., [181]).

The production of Al is actually another rather complicated process, due to the temperature dependence of the Mg–Al chain as well as the fragility of ^{26}Al to neutron capture, for details see [113]. Recall that there may be three neutron sources active in the regions where ^{26}Al is present: The secondary ^{13}C from CN cycling in the H-burning shell, a possible primary ^{13}C pocket from partial mixing of protons downward during the third dredge-up phase, and the ^{22}Ne source in the ISCZ.

The production (and destruction!) of ^{26}Al is summarized in Figure 2.27, where we show five regions after [113]. Firstly, Region A represents the small amount of mass ($\approx 0.001\,M_\odot$) between the outermost tip of the ISCZ and the H-burning shell (effectively the bottom of the convective envelope). Region B indicates the material processed by the H-burning shell since the last pulse, usually referred to as ΔM_H and up to a few $\times 10^{-2} M_\odot$ in extent. Region C represents the ^{13}C pocket (which has now been transformed into the s-process pocket). Region D is the overlap between the current ISCZ and the previous ISCZ, and contains material subjected to He-burning at the previous pulse. It is also up to a few $\times 10^{-2}\,M_\odot$ in extent. Region E is the new ISCZ, which mixes together Regions B, C, and D. Within Region A the H-shell has essentially burned all ^{25}Mg into ^{26}Al by proton captures. Region B also contains ^{26}Al produced by the H-shell, but with a half-life comparable to the interpulse period there will have been some decay into ^{26}Mg at the bottom of Region B, where the ^{26}Al has been present for the longest time. It is also Region B that sees a neutron flux from the ^{13}C produced in the H-shell. The neutron capture cross-section for ^{26}Al is quite large, so that substantial ^{26}Al is destroyed via ^{26}Al(n, p)^{26}Mg. There is indeed fierce competition for these neutrons, with ^{14}N, the iron group, and the s-process elements all involved. The small ^{13}C pocket in Region C (of order a few $\times 10^{-4}\,M_\odot$) has already experienced neutron captures to form the s-process elements. Region D contains the material left over from the last pulse, which is mostly ^{12}C-rich but there may be other relevant species (especially the heavy Mg isotopes, if the temperature at the base of the ISCZ exceeds about 300 million K). Finally, Region E mixes Regions B, C, and D together and subjects the whole mixture to high temperatures in an α-rich environment. During this phase there may be further destruction of ^{26}Al via neutron capture if the ^{22}Ne neutron source is activated. If this is the case, then ^{25}Mg and ^{26}Mg will also be produced.

If the temperature in the H-burning shell exceeds about 65 million K, then Region B will show the transformation of ^{24}Mg into ^{25}Mg, and ^{26}Al into ^{27}Al that is characteristic of the Mg–Al chain at higher temperatures. This can be expected for more massive or more metal-poor stars, or the later pulses of lower-mass stars. The whole calculation must be performed self-consistently, because not only do the temperatures of the various regions change with evolution, but so do their sizes.

2.5.11 Hot Bottom Burning

Hot bottom burning (hereafter HBB) is the colorful name given to the circumstance where the bottom of the convective envelope reaches temperatures high enough for nuclear processing; it is also sometimes called "convective envelope burning." Another way of thinking of it is that the convective envelope reaches into the top of the hydrogen burning shell. Perhaps the earliest self-consistent calculation showing the occurrence of HBB was the evolution of a $7\,M_\odot$ model by Iben [76]. This was soon followed by calculations by Sackmann, Smith, and Despain [122], who assumed an HBB structure and looked at the resulting nucleosynthesis. Many envelope structures were determined by Scalo, Despain, and Ulrich [125]. Crucially, these latter two sets of calculations included an approximation for time-dependent mixing, and they showed that Li could be produced in large quantities.

Interest was renenewed in HBB when Smith and Lambert discovered Li-rich giants in the Magellanic Clouds [134, 135]. Many models were calculated beginning in the early 1990s (e.g., [17, 21, 96]). These showed that for intermediate-mass stars the temperature at the base of the convective envelope could reach up to 100 million K. We now know that the minimum mass for the onset of HBB depends on the composition, ranging from about $5\,M_\odot$ at solar metallicity to maybe as low as $2\,M_\odot$ for $Z = 0$ stars [131].

Figure 2.28 shows the evolution of the temperature at the bottom of the convective envelope during the AGB phase of evolution for a $6\,M_\odot$ model

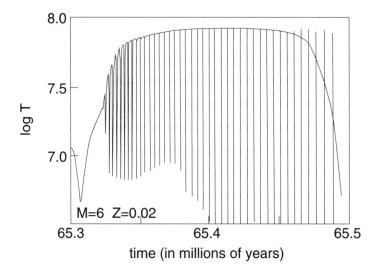

Fig. 2.28. The temperature at the bottom of the convective envelope during the AGB phase of evolution for a $6\,M_\odot$ model with $Z = 0.02$. Note that HBB ceases, due to a reduced envelope mass, near the end of the AGB evolution

with $Z = 0.02$. Note that the temperature rapidly rises, reaching almost 100 million K. As the evolution proceeds, and mass loss reduces the envelope, the HBB is eventually shut off when the temperature declines again. The details of when this happens will depend on the mass-loss formula used (the calculations quoted here used the formula in [154]).

The temperature distribution near the bottom of the envelope is shown in Figure 2.29. The shaded region is the convective envelope, and the steep temperature gradient at the bottom of the envelope is clearly visible. Although the temperatures are quite high, the region exposed to these temperatures is quite small in mass, as shown on the enlarged scale in Figure 2.30. Typically, the region seeing $T \gtrsim 10$ million K is $\approx 10^{-4}\,M_\odot$ in extent, and the region with $T \gtrsim 30$ million K is ten times smaller ($\approx 10^{-5}\,M_\odot$). The extent in radius is typically a few $\times 10^{-2}\,R_\odot$, and the density is $\approx$ few $\mathrm{g\,cm^{-3}}$.

Material in an HBB envelope is transported through the small, hot region many times between pulses. If we take a typical convective velocity as $v_{\mathrm{conv}} \approx 10^5\,\mathrm{cm\,s^{-1}}$ and a typical stellar radius as several hundred $R_\odot$, then we can estimate a convective turnover time as roughly $\tau \approx 500\,R_\odot/v_{\mathrm{conv}} \approx$ few years. Given that the interpulse period is a few thousand years, then each nucleus in the envelope is exposed to the high temperatures at the bottom of the envelope $\approx 10^3$ times between each pulse.

Most stellar convective zones have turnover times that are much less than the nuclear burning time scale, and hence the composition in these zones is

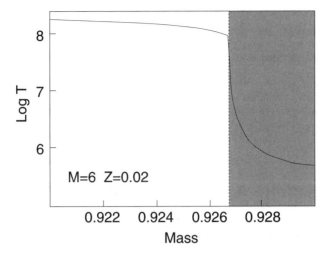

Fig. 2.29. Variation of temperature near the bottom of the convective envelope in a $6\,M_\odot$ model with $Z = 0.02$. The shaded region is convective, and the total mass is $5.623\,M_\odot$ at this stage

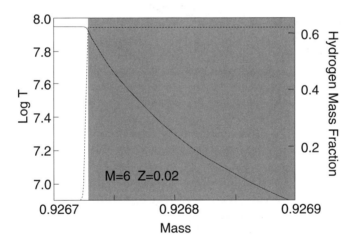

Fig. 2.30. Same as Figure 2.29 but at higher resolution, and also with the H-profile shown dashed

homogeneous to high accuracy. That is not true in the present case, at least for some species. For this reason it is crucial that the composition in the HBB envelope be calculated with some approximation for time-dependent mixing: Usually a diffusion equation is solved for the mixing process.

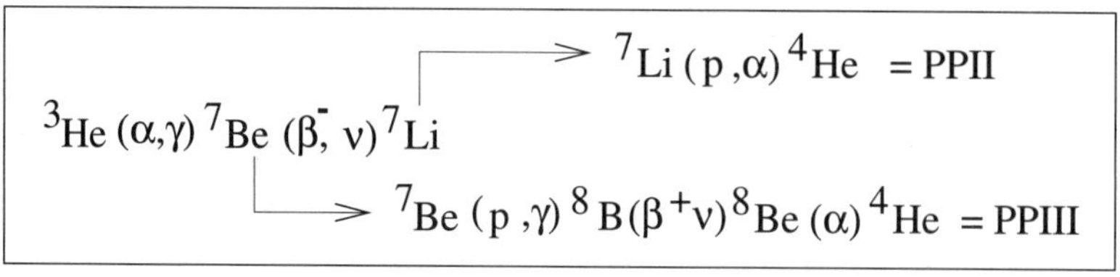

$$^3\text{He}(\alpha,\gamma)^7\text{Be}(\beta^-,\nu)^7\text{Li} \begin{cases} \longrightarrow {}^7\text{Li}(p,\alpha)^4\text{He} = \text{PPII} \\ \longrightarrow {}^7\text{Be}(p,\gamma)^8\text{B}(\beta^+\nu)^8\text{Be}(\alpha)^4\text{He} = \text{PPIII} \end{cases}$$

Fig. 2.31. Summary of the Cameron–Fowler mechanims for producing ^{7}Li

Before proceeding with the nucleosynthesis aspects we discuss briefly another important aspect of HBB: It dramatically alters the core mass–luminosity relation on the AGB, as discussed earlier in Section 2.3.2. The linear relation between the mass M_H of the H-exhausted core and the surface luminosity L (the Paczynski relation) has been given above. Later authors extended this equation to stars of lower masses, and derived similar if slightly different relations, with some composition dependence (e.g., [19, 94, 175]). These prompted attempts to understand the physical basis of the relations, using semianalytical arguments and homology relations, such as in [91]. It was shown in [144] that the crucial requirement was a radiative zone separating the convective envelope from the degenerate core. It is this radiative "buffer" zone that permits the evolution to be largely independent of the envelope mass, and results in the relation between the core mass and the luminosity. However, for more massive AGB stars this zone does not exist, because the convective envelope reaches into the top of the H-burning shell. Hence the very existence of HBB precludes the usual relation between core mass and luminosity [17, 21]. The HBB models show luminosities much higher than predicted by the core mass–luminosity relations, and hence estimates based on these formulae are invalid for the more massive models. In particular, since the limiting AGB core mass is the Chandrasekhar mass, this has sometimes been used in conjunction with a core mass–luminosity relation to provide a luminosity limit for AGB stars. This is not a valid procedure, because HBB stars can exceed this luminosity.

2.5.12 Hot Bottom Burning: Li Production

Perhaps the first consequence of HBB is the rapid production of ^{7}Li. Cameron [28] outlined the reactions involved, and Cameron and Fowler [29] suggested that AGB stars could be the site of the reactions. The reactions involved are shown in Figure 2.31. At the bottom of the convective envelope some ^{3}He (created earlier in the evolution) captures a ^{4}He nucleus to create ^{7}Be. If this ^{7}Be is exposed to high temperatures, then it can capture a proton to complete the PPIII chain. Alternatively, if the ^{7}Be can capture an electron, it can produce ^{7}Li. Of course, this ^{7}Li is itself subject to proton captures, and

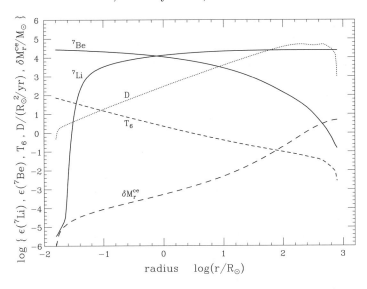

Fig. 2.32. Envelope structure and compositon for a 6 M$_\odot$ model with $Z = 0.02$, near the time of maximum ^{7}Li content. From [121]

these would lead to completion of the PPII chain. The requirement for ^{7}Li production is therefore a high-temperature region that can produce the ^{7}Be, and then some mixing to transport the ^{7}Be away to cooler regions where it can capture an electron to make the ^{7}Li. This is exactly what occurs in stars that have HBB in their envelopes.

We here summarize the comprehensive account of ^{7}Li production by Sackmann and Boothroyd [121]. Figures 2.32 and 2.33 show some important structure and composition information for a 6 M$_\odot$ model with $Z = 0.02$, taken from [121], at the time of maximum ^{7}Li content. Figure 2.32 shows the ^{7}Be and ^{7}Li content, as well as the temperature (in 10^6 K) and the diffusion coefficient used in the mixing algorithm $D = v_{\rm conv}\ell$, where ℓ is the mixing length. Also shown is the mass $\delta M^{\rm ce}$ as measured from the bottom of the convective envelope. In Figure 2.33 are shown various time scales throughout the envelope at the same time as in Figure 2.32. The convective mixing time scale $\tau_{\rm conv}$ is defined as $\ell/v_{\rm conv}$, and the nuclear time scales $\tau(x)$ are the time scales for the reactions destroying nucleus x via all relevant reactions, unless otherwise indicated.

The crucial requirement is time-dependent mixing: if one makes The usual assumption of instantaneous mixing within convective zones (i.e., that the mixing time scale $\tau_{\rm conv}$ is much less than the nuclear time scale $\tau_{\rm nuc}$) then the ^{7}Li abundance decreases during AGB evolution, which explains the low ^{7}Li abundances found in [76] despite a hot envelope. As shown in Figure 2.33

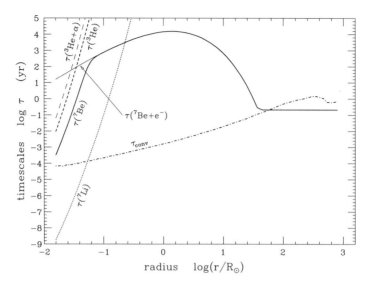

Fig. 2.33. Some important time scales in the envelope of a $6\,M_\odot$ model with $Z = 0.02$, near the time of maximum ^{7}Li content. From [121]

the time scale for ^{3}He destruction is always much longer than $\tau_{\rm conv}$, so we expect (and find) the ^{3}He abundance to be homogeneous within the envelope. Contrast this with ^{7}Be and ^{7}Li, which are far from constant within the convective zone. Near the surface the time scale for ^{7}Be destruction by electron capture is shorter than $\tau_{\rm conv}$, and hence the ^{7}Be content decreases substantially here. In the inner part of the envelope we have $\tau(^7{\rm Li}) < \tau_{\rm conv}$, which is why the ^{7}Li decreases in this region. In other words, the ^{7}Be is made at the bottom of the envelope and mixed outward. When $\tau(^7{\rm Be})$ decreases to a value near (or below) $\tau_{\rm conv}$ (i.e., near the surface), then the ^{7}Be forms ^{7}Li via electron capture. This ^{7}Li is then mixed downwards until it reaches a point where $\tau(^7{\rm Li})$ decreases to a value near (or below) $\tau_{\rm conv}$, with consequent ^{7}Li destruction through the PPII reactions.

Figure 2.34 shows the variation of the surface content of ^{7}Li for the $6\,M_\odot$ model shown also in Figures 2.28–2.30. The ^{7}Li content increases quite quickly, but after reaching a peak of $\log \epsilon(^7{\rm Li}) \approx 4$ it starts to decrease again. From the previous paragraph, the reasons are clear. The ^{7}Li is destroyed on each passage through the bottom of the convective envelope, by proton captures (see Figure 2.31), completing the PPII chain. However, the ^{7}Li is replenished by the ^{7}Be produced at the bottom of the envelope via the ^{3}He + ^{4}He reactions occurring there. This can continue only as long as there is ^{3}He present, because this is exhausted first. The resulting maximum surface

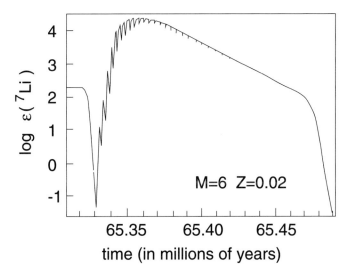

Fig. 2.34. Surface ^{7}Li content during the AGB evolution of a $6\,M_\odot$ model with $Z = 0.02$

abundances are typically $\log \epsilon(^7\mathrm{Li}) \approx 4$ and are in very good agreement with the observed super Li-rich giants seen in the Magellanic Clouds (e.g., [1, 134]).

Finally, we note that the details of the ^{7}Li production depend on the convection model used, through quantities such as v_{conv}. A comparison between the standard mixing-length theory and the Canuto and Mazzitelli [30, 31] "full spectrum of turbulence" (FST) theory is given in [39, 104, 156]. FST favors HBB (and ^{7}Li production) at lower masses than the mixing-length theory. Which theory is most favored by the observations is still uncertain at present.

2.5.13 Hot Bottom Burning: F and the CNO Cycles

2.5.13.1 The CNO cycles

The CNO cycles are summarized in Figure 2.35. The most important, for our purposes, is CNO$_\mathrm{I}$, often simply called the CN cycle. Due to the dominance of $^{15}\mathrm{N}(p,\alpha)$ over $^{15}\mathrm{N}(p,\gamma)$ this chain is a true cycle, operating about 1000 times faster than CNO$_\mathrm{II}$, and coming into equilibrium very quickly. The main result, apart from fusing H into ^{4}He, is the conversion of most nuclei into ^{14}N. This is due to the bottleneck created by the slow $^{14}\mathrm{N}(p,\gamma)^{15}\mathrm{O}$ reaction. The ratio $^{12}\mathrm{C}/^{13}\mathrm{C}$ reaches an equilibrium value of about 3–4, depending on the temperature.

There are three possible branches, at nuclides ^{15}N, ^{17}O, and ^{18}O (and the possibility of breakout to the Ne–Na cycle via ^{19}F). Note that ^{17}O is

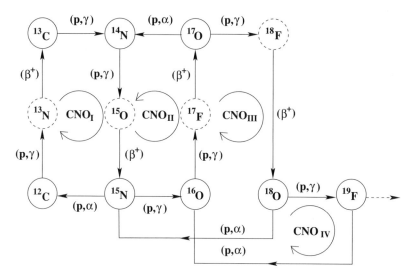

Fig. 2.35. Summary of the CNO cycles. Unstable isotopes are shown in dashed circles, and the dashed line shows leakage out of the chains. After [6]

produced at low temperatures but destroyed by temperatures above about 20 million K. The branching at ^{17}O is very sensitive to temperature, and although quite uncertain, this will not hinder our discussion very much. The "hot CNO cycles," which are defined by proton captures on the radioactive nuclei ^{13}N, ^{17}F, and ^{18}F are not important in AGB stars.

2.5.13.2 Hot Bottom Burning

We can make a rough estimate of the conditions experienced by material at the bottom of the convective envelope, and hence its consequences for the operation of the CNO cycles. From Figure 2.32 we see that the thickness of the burning region is about $0.1\,R_\odot$, and if we take a typical convective velocity as $10^5\,\mathrm{cm\,s^{-1}}$, then a typical time to traverse the burning region is $\approx 10^{-3}$ yr. Given that material circulates through the envelope a few thousand times between pulses, then material will spend a few years, on average, in the high-temperature region between pulses. But at these temperatures the time scale for the CN cycle (i.e., CNO$_\mathrm{I}$) to come into equilibrium is also of order years. Hence we may expect the cycle to reach equilibrium over the entire envelope in just a few thermal pulse cycles. Thus the surface value of the ratio ^{12}C/^{13}C should rapidly approach the equilibrium value of 3–4. This indeed does occur, as shown in Figure 2.36 for the $6\,M_\odot$ model we have considered earlier. Very rapidly the ratio decreases towards its equilibrium value, during the first few pulses. As the dredge-up deepens, we see an increase in the ratio due to the

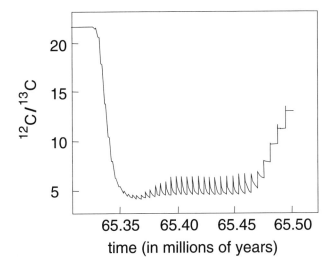

Fig. 2.36. The variation of the surface ratio of $^{12}C/^{13}C$ during the AGB evolution of a 6M$_\odot$ model with $Z = 0.02$

addition of ^{12}C, but then the ratio decreases again as the HBB processes the added ^{12}C into ^{13}C. Toward the end of the evolution the decrease in the envelope mass via mass loss results in the HBB ceasing while the dredge-up continues: Hence we see the increase again in $^{12}C/^{13}C$ toward the end of the evolution.

Figure 2.37 shows the surface abundances of ^{12}C, ^{13}C, and ^{14}N during the AGB phase of the 6M$_\odot$ model, also shown in Figure 2.36. In this star the HBB begins immediately, and we see the ^{12}C content of the envelope decrease as it is burned into ^{13}C. This occurs at the same time that the star produces large amounts of ^7Li, as discussed in Section 2.5.12 and shown in Figure 2.34. Once the ^{13}C abundance reaches its peak, it begins to decrease in concert with the ^{12}C as the CN cycle continues to operate in equilibrium, when averaged over the interpulse period. Each third dredge-up event increases the ^{12}C content, shown in the figure as essentially instantaneous increases. HBB then burns this to ^{13}C before the next pulse occurs. A consequence of this CN cycling is the progressive increase in the amount of ^{14}N present in the envelope, as clearly shown in Figure 2.37. Note that this is *primary* ^{14}N, since it is made from the primary ^{12}C produced in the stellar interior by the triple-alpha reactions.

But what are the observational consequences of this HBB? It is repeated additions of ^{12}C to the envelope via the third dredge-up that lead to the formation of C stars. So if HBB converts this ^{12}C mostly into ^{14}N, then this can prevent the star from becoming a C star, despite the presence of effi-

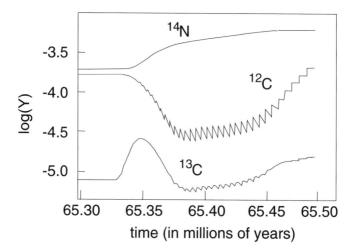

Fig. 2.37. Surface abundances of ^{12}C, ^{13}C, and ^{14}N during the AGB evolution of a 6 M$_\odot$ model with $Z = 0.02$

cient dredge-up [22]. We summarize the situation in Figure 2.38. This figure shows the C/O ratio on the AGB for $M = 4$, 5, and 6 M$_\odot$, each at $Z = 0.02$, 0.008, and 0.004. The 4 M$_\odot$ models (left column) do not experience HBB and succeed in adding enough C to their envelope to become C stars. Note that the resulting values of C/O are higher for lower metallicities: This is easily understood as a consequence of the low initial O abundance (scaling roughly with Z), whereas the amount of ^{12}C added is roughly constant (in fact, the efficiency of the third dredge-up increases at lower Z, further enhancing this effect). Next examine the 6 M$_\odot$ cases, which all show HBB: Hence the C/O ratio declines from the start of the AGB. (The increase toward the end is due to mass loss and will be discussed below). The intermediate case of 5 M$_\odot$ is interesting. First consider the $M = 5$ M$_\odot$, $Z = 0.004$ case. Here the C/O ratio begins to increase due to third dredge-up, and then HBB begins and the C/O ratio decreases rapidly. The same is seen in the $M = 5$ M$_\odot$, $Z = 0.008$ case, except that here the model almost becomes a C star before the decline. Changes in the mass loss rate or convection parameters could allow this star to (temporarily) become a C star. Finally, the $M = 5$ M$_\odot$, $Z = 0.02$ case starts burning C via HBB just when the decrease in envelope mass shuts off the HBB again, and the C/O ratio increases. A similar situation is seen in the $M = 4$ M$_\odot$, $Z = 0.004$ case.

Figure 2.38 nicely illustrates the interplay between the third dredge-up and HBB and how they determine whether a star has C/O > 1 or < 1 on the AGB. It also illustrates the critical mass for the appearance of HBB, and how this decreases with decreasing metallicity, reflecting the well-known

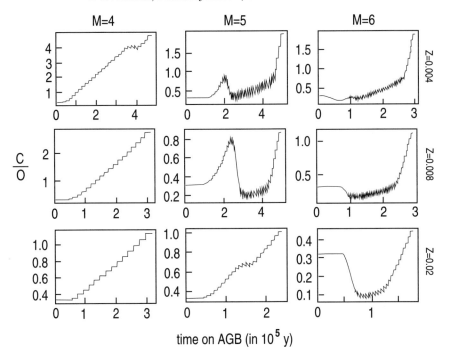

Fig. 2.38. The C/O ratio during the AGB evolution for nine stellar models during the AGB phase. The **columns** give the three masses 4, 5, and 6 M$_\odot$, and the three **rows** correspond to $Z = 0.004$ (roughly the composition of the Small Magellanic Cloud), $Z = 0.008$ (similar to the Large Magellanic Cloud), and $Z = 0.02$ (solar composition)

fact that stars of lower metallicity burn at higher temperatures. AGB stars more luminous than $M_{bol} \approx -6.4$ should not be C, stars due to HBB [22]. ^{7}Li is produced when the temperature at the base of the convective envelope exceeds 50×10^6 K, but ^{12}C destruction requires at least 80×10^6 K. Hence there is a small range of parameters where a star may be both a C star and a Li-rich star; see also [156]. We show the ratio ^{12}C/^{13}C for various masses and compositions in Figure 2.39, which again shows how this ratio is usually a good indicator for the presence of HBB.

In summary, intermediate-mass stars that show HBB should become ^{14}N-rich giants (with low values of ^{12}C/^{13}C and possible enhancements of ^{7}Li) rather than C stars. Lower masses, which do not show HBB, are expected to form the majority of the known C stars.

One final feature in Figure 2.38 is of interest. HBB requires a deep envelope, and this is why it is not present in low-mass stars. Mass loss on the AGB will reduce the envelope mass, and eventually this can stop HBB from

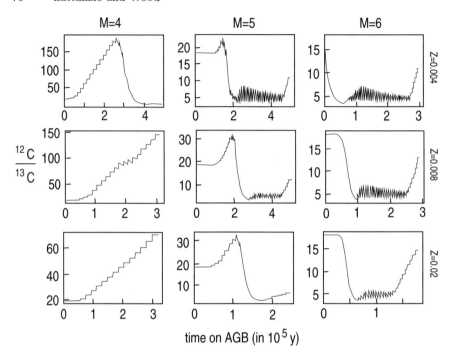

Fig. 2.39. Same as Figure 2.38 but for $^{12}C/^{13}C$

operating. In addition, the third dredge-up is favored at high envelope masses (e.g., [164]). As the envelope mass decreases we face the important question as to which will stop first, dredge-up or HBB. Figure 2.38 shows clearly that dredge-up continues after HBB stops, as shown in the rise in the C/O ratio toward the end of the evolution of (for example) the $5\,M_\odot$ model with $Z = 0.004$. Indeed, this model becomes a C star right near the end of its life, as do all the $6\,M_\odot$ models in the figure, as well as the $5\,M_\odot$ model with $Z = 0.008$. Thus, in these cases HBB delays, but does not fully prevent, the formation of C stars. These models clearly predict the existence of luminous C stars that are very near the end of their AGB lifetime and have $M_{bol} \approx -7$. This luminosity is brighter than that of optically visible C stars. However, at this stage of evolution, the mass loss rate is a few $\times 10^{-5}\,M_\odot\,yr^{-1}$, and hence these stars will be obscured by their thick circumstellar envelope: They will be visible only at infrared wavelengths. Optically obscured, infrared-bright C stars have been identified in the Magellanic Clouds [151, 152], and the models above seem to provide a natural explanation for them; see [56] for details.

Another aspect of HBB nucleosynthesis in the CNO cycles is the modification of the three oxygen isotopes. Refer again to Figure 2.35, where we see that CNO_{II} can produce ^{17}O, and that CNO_{III} destroys both ^{17}O and

^{18}O (see also [6]). The variation of the oxygen isotopes during the various dredge-up events and HBB has been studied in detail [23, 24]. Briefly, the ratio ^{17}O/^{16}O increases slightly at the first dredge-up, but is hardly affected by the second or third dredge-up, remaining at ^{17}O/^{16}O $\approx 10^{-2}$ to 10^{-3} (depending on the mass). This is because the material dredged to the surface contains little oxygen of any kind. This is not true if there is substantial overshoot at the boundaries of the ISCZ (see [65]). The ratio ^{18}O/^{16}O decreases due to the first dredge-up but is largely unaffected by the second and third dredge-ups, maintaining its value of $\approx 10^{-2}$. When HBB starts, for the more massive stars, we find the rapid and almost complete destruction of the ^{18}O, so that the ^{18}O/^{16}O $< 10^{-6}$, while there is only a small change in the ^{17}O/^{16}O ratio, which remains at $\approx 10^{-1}$–10^{-3}. These results have been compared with oxygen isotopic ratios in presolar meteorite grains leading to the conclusion that some form of extra-mixing is required, often called "cool bottom processing," on the first giant branch and possibly also the AGB [24, 114]. Note that this mechanism is also assumed responsible for the low observed values of ^{12}C/^{13}C seen on the first giant branch (see Section 2.5.1).

Finally, in this section, we return to ^{19}F and how it is affected by HBB. We have seen in Section 2.5.7 that AGB stars produce ^{19}F by a complicated series of reactions. At the high envelope temperatures characteristic of HBB we find that the ^{19}F is efficiently destroyed by ^{19}F(p, α)^{16}O. There is a short period at the start of HBB when there may be a small production of ^{19}F from proton captures on ^{18}O, but this is very soon overcome by proton captures on the ^{19}F itself [111]. As an example we show the surface ^{19}F content in a 5 M$_\odot$ model with $Z = 0.004$ in Figure 2.40. Typically, HBB reduces the ^{19}F content to $Y(^{19}\text{F}) \approx 10^{-9}$–$10^{-10}$, as seen in the figure.

2.5.14 Hot Bottom Burning: the Ne–Na and Mg–Al Chains

HBB facilitates many proton capture reactions, and hence the Ne–Na and Mg–Al cycles will be active. The first effect of the Ne–Na cycle is the transformation of ^{22}Ne into ^{23}Na. This ^{22}Ne was produced in the ISCZ by successive α-captures on the abundant ^{14}N, which in turn was produced (mostly) by CN cycling in the H-burning shell. Hence it has both a *primary* and a *secondary* component. The primary one comes from the ^{12}C produced in the star, while the secondary component is from the initial CNO nuclei in the star; both are transformed into ^{14}N by CN cycling.

Overall, the surface ^{22}Ne shows complicated behavior. Stars without HBB can show an increase of ^{22}Ne due to third dredge-up, typically up to 1.0 dex. For higher masses the ^{22}Ne is burnt in the ISCZ into the heavy Mg isotopes, and hence there is little ^{22}Ne dredged to the surface. Rather, since these masses usually exhibit HBB, there are proton captures on the ^{22}Ne to produce ^{23}Na. This is seen in Figure 2.41. The ^{22}Ne initially present is burnt to ^{23}Na, and hence the ^{22}Ne abundance decreases, while the ^{23}Na content increases. The addition of ^{22}Ne following each dredge-up event is clearly seen, although

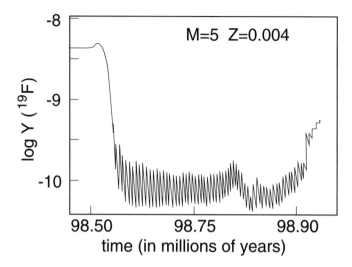

Fig. 2.40. Surface ^{19}F composition during the AGB evolution of a $5\,M_\odot$ model with $Z = 0.004$

it is relatively small in amount. We also see the destruction of this ^{22}Ne as it is transformed into ^{23}Na by HBB during each interpulse phase. There is a gradual increase in the surface abundance of ^{22}Ne, but when mass loss terminates HBB we see a substantial amount of ^{22}Ne dredged to the surface following the final few pulses.

^{21}Ne never reaches a significant abundance, being efficiently destroyed by HBB to the level of $Y(^{21}\text{Ne}) \approx 10^{-9}$ (see Figure 2.42). Although we expect the Ne–Na cycle to begin destroying ^{20}Ne at high temperatures, this is offset by ^{23}Na(p, α)^{20}Ne (which dominates over ^{23}Na(p, γ)^{24}Mg at these temperatures) so that the changes in ^{20}Ne are totally negligible.

At the high temperatures seen during HBB we expect some leakage from the Ne–Na chain into the Mg–Al chain via ^{23}Na(p, γ)^{24}Mg (see Figure 2.23). However we see in Figure 2.41 that the ^{24}Mg content actually decreases slightly. This is efficient only at high temperatures, such as seen in HBB in stars of higher mass or lower metallicity, and is due to the Mg–Al reactions. Note that both heavy Mg isotopes increase due to the third dredge-up, as we saw previously in Figure 2.26. This increase can be over an order of magnitude, and can produce very nonsolar Mg isotopic ratios [108]. This reflects the (α, n) and (α, γ) reactions occurring on ^{22}Ne in the ISCZ. Nevertheless, once ^{25}Mg and ^{26}Mg are in the envelope, they suffer proton captures to make ^{26}Al and ^{27}Al, as also seen Figures 2.41 and 2.42. Indeed, we can produce ^{26}Al up to the level of about 10% of the ^{27}Al present. This ^{26}Al will later decay to ^{26}Mg on a time scale of order one million yr. The increase in ^{27}Al is more

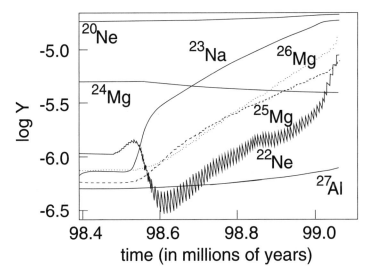

Fig. 2.41. Surface abundances for various species during the AGB evolution of a 5 M$_\odot$ model with $Z = 0.004$

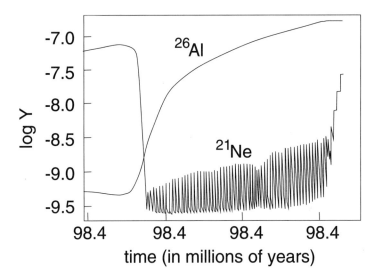

Fig. 2.42. Surface abundances for various species during the AGB evolution of a 5 M$_\odot$ model with $Z = 0.004$

modest, reaching maybe 0.3 dex or so. The resulting ^{26}Al/^{27}Al ratios are in broad agreement with those measured in presolar grains (e.g., [98, 180]).

2.5.15 AGB Star Yields

Having discussed the main nucleosynthesis occurring in AGB stars, it is natural to look at their contribution to the inventory of the various nuclides within the Galaxy. For this we need the "yield" of each element as a function of mass and composition. Such estimates are, of course, limited by the uncertainties in the calculations, and there are many. These include the depth and occurrence of the third dredge-up, the mass-loss formula, and many of the reaction rates themselves. These are the most serious problems at present, and they limit the reliability of the sort of detailed models presented here. For example, to determine whether HBB enables intermediate-mass AGB stars to contribute to the overall ^{7}Li content of the Galaxy we need to know precisely when the high mass-loss phase occurs. As we saw, the large overabundances of ^{7}Li are temporary, so that if the ^{7}Li is destroyed before the super-wind sets in, then the stars will not contribute to the overall ^{7}Li content of the Galaxy and Universe; on the other hand, if the mass loss is substantial while the envelope is rich in ^{7}Li, then these stars could indeed contribute substantial amounts of ^{7}Li (see [143]).

Until recently the huge amount of computer time needed for detailed models has precluded direct calculations of AGB star yields. Hence the yields presented in the literature until now have resulted from "synthetic" evolution models (see Chapter 3 in this volume), using parameterized fits to detailed models and related approximations. These must be used with care and caution: Sometimes the synthetic procedures can be misleading. Yields from detailed models covering a wide range of masses and compositons are only now becoming available (Karakas and Lattanzio, in press).

2.6 Variability

Variability is a very common feature of AGB stars. Its cause is mainly pulsation, although there is some evidence for variations caused by episodic ejections of dust shells or orbiting dust clouds. Helium shell flashes also lead to longer-term variability, which can interact with the shorter-term pulsation. In this chapter, all these causes of variability are discussed.

2.6.1 Variable Types and Light Curves

The red giant variable stars listed in the *General Catalog of Variable Stars* [90] fall into three broad groups: The Mira variables, the semiregular variables, and the irregular variables. The semiregular variables are further divided into subgroups SRa to SRd. By definition, the Mira variables are of large amplitude (> 2.5 mag in V), and their variations are relatively regular; the semiregular variables are of smaller amplitude than the Miras (<2.5 mag in V) with some definite periodicity; and the irregular variables show

little periodicity, although this is often due to poorly studied light curves rather than lack of intrinsic periodicity. The high-quality data produced by microlensing surveys such as MACHO [4], EROS [9], and OGLE [146] provide fine examples of light curves of all classes of these variables, and some examples from the MACHO database[7] are shown in Figure 2.43. Web-based visual light curves for visually-bright nearby AGB variables can be obtained using the American Association of Variable Star Observers (AAVSO) light curve generator at
http://www.aavso.org/adata/curvegenerator.shtml,
or from VSNET at
http://vsnet.kusastro.kyoto-u.ac.jp/vsnet/gcvs/,
where links to other data sources can be found.

Although the amplitudes of the visual light curves of the Mira variables can be 6 mag or more, the bolometric light curve amplitudes are much smaller, typically 1 mag and with amplitude increasing with period [49, 159]. The reason for this is that most of the energy emitted by AGB variables comes out in the IR, where the light curve amplitude is much smaller than in the visible. This is demonstrated in Figure 2.44, which shows the light curves of the Mira variable RR Sco (P = 281 d, spectral type M6e) in the bandpasses $UBVRIJHKL$ (wavelengths from 0.36 to 3.6 µm). The large amplitude in the shorter wavelength regions is a result of two effects [132]: (1) absorption by TiO, which varies strongly throughout the pulsation cycle, and (2) a large change in monochromatic flux with effective temperature at wavelengths shortward of the peak in the overall flux distribution (this is a property of blackbody flux distributions, which are reasonable approximations to stellar flux distributions: The peak in the flux distribution occurs around 1 µm in Mira variables). Figure 2.44 also shows the lag in phase of 0.1 to 0.2 that exists between the visual light curve maximum and the infrared (or bolometric) light curve maximum [102].

An important class of AGB variables not found in the GCVS consists of the dust-enshrouded infrared variables that are too faint to have been discovered in the visible (in some cases, visual light curves have been obtained after discovery). The oxygen-rich versions of these stars have been found in two ways: Firstly, by their 1612 MHz OH maser emission, in which case they are known as OH/IR stars; and secondly, by infrared surveys. Herman and Habing [63] show light curves for the OH flux variations in a sample of OH/IR stars with periods up to ≈ 2000 d. The carbon-rich analogues of these

---

[7] This chapter utilizes public-domain data obtained by the MACHO Project, jointly funded by the US Department of Energy through the University of California, Lawrence Livermore National Laboratory under contract No. W-7405-Eng-48, by the National Science Foundation through the Center for Particle Astrophysics of the University of California under cooperative agreement AST-8809616, and by the Mount Stromlo and Siding Spring Observatory, part of the Australian National University.

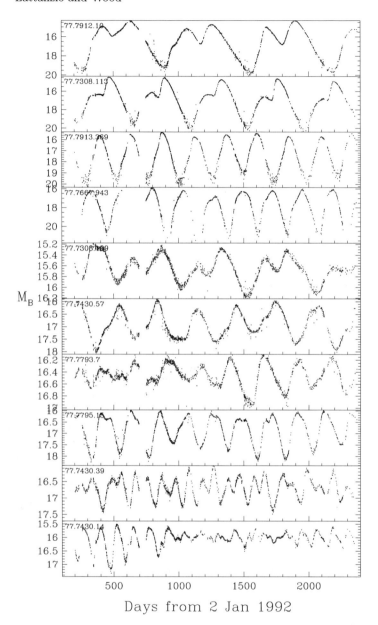

Fig. 2.43. Examples of light curves of Mira variables (**top four** panels) and semiregular variables (**bottom six** panels) in the LMC from the MACHO database. The bandpass of the lightcurves is MACHO blue, which is centered near 0.53 μm, similar to the visual bandpass. All red giants in the MACHO database seem to show distict periodicities at some times, but without high-quality light curves such as these, they would possibly be classified as irregular

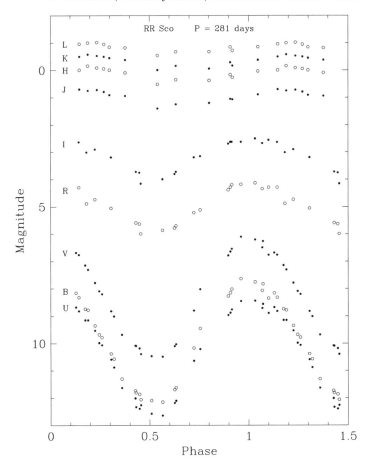

Fig. 2.44. Light curves for the Mira variable RR Sco in the bandpasses $UBVRIJHKL$. The $UBRVI$ light curves come from Eggen [44], while the $JHKL$ light curves are from Catchpole et al. [34]

stars have been found mainly by infrared surveys, since carbon stars do not produce OH maser emission. In both cases, the stars generally pulsate with larger amplitudes (δK up to 3 mag) and longer periods ($\gtrsim 600$ d) than the optical Mira variables. Examples of K light curves can be seen in [160, 172], and light curves for a small number of these stars in various bandpasses from 1 to 20 μm may be found in [100]. The dust-enshrouded infrared variables are in a more advanced evolutionary state than the Mira variables, and they have stellar winds with higher mass-loss rates than the Miras. The less extreme examples of stars in this class are readily seen in the near-infrared part of the spectrum. Examples of light curves of some LMC variables of this class are

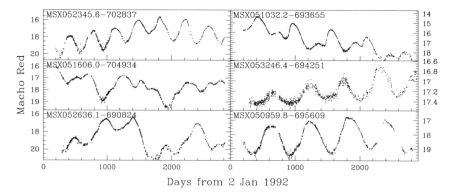

Fig. 2.45. MACHO red light curves for some dust-enshrouded AGB stars found in the LMC by the MSX satellite

shown in Figure 2.45, using the red MACHO bandpass centered near 0.70 µm. The light curves of these stars are far from regular, showing brightening and dimming over time scales of ≈ 10 yr, rapidly changing amplitudes, and multiperiodicity.

2.6.2 Intrinsic Parameters

2.6.2.1 Masses

Variable AGB stars occur over the whole mass range occupied by AGB stars, from globular clusters (e.g., [106]) where the turnoff mass is $\approx 0.85\,M_\odot$ to the upper mass limit of AGB stars of 6–8 $M_\odot$: The latter variables have been identified in the LMC [170], and they often have luminosities near the upper limit for AGB stars. In the solar vicinity, the kinematics and scale height out of the Galactic plane of Mira and semiregular variables have been studied to find the mass of the typical local population [8, 47, 50, 51, 86, 87, 89]. For the Mira variables, the average progenitor is an old disk star whose mass increases slightly with period, being $\gtrsim 1.1\,M_\odot$ for $P > 300$ d and $\lesssim 1.1\,M_\odot$ for $P < 300$ d. The Miras with periods near 200 d have the kinematics of Galactic halo stars, and they are found in the more metal-rich globular clusters, so they are clearly of initial mass $\approx 0.85\,M_\odot$. The local semiregular variables generally have masses similar to those of the Mira variables with periods $\gtrsim 300$ d. However, semiregular variables are also found in globular clusters (of all metallicities), so the mass range of the semiregulars is almost certainly as broad as that of the Miras.

2.6.2.2 Position in the HR-Diagram

The AGB variables lie on the coolest, most luminous part of the AGB. In the solar vicinity, Eggen [43] studied individual local red giants for small-amplitude variability and came to the conclusion that all giants redder than $R - I = 0.9$ were variable. Data from experiments such as MACHO now allow this type of study to be carried out on very large numbers of stars with more reliable results and quantitative estimates of the variable star fractions in different parts of the HR-diagram.

The **left** panel of Figure 2.46 shows the HR-diagram for all stars in the 0.5×0.5 degree region of the LMC bar [169]. All stars above and to the right of the dashed lines in Figure 2.46 were searched for variability down to a detection limit of 0.02 mag. The **middle** panel shows the nonvariable stars, while the **right** panel shows the variable stars. Confining discussion to the red giant branch and ignoring the Cepheid variables, it is found that > 90% of stars above the tip of the first red giant branch (RGB) are variable and that the major group of variable stars extends down to the minimum luminosity experienced by stars of roughly solar mass on the thermally pulsing AGB (TP-AGB). The conclusion to be drawn from Figure 2.46 is that 90% or more of stars on the TP-AGB are variable at the level of a few percent or more.

Finally, note that there are a small number of stars below the TP-AGB minimum luminosity in Figure 2.46: Such stars must be on the RGB. They have exceptionally regular, small-amplitude light curves and they are almost certainly binary stars or perhaps rotating spotted stars rather than pulsators (see [169]). These stars make up about 0.5% of the RGB stars near the tip of the RGB.

2.6.3 Period–Luminosity Relations

Early attempts to look at period–luminosity (PL) relations for Mira and semiregular variables relied on statistical parallax methods for providing distances and hence absolute magnitudes for local variables (e.g., [119]). For small numbers of local variables, Hipparcos distances have been used to look at period–luminosity relations [13, 150, 158], but the errors in individual distance are large. By far the most revealing studies of the PL relations for AGB variables come from studies of AGB stars in the LMC, where the distance is known and interstellar reddening is small. The results of these studies are utilized here.

Figure 2.47 shows the PL relation for optically visible semiregular and Mira variables in the LMC. It is clear from this figure that there are four[8] parallel period–luminosity sequences (A–D) lying above $K_0 = 12.9$, which

---

[8] Very recent work (Keller, S.C., Cook, K.H. and Wood, P.R., in preparation) shows that sequence B splits into two sequences B_1 and B_2.

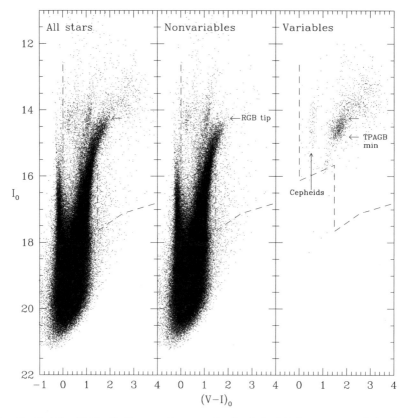

Fig. 2.46. The $(I, V-I)$ diagram for stars in the 0.5×0.5 degree area of the LMC searched for variable stars [169]. The left panel shows all stars in the search area; the middle panel shows the nonvariable stars; and the right panel shows the variable stars. Variables were searched for only in the region lying above and to the right of the dashed line. The positions of the RGB tip and the minimum luminosity for thermally pulsing AGB (TP-AGB) stars are indicated. The main sequence can be seen extending almost vertically at $V-I \approx -0.1$; the red giant branch [combined RGB and AGB] for low-mass ($M < 2.25\,M_\odot$) stars is the thick sequence on the right; the almost vertical sequence near $V-I \approx 1.2$ and $I \approx 14$ is the red giant branch for intermediate-mass ($2.25 < M/M_\odot < 8$) stars; and the stars with $14 < I < 15$ and $0 < V-I < 1$ are intermediate-mass stars on blue He-core burning loops

corresponds to the minimum luminosity for thermally pulsing AGB stars. Another sequence (E) extends downward to $K_0 \approx 14.3$ (the lower magnitude limit of the search for variability in [169]) and consists mostly of stars on the RGB. Examination of the light curves of the stars in Figure 2.47 shows that the Mira variables (i.e., the stars with visual light curve amplitudes >2.5

2 Evolution, Nucleosynthesis, and Pulsation of AGB Stars   87

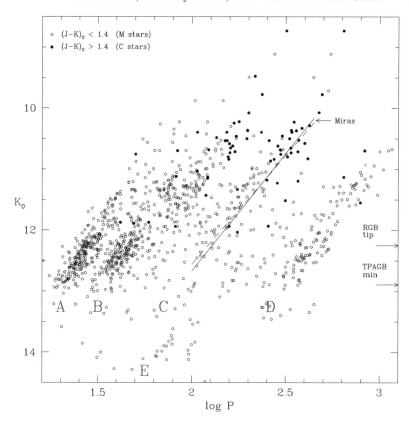

Fig. 2.47. The period–luminosity relations for optically visible red variables in a 0.5×0.5 degree area of the LMC plotted in the (K_0, $\log P$) plane. The letters A–E are used to label the five sequences visible in the diagram: Very recent work shows that the B sequence splits into two sequences B_1 and B_2. The positions of the tip of the first red giant branch (RGB) and the minimum luminosity for thermally pulsing AGB (TP-AGB) stars with $M \approx 1\,M_\odot$ are indicated by arrows. The solid and dashed lines are the $K - \log P$ relations from [74] and [48], respectively. Solid circles correspond to stars with $J-K > 1.4$; spectra show such stars to be mostly carbon stars, although the few longer-period, fainter stars are dust-enshrouded Mira variables that could be oxygen-rich. Open circles correspond to stars with $J-K < 1.4$, and they are assumed to be oxygen-rich M or K stars. From [167]

mag) lie at the upper end of sequence C. Sequence C also coincides with the $K - \log P$ relation found for large-amplitude AGB variables in the LMC [48, 74].

The semiregular variables (i.e., those with visual light curve amplitudes <2.5 mag) occupy sequences A and B and the lower half of sequence C. All the semiregular light curves shown in Figure 2.43 are from sequence C, while

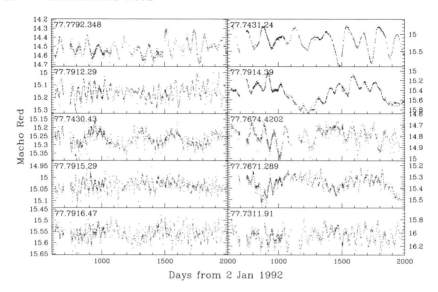

Fig. 2.48. Examples of light curves of stars on sequence A (**left** column) and sequence B (**right** column). The long secondary periods evident in some of the light curves (for example, the ≈ 700 d period in star 77.7914.39) belong on sequence D

some light curves of semiregular variables on sequences A and B are shown in Figure 2.48. In the solar vicinity, [13] used Hipparcos distances to show that local semiregular (SR) variables lie on both sequence C (along with the Miras) and sequence B, which was previously identified in the LMC by [173]. The obvious interpretation of the sequences A, B, and C in Figure 2.47 is that they represent pulsation in different modes (mode identifications will be made in Section 2.6.5). Thus, semiregular variables can pulsate in a number of modes, often simultaneously, while the Mira variables are confined to the single pulsation mode corresponding to sequence C.

Some examples of light curves of stars belonging to sequences D and E are shown in Figure 2.49. The sequence E stars lie below the TP-AGB minimum luminosity (see Section 2.6.2.2), and they have quite regular light curves, indicative of some sort of binary or rotational origin.

The sequence D stars are invariably multiperiodic, having one period that lies on sequence D and another that lies on sequence A, B, or (occasionally) C. The origin of the period associated with sequence D is currently a mystery. Possibilities include some currently unknown form of pulsation (radial or nonradial), an effect of rotation in a merged binary (ellipsoidal variations), episodic dust formation, or orbiting dust clouds [169].

The $K-\log P$ diagram shown in Figure 2.47 contains only optically detectable LPVs from a small 0.5×0.5 degree area of the LMC. Toward the end

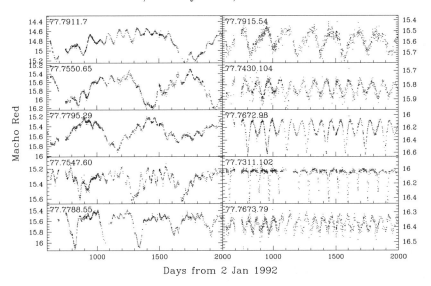

Fig. 2.49. Examples of light curves of stars on sequence D (**left** column) and sequence E (**right** column)

of the Mira phase of evolution, an AGB star becomes enveloped in a dusty circumstellar shell resulting from the high mass-loss rate. This dusty phase is very short, so that such stars are very rare. However, hundreds of such stars have recently been revealed in the LMC by the infrared MSX satellite. About one-half to one-third of these sources were seen by MACHO, so that pulsation periods can be derived (see Figure 2.45 for examples of light curves). These stars are shown in Figure 2.50, which is just Figure 2.47 with the infrared stars overplotted. Note that the area from which the MSX-selected variables comes is about 50 times larger than the area from which the optically selected stars were derived, showing the rareness of the infrared stars.

Clearly, most of the dust-enshrouded stars have evolved from the end of the Mira sequence, declining dramatically in K magnitude when their dust shells become so thick and cool that their flux is emitted well longward of the K band. It needs to be stressed that the MSX-selected variables shown in Figure 2.50 were all seen by MACHO. This is almost certainly a powerful selection against the dustiest, most-evolved AGB stars. Thus the maximum period shown is probably an underestimate. Infrared monitoring is currently being carried out to find periods for the redder objects.

A small number of the MSX-selected sources in Figure 2.50 lie above the Mira sequence. These are more massive and luminous AGB stars that have not yet become dust-enshrouded but that still have high mass-loss rates as a result of their high luminosity.

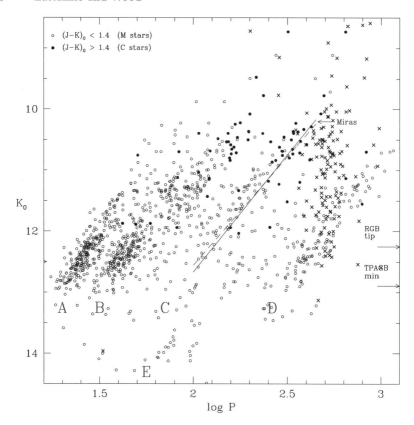

Fig. 2.50. Same as Figure 2.47 but with addition of MSX-selected infrared sources from a 3×3.5 degree arc of the LMC (crosses). The periods for these stars came from the MACHO database. From [168]

2.6.4 Linear Pulsation Models of AGB Stars

Linear radial pulsation models for AGB stars yield periods and growth rates. Extant computations [53, 59, 115] have generally used relatively simple computer programs that do not take detailed account of the interaction of pulsation and convection; calculations that do make an attempt to model the dynamical and thermal coupling between pulsation and convection have recently been performed [176].

The radial pulsation modes of AGB stars are essentially confined to the large convective envelope, with the central nuclear-burning core acting as a point source of gravity and luminosity only. It needs to be remembered that energy transport, which determines the stability and growth rates of pulsation, is dominated by convection. Furthermore, convection is usually treated in the simplifying approximation of mixing-length theory, so that

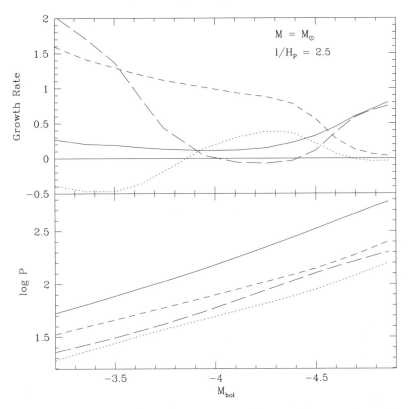

Fig. 2.51. The period P (in d) and the growth rate (fractional increase in amplitude per year) for a 1 M$_\odot$ star plotted against luminosity on the AGB. Convective energy transport is treated using mixing-length theory with a mixing length of 2.5 pressure scale heights. The solid line shows the fundamental mode; the short dashed line shows the first overtone; the long dashed line shows the second overtone; and the dotted line shows the third overtone

the derived growth rates need to be treated with great caution, although the periods are probably reasonably accurate (but see Section 2.6.6). The general character of the eigenmodes, the regions causing driving, and the important regions of the star for the determination of the pulsation periods will be described here (see also [53]); for a description of the computer code see [173], note that this code computes a variation in convective flux through the pulsation cycle, but it does not include the effects of turbulent pressure.

Figure 2.51 shows the period for linear modes in a 1 M$_\odot$ AGB star as its luminosity increases up the AGB. As expected, the period increases with luminosity (stellar radius) up the AGB, since $P \propto R^\alpha M^{-\beta}$, where $\alpha \approx 1.5$–2.5 and $\beta \approx 0.5$–1.0 [53, 165]. Turning to growth rates of pulsation, Figure 2.51

shows that at low luminosity, the second overtone has the highest growth rate; then the first overtone has the highest growth rate at intermediate luminosities, and the fundamental mode has the highest growth rate at high luminosities. Although a particular mode having the highest growth rate does not guarantee that the star will pulsate in that mode, it is a reasonable indicator. This series of models demonstrates a general characteristic of pulsation on the AGB. As a star evolves to higher luminosity, it first becomes unstable in the second or third overtone, then switches to lower overtones and eventually the fundamental mode at the highest luminosities. At some luminosities, different modes have similar positive growth rates, suggesting that multimode oscillations may occur, a situation that is observed to be quite common [92, 169].

Some properties typical of pulsating AGB stars are shown in Figure 2.52. The top panel clearly demonstrates the complete dominance of convective energy transport through most of the envelope of the star. The bottom panel shows that as in nearly all stars, the amplitude of the radial fundamental is roughly proportional to R, while the overtones in these stars are concentrated near the surface, so that considerable care must be taken to treat the surface layers correctly. The partial work integral W_r (third panel) shows the work done over one pulsation cycle by all matter interior to each radius. Regions where W_r decreases with r are damping regions, while regions where W_r increases with r contribute to excitation of pulsation. Regions of the star interior to the H and first He ionization zones are clearly damping regions: This effect is analagous to the radiative damping that occurs deep in nonconvective pulsating stars as a result of the Kramers-like opacity law (e.g., Cox and Giuli [38]). Driving of the pulsation occurs throughout the H and first He ionization zones. In the surface layers where $T < 8000\,\mathrm{K}$, the opacity is low, and these regions make very little contribution to either driving or damping of pulsation, even in overtone pulsators. Finally, the partial integral $J\Sigma_r^2$ allows one to see which parts of the envelope contribute most to determining the pulsation period of the star. Note that if a time dependence $\exp(i\sigma t)$ is assumed, then $\sigma^2 \propto J\Sigma_R^2/J$, where J is a real, positive integral. Regions where $J\Sigma_r^2$ varies the most rapidly with r contribute most to determining the pulsation period. The second panel of Figure 2.52 shows that most of the layers below the top of the H ionization zone at $T \approx 8000\,\mathrm{K}$ contribute to the determination of the fundamental mode period, whereas all layers, including the surface layers, contribute to determining the first (and higher) overtone period. For a more detailed description see [53].

2.6.5 Comparison of Linear Models with Observations

The MACHO observations of AGB variables in the LMC provide an excellent opportunity for comparing theory with observation. A comparison of the linear pulsation models with observations is shown in Figure 2.53, adapted from [169]. A good fit requires that at a given luminosity, the models have the

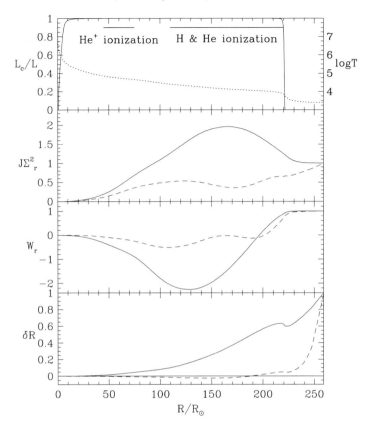

Fig. 2.52. Various properties of a model Mira variable with $L = 5000 \, L_\odot$ and $M = 1 \, M_\odot$ and solar metallicity plotted against radius within the star. The fundamental mode period of the model is 333 d, similar to that of the prototypical Mira o Ceti. **Top** panel: $\log T$ (dotted line) and the fraction of the energy flux carried by convection (solid line); **second** panel: the partial integral $J\Sigma_r^2$ for the fundamental mode (solid line) and the first overtone (dotted line); **third** panel: the partial work integral W_r for the fundamental mode (solid line) and the first overtone (dotted line); **bottom** panel: the real part δR of the eigenfunction for the fundamental mode (solid line) and the first overtone (dotted line), where the eigenfunction is defined to have complex amplitude $(1.0, 0.0)$ at the stellar surface

observed temperature of the AGB and the observed pulsation period. In order to satisfy the first requirement, the mixing length was adjusted so that at $L = 5000 \, L_\odot$, a $1 \, M_\odot$ AGB star with metallicity $Z = 0.008$ and helium abundance $Y = 0.3$ lies on the observed AGB. The mixing length to pressure scale height ratio (3.2) was then held constant for all other models, and the mass was adjusted at each luminosity until the observed AGB effective temperature was attained. The resulting masses and the theoretical AGB are shown in the bottom panel of Figure 2.53. Higher masses were required for pulsating

AGB stars at higher luminosities, which is reasonable if higher masses evolve to higher luminosities on the AGB. The masses shown are also consistent with masses estimated from the kinematics of Galactic Miras [47]. Given the mass, temperature, and luminosity, the pulsation period can then be computed from the period–mass–radius (PMR) relation $P = P(M, R)$, which can be written as $P = P(M, T_{eff}, L)$ using the definition of effective temperature $L = 4\pi\sigma_R R^2 T_{eff}^4$. Pulsation periods for various modes are plotted in the top panel of Figure 2.53. In this panel, luminosity is represented by the quantity

$$I_W = I_0 - 1.38(V - I)_0,$$

which was chosen (a) because it is almost reddening free and (b) because the theoretical period–luminosity–color relation for LPVs suggests that stars pulsating in a given mode should lie on almost the same $(I_W, \log P)$ relation regardless of mass; i.e., the color term is very small in the $(I_W, \log P)$ plane.

It is clear that fundamental mode pulsation is able to accurately reproduce the P-L relation of sequence C, which contains the Miras and some SR variables. The SR variables on sequences A and B (which splits into B_1 and B_2) become first, second, or third overtone pulsators. However, the definitive test for the mode of pulsation comes from the period *ratios* rather than the periods themselves: With the mode assignment shown, the observed period ratios and the theoretical ones are in reasonable agreement, given the limitations of mixing-length theory. On the other hand, if sequence C is assumed to correspond to the first overtone mode, the observed and theoretical period ratios show absolutely no correlation. This result is probably the strongest test known for assigning fundamental mode pulsation as the mode of the Miras.

It is interesting to ask why pulsation modes higher than the third overtone are not seen in AGB stars. This is probably because for modes higher than the third overtone, traveling waves incident at the surface from the interior travel out into the photosphere rather than being reflected back as assumed in the pulsation equations [53]. In such a situation, any interior driving of the pulsation will be damped by energy and momentum loss from the surface by traveling waves, and the corresponding mode will be stable. The dotted line in the top panel of Figure 2.53 corresponds to the shortest pulsation period for which waves incident at the stellar surface will be reflected back (the pulsation frequency at this point is known as the acoustic cutoff frequency). This line corresponds reasonably well with the highest observed mode of pulsation.

Using all the above results, we are now in a position to speculate on the evolution of pulsation properties of an AGB star. As the star evolves up the AGB, it will presumably first pulsate at low amplitude on sequence A. Then with further evolution, the sequence B mode will become unstable, and its amplitude will increase, while the sequence A mode amplitude will decrease, so the star will be a multimode pulsator having pulsation periods

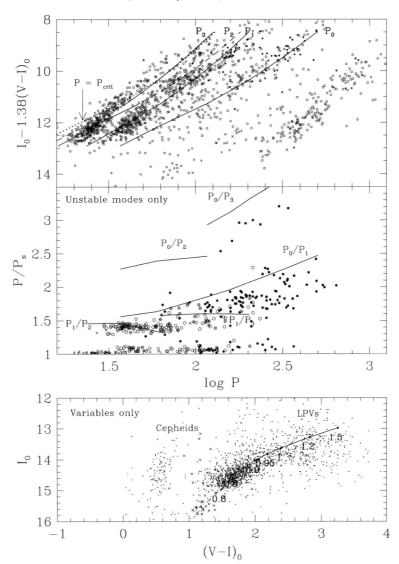

Fig. 2.53. Top two panels: The luminosity index I_W (see text) and the period ratio plotted against P. The solid (dashed) lines correspond to the unstable (stable) radial pulsation modes P_0, P_1, P_2 and P_3, as indicated. In the P-ratio plot, a line is shown only if both modes are unstable. The dotted line in the top panel is the shortest period of oscillation for which waves incident at the stellar surface from the interior are reflected back rather than running out into the circumstellar matter (see text). **Bottom panel**: the theoretical AGB in the $I_0, (V-I)_0$ diagram. The mass ($M_\odot$) used at each luminosity is displayed. Crosses represent stars with light curves like contact binaries. From [169].

lying on sequences A and B. While this is happening, the pulsation period of each mode will slowly increase as the stellar radius increases. With further evolution, the star will develop an unstable fundamental mode on sequence C. It will remain a multimode, semiregular variable with pulsation periods belonging to sequences B and C, and possibly A. It will then evolve up sequence C, increasing in amplitude and become a Mira variable. Finally, it will end up as one of the large-amplitude, long-period, dust-enshrouded AGB stars (still fundamental mode pulsators) that eject all their envelope mass on the way to the PN phase. In the event of a helium shell flash occurring, the stellar luminosity will decline, and the above transitions will be temporarily reversed before the star once again increases in luminosity and retraces the above changes in pulsation mode.

2.6.6 Large-Amplitude Pulsation Models

In order to get a proper theoretical understanding of the mass-loss processes that terminate evolution on the AGB, it will be necessary to have accurate models for the large-amplitude, nonlinear pulsation that occurs at this time. Such models will also allow the derivation of intrinsic properties (M, L, T_{eff}) of ordinary Miras by the modeling of light-curve shapes. For example, many of the Miras in the LMC show double-peaked light curves analagous to the bump Cepheids (see Figure 2.43), and reproducing such light curves requires a sensitive adjustment of stellar parameters. The fact that such light curves are rare among Miras in the solar vicinity (R Cen and R Nor are two examples) is a clear indication of differences between the intrinsic properties of solar neighborhood and LMC Miras.

Unfortunately, making full-amplitude, nonlinear models for Mira variables is not generally possible at the present time. Because growth rates in extant models seem to be much higher than those in real Miras (a result of our inability to model convective energy transport correctly), the model amplitudes grow well beyond the observed amplitudes, and the stars tend to blow apart, leading to convergence failure in the pulsation codes [145, 161]. Models with stable limiting amplitudes can be made, provided that the envelope and total mass are sufficiently large and/or the luminosity is not too large (typically, $M \gtrsim 1.5\,M_\odot$ for a typical luminosity of $\approx 5000\,L_\odot$ yields a model with a stable limiting amplitude). Alternatively, some scheme to artificially limit the driving of the models can be introduced [14]. At this time, we still await satisfactory full-amplitude, nonlinear pulsation models for AGB stars.

The detailed models that examine the mass-loss processes in the exterior layers (see Chapter 6) get around the lack of theoretical interior pulsation models by replacing the stellar interior by an inner boundary condition consisting of a sinusoidally moving piston situated a few pressure scale heights below the photosphere. In such models, the amplitude of the driving piston is arbitrary, and some assumption has to be made about how the luminosity varies (for example, T_{eff} can be assumed constant, or the luminosity at the piston can be held constant). Proper interior models are required to provide realistic interior boundary conditions for the mass-loss models.

The main purpose for which nonlinear pulsation models have been constructed so far is to try to determine the pulsation mode of the Mira variables. Note that to first order, the pulsation period satisfies the relation $P = QR^{\frac{3}{2}}M^{-\frac{1}{2}}$ for a given pulsation mode. In a Mira, the pulsation constant Q for the fundamental mode is roughly twice as large as Q for the first overtone mode. Therefore, for a given stellar mass, the radius required to produce fundamental mode pulsation at a given observed P is considerably smaller than the radius required to produce the same period with first overtone pulsation.

The problem of pulsation mode determination for Mira variables has a long history. Firstly, Wood [162] showed that photometric radii for Miras (i.e., radii derived from the definition $L = 4\pi\sigma_R R^2 T_{eff}^4$ when L and T_{eff} are derived from photometry) fitted the period–mass–radius relation for first overtone pulsators. A few years later, Hill and Willson [67] constructed pulsation models for the surface layers that indicated that the observed pulsation velocity amplitudes in Miras could not be reproduced by first overtone pulsation models because of their large surface radii and low surface gravities. On the other hand, the more compact, higher-gravity stars that result from the assumption of fundamental mode pulsation were able to produce the observed pulsation velocities.

Many direct measurements of Mira radii have been made using interferometric techniques (e.g., [61, 149]). The conversion of the interferometric observations (the visibility curves) to some identifiable radius position in a Mira variable (for example, the radius at Rosseland mean optical depth unity) requires a model for the pulsating outer layers of the Mira star. One of the main effects of pulsation is to extend the outer layers [163] so that models of static stars are inadequate for this purpose. Limb-darkened model atmospheres and associated visibility curves for Mira variables have been computed [73] by taking the density structure from nonlinear pulsation models and using this structure in full model atmosphere calculations. Comparison of observed visibility curves with these models yields the stellar angular diameter. With the further adoption of a distance, radii can be derived and compared to radii expected from the theoretical PMR relation, which is usually derived from linear pulsation models. The results of such comparisons have almost always shown that the observed radii are consistent with the radii expected if the Miras are first overtone pulsators (e.g., [61, 71, 149]). However, a few very recent measurements of Mira radii in the near-infrared have yielded values consistent with fundamental mode pulsation [72]. More measurements of this type should be available soon, and it will be interesting to see whether they continue to yield the smaller photospheric radii needed for fundamental mode pulsation.

Returning to pulsation velocity amplitudes of Miras, we note that many additional measurements of Mira pulsation velocities [69, 70] have been made since the initial results [67]. In addition, the factor required for converting observed to pulsation velocity has been examined in detail [128]. These more recent results continue to support the requirement for fundamental mode pulsation [128].

A definitive determination of the Mira pulsation mode from radius and velocity measurements is still to be made. However, we believe that the results of Section 2.6.5 unambiguously show that the Miras are fundamental mode pulsators: In particular, the period ratios observed in the LMC pulsators can be explained only if the Miras are fundamental mode pulsators. In this situation, we need to ask why the observed Mira radii are much larger than the radii predicted by the PMR relation derived from linear fundamental-mode pulsation models.

One possibility is that there is much more scattering of light in the upper layers of Mira atmospheres than is currently accounted for in the model atmospheres used for converting observed to model radii. In this case, the radii derived from observations will be larger than the true photospheric diameters. It is unlikely that dust could supply the necessary photon scattering [12]. If scattering is the source of the large Mira sizes, then the scattering source remains to be identified: H_2O could play a role [12], but also strong scattering by H_2 occurs in Mira atmospheres [68]. A very important result in this connection is the identification in Mira variables of "hot" and "cool" layers of H_2O with large optical depth, and at radii from 1 to 2 times the photospheric radius. Scattering of photons in these layers could clearly lead to observers estimating radii for Mira variables much larger than the true photospheric radius unless the models used to convert observed visibilities to radii include these scattering layers.

Nonlinear pulsation models constructed by [177] offer a possible way of obtaining fundamental mode pulsation with relatively large radii. In these models, if a Mira model is kept pulsing for a very long time (several hundred years), its internal structure changes in such a way that the period decreases significantly without a large change in the surface radius. This could allow agreement between the observed radii and the radii computed from the nonlinear PMR relation [178]. However, this explanation needs further investigation: a similar calculation by [166] did not show the long-term period decline. In fact, the model expanded, and the pulsation period in the relaxed model was longer than in the static starting model!

Overall, the nonlinear pulsation of AGB variables is a relatively unexplored area, largely because of the great uncertainties involved in treating energy transport by convection and the consequent large uncertainty in the limiting amplitude of the pulsation models.

2.6.7 The Interaction of Pulsation and Evolution

Changes in pulsation period provide a sensitive measure of slow evolutionary changes. It has long been known that the two Mira variables R Hya and R Aql have rapidly decreasing periods: The period of R Hya has changed from ≈ 500 d in 1700 to ≈ 387 d in 2000, while the period of R Aql has decreased from ≈ 320 d in 1915 to ≈ 267 d in 2000. Wood and Zarro [175] showed that the time scale of this change is consistent with the change expected in an AGB star whose surface luminosity is declining rapidly (over several hundred years) from the peak associated with a helium shell flash. The period changes in these two stars thus provide direct evidence for the existence of helium shell flashes, and they identify two stars that have undergone a shell flash with the last 300–500 yr. A number of other Mira variables with less dramatic period changes have been suggested as helium shell flash candidates: W Dra [175], T UMi [57], S Sex [107], and R Cen [62].

2.7 Conclusions and Outlook

Much is now known about the interior workings of AGB stars, but it is hard to think of *any* aspects of the interior where we are confident that we can compute *quantitatively* accurate models. This is due almost entirely to our inability to accurately model stellar convection, whose influence dominates so much of AGB evolution. For example, in the stellar envelope, convective energy transport is the main mechanism by which energy is transported, and uncertainties here mean that we are currently unable to reliably compute the pulsational stability of AGB stars, and to a lesser extent we are uncertain whether computed radii and effective temperatures are realistic. At the same time, uncertainties in convective transport processes in the stellar envelope mean that nucleosynthesis by hot bottom burning must remain uncertain. Perhaps the greatest uncertainty lies in the amount of mixing that occurs at convective boundaries due to "overshoot" or some other mixing process. This affects dredge-up and C star formation, as well as the depth of first and second dredge-ups. Overshoot at the boundaries of the intershell convection zone affects the composition of this zone and hence the composition of dredged-up material. Finally, the ^{13}C-pocket may well owe its existence to a partially mixed overshoot region at the base of the convective envelope during third dredge-up. Hence, the production of s-process elements in AGB stars is highly dependent on the amount of convective overshoot. Clearly, progress in the theoretical understanding of AGB star interiors awaits improvements in the treatment of stellar convection and hydrodynamics.

Acknowledgments: We would like to thank Roberto Gallino for a careful reading of the manuscript, and Amanda Karakas for providing results ahead of publication.

References

1. Abia, C., Boffin, H. M. J., Isern, J., and Rebolo, R. *A&A*, 245, L1, 1991.
2. Abia, C., Busso, M., Gallino, R., et al. *ApJ*, 559, 1117, 2001.
3. Abia, C., Domínguez, I., Gallino, R., et al. *ApJ*, 579, 817, 2002.
4. Alcock, C., Allsman, R. A., Axelrod, T. S., et al. *AJ*, 109, 1653, 1995.
5. Arlandini, C., Käppeler, F., Wisshak, K., et al. *ApJ*, 525, 886, 1999.
6. Arnould, M., Goriely, S., and Jorissen, A. *A&A*, 347, 572, 1999.
7. Arp, H. C. *AJ*, 60, 317, 1955.
8. Aslan, Z. *MNRAS*, 165, 337, 1973.
9. Aubourg, E., Bareyre, P., Brehin, S., et al. *Nature*, 365, 623, 1993.
10. Becker, S. A. *ApJ*, 248, 298, 1981.
11. Becker, S. A. and Iben, I. *ApJ*, 237, 111, 1980.
12. Bedding, T. R., Jacob, A. P., Scholz, M., and Wood, P. R. *MNRAS*, 325, 1487, 2001.
13. Bedding, T. R. and Zijlstra, A. A. *ApJ*, 506, L47, 1998.
14. Bessell, M. S., Scholz, M., and Wood, P. R. *A&A*, 307, 481, 1996.
15. Blöcker, T. *A&A*, 297, 727, 1995.
16. Blöcker, T. *A&A*, 299, 755, 1995.
17. Blöcker, T. and Schoenberner, D. *A&A*, 244, L43, 1991.
18. Boothroyd, A. I. and Sackmann, I.-J. *ApJ*, 328, 632, 1988.
19. Boothroyd, A. I. and Sackmann, I.-J. *ApJ*, 328, 641, 1988.
20. Boothroyd, A. I. and Sackmann, I.-J. *ApJ*, 328, 653, 1988.
21. Boothroyd, A. I. and Sackmann, I.-J. *ApJ*, 393, L21, 1992.
22. Boothroyd, A. I., Sackmann, I.-J., and Ahern, S. C. *ApJ*, 416, 762, 1993.
23. Boothroyd, A. I., Sackmann, I.-J., and Wasserburg, G. J. *ApJ*, 430, L77, 1994.
24. Boothroyd, A. I., Sackmann, I.-J., and Wasserburg, G. J. *ApJ*, 442, L21, 1995.
25. Buonanno, R., Corsi, C. E., Buzzoni, A., et al. *A&A*, 290, 69, 1994.
26. Burbidge, E. M., Burbidge, G. R., Fowler, W. A., and Hoyle, F. *Reviews of Modern Physics*, 29, 547, 1957.
27. Busso, M., Gallino, R., and Wasserburg, G. J. *ARA&A*, 37, 239, 1999.
28. Cameron, A. G. W. *ApJ*, 121, 144, 1955.
29. Cameron, A. G. W. and Fowler, W. A. *ApJ*, 164, 111, 1971.
30. Canuto, V. M. and Mazzitelli, I. *ApJ*, 370, 295, 1991.
31. Canuto, V. M. and Mazzitelli, I. *ApJ*, 389, 724, 1992.
32. Castellani, V., Giannone, P., and Renzini, A. *Ap&SS*, 10, 340, 1971.
33. Castellani, V., Giannone, P., and Renzini, A. *Ap&SS*, 10, 355, 1971.
34. Catchpole, R. M., Robertson, B. S. C., Lloyd-Evans, T. H. H., et al. *South African Astronomical Observatory Circular*, 1, 61, 1979.
35. Charbonnel, C. *A&A*, 282, 811, 1994.
36. Charbonnel, C. *ApJ*, 453, L41, 1995.
37. Clayton, D., Fowler, W., Hull, T., and Zimmerman, B. *Ann. Phys.*, 12, 331, 1961.
38. Cox, J. P. and Giuli, R. T. *Principles of Stellar Structure*. Gordon and Breach: New York, 1968.
39. D'Antona, F. and Mazzitelli, I. *ApJ*, 470, 1093, 1996.
40. Dearborn, D. *Phys. Rep.*, 210, 367, 1992.
41. Dopita, M. A., Vassiliadis, E., Meatheringham, S. J., et al. *ApJ*, 460, 320, 1996.

42. Dray, L. M., Tout, C. A., Karakas, A. I., and Lattanzio, J. C. *MNRAS*, 338, 973, 2003.
43. Eggen, O. J. *ApJ*, 184, 793, 1973.
44. Eggen, O. J. *ApJS*, 29, 77, 1975.
45. El Eid, M. F. *A&A*, 285, 915, 1994.
46. El Eid, M. F. and Champagne, A. E. *ApJ*, 451, 298, 1995.
47. Feast, M. W. *MNRAS*, 125, 367, 1963.
48. Feast, M. W., Glass, I. S., Whitelock, P. A., and Catchpole, R. M. *MNRAS*, 241, 375, 1989.
49. Feast, M. W., Robertson, B. S. C., Catchpole, R. M., et al. *MNRAS*, 201, 439, 1982.
50. Feast, M. W. and Whitelock, P. A. *MNRAS*, 317, 460, 2000.
51. Feast, M. W., Woolley, R., and Yilmaz, N. *MNRAS*, 158, 23, 1972.
52. Forestini, M., Goriely, S., Jorissen, A., and Arnould, M. *A&A*, 261, 157, 1992.
53. Fox, M. W. and Wood, P. R. *ApJ*, 259, 198, 1982.
54. Frogel, J. A. and Elias, J. H. *ApJ*, 324, 823, 1988.
55. Frogel, J. A., Mould, J., and Blanco, V. M. *ApJ*, 352, 96, 1990.
56. Frost, C. A., Cannon, R. C., Lattanzio, J. C., Wood, P. R., and Forestini, M. *A&A*, 332, L17, 1998.
57. Gal, J. and Szatmary, K. *A&A*, 297, 461, 1995.
58. Gallino, R., Arlandini, C., Busso, M., et al. *ApJ*, 497, 388, 1998.
59. Gautschy, A. *A&A*, 349, 209, 1999.
60. Goriely, S. and Siess, L. *A&A*, 378, L25, 2001.
61. Haniff, C. A., Scholz, M., and Tuthill, P. G. *MNRAS*, 276, 640, 1995.
62. Hawkins, G., Mattei, J. A., and Foster, G. *PASP*, 113, 501, 2001.
63. Herman, J. and Habing, H. J. *A&AS*, 59, 523, 1985.
64. Herwig, F. *A&A*, 360, 952, 2000.
65. Herwig, F., Blöcker, T., Schoenberner, D., and El Eid, M. *A&A*, 324, L81, 1997.
66. Herwig, F., Schoenberner, D., and Blöcker, T. *A&A*, 340, L43, 1998.
67. Hill, S. J. and Willson, L. A. *ApJ*, 229, 1029, 1979.
68. Hinkle, K. H., Aringer, B., Lebzelter, T., Martin, C. L., and Ridgway, S. T. *A&A*, 363, 1065, 2000.
69. Hinkle, K. H., Hall, D. N. B., and Ridgway, S. T. *ApJ*, 252, 697, 1982.
70. Hinkle, K. H., Scharlach, W. W. G., and Hall, D. N. B. *ApJS*, 56, 1, 1984.
71. Hofmann, K.-H., Balega, Y., Scholz, M., and Weigelt, G. *A&A*, 353, 1016, 2000.
72. Hofmann, K.-H., Beckmann, U., Blöcker, T., et al. *New Astronomy*, 7, 9, 2002.
73. Hofmann, K.-H., Scholz, M., and Wood, P. R. *A&A*, 339, 846, 1998.
74. Hughes, S. M. G. and Wood, P. R. *AJ*, 99, 784, 1990.
75. Iben, I. *ApJ*, 154, 581, 1968.
76. Iben, I. *ApJ*, 185, 209, 1973.
77. Iben, I. *ApJ*, 196, 525, 1975.
78. Iben, I. In Iben, I. and Renzini, A., editors, *Physical Processes in Red Giants*, page 3. D. Reidel Publishing Company: Dordrecht, 1981.
79. Iben, I. *ApJ*, 277, 333, 1984.
80. Iben, I., Kaler, J. B., Truran, J. W., and Renzini, A. *ApJ*, 264, 605, 1983.
81. Iben, I. and Renzini, A. *ApJ*, 259, L79, 1982.
82. Iben, I. and Renzini, A. *ApJ*, 263, L23, 1982.

83. Iben, I. and Renzini, A. *ARA&A*, 21, 271, 1983.
84. Johnson, H. R. *ApJ*, 260, 254, 1982.
85. Jorissen, A., Smith, V. V., and Lambert, D. L. *A&A*, 261, 164, 1992.
86. Jura, M. and Kleinmann, S. G. *ApJS*, 79, 105, 1992.
87. Jura, M., Yamamoto, A., and Kleinmann, S. G. *ApJ*, 413, 298, 1993.
88. Kaeppeler, F., Gallino, R., Busso, M., Picchio, G., and Raiteri, C. M. *ApJ*, 354, 630, 1990.
89. Kerschbaum, F. and Hron, J. *A&A*, 263, 97, 1992.
90. Kholopov, P. N. *General Catalogue of Variable Stars, 4th edition*. Nauka publishing House: Moscow, 1985.
91. Kippenhahn, R. *A&A*, 102, 293, 1981.
92. Kiss, L. L., Szatmáry, K., Cadmus, R. R., and Mattei, J. A. *A&A*, 346, 542, 1999.
93. Langer, N., Heger, A., Wellstein, S., and Herwig, F. *A&A*, 346, L37, 1999.
94. Lattanzio, J. C. *ApJ*, 311, 708, 1986.
95. Lattanzio, J. C. *ApJ*, 344, L25, 1989.
96. Lattanzio, J. C. *Proceedings of the Astronomical Society of Australia*, 10, 120, 1992.
97. Lattanzio, J. C. and Boothroyd, A. I. In Bernatowicz, T. J. and Zinner, E., editors, *Astrophysical Implications of the Laboratory Study of Presolar Materials*, page 85. AIP: Woodbury N.Y., 1997.
98. Lattanzio, J. C., Frost, C. A., Cannon, R. C., and Wood, P. R. In Wing, R. F., editor, *IAU Symp. 177: The Carbon Star Phenomenon*, page 449. Kluwer Academic Publishers: Dordrecht, 2000.
99. Lauterborn, D., Refsdal, S., and Weigert, A. *A&A*, 10, 97, 1971.
100. Le Bertre, T. *A&AS*, 97, 729, 1993.
101. Little-Marenin, I. R. and Little, S. J. *AJ*, 84, 1374, 1979.
102. Lockwood, G. W. *ApJS*, 24, 375, 1972.
103. Luck, R. E. and Bond, H. E. *ApJS*, 77, 515, 1991.
104. Mazzitelli, I., D'Antona, F., and Ventura, P. *A&A*, 348, 846, 1999.
105. McClure, R. D. *ApJ*, 280, L31, 1984.
106. Menzies, J. W. and Whitelock, P. A. *MNRAS*, 212, 783, 1985.
107. Merchán Benítez, P. and Jurado Vargas, M. *A&A*, 353, 264, 2000.
108. Messenger, B. and Lattanzio, J. *Nuc. Phys. A*, 688, 405, 2001.
109. Meyer, B. S. *ARA&A*, 32, 153, 1994.
110. Mowlavi, N. *A&A*, 350, 73, 1999.
111. Mowlavi, N., Jorissen, A., and Arnould, M. *A&A*, 311, 803, 1996.
112. Mowlavi, N., Jorissen, A., and Arnould, M. *A&A*, 334, 153, 1998.
113. Mowlavi, N. and Meynet, G. *A&A*, 361, 959, 2000.
114. Nollett, K. M., Busso, M., and Wasserburg, G. J. *ApJ*, 582, 1036, 2003.
115. Ostlie, D. A. and Cox, A. N. *ApJ*, 311, 864, 1986.
116. Paczynski, B. *Acta Astronomica*, 20, 47, 1970.
117. Paczynski, B. *Acta Astronomica*, 21, 417, 1971.
118. Paczynski, B. *ApJ*, 202, 558, 1975.
119. Robertson, B. S. C. and Feast, M. W. *MNRAS*, 196, 111, 1981.
120. Sackmann, I.-J. *ApJ*, 241, L37, 1980.
121. Sackmann, I.-J. and Boothroyd, A. I. *ApJ*, 392, L71, 1992.
122. Sackmann, I. J., Smith, R. L., and Despain, K. H. *ApJ*, 187, 555, 1974.
123. Sandage, A. R. *AJ*, 58, 61, 1953.

124. Sandage, A. R. and Walker, M. F. *ApJ*, 143, 313, 1966.
125. Scalo, J. M., Despain, K. H., and Ulrich, R. K. *ApJ*, 196, 805, 1975.
126. Schoenberner, D. *A&A*, 79, 108, 1979.
127. Schoenberner, D. *A&A*, 103, 119, 1981.
128. Scholz, M. and Wood, P. R. *A&A*, 362, 1065, 2000.
129. Schwarzschild, M. and Härm, R. *ApJ*, 142, 855, 1965.
130. Seeger, P. A., Fowler, W. A., and Clayton, D. D. *ApJS*, 11, 121, 1965.
131. Siess, L., Livio, M., and Lattanzio, J. *ApJ*, 570, 329, 2002.
132. Smak, J. *ARA&A*, 4, 19, 1966.
133. Smith, V. V. and Lambert, D. L. *ApJ*, 311, 843, 1986.
134. Smith, V. V. and Lambert, D. L. *ApJ*, 345, L75, 1989.
135. Smith, V. V. and Lambert, D. L. *ApJ*, 361, L69, 1990.
136. Straniero, O., Chieffi, A., Limongi, M., et al. *ApJ*, 478, 332, 1997.
137. Straniero, O., Gallino, R., Busso, M., et al. *ApJ*, 440, L85, 1995.
138. Sugimoto, D. and Nomoto, K. *PASJ*, 27, 197, 1975.
139. Sweigart, A. V. and Mengel, J. G. *ApJ*, 229, 624, 1979.
140. Takahashi, K. and Yokoi, K. *Atomic Data and Nuclear Data Tables*, 36, 375, 1987.
141. Thomas, H.-C. *Zeitschrift Astrophysics*, 67, 420, 1967.
142. Travaglio, C., Galli, D., Gallino, R., et al. *ApJ*, 521, 691, 1999.
143. Travaglio, C., Randich, S., Galli, D., et al. *ApJ*, 559, 909, 2001.
144. Tuchman, Y., Glasner, A., and Barkat, Z. *ApJ*, 268, 356, 1983.
145. Tuchman, Y., Sack, N., and Barkat, Z. *ApJ*, 234, 217, 1979.
146. Udalski, A., Szymanski, M., Kaluzny, J., et al. *Acta Astronomica*, 43, 289, 1993.
147. Ulrich, R. In Schramm, D. N. and Arnett, W. D., editors, *Explosive nucleosynthesis*. University of Texas Press: Austin, 1973.
148. Uus, U. *Nauch. Infor.*, 17, 3, 1970.
149. van Belle, G. T., Dyck, H. M., Benson, J. A., and Lacasse, M. G. *AJ*, 112, 2147, 1996.
150. van Leeuwen, F., Feast, M. W., Whitelock, P. A., and Yudin, B. *MNRAS*, 287, 955, 1997.
151. van Loon, J. T., Zijlstra, A. A., Whitelock, P. A., et al. *A&A*, 325, 585, 1997.
152. van Loon, J. T., Zijlstra, A. A., Whitelock, P. A., et al. *A&A*, 329, 169, 1998.
153. VandenBerg, D. A., Swenson, F. J., Rogers, F. J., Iglesias, C. A., and Alexander, D. R. *ApJ*, 532, 430, 2000.
154. Vassiliadis, E. and Wood, P. R. *ApJ*, 413, 641, 1993.
155. Vassiliadis, E. and Wood, P. R. *ApJS*, 92, 125, 1994.
156. Ventura, P., D'Antona, F., and Mazzitelli, I. *ApJ*, 524, L111, 1999.
157. Weight. A. *Zeitschrift Astrophysics*, 64, 395, 1966.
158. Whitelock, P. and Feast, M. *MNRAS*, 319, 759, 2000.
159. Whitelock, P., Marang, F., and Feast, M. *MNRAS*, 319, 728, 2000.
160. Whitelock, P., Menzies, J., Feast, M., et al. *MNRAS*, 267, 711, 1994.
161. Wood, P. R. *ApJ*, 190, 609, 1974.
162. Wood, P. R. *MNRAS*, 171, 15P, 1975.
163. Wood, P. R. *ApJ*, 227, 220, 1979.
164. Wood, P. R. In Iben, I. and Renzini, A., editors, *Physical Processes in Red Giants*, page 135. D. Reidel Publishing Company: Dordrecht, 1981.

165. Wood, P. R. In Mennessier, M. and Omont, A., editors, *From Miras to Planetary Nebulae: Which Path for Stellar Evolution?*, page 67. Editions Frontières: Gif sur Yvette, 1990.
166. Wood, P. R. In Stobie, R. and Whitelock, P., editors, *IAU Colloq. 155: Astrophysical Applications of Stellar Pulsation*, page 127. ASP: San Francisco, 1995.
167. Wood, P. R. *Publ. of the Astron. Soc. of Australia*, 17, 18, 2000.
168. Wood, P. R. In Nakada, Y., Honma, M., and Seki, M., editors, *Mass-Losing Pulsating Stars and their Circumstellar Matter*, page 3. Kluwer Academic Publishers: Dordrecht, 2003.
169. Wood, P. R., Alcock, C., Allsman, R. A., et al. In Le Bertre, T., Lèbre, A., and Waelkens, C., editors, *IAU Symp. 191: Asymptotic Giant Branch Stars*, page 151. ASP: San Francisco, 1999.
170. Wood, P. R., Bessell, M. S., and Fox, M. W. *ApJ*, 272, 99, 1983.
171. Wood, P. R. and Faulkner, D. J. *ApJ*, 307, 659, 1986.
172. Wood, P. R., Habing, H. J., and McGregor, P. J. *A&A*, 336, 925, 1998.
173. Wood, P. R. and Sebo, K. M. *MNRAS*, 282, 958, 1996.
174. Wood, P. R., Whiteoak, J. B., Hughes, S. M. G., et al. *ApJ*, 397, 552, 1992.
175. Wood, P. R. and Zarro, D. M. *ApJ*, 247, 247, 1981.
176. Xiong, D. R., Deng, L., and Cheng, Q. L. *ApJ*, 499, 355, 1998.
177. Ya'Ari, A. and Tuchman, Y. *ApJ*, 456, 350, 1996.
178. Ya'Ari, A. and Tuchman, Y. *ApJ*, 514, L35, 1999.
179. Zahn, J.-P. *A&A*, 265, 115, 1992.
180. Zinner, E. K. In Busso, M., Raiteri, C., and Gallino, R., editors, *Nuclei in the Cosmos III*, page 567. AIP: Woodbury, N.Y., 1995.
181. Zinner, E. K. In Bernatowicz, T. J. and Zinner, E., editors, *Astrophysical Implications of the Laboratory Study of Presolar Materials*, page 3. AIP: Woodbury, N.Y., 1997.

3 Synthetic AGB Evolution

Martin A.T. Groenewegen[1] and Paula Marigo[2]

[1] Instituut voor Sterrenkunde, Katholieke Universiteit Leuven
[2] Department of Astronomy, University of Padova

3.1 The Role of Synthetic Evolutionary Models

Full stellar evolution calculations are computer-time-consuming, especially when they reach the AGB phase, reflecting the complexity of the inner structure of AGB stars and of the temporal evolution due to thermal pulses (see Chapter 2). Therefore, if one is interested in distribution functions (for example, the luminosity function of AGB stars in a galaxy, or the fraction of AGB stars in a galaxy that are obscured by dust, or population synthesis in general), this is hardly achievable from calculating and integrating over a set of tens to hundreds of full stellar evolutionary models. In addition, full stellar models depend critically on a number of uncertain parameters, like mass loss, mixing length, and convective dredge-up of nucleosynthesis products. Clearly, it is computationally prohibitive to calculate a grid of stellar evolution models (including the AGB phase) sufficiently spaced in mass, and for different choices of initial metallicities, mass loss, and other parameters.

A complementary approach to *full evolutionary models* is provided by *synthetic evolutionary models*, with the aim to supply simplified descriptions of stellar evolution, or at least of the most important observable parameters, like the time-evolution of the total mass, atmospheric abundances, surface luminosity and effective temperature. The recipes and descriptions used in synthetic evolution are based on the results of full evolutionary models, and/or derived from empirical calibrations. Therefore, the physical correctness of synthetic evolution models can never be better than the full models they are based on. The main limitation of this approach is the unavoidable loss of detail in going from full models to synthetic models, which, in general, do not provide all the details on the internal structure of AGB stars.

On the positive side, the advantages of synthetic models are clear. First of all, the evolution of a star through the AGB phase is calculated more quickly. Hence, the enormous gain in computer time makes population synthesis readily feasible. Furthermore, it should be remarked that in looking at distribution functions of properties of AGB stars in external galaxies, the details—which are lost in synthetic evolution—are almost always washed away by observational error or are not resolved in the first place. Second, the influence of parameters like dredge-up, and mass loss, can be much more easily investigated. Third, predictions can be readily compared with observations, the reproduction of which—through calibration of key-model parameters—may give useful indications on related physical processes, and hence provide feedback to the developers of full models.

In this chapter we will provide a comprehensive overview of how a synthetic evolution program might look like. In fact, using the recipes provided, the reader may construct a workable synthetic evolution program of his own. Suggestions for improvements or extensions are included. Specifically, the organization of this chapter is the following. In Section 3.2 we give a brief overview of published AGB synthetic models, from pioneer to present works. Section 3.3 illustrates the necessary ingredients for developing a synthetic model for evolution of a single AGB star. As a next step, the basic information necessary to construct synthetic populations of AGB stars is given in Section 3.5. Finally, Section 3.6 is dedicated to presenting a few relevant observational constraints that synthetic models should reproduce, discussing them on the basis of recent results.

3.2 A Historical Overview

The first significant AGB synthetic model was developed by [50] (and references therein). They were primarily interested in the s-process nucleosynthesis occurring in AGB stars (see Chapter 2). The most influential study was undoubtedly that of [87]. Some of the recipes they introduced are sometimes still referred to today, even if the paper is outdated by now. In this respect, it should be remarked that the [87] model was *uncalibrated*; i.e., the parameters of the relevant process (like mass loss and dredge-up) were not constrained to observations, but rather derived from then current theoretical indications. However, that work had the merit of outlining the structure and basic ingredients of all synthetic evolution models that came later. It also indicated the kind of analysis that can be done with synthetic models: For instance, their results were used to compare the theoretical luminosity function of carbon stars with the observed one in the LMC and to compare the predicted abundances in the ejecta of AGB stars with the observed abundances in planetary nebulae. Moreover, [87] calculated the amount and chemical compositions of matter returned to the interstellar medium, thus providing the key ingredients to Galactic chemical evolution models. Other studies from that epoch include [2] and [14].

In most cases the formulae used in these studies were based on full evolutionary AGB calculations for massive stars ($3 \lesssim M/M_\odot \lesssim 8$). These results were then extrapolated to less-massive stars. This was one limitation of these early works. Other important aspects that had been neglected in almost all early studies were the metallicity dependence of the evolutionary algorithms used, the time variation of the luminosity (and other parameters) during the thermal-pulse cycle, and the fact that the first few pulses are not yet at full amplitude. Finally, the breakdown of the core mass-luminosity relation caused by the occurrence of hot bottom burning in the most massive AGB stars was not accounted for, since that was a later finding of full stellar models (i.e., [6]).

A synthetic evolution model that tried to incorporate the effects of metallicity dependence, time-variation over the thermal-pulse cycle, and the low amplitude of the first few pulses, and also made use of the full evolutionary models for low-mass AGB stars that appeared in the late 1980s (by Lattanzio, Boothroyd and Sackmann, and Hollowell; see Chapter 2) was the model by [32]. Using this new synthetic model, the luminosity function of carbon stars in the LMC was fitted to determine the dredge-up parameters and efficiency of Reimers mass-loss law. In this sense the model was *calibrated*.

In other papers the model was extended to compare the abundances in the ejecta of AGB stars to those observed in planetary nebulae (PNe) in the LMC [33], compare period distributions of Mira stars in the LMC [34], check the influence of different mass-loss prescriptions [35], study the formation and evolution of Galactic carbon stars [38], and calculate the stellar yields that are needed in chemical evolution models [101].

A model that is largely based on that of [32], but that includes a much more detailed description of the nucleosynthesis, is the calibrated model by [66]. In that work the changes in the surface abundances caused by hot bottom burning are not calculated from simple approximations, but instead by solving a nuclear network for the outer envelope. This approach is more costly in computer time (though much less than full AGB stellar calculations), but is a correct procedure for a detailed analysis of hot bottom burning phenomenon.

Both the [32] and the [66] models were significant improvements over the models from the 1980s, yet the issue remained that the recipes were not derived from a uniformly calculated grid of full stellar models, and that not all recipes included a metallicity dependence. In this perspective, [107] calculated an extensive grid of full stellar models, from which [108] derived detailed recipes useful for synthetic AGB evolution, yet not for the possible changes in the surface abundances. Some of the prescriptions in [108] have already been included in some calculations, such as those presented by [63], and [68].

These latter works mark further improvements of synthetic AGB models. A method, based on envelope integrations, was first developed by [67] and [63] to account for the breakdown of the core mass–luminosity relation as a consequence of hot bottom burning in the most massive AGB stars ($M \gtrsim 3.5 - 4\ M_\odot$). The second notable improvement concerns the treatment of the third dredge-up. In [68] a criterion, based on envelope integrations, was adopted to determine whether and when the third dredge-up occurs in a star of given mass and metallicity. Based on that, the carbon star luminosity functions in both Magellanic Clouds were reproduced, yielding a metallicity-dependent calibration of the dredge-up parameters.

3.3 The Main Ingredients of a Synthetic AGB Model

Synthetic AGB models are constructed on the basis of the results from full evolutionary calculations, which can be often "summarized" in the form of fitting analytical relations, and/or prescriptions derived from observations (e.g., mass loss, dredge-up efficiency). The approach can be purely analytical, or semianalytical when analytical prescriptions are coupled to numerical integrations of structural models (e.g., convective envelope models, "extended" envelope models including the nuclear burning shell(s)).

Synthetic models should easily predict the average evolution of the main physical quantities of an AGB star, such as the surface luminosity and effective temperature, the growth of the core mass during quiescent inter-pulse periods and its reduction when dredge-up occurs at thermal pulses (TPs), the changes in the surface chemical composition, the amount of material lost by stellar winds, and hence the final mass of the remnant.

To this end, the following sections are dedicated to presenting the main necessary ingredients, which should be considered as a "shopping list" for anyone intending to follow the evolution of a star on the AGB by means of a synthetic model. We also recommend that the reader refer to Chapter 2 for an exhaustive analysis of AGB stellar structure and evolution as predicted by full stellar models.

3.3.1 The Evolution Prior to the First Thermal Pulse

In synthetic AGB calculations, the evolution prior to the first TP should be taken into account, essentially to determine the physical conditions of a star at the onset of the thermally pulsing regime, namely, the core mass, the envelope mass possibly reduced by previous mass loss, the surface luminosity, and the envelope chemical composition.

3.3.1.1 Up to the Core Helium Ignition

Low-mass stars that experience the core helium flash ($M_i < M_{\mathrm{HeF}} \approx 1.7$–$2.5$ $M_\odot$, depending on metallicity and model details) may lose a non negligible amount of mass on the Red Giant Branch (RGB). This can be easily estimated by directly applying the classical Reimers law [84] ($\dot{M} = 4 \times 10^{-13} \eta L_* R_*/M_*$ $M_\odot$ yr^{-1}, where the stellar values are in solar units) for mass loss to the RGB evolutionary tracks of some stellar library (e.g., [21, 30, 80, 92]). In the standard procedure, the efficiency parameter η in the Reimers prescription should be calibrated (usually $\eta \approx 0.35$–0.40) in order to reproduce the observed morphology of horizontal branches in galactic globular clusters (see [86]).

Predictions of the surface chemical changes produced by the *first dredge-up*, occurring at the base of the RGB, can be taken as well from full stellar models, as a function of the stellar mass and chemical composition. Tables are nowadays available for sufficiently dense grids of M and Z (e.g., [30]), so that simple interpolations in the two variables can be made.

3.3.1.2 The Early-AGB (E-AGB)

Similarly to the RGB, the evolution along the E-AGB and the related mass loss by stellar winds can be easily derived from available sets of full stellar models—usually calculated at constant mass—to which one should then apply some mass-loss prescription. The procedure is quite simple, so that analytical fitting relations describing the evolution of the stellar parameters on the E-AGB are not given here. An analytical description of the E-AGB can be found in [108].

The expected changes in the chemical composition of the envelope due to the *second dredge-up*, occurring at the base of the E-AGB for stars with $M_i \gtrsim 3.5 \, M_\odot$, can also be obtained from available stellar libraries.

3.3.2 The Thermally Pulsing (TP) AGB Phase: Some Definitions

Let us first define some quantities that are used throughout the rest of this chapter.

A *thermal pulse cycle* (TPC) is the time interval from a local maximum of helium burning, through quiescent hydrogen burning, up to the next TP.

The *core* is defined as the part of a star inside the location where the local hydrogen content reaches half the photospheric value. Since the H-burning shell is extremely narrow (both in radius and relative mass units), this essentially coincides with all other definitions of the core mass in the literature. The *core growth* ΔM_c is defined as $\int \max\{\frac{dM_c}{d\tau}, 0\} d\tau$. Usually, $\Delta M_c = M_c(\tau) - M_{c,0}$, where $M_{c,0}$ is the core mass at the first TP.

The *first TP* is the one in which the maximum (integrated) luminosity produced by He-burning exceeds the maximum H-burning luminosity prior to this for the first time. Before the first TP there may be *prepulses*.

We refer to *asymptotic phase* or *full-amplitude regime* when the global quantities of the TP-AGB evolution have reached their asymptotic behavior, usually after a few (pre-)pulses.

3.3.3 The Initial Conditions at the First TP

The quantities that govern the physical condition of a star at the beginning of its TP-AGB phase are the total mass M, the core mass M_c, the surface luminosity L, the effective temperature T_{eff}, and the chemical composition of the envelope.

The core mass at the first thermal pulse, $M_{c,0}$, is an important quantity, since it fixes a lower limit to the remnant mass, and therefore it is closely linked to the initial–final mass relation (see Section 3.6.1).

Table 3.1. The coefficients of Equation 3.1

Z	p_1	p_2	p_3	p_4	p_5	p_6	p_7
0.02	0.0294	1.478	0.550	0.0634	0.572	3.193	0.260
0.0001	0.0213	2.589	0.592	0.0324	0.790	2.867	0.260

We report from [108] the following equations:

$$M_{c,0}(M_i) = \left(-p_1(M_i - p_2)^2 + p_3\right) f + (p_4 M_i + p_5)(1 - f) \qquad (3.1)$$

where

$$f(M_i) = \left(1 + e^{\frac{M_i - p_6}{p_7}}\right)^{-1},$$

which are fitting relations to the results of full stellar calculations. The parameters p_1 to p_7 can be found in Table 3.1, and can be interpolated linearly in $\log Z$. The luminosity at the first TP is found by inserting the core mass at the first TP into the core mass–luminosity relation presented in Section 3.3.4.5.

It should be remarked that the fitting relations quoted above are derived from classical models; i.e., the convective boundaries are determined via the Schwarzschild criterion. Stellar models adopting convective overshoot would predict a similar trend for $M_{c,0}(M_i)$, except that at given stellar mass the core mass is larger than in classical models (e.g., [30, 92]).

3.3.4 The Luminosity Evolution on the TP-AGB

In the following we briefly recall the main features of the luminosity evolution during a pulse cycle, which is depicted in Figure 3.1. This latter has been obtained by adopting the analytical relations presented by [108], which one should refer to for all the details (see also Section 3.3.4.5).

3.3.4.1 The Core Mass–Luminosity Relation (CMLR)

On the basis of stellar evolution calculations, [76] first discovered the existence of a simple relationship between the core mass M_c of a TP-AGB star and its quiescent interflash luminosity (part of the lightcurve around L_A in Figure 3.1). He derived the fitting formula in the range 0.57 $M_\odot < M_c <$ 1.39 $M_\odot$:

$$L = 59250(M_c - 0.522), \qquad (3.2)$$

where masses and luminosities are in solar units.

Over the years, continuous upgrading of the input physics and detailed calculations of the AGB phase over wider ranges of stellar masses and chemical compositions resulted into a flourishing of several $M_c - L$ relationships [9, 41, 46, 50, 59, 108, 120]. Many of them are displayed in Figure 3.2.

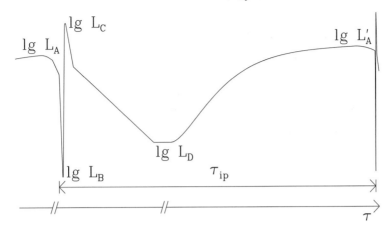

Fig. 3.1. Typical evolution of the surface luminosity during a thermal pulse cycle. The time axis has been stretched around the first TP shown. From [108]

From the most recent analysis it turns out that the CMLR contains a non negligible dependence upon the surface chemical composition; i.e., for a given core mass the luminosity is higher at increasing metal and helium contents (hence higher mean molecular weight). For instance, for core masses in the range $0.5\ M_\odot < M_c < 0.66\ M_\odot$, [9] derived

$$L = 238000\mu^3 (Z_{\rm CNO})^{0.04} (M_c^2 - 0.0305 M_c - 0.1802), \qquad (3.3)$$

where $\mu = 4/(5X+3-Z)$ is the mean molecular weight under the assumption of a fully ionized gas, and $Z_{\rm CNO}$ is the fractional surface abundance (by mass) of the CNO elements. It follows that at given core mass, stars of solar composition ($Z = 0.02$, $\mu \approx 0.62$) are $\approx 25\%$ more luminous than metal-poor stars ($Z = 0.001$, $\mu \approx 0.598$).

The existence of the CMLR can be explained from first principles, under specific structural conditions that make the thermal evolution of the core be essentially decoupled from that of the outer envelope. The reader can find transparent demonstrations of this in [54, 76, 82, 99].

It should be emphasized that the structure of a TP-AGB star fulfills the conditions necessary for a CMLR to exist only in particular stages of the evolution, i.e., during the quiescent part of the pulse cycle (when the H-burning shell supplies most of the stellar energy), provided that the full-amplitude regime has been attained, and hot bottom burning (HBB, see later in this section) is not operating.

Out of these specific phases, the luminosity evolution of a TP-AGB star does not follow the CMLR listed above. The main deviations, which should be included in synthetic calculations, are briefly recalled and discussed in the following.

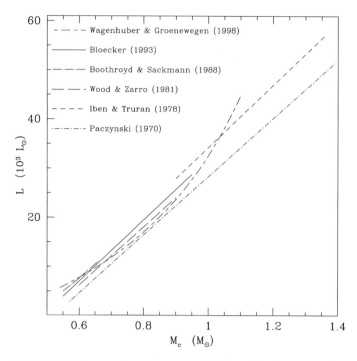

Fig. 3.2. CMLRs from various authors, each plotted in the range of M_c considered to obtain the analytical fit to the model results. Where a composition dependence is present, the solar case is considered, i.e., ($Z = 0.02, Y = 0.28$). The relation by Iben & Truran [50] refers to a 7 $M_\odot$ model

3.3.4.2 The Subluminous Prepulses

The onset of thermal pulses usually takes place at lower luminosities than predicted by the standard CMLR for the same core mass. Moreover, the initial evolution of the quiescent interpulse luminosity as a function of M_c approaches the asymptotic regime of full amplitude with a steeper slope than that of the CMLR (e.g., [9]). Recently, [44] have reported that the occurrence of the third dredge-up already at the first thermal pulses may alter the approach toward the full amplitude regime, possibly preventing the settlement onto the CMLR (but see also [70]).

3.3.4.3 Flash-Driven Luminosity Variations: Peaks and Dips

The occurrence of a TP may cause significant variations of the surface luminosity as long as the thermal instability is removed and quiescent H-burning recovered. The most significant features following a TP are shown in Figure 3.1, namely, (i) the rapid dip (fast luminosity drop indicated by L_B), (ii) the post-thermal-pulse luminosity peak (rapid rise of light curve up to the maximum denoted by L_C), and (iii) the slow dip (subluminous part of the pulse cycle labeled L_D). These features have amplitudes and durations that depend on stellar parameters, mainly envelope mass and metallicity (see [8] for a detailed analysis).

It should be remarked that the subluminous slow dip is that of the greatest significance, since it may produce observable and testable effects, especially during the evolution of the lowest-mass stars. These can spend as much as 20%–30% of the interpulse period at a luminosity a factor of two lower than predicted by the CMLR for the same core mass. As a consequence, the slow dip is found to importantly affect the faint wing of the carbon star luminosity function (see Section 3.6.2), and may explain the observed mixture of carbon- and oxygen-rich stars at the same luminosities (e.g., [69]).

3.3.4.4 Over-Luminosity Produced by Hot Bottom Burning (HBB)

In the most luminous and massive TP-AGB stars ($M_i \gtrsim 3.5$ M$_\odot$) the base of the convective stellar envelope can deepen into high-temperature regions ($T > 20 - 40 \times 10^6$ K) so that H-burning via the CNO cycle can occur. This process is called hot bottom burning (HBB) or envelope burning. The most notable signature of its occurrence is the breakdown of the CMLR toward higher luminosities, as first pointed out by [6], and confirmed by subsequent studies (e.g., [7, 12, 20, 61, 106]).

The violation of the CMLR produced by HBB should be included in modern AGB synthetic calculations, provided that the results of full AGB models can be reproduced. To this end, substantial efforts have been recently performed by [63, 67, 108]. In the former work a solution method, based on envelope integrations, has been developed in order to account for the production of nuclear energy at the base of the convective envelope; the latter study presents analytical fitting relations derived from complete calculations of AGB models with HBB.

Both works successfully reproduce the expected luminosity evolution of a massive TP-AGB star with HBB, as a function of model parameters (i.e., mass-loss rates, mixing length parameter), and stellar parameters (i.e., current envelope mass, metallicity). An example of synthetic calculations (from [63]) is illustrated in Figure 3.3, showing the luminosity evolution of a ($5\,\text{M}_\odot$, $Z = 0.008$) model as a function of the core mass. Mass loss is included according to the [106] formalism.

We can notice that HBB develops since the first subluminous interpulse periods, so that when thermal pulses attain the full-amplitude regime the star does not settle on the CMLR (dashed line), but quickly reaches higher and higher luminosities, with a brightening rate much greater than expected

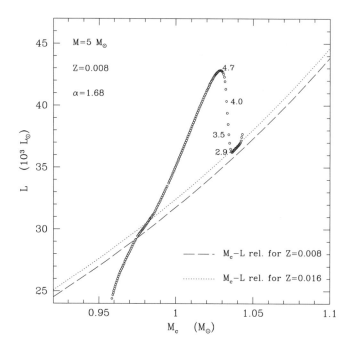

Fig. 3.3. Luminosity evolution of a $5.0 \, M_\odot$, $Z = 0.008$ star experiencing HBB. The open circles correspond to the maximum quiescent luminosity before each thermal pulse. The numbers along the curve indicate the current stellar mass in solar units. The dashed and dotted lines represent the reference CMLR relation [108] for $Z = 0.008$ and $Z = 0.016$, respectively. From [63]

from the slope of the standard relation. The luminosity increases as long as the stellar envelope is massive enough to maintain high temperatures at its bottom. The luminosity maximum and subsequent decline are determined by the onset of the superwind, when the star starts to rapidly lose mass at significant rates (i.e., $\dot{M} \approx 10^{-5}$ to $10^{-4} \, M_\odot \, \text{yr}^{-1}$). Due to the drastic reduction of the stellar envelope mass, HBB weakens and eventually extinguishes. As a consequence, the star approaches again the CMLR, where it lies till the end of the evolution.

It is worth remarking that owing to the dependence of the CMLR on the chemical composition, the star finally recovers a CMLR consistent with an increased stellar envelope metallicity due to the combined effect of the third dredge-up and HBB (see Figure 3.3).

3.3.4.5 Useful Analytical Relations

Analytical formulas describing secular light curves of TP-AGB stars, based on extensive grids of evolutionary calculations for stars with initial masses in the range $0.8\,M_\odot \leq M_i \leq 7\,M_\odot$, and metallicities $Z = 0.0001$, $Z = 0.008$, and $Z = 0.02$, have recently been presented by [108]. The maximum luminosity during quiescent H-burning is given as the sum of different terms:

$$L = (18160 + 3980\log\tfrac{Z}{0.02})(M_c - 0.4468) \tag{4a}$$
$$+ \left[\,10^{2.705+1.649 M_c}\right. \tag{4b}$$
$$\left.\times\,10^{0.0237(\alpha-1.447)M_{c,0}^2 M_{\text{env}}^2 (1-e^{-\Delta M_c/0.01})}\,\right] \tag{4c}$$
$$-10^{3.529-(M_{c,0}-0.4468)\Delta M_c/0.01} \tag{4d}$$

The first term (4a) represents the classical linear $M_c - L$ relation, giving the quiescent luminosity for stars already in the full-amplitude regime, with core masses in the range $0.6\,M_\odot \lesssim M_c \lesssim 0.95\,M_\odot$. The second term (4b) provides a correction, becoming significant for high values of the core mass $M_c > 0.95\,M_\odot$. The third term (4c) accounts for the extra luminosity produced by HBB, as a function of the stellar envelope mass M_{env}. A dependence on the mixing-length parameter, α, is included to reproduce the strength of HBB on the basis of calculations carried out by [107] with $\alpha = 1.5$, and other authors, namely, [7] with $\alpha = 2.1$, [20] with $\alpha = 2.75$. Finally, the fourth term (4d) gives a negative correction to the luminosity in order to mimic the subluminous and steep evolution typical of the first pulses.

We recall that in [108] the reader can find many other useful analytical relations, e.g., describing the flash-driven variations already mentioned in Section 3.3.4.3.

3.3.5 The Time Evolution on the TP-AGB

The time evolution of a TP-AGB star is essentially determined by the outward advance of the H-burning shell, which lasts for most of the interpulse period, τ_{ip}. After this, the next thermal pulse occurs, possibly followed by a dredge-up event that reduces the core mass, and a new pulsecycle starts.

3.3.5.1 The Rate of Core Growth During the Interpulse Period

The equation that describes the rate of core growth reads

$$\frac{\mathrm{d}M_c}{\mathrm{d}t} = \frac{q}{X}L_H, \tag{5}$$

where X is the hydrogen abundance (in mass fraction) in the stellar envelope, L_H is the luminosity produced by the H-burning shell, and q is the mass burnt per unit of energy released. Following [108], it turns out that q is slightly metallicity dependent due to the contribution from the pp-cycle for low Z. The calculations with a nuclear network yield the mean value $q = [(1.02 \pm 0.04) + 0.017(Z/0.02)]10^{-11}\,M_\odot\,L_\odot^{-1}\,\text{yr}^{-1}$.

The luminosity produced by H-burning (L_H) that enters Equation 5 is lower than the total luminosity, since gravitational energy release due to the core shrinking and the flow of burnt envelope material down to the core, together with relic He-burning, additionally contribute during quiescent H-burning. According to [108], the relative contribution from H-burning to the inter-pulse surface luminosity as given by their equation (1.4) (part of the light curve labeled L_A in Figure 3.1) is

$$\lg(L_H/L) = -0.012 - 10^{-1.25-113\Delta M_c} - 0.0016 M_{env}. \tag{6}$$

This equation has been derived from complete AGB models without HBB, i.e., with the term (4c) set to unity.

3.3.5.2 The Core Mass–Interpulse Period Relation

The time-scale characteristic of a thermal pulse, τ_{ip}, which is the time interval between two subsequent pulses, is a decreasing function of the core mass as first pointed out by [77],

$$\log \tau_{ip} = 4.5\,(1.678 - M_c), \tag{7}$$

and subsequently confirmed by complete AGB calculations (e.g., [10, 120]). In the work of [10] a strong dependence on composition was found: At the same core mass, τ_{ip} for the low-metallicity case (e.g., $Z = 0.001$) is found to be nearly twice as long as in the high-metallicity case (e.g., $Z = 0.02$).

The recent relation presented by [108] is

$$\log \tau_{ip} = (-3.628 + 0.1337 \, \log \tfrac{Z}{0.02})\,(M_c - 1.9454) \tag{8a}$$
$$-10^{-2.080-0.353 \, \log \tfrac{Z}{0.02}+0.200(M_{env}+\alpha-1.5)} \tag{8b}$$
$$-10^{-0.626-70.30\,(M_{c,0}-\log \tfrac{Z}{0.02})\,\Delta M_c}. \tag{8c}$$

Three components can be distinguished: The term (8a) expresses τ_{ip} as a decreasing exponential function of M_c with some dependence on the metallicity; the term (8b) gives a negative correction to include the effect of hot bottom burning in somewhat reducing the interpulse period; and term (8c) reproduces the initial increase of τ_{ip} starting from values that are shorter by almost a factor of two compared to those derived from the asymptotic relation for the same core mass.

3.3.6 The Effective Temperature

The effective temperature, T_{eff}, can be determined in two possible ways, either (1) using semiempirical calibrations, or (2) with the aid of stellar envelope integrations.

Empirical prescriptions can be derived, for instance, by coupling observations of AGB variables with measured periods and light curves (e.g., Miras) to pulsation model atmospheres, in which the effective temperature is made to vary until good spectral and light-amplitude fits are obtained (e.g., [4, 13]). This allows one to get a relation between the bolometric luminosity and effective temperature, including some dependence on the stellar mass and metallicity (e.g., [118]).

The purely theoretical approach, based on stellar envelope models, requires that four boundary conditions be specified in order to integrate the four equations of stellar structure for L_r, T_r, P_r (or ρ_r), and r as a function of the Lagrangian mass coordinate M_r. Let us denote by M, L, P, R, $T_{\rm eff}$ the photospheric values, and assume that M, $M_{\rm c}$, and the chemical composition of the homogeneous convective envelope are given.

Two outer boundary conditions are immediately available: The effective stellar surface radius is determined by the Stefan-Boltzmann law, and the photospheric pressure can be derived as $P = P(L, M, T_{\rm eff})$ from the integration of a stellar atmosphere model; see Chapter 4.

The third condition can be derived, for instance, considering that the transition to the core is very well defined, due to the extreme steepness of the structural gradients in a very thin region below the base of the convective envelope. In other words, we can safely assume that the Lagrangian mass coordinate of the base of the envelope coincides with that of the core.

These three conditions are sufficient to solve the stellar envelope structure up to the photosphere (hence to determine $T_{\rm eff}$) if no energy source is present inside the envelope, i.e., when HBB is not operating. In this circumstance the integration of the equation of thermal balance can be skipped, just setting $L_r = L$, i.e., the luminosity is constant throughout the envelope. We recall that during quiescent H-burning L is determined by the CMLR (see Section 3.3.4.1).

Differently, in massive AGB stars with HBB, not only does the surface luminosity depend on $M_{\rm c}$, but it is also affected by the physical properties of the stellar envelope itself (e.g., envelope mass, temperature, and density stratification) as well as by model prescriptions (e.g., mixing-length parameter, overshooting). All these factors actually affect the efficiency of nuclear rates in the deepest layers of the envelope. The fourth boundary condition is then needed. Possible boundary conditions can be figured out with the aid of complete calculations of the AGB phase. For instance, one could adopt the *core mass–core radius relation* for AGB stars to determine the radial coordinate below which a mass equal to $M_{\rm c}$ should be contained. Another alternative is to constrain the local energy flux at the base of the convective envelope to be the sum of the energy contributions from the gravitational contraction of the core, He-burning shell, and radiative layers of the H-burning shell. Physical and technical details on envelope integrations are extensively described in [63, 66, 87, 90, 108, 120].

Stellar envelope integrations in synthetic AGB models are also useful for quickly exploring the sensitivity of the results to different input prescriptions, such as the mixing-length parameter α, nuclear reaction rates, and gas opacities. For instance, [65] has recently investigated the remarkable effects produced by molecular opacities if properly coupled to the actual carbon and oxygen surface abundance, hence C/O ratio. This point is discussed in Section 3.6.3.

3.3.7 Mass Loss by Stellar Winds

Mass loss by stellar winds plays a crucial role in the evolution of an AGB star, since it not only determines the final mass at the end of the AGB and hence the final luminosity, but it also affects its internal structure and nucleosynthesis. Many mass-loss prescriptions for the AGB phase have been suggested over the years, based on both theoretical arguments (e.g., [1, 7, 13, 24]) and/or (semi-) empirical calibrations (e.g., [106]), which have proved to give a better description of mass-loss rates than the classical Reimers law [84].

This latter was originally designed to account for mass loss along the RGB (see Section 3.3.1.1). The adoption of the Reimers law does not account for the highest mass loss rates observed in stars close to the tip of the AGB ($\dot{M} \approx 10^{-5} - 10^{-4} M_\odot$ yr^{-1}), being also inadequate to explain the ejection of planetary nebulae at the observed luminosities. A sudden transition to the so-called *superwind* regime should then be invoked to reconcile predictions to observations [87]. High mass-loss rates can be obtained with Reimers-like prescriptions only by adopting quite large values of the efficiency parameter (e.g., $\eta = 5$ as in [32]). A natural development of the super wind is instead allowed by more recent prescriptions for mass loss (e.g., [24, 106]) that consider the combined role of radiation pressure, stellar pulsation, and dust formation in driving mass loss on the AGB. Moreover, it turns out that mass loss depends on metallicity, roughly as $\approx Z^{0.5}$, as suggested from both theoretical [57, 103] and observational [37] evidence. In addition, since mass loss seems to correlate with luminosity, and the luminosity varies over the TP-cycle, mass loss variations on a time scale of 10^4 years are implied (see Section 3.6.6).

The effect of different mass-loss prescriptions on the predicted initial–final mass relation and its dependence on the chemical composition will be discussed in Section 3.6.1, on the basis of synthetic AGB calculations.

3.3.8 Surface Chemical Changes During the TP-AGB

We will illustrate here the basic schemes for describing the main processes that may change the surface chemical composition of an AGB star, namely, the third dredge-up and hot bottom burning.

3.3.8.1 The Third Dredge-Up

A synthetic treatment of the third dredge-up requires three basic inputs:

- the minimum core mass $M_c^{\min}$ for dredge-up to occur;
- the efficiency λ;
- the chemical composition of the material being dredged up.

A TP-AGB star can experience the third dredge-up if its core mass has grown over a critical value $M_c^{\min}$, as indicated by theoretical analysis of thermal pulses (e.g., [11, 49, 60, 117]).

The quantity λ expresses the dredge-up efficiency, since it measures the extent of the inward penetration of the convective envelope. It is defined as the fraction of the core mass increment, ΔM_c, during the preceding inter-pulse period that is dredged up to the surface by envelope convection:

$$\lambda = \frac{\Delta M_{\text{dredge}}}{\Delta M_c}, \qquad (9)$$

where ΔM_{dredge} is the mass of the dredged-up material.

Finally, the chemical composition of the dredged-up material coincides with that of the convective intershell that develops during a thermal pulse, mainly containing ^{4}He and the nuclear products of alpha-captures (C and O).

Once the three parameters are specified, the new surface abundance of a given element Y_i^{new} soon after a dredge-up event can be calculated,

$$Y_i^{\text{new}} = \frac{Y_i^{\text{old}} M_{\text{env}} + Y_i^{\text{cs}} \Delta M_{\text{dredge}}}{M_{\text{env}} + \Delta M_{\text{dredge}}}, \qquad (10)$$

where Y_i^{old} and M_{env} are the stellar envelope abundance and mass before the dredge-up, respectively; Y_i^{cs} is the elemental abundance in the convective intershell, hence in the dredged-up material.

In principle, the aforementioned parameters in synthetic AGB models could be specified following the predictions of full AGB calculations. From doing so in past works (e.g., [87]) a clear discrepancy between observations and predictions emerged: Theory predicted too few faint (single) carbon stars, an aspect of the so-called *carbon star mystery* first pointed out by [47].

In this context, [32], [66], and [68] have moved from another perspective: The dredge-up parameters are adopted as free parameters to be calibrated on the observed luminosity functions of carbon stars in the Magellanic Clouds. The discrepancy could be removed assuming, on average, lower $M_c^{\min}$ and larger λ than found in full evolutionary calculations.

In recent years the detailed treatment of the third dredge-up has been improved, leading, in some cases, to an earlier occurrence and a greater efficiency than in previous calculations (e.g., [43, 44, 94]).

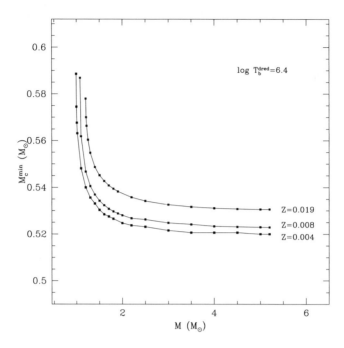

Fig. 3.4. Predicted behavior of the minimum core mass for the third dredge-up, $M_c^{\min}$, as a function of the actual stellar mass, for three values of the initial metallicity, and assuming the specified value for the minimum temperature at the base of the convective envelope (T_b^{dred}). From [68]

The real values of $M_c^{\min}$ are still a matter of debate, since this quantity heavily depends on the delicate interplay between model prescriptions (e.g., opacities, nuclear reaction rates, mixing-length parameter, mass loss, time and mass resolution) and intrinsic properties of the star (e.g., chemical composition, stellar envelope mass, strength of the pulse). Nevertheless, a certain trend can be extracted from full stellar models; i.e., the onset of the third dredge-up is favored (i.e., $M_c^{\min}$ is lower) at higher stellar mass M, lower metallicity Z, and greater efficiency of envelope convection (e.g., obtained by increasing the mixing-length parameter).

The estimation of $M_c^{\min}$ in synthetic models has been recently improved by [68] following the scheme proposed by [117]. The usual criterion based on a constant $M_c^{\min}$ parameter is replaced by one on the minimum temperature at the base of the convective stellar envelope, T_b^{dred}, at the stage of the post-thermal-pulse luminosity maximum. Envelope integrations allow one then to determine $M_c^{\min}$ as a function of stellar mass, metallicity, and pulse strength, thus implying whether and when dredge-up first occurs. Predictions for $M_c^{\min}$ are shown in Figure 3.4, as a function of stellar mass and metallicity. It turns out that the onset of the third dredge-up is favored in stars with lower metallicity and larger mass. The $T_\mathrm{b}^\mathrm{dred}$ criterion can also predict the possible shutdown of the process, e.g., in the last stages of the AGB evolution when the envelope mass is drastically reduced and the minimum base temperature is no longer attained.

The question of the real values of λ is also troublesome, since it is crucially affected by the treatment (both physical and numerical) of the instability of matter against convection (see, e.g., [27, 72]). It should be remarked that the difficulty of earlier complete AGB models in finding significant dredge-up (e.g., [28, 49, 117]), particularly in faint low-mass stars, still affects many recent calculations (e.g., [106, 107]). However, recent analysis would indicate that such a problem can be overcome, either on the basis of technical improvements [94], or by allowing for deep extra mixing beyond the formal convective boundaries [44, 72]. However, these latter AGB calculations still need to be extended to the lowest stellar masses and over a larger range of metallicities, and then tested through a full comparison with observations of carbon stars.

Conversely, a limitation still affecting AGB synthetic calculations is that λ is often assumed a constant parameter, regardless of current mass and metallicity. On the basis of recent complete AGB calculations with deep extra mixing, [72] suggests an analytical fitting formula

$$\lambda = 7 \times (M_\mathrm{c} - 0.571) \tag{11}$$

expressing the efficiency of the third dredge-up as a function of the core mass. The correctness of this law has still to be tested through systematic synthetic calculations.

The chemical composition of the dredge-up material is also affected by model details. According to full calculations of the third dredge-up and related nucleosynthesis, after the first peculiar pulses the intershell approaches a typical chemical composition, in which the distribution of the elements seems to be crucially dependent on the treatment of the convective boundaries, with $[X(^4\mathrm{He}) \approx 0.76;\ X(^{12}\mathrm{C}) \approx 0.22;\ X(^{16}\mathrm{O}) \approx 0.02]$ according to classical models based on the Schwarzschild criterion (e.g., [10]); and $[X(^4\mathrm{He}) \approx 0.25;\ X(^{12}\mathrm{C}) \approx 0.50;\ X(^{16}\mathrm{O}) \approx 0.25]$ following the recent results by [44] based on the adoption of a deep overshooting.

For all the reasons mentioned, the treatment of the third dredge-up in synthetic models is preferably a parametric one; that is, the parameters are not assumed a priori, but calibrated on observations (i.e., carbon star luminosity functions). This important application will be discussed in detail in Section 3.6.2, on the basis of recent AGB synthetic calculations.

3.3.8.2 Nucleosynthesis During HBB

The changes in surface abundances (e.g., H, He, C, N, O) caused by HBB during the interpulse evolution of the most massive AGB stars should be taken into account in synthetic AGB models. This can be done either (i) by directly solving a specified nuclear network (see, for instance, the works by [67, 87, 90]), or (ii) by adopting a parametric scheme like that suggested by [32], to which the reader is referred for a description.

In case (i), static stellar envelope integrations can be used to determine, at each time step, the temperature and density stratification across the envelope. Then, the changes in elemental abundances can be obtained by directly solving the system of differential equations, of the kind

$$\frac{dY_i}{dt} = -[ij]\ Y_i Y_j + [rs]\ Y_r Y_s, \quad i = 1, ..., N_{\rm el}, \qquad (12)$$

where $N_{\rm el}$ is the number of elemental species involved; $Y_i = X_i/A_i$, X_i, and A_i denote the abundance by number (mole g^{-1}), the abundance by mass, and the atomic mass of the elemental species i, respectively. On the right-hand side of equation (12), $[ij]$ stands for the rate of the generic reaction that converts the element i into another nucleus because of the interaction with the element j, whereas $[rs]$ is the rate of the generic reaction transforming nuclei s and r into the element i.

The system of Equations (12) can be solved locally, i.e., in each layer of the stellar envelope. Finally, the new chemical composition of the envelope can be determined by mass-averaging the new local abundances in each layer throughout the envelope; i.e., it is equivalent to assume that convective currents completely homogenize the envelope composition.

3.3.8.3 Some Exemplifying Calculations

As mentioned in Section 3.3.8.1, the main processes that may alter the surface abundances of TP-AGB stars are (i) the third dredge-up, and (ii) H-burning via the CNO cycle (HBB) in the hottest stellar envelope layers of the most massive AGB stars. The former essentially determines the enrichment in primary ^{12}C, ^{16}O, ^{4}He, and s-process elements. The latter mainly leads to the increase of ^{13}C abundance at the expense of that of ^{12}C and, if reactions of the CNO cycle proceed further on, so as to allow an efficient synthesis of ^{14}N.

Figures 3.5 and 3.6 exemplify the typical evolution of the CNO surface abundances during the TP-AGB phase caused by (1) sole third dredge-up as in the 2.5 M$_\odot$ model; (2) third dredge-up *and* HBB as in the 5 M$_\odot$ model.

Case of Sole Third Dredge-Up
As far as the 2.5 M$_\odot$ model is concerned (Figure 3.5), the following points are worthy of notice: The transition to carbon star (as soon as C/O $>$ 1) after the first few thermal pulses, the increase of the ^{12}C/^{13}C ratio, the modest increase of the ^{16}O abundance, and the concomitant small decrease of the N/O ratio. Of course, these results are highly dependent on the adopted chemical distribution of the intershell, and the efficiency of envelope penetration, expressed by the parameter $\lambda = 0.5$, which has been calibrated in this

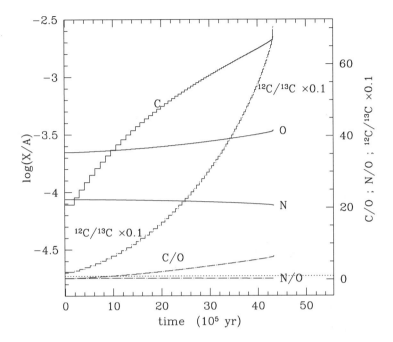

Fig. 3.5. Evolution of CNO surface abundances (by number, mole gr^{-1}) and ratios from the first thermal pulse until the complete ejection of the stellar envelope for $Z = 0.008$ TP-AGB model with initial mass of $2.5\,M_\odot$. The efficiency parameter for the third dredge-up is assumed to be $\lambda = 0.5$; the mixing-length parameter is $\alpha = 2.0$. Based on synthetic calculations by [63]

case on the luminosity function of carbon stars in the LMC (cf. Section 3.6.2).

Case of Third Dredge-Up and HBB

The saw-like trend of the C and ^{12}C/^{13}C curves for the $5\,M_\odot$ model (Figure 3.6) is caused by the combined effect of the convective dredge-up events at each thermal pulse, clearly visible in the sudden spikes, and HBB during the subsequent interflash period, which quickly converts newly dredged-up ^{12}C first into ^{13}C and then into ^{14}N.

The initial increase of both C abundance and ^{12}C/^{13}C ratio, due to the sole effect of the third dredge-up, is followed by a rapid drop as soon as HBB has become efficient enough to convert carbon isotopes into nitrogen via the first reactions of the CN cycle. *The ^{12}C/^{13}C ratio attains and keeps its equilibrium value of ≈ 3.4*, until the last stages of the evolution, when the

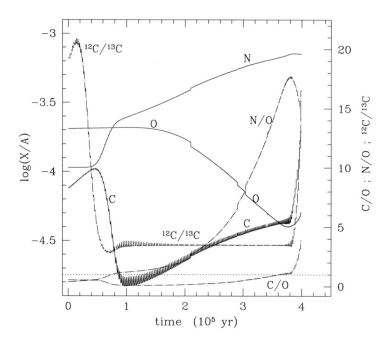

Fig. 3.6. The same as in Figure 3.5 but for a $Z = 0.008$ TP-AGB model with initial mass of $5.0\,M_\odot$

drastic reduction of the stellar envelope mass causes a sudden extinction of nuclear reactions.

Note that for most of the TP-AGB lifetime the star is expected to be oxygen-rich; i.e., the C/O ratio is below unity. In other words, *HBB may delay or even prevent the formation of carbon stars at high luminosities*. In this case, the transition to a carbon star eventually happens when HBB weakens and ^{12}C enrichment due to convective dredge-up becomes dominant. Observationally, this would correspond to the so-called obscured carbon stars [105] being optically invisible because of the dense dusty circumstellar envelope ejected during the superwind regime (cf. [26]).

The oxygen abundance remains practically unchanged for nearly half the TP-AGB lifetime. This holds as long as the base temperature, T_b, of the convective envelope has not become high enough to activate the ON cycle (^{16}O has the longest nuclear lifetime against proton captures). When this happens, for $T_b \gtrsim 80 \times 10^6$ K, ^{16}O is efficiently destroyed.

The main product of the CNO bicycle is ^{14}N, as illustrated by the rising of the nitrogen curve. The decrease of oxygen and the rise of nitrogen both concur to significantly increase the N/O ratio up to a factor of ≈ 40. In general, HBB is expected to be an important source of nitrogen, a prediction that may be relevant for the interpretation of the overabundance of this element measured in the atmospheres of some giant stars and PNe, as well as of its observed distribution in galaxies of different metallicities.

3.3.8.4 Secondary and Primary Synthesis of CNO Nuclei

During the TP-AGB evolution of stars that undergo HBB, at each dredge-up event some primary ^{12}C and ^{16}O nuclei (produced by α-capture reactions) are injected into the stellar envelope, where they can partake in the CNO cycle for the conversion of hydrogen into helium. As a consequence, the surface distribution of the CNO nuclei consists of two components:

- a *secondary component* produced by nuclear reactions involving only seed metals originally present in the star at the epoch of its formation;
- a *primary component* synthesized by a chain of nuclear burnings beginning from H and He and hence independent of the original metal content.

Hence, the surface abundance (by number) Y_i of the element i belonging to the CNO group at an instant t, can be expressed as $Y_i = Y_i^S + Y_i^P$, where Y_i^S and Y_i^P are the secondary and primary fractions, respectively.

In Figure 3.7 the evolution of secondary and primary ^{12}C, ^{13}C, ^{14}N, ^{15}N, ^{16}O, ^{17}O, and ^{18}O surface abundances is plotted as a function of time for a $4.5\,M_\odot$ TP-AGB star with $Z = 0.004$. It can be noticed that in this case, for all CNO nuclei the primary component (which is initially equal to zero) of the surface abundance grows such that it overcomes the secondary one.

3.4 Stellar Yields

Low- and intermediate-mass stars play an important role in galactic chemical evolution, thanks to the ejection into the interstellar medium of material containing newly synthesized nuclear products. These stars contribute to the chemical enrichment of the ISM essentially during the RGB and AGB phases, both characterized by the occurrence of both mass loss and events of surface chemical pollution. The most relevant contribution concerns ^{4}He, ^{12}C, and ^{14}N, but also the *s*-process elements (see [98] and references therein).

In the following, first we briefly recall the basic equations for the calculation of the stellar yields, and then show the predictions for the main CNO isotopes, based on recent synthetic AGB models.

3.4.1 Derivation of the Wind Contributions

According to the classical definition by [95], the *stellar yield*, $p_k(m)$, of a given chemical element k is the mass fraction of a star with initial mass m

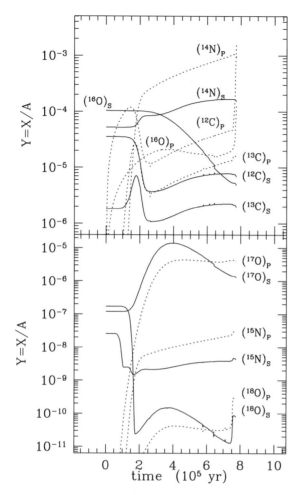

Fig. 3.7. Secondary and primary surface abundances (mole g^{-1}) of CNO isotopes for a (4.5 M$_\odot$, Z = 0.004) model evolving on the TP-AGB phase. The adopted mixing length parameter is α = 2.0, and the third dredge-up efficiency is λ = 0.65. Taken from synthetic AGB models by [63]

that is converted into the element k and returned to the ISM during its entire lifetime, $\tau(m)$. It follows that the quantity

$$M_y(k) = m\, p_k(m) \tag{13}$$

gives the mass of the ejected matter in the form of the newly synthesized element k, which can be calculated by

$$M_y(k) = \int_0^{\tau(m)} [X_k(t) - X_k^0]\, \dot{M}(t)\, \mathrm{d}t \tag{14}$$

where $\dot{M}(t)$ represents the mass-loss rate at the current time t, and $X_k(t)$, X_k^0 denote the current and initial surface abundances of the element k, respectively. The calculation of total wind contributions should account for any

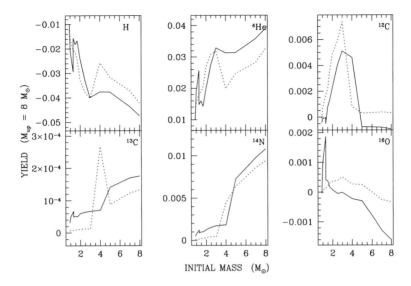

Fig. 3.8. Stellar yields, $p_k(m)$, from low- and intermediate mass stars, taken from calculations by [101] with $M_{\rm up} = 8 M_\odot$, for initial metallicities $Z = 0.02$ (solid line) and $Z = 0.004$ (dotted line)

possible change of elemental species caused, for instance, by the first, second, third dredge-up, and HBB. According to Equation 14, *negative* p_k correspond to those elemental species that are prevalently *destroyed* and diluted in the stellar envelope, so that their abundances in the ejected material are lower with respect to the main sequence values. In contrast, *positive* p_k correspond to those elements that are prevalently *produced*, so that a net enrichment of their abundances in the ejecta is predicted.

3.4.2 Yields as a Function of M_i and Z

Figures 3.8 and 3.9 show the fractional yields p_k as a function of the initial stellar mass and two choices of the metallicity, taken from two sets with different values of the upper mass limit $M_{\rm up}$ (i.e., [64, 101]). Despite some differences, in both works the predicted yield of a given elemental species displays a similar trend with the initial mass and metallicity.

- **H and ^{4}He.** The net yields of these elements have mirror-like behaviors, corresponding to a destruction of H during the synthesis of ^{4}He. The peak in ^{4}He production at around 2–3 $M_\odot$ (depending on metallicity) is mainly due to the fact that stars in this mass range have the longest TP-AGB lifetimes, so that they suffer the largest number of dredge-up events. The increase of ^{4}He yields toward higher masses, $M_i \geq 3.5 M_\odot$, is favored by the

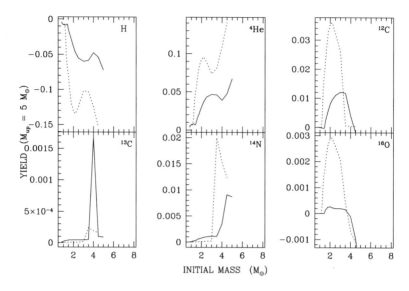

Fig. 3.9. The same as in Figure 3.8, but data taken from [64], with $M_{\rm up} = 5\,M_\odot$

occurrence of both dredge-up and HBB, depending on both the efficiency and duration of the processes. At decreasing metallicity, the synthesis of new ^{4}He is favored by the larger efficiency of HBB in the models by [64], whereas the shorter TP-AGB lifetimes produce the opposite effect in the results of [101].

- ^{12}C: The net positive yield increases with the initial mass, with a production peak at about 2–3 $M_\odot$, which corresponds to stars experiencing the largest number of dredge-up episodes on the TP-AGB. The subsequent decline at increasing stellar mass is first determined by the decrease of the TP-AGB lifetimes, and then by the prevailing effect of HBB, so that the ^{12}C yield from the most massive stars finally becomes negative. It is worth noticing that both in the [64] and [101] results the peak of ^{12}C production moves toward lower initial masses for decreasing metallicities, reflecting the longer TP-AGB duration and greater efficiency of HBB.
- ^{13}C: The positive ^{13}C yields result mainly from the first dredge-up in lower-mass stars, and mild HBB in the most massive AGB models as long as the creation of ^{13}C via the reaction ^{12}C$(p,\gamma)^{13}$C prevails over destruction via the reaction ^{13}C$(p,\gamma)^{14}$N. The results also depend on the interplay between the efficiency of HBB and stellar winds. A suitable tuning between the strength of HBB and mass loss can occasionally create favorable conditions for the synthesis of ^{13}C, giving a peak of the related yields around $4\,M_\odot$ for $Z = 0.02$ in [101], and for $Z = 0.004$ in [64]. In these cases, the efficiency of CNO-cycle reactions allows the synthesis

of ^{13}C for a long time before the drastic reduction of the envelope mass causes the extinction of nuclear burning.
- ^{14}N: The surface abundance of ^{14}N can increase as a consequence of the second dredge-up and HBB, the latter process being the major cause of the synthesis of this element in intermediate-mass stars. It turns out that at decreasing metallicities, nitrogen production by HBB is favored by the larger nuclear reaction rates and/or longer TP-AGB lifetimes. It is worth remarking that low- and intermediate-mass stars play an important role in the galactic nucleosynthesis of nitrogen. As already mentioned in Section 3.3.8.4, AGB stars with HBB ($M_i \gtrsim 3.5\,M_\odot$) provide a source of primary ^{14}N, which is synthesized as newly dredged-up ^{12}C is injected into the CNO cycle. This primary contribution could be relevant in explaining the observed lack of a correlation between the [N/Fe] and [Fe/H] ratios in systems of very low metallicities, where a large scatter is observed (see the review by [71]).
- ^{16}O: The third dredge-up may determine a positive net yield of this element for stars with $0.8\,M_\odot \lesssim M_i \lesssim 3.5\,M_\odot$, whereas at larger masses HBB produces an opposite effect. The results are found to significantly depend on metallicity. At decreasing metallicities larger positive yields of oxygen are expected, reflecting the longer duration of the TP-AGB (and hence the greater number of dredge-up episodes). As expected, the trend of ^{16}O is similar to that of ^{12}C.

3.5 From One Star to Population Synthesis

With the recipes described in Section 3.3, the TP-AGB evolution of a single star with given initial mass and metallicity can be calculated. The basic structure of a synthetic TP-AGB model for single-star evolution is outlined in Section 3.5.1.

Then, if one is interested in reproducing and interpreting the observed distribution functions of specific properties of the cool giants belonging to a stellar cluster or galaxy, the next step is to model a population of AGB stars. To this end, it is necessary to integrate the properties of single stars over the range of relevant masses. A suitable population synthesis approach is described in Section 3.5.2.

3.5.1 A Simple Scheme to Calculate the Synthetic Evolution of a Single TP-AGB Star

1. Set up the initial conditions at the first thermal pulse (M, M_c, L, T_{eff}, envelope chemical composition), and specify the basic model parameters (e.g., mass-loss law, dredge-up efficiency λ).
2. Set up counters that keep track of the number of thermal pulses and the phase within the current TPC.

3. The main loop on the sequence of TPCs starts.
4. The secondary loop on the current TPC starts.
5. Test the criterion for the third dredge-up to take place (e.g., $M_c > M_c^{min}$): If satisfied, the core mass is reduced and the stellar envelope mass is increased by the amount $\lambda \Delta M_c$, and the envelope chemical abundances are changed according to Equation 10.
6. Calculate the interpulse period, e.g., via the M_c-τ_{ip} relation described in Section 3.3.5.2. The inner loop on the quiescent interpulse period τ_{ip} starts.
7. Determine the time step. There are three relevant time scales: the growth of the core, mass loss from the star, and the changes in the luminosity over the TPC. No general formula can be given, and several recipes can be tried out. For example, one would like to resolve each TPC with N steps, which would lead to one estimate of the time step. Another estimate could come from the condition that mass loss cannot reduce the total mass by more than a certain amount. The time step to be chosen is then the smallest of those considered. Then, for each time step Δt:
   a) Calculate the increment of the core mass, ΔM_c, according to Equation 5.
   b) Derive the variation of the luminosity as a function of the TPC phase according to Section 3.3.4.
   c) Estimate the possible changes in the envelope chemical composition due to HBB nucleosynthesis for the more massive models, as outlined in Section 3.3.8.2.
   d) Single out the variation of the effective temperature as described in Section 3.3.6, via either analytical recipes or stellar envelope integrations.
   e) Calculate the amount of ejected mass, $\Delta M_{ej} = \dot{M} \Delta t$, by stellar winds, so that the current stellar mass becomes $M = M - \Delta M_{ej}$.
   f) Update all variables at the current time $t = t + \Delta t$.
   g) Test the criterion for determining the AGB termination: If satisfied go to step 8.
   h) Go back to step 7 if the current TPC has not ended yet. Otherwise, a new thermal pulse occurs, hence update the corresponding number pulse, and go back to step 4.
8. End of the main loop started at step 3, corresponding to the termination of the AGB evolution. The ending conditions are reached either (i) when the core mass has grown to the Chandrasekhar limit of about $1.4\,M_\odot$, which would imply explosive carbon ignition (this might happen for massive stars with low mass-loss rates), or (ii) when the stellar envelope mass has been reduced to some prescribed small value, of order $0.01\,M_\odot$ or less.

3.5.2 A Simple Scheme to Construct a Synthetic AGB Population

Stars that are presently on the AGB must have been born an appropriate time ago, so that their mass-distribution function $N(M)$ (in number of stars per unit mass interval) may be written as

$$N(M)\mathrm{d}M \propto \int_0^{t_{\mathrm{AGB}}(M,Z)} \phi(M)\,\psi(T_G - \tau(M,Z) - x)\,\mathrm{d}x,$$

where ϕ (in $M_\odot^{-1}$) is the initial mass function (IMF, usually written in the form of a power law $\sim M^{-\alpha}$), ψ (in $M_\odot\,\mathrm{yr}^{-1}$) is the star formation rate (SFR), t_{AGB} is the lifetime of a star on the AGB, T_G is the age of the system, and $\tau(M)$ is the pre-AGB lifetime of a star with mass M.

Realizing that $t_{\mathrm{AGB}} \ll \tau$ for all masses leads to

$$N(M)\mathrm{d}M \propto \phi(M)\psi(T_G - \tau(M,Z))t_{\mathrm{AGB}}(M,Z) \qquad (15)$$

in the range $M_{\mathrm{low}} < M < M_{\mathrm{up}}$, where M_{low} and M_{up} are the lower and upper mass limits, respectively, of stars that can potentially evolve through the AGB. These are constrained, respectively, by the age of the galaxy [i.e., $\tau(M_{\mathrm{low}}) \approx T_G$), and the maximum mass for a star to develop a degenerate C-O core, with typical values of about $0.9\,M_\odot$ and $8\,M_\odot$ (lowered to 5–6 $M_\odot$ in the case of models with convective overshoot). More restricted mass intervals could be specified according to particular ages of interest.

The other relevant ingredient is the age–metallicity relation. Low-mass stars, born a long time ago, have main-sequence metallicities that are lower than those of stars born today. Examples of age–metallicity and star-formation rate relations for our Galaxy can be found in [16, 88, 89] or [100]; for the Magellanic Clouds in [45] and [78]. In order to use such a relation one needs as well a relation between τ and mass (and metallicity). Such a relation can be found in [48], which unfortunately is independent of metallicity, or, by interpolation, in any published set of evolutionary tracks.

Finally, one needs to know t_{AGB}. Because this depends on the evolution of the stars in the population under study, its value can be determined only iteratively. Since in Equation 15 only the relative ages are of importance, the dependence of the distribution of stars over mass does not depend very much on t_{AGB} (see Appendix B in [32]). One viable approach would be to start with (1) a constant value for t_{AGB}, (2) run the model and adjust the parameters to fit the observations, (3) determine $t_{\mathrm{AGB}}(M)$ for that set of parameters, and (4) use this new estimate, and repeat the calculations until differences in the results fall below a predetermined tolerance.

Here we outline the logical path of steps to develop a synthetic AGB population model, based on the Monte Carlo simulation method. It is worth recalling that as an alternative, the AGB could be populated analytically, using the basic equations of population synthesis (see, e.g., [66]).

1. Read in, or set up: The initial mass function, star formation rate, age–metallicity relation, (initial mass)–(pre-AGB-lifetime) relation. Specify the mass (hence age) interval of interest.
2. Randomly draw a mass within the specified interval, according to the probability function that a star of a given mass is on the AGB at present (see Equation 15). Determine the corresponding main sequence metallicity. This may require an iteration scheme, since the pre-AGB lifetime depends on mass but also on metallicity. One solution would be to choose an initial metallicity, then calculate the pre-AGB lifetime for that mass and metallicity, and subsequently use the age–metallicity relation to get a better estimate for the metallicity.
3. Run the routine described in Section 3.5.1 to predict the synthetic TP-AGB evolution of the specified single star. At each time step, update the distribution functions (of, e.g., luminosity, effective temperature, mass loss, final mass, chemical properties) you are keeping track of.
4. The procedure ends when a predetermined number of stars is randomly picked up and the distribution functions can be built.

3.6 Observational Constraints

3.6.1 The Initial–Final Mass Relation (IFMR)

The final fate of an AGB star is determined by the competition between the growth of the core mass and the mass loss by stellar winds. If the core mass is allowed to grow up to the Chandrasekhar limit ($M_c \approx 1.4\,M_\odot$), then explosive carbon ignition would occur (type I-1/2 supernova; see [49]). Conversely, if this circumstance is prevented by the the complete shedding of the stellar envelope, then the end-product of the AGB evolution is a C-O white dwarf (WD).

Establishing the relation between the final mass, M_f, left at the end of the AGB, and the initial mass, M_i, of the progenitor star is of paramount importance for various aspects related to both stellar evolution and the history of the host galaxy. The age of a coeval stellar system can in principle be derived once the mass of the detected WDs is empirically estimated. In particular, the initial–final mass relation (IFMR) for low- and intermediate-mass stars is an important tool for investigating the history of the Galaxy, e.g., to estimate the age of the local disk [48, 116]. Moreover, the IFMR provides information on the total amount of mass that each star loses in the course of its entire evolution and hence on the evolution of the mass budget of stellar populations and galaxies [85]. Finally, assessing the upper mass limit for WD progenitors is a crucial aspect, since it affects the expected rate of type II supernovae events with important implications for the star formation history, the chemical enrichment of the interstellar medium, and the birth rate of neutron stars [112].

In Figure 3.10, a few empirical IFMRs for the solar neighborhood are displayed (top left-hand side panel). Note that the most recent studies (e.g., [42, 53, 56, 83, 111]) have notably changed the shape of the relation with respect to the earlier determination by [109], as recently discussed by [110] and [111]. The IFMR presented by [42], for instance, presents a flat slope up to Hyades location ($M_i \approx 3\,M_\odot$, $M_f \approx 0.7\,M_\odot$), followed by a steeper rise, and a final flattening toward higher initial masses ($M_i > 4\,M_\odot$). Another important point concerns the expected value for M_{WDpro}, i.e., the maximum initial mass for the WD progenitors, which is still a matter of debate. For the same open cluster NGC 2516 [56] and [83] derive $M_{\mathrm{WDpro}} \approx 8\,M_\odot$, whereas [53] estimates $M_{\mathrm{WDpro}} \approx 5\text{--}6\,M_\odot$. The former value of the limiting mass would be consistent with classical evolutionary models, the latter supporting stellar models with overshooting.

As far as the theoretical predictions of the IFMR are concerned, the following remarks should be made. Thanks to the computational agility of synthetic AGB models, these are suitable tool for a systematic investigation of the IFMR through its dependence on various parameters (e.g., mass loss prescription, dredge-up law, luminosity evolution, metallicity).

Past synthetic AGB models (e.g., [49, 87]) have predicted IFMRs in clear disagreement with the observed calibrations, i.e., a sizable excess of WDs more massive than $0.7\,M_\odot$, with the most massive stars building C-O cores up to the Chandrasekhar value of $1.4\,M_\odot$. We can notice the quick divergency at increasing mass that characterizes the relation predicted by [49] (top right-hand side panel in Figure 3.10). Actually, the WDs belonging to the youngest open clusters in the solar neighborhood have estimated masses not larger than about $1.1\,M_\odot$ [53, 56, 109].

The main reason for such a discrepancy in earlier theoretical studies resides in the adoption of unsuitable mass-loss prescriptions for the AGB, i.e., the straightforward application of the Reimers law [84] originally calibrated on RGB stars. In this respect we recall that as already noticed by [87], the Reimers formula (with $\eta = \frac{1}{3}$ to 1) is not able to predict the "superwind" mass loss rates ($\dot{M} \approx 10^{-5}$ to $10^{-4}\,M_\odot\,\mathrm{yr}^{-1}$) required to account for the typical masses (and expansion velocities) of observed PNe. Since then, high mass-loss rates, consistent with superwind values, have actually been measured in the brightest AGB stars (e.g., OH/IR and Mira variables) of the Galaxy (cf. [55]) and Magellanic Clouds (cf. [105, 119]); see Chapter 7.

In more recent years, several prescriptions have been proposed to specifically describe mass loss during the AGB (e.g., [7, 13, 24, 106]). All these studies rely (either from theoretical arguments and/or empirical evidence) on the fact that radiation pressure, pulsation, and dust should play an essential role in driving mass loss on the AGB; see Chapters 4 and 6. The reader can refer to [35] and [29] for an exhaustive comparison of the mostly used laws, based on various observational constraints in addition to the observed IFMR (e.g., the carbon star luminosity function, the number ratio of carbon-

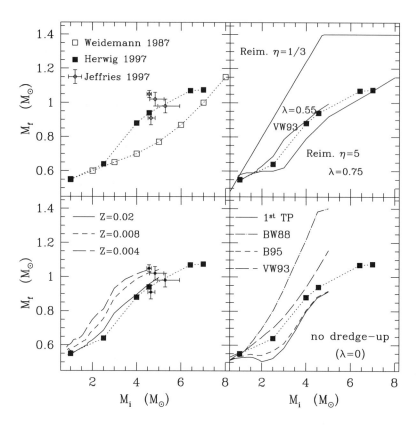

Fig. 3.10. Initial–final mass relation for low- and intermediate-mass stars. Top left-hand side panel: Empty and filled squares correspond to the semiempirical relation for the solar neighborhood by [109] and [42], respectively. Data from [53] are shown by points with error bars. Top right-hand side panel: Theoretical predictions for models with solar metallicity and various mass-loss prescriptions, i.e., according to [49] with the Reimers law ($\eta = \frac{1}{3}$); according to [64] with mass loss from [106]; [101] with the Reimers law ($\eta = 5$). For the two latter models, the adopted dredge-up efficiency parameter is specified. Bottom left-hand side panel: Theoretical predictions for various metallicities [64]. Bottom right-hand side panel: Theoretical predictions by [29] for models with solar metallicity, according to various mass-loss laws, i.e., [7, 13, 106]. In these calculations no dredge-up is allowed during the TP-AGB phase. The core mass at the first thermal pulse is also shown (solid line)

rich to oxygen-rich stars, chemical abundances in PNe, integrated V-K colors in LMC clusters).

The relevant point to be stressed here is that, with these new formalisms adopted in synthetic AGB models, the expected IFMRs are in better agreement with observations, if compared to those predicted by the Reimers law. This can be seen in Figure 3.10, which displays several theoretical IFMRs.

From these calculations we can derive a first important result. At least for not extremely low metallicities (i.e., $Z > 0.001$), even the most massive AGB stars are likely to lose all the residual stellar envelope before their C-O cores reach the Chandrasekhar mass. This would prevent the occurrence of the so-called type I-1/2 SNe, first introduced by [49].

Another relevant point is the dependence of the IFMR on the metallicity of the stellar generation. Stars with the same initial masses build up final C-O cores of larger masses at decreasing metallicity (see the bottom left-hand side panel of Figure 3.10). In a simple picture, this result can be explained by two concurring effects expected at lower metallicities in stars of given initial masses: (i) Stars already enter the TP-AGB phase with larger C-O cores left at the end of the He-burning phase; (ii) the location of the Hayashi lines in the HR diagram is characterized by higher effective temperatures, thus yielding lower mass-loss rates.

In addition to the mentioned points, there are additional factors that may, however, affect the resulting IFMR, namely, (iii) the dredge-up process taking place at thermal pulses, and (iv) the overluminosity produced by HBB in the most massive AGB stars.

Every time a dredge-up event occurs, it is assumed to instantaneously reduce the core mass by the quantity $\Delta M_{\mathrm{dredge}} = \lambda \Delta M_{\mathrm{c}}$, where ΔM_{c} is the increment of the core mass over the previous interpulse period (see Equation 9). Following recent full calculations of thermal pulses [44] it turns out that λ could be even larger than unity, so that in this case the core mass is effectively reduced. The impact of this new scenario on the resulting IFMR deserves, however, further calculations as well as observational tests. Another effect produced by the third dredge-up on the IFMR is indirectly related to the changes in the surface chemical composition. The enrichment in carbon (and oxygen) may affect the relevant molecular absorption and dust features in the stellar atmosphere, and hence the efficiency of mass loss.

As far as the occurrence of HBB is concerned, it turns out that the steeper luminosity increase with the core mass (if compared with the classical CMLR) anticipates the transition to the superwind regime, concurring to produce remnants with lower masses. This circumstance could help to explain the flattening of the empirical IFMR toward larger initial masses. Finally, we should remark that the IFMR at larger initial masses may be influenced by HBB nucleosynthesis, since the changes in the stellar envelope abundances (mainly the increase of nitrogen) may affect the effective temperature, and hence the mass-loss rate.

The sensitivity of the IFMR to the changes in the surface chemical composition of stars with different initial masses (and chemical evolution) is an important aspect, which, however, still deserves future theoretical analysis.

3.6.2 The Carbon Star Luminosity Function

The observed carbon star luminosity function (CSLF) represents a fundamental constraint to AGB models, since a sole observable contains a number of implications, concerning both the evolution of single stars (e.g., evolutionary rates, mass-loss efficiency, lifetimes, initial–final mass relation, dredge-up law), and integrated properties of the host systems (e.g., star formation history, age–metallicity relation).

The need for a systematic analysis of all these factors makes synthetic AGB models the suitable tool to investigate the CSLF. Essentially, two theoretical approaches can be used to derive the predicted number, N_k, of carbon stars in each $k^{\rm th}$ luminosity bin, namely:

1. analytical method (cf. [68]);
2. Monte Carlo simulation method (cf. [32]).

Clearly, both methods should produce the same results for large numbers of carbon stars.

In the former scheme, denoting by $N_k(M_i)$ the number of carbon stars with a progenitor mass M_i and a present luminosity within the $k^{\rm th}$ bin, the number N_k of carbon stars populating the $k^{\rm th}$ bin is simply given by integration of $N_k(M_i)$ over all masses M_i:

$$N_k = \int_{M_i^{\rm min}}^{M_i^{\rm max}} N_k(M_i) {\rm d}M_i$$

$$\propto \int_{M_i^{\rm min}}^{M_i^{\rm max}} \psi(M_i)\, \phi[T_{\rm G} - \tau(M_i)]\, \Delta\tau_k(M_i)\, {\rm d}M_i, \qquad (16)$$

where $\psi(M_i)$ is the IMF, $T_{\rm G}$ is the age of the galaxy, $\tau(M_i)$ is the stellar lifetime, $\phi[T_{\rm G} - \tau(M_i)]$ corresponds to the star formation rate evaluated at the birth epoch of the stars of initial mass M_i, and $\Delta\tau_k(M_i)$ is the time each carbon star of mass M_i spends in the $k^{\rm th}$ luminosity interval.

The latter scheme is a probabilistic one in which, given a selected sample of AGB stars, the corresponding initial masses are randomly assigned according to the distribution function in Equation 15. Once the present status (i.e., whether it is a carbon star or not, and its luminosity) of every sample star is determined, the luminosity distribution can be derived by summing up the number of carbon stars in each luminosity bin. Both methods require that the minimum ($M_{\rm min}$) and maximum ($M_{\rm max}$) initial masses for a star to become a carbon star be specified.

These methods have been successfully employed in recent synthetic AGB models (i.e., [32, 68]) aimed at reproducing the observed CSLFs in the Magellanic Clouds, via the calibration of the dredge-up parameters ($M_c^{\min}$ and λ; see Section 3.3.8.1). Both works conclude that the third dredge-up should occur already in low-mass stars (with $M_i \gtrsim 1.2$–$1.4\,M_\odot$) and be, on average, more efficient ($\lambda \approx 0.5$–0.6), than usually predicted by most full AGB calculations.

Figure 3.11 shows the best fits to the observed CSLFs in the Large and Small Magellanic Clouds, as obtained from [68], to which the reader is referred for more details. According to that analysis, the most remarkable differences between the two observed distributions (i.e., peak location and shape of the tails) are mainly the expression of different initial metallicities, characterizing the stellar populations in the two host galaxies (i.e., on average, $Z = 0.008$ for the LMC, and $Z = 0.004$ for the SMC).

In particular, the metallicity-dependence of the third dredge-up is found to play the dominant role. At decreasing metallicity the minimum mass, $M_{\min}^{\text{carb}}$, for a star to experience convective dredge-up of carbon decreases (i.e., $\approx 1.3\,M_\odot$ for the LMC, and $\approx 1.2\,M_\odot$ for the SMC), and the efficiency of the third dredge-up increases (i.e., on average, $\lambda \approx 0.55$ for the LMC, and $\lambda \approx 0.65$ for the SMC). Moreover, for a given stellar mass the onset of the third dredge-up is expected to be favored at lower metallicity, i.e., the lower Z is, the lower $M_c^{\min}$ is (see Figure 3.4). In general, both distributions would involve stellar progenitors with initial masses in the range from $M_{\min}^{\text{carb}}$ up to about 4.5–$5.0\,M_\odot$, corresponding to ages from a few 10^9 down to about 10^8 yr.

3.6.2.1 The Low-Luminosity Tail

The faint tail of observed CSLFs deserves some important remarks. First of all, synthetic AGB models (cf. [32, 49, 66, 68]) have clearly shown that the well-extended and long-lived low-luminosity dip in the light curve of lower-mass AGB stars (cf. Section 3.3.4.3) has an important effect in shaping the tail of the CSLFs towards fainter luminosities. Just recall, for instance, that a low-mass AGB star can spend as much as ≈ 20–30% of an interpulse period at a luminosity quite lower than expected by the M_c-L relation. Moreover, we should consider that the initial stages of the TP-AGB evolution occur at lower luminosities, and with a steeper rate of luminosity increase, than predicted by the M_c-L relation for the same core mass (see Section 3.3.4.2).

Another factor that may affect the low-luminosity extension of the CSLF is the contamination by faint carbon stars formed in closed binary systems, in which the surface of a low-mass star has been contaminated by the ejecta of a former AGB star companion (e.g., [25], Chapter 9). It was shown by [113] that the CSLF in the SMC is characterized by an extended tail down to $M_{\text{bol}} \approx -1.8$. These carbon stars (e.g., J-type stars, dwarf carbon stars, R stars) are indeed too faint to be explained as the result of single-star evolution.

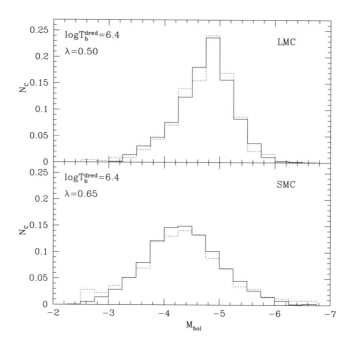

Fig. 3.11. Observed CSLFs (dotted lines) in the LMC ([19], top panel) and SMC ([81], bottom panel), and the theoretical best fits (solid lines, [68]). The first bin of the observed CSLF in the SMC contains all stars fainter than $M_{bol} = -2.5$. The quantity T_b^{dred}, a parameter related to M_c^{min}, represents the minimum temperature at the base of the convective envelope for the occurrence of the third dredge-up, as derived from stellar envelope integrations

However, it should be noted that in the case of the SMC the fraction of the faintest carbon stars amounts to less than 3 % of the total sample [81], so that the calibration procedure applied by [68] should not be affected.

3.6.2.2 The High-Luminosity Tail

The luminosity distribution of carbon stars in the brightest bins is mainly affected by two factors: mass loss and hot bottom burning. The former controls the luminosity excursion of the most massive carbon stars (since it determines the end of the AGB as soon as the residual stellar envelope is ejected); the latter prevents the most massive and luminous AGB stars from becoming carbon stars.

Unfortunately, the efficiency of both processes is still affected by large uncertainties. However, important indications can be indirectly derived from other observables, e.g., the initial–final mass relation (related to mass loss; cf. Section 3.6.1), the maximum luminosities of observed AGB stars (related to the overluminosity produced by HBB; see Section 3.3.4.4), the chemical composition of most luminous AGB stars (related to the efficiency of HBB; see Section 3.3.8.3).

3.6.2.3 Number Ratios N_C/N_M

As for the carbon stars, an important additional constraint to the models would be a luminosity function of the M-type AGB stars. However, no reliable LF exists, since most investigators have focused on carbon stars. A constraint that is available is the observed number ratio of carbon to M-type stars (N_C/N_M). One source of uncertainty is that not all M-type stars should be on the TP-AGB. Such uncertainty is reflected in the fact that the N_C/N_M ratio strongly depends on the spectral type that is included. For the LMC, [5] gives $N_C/N_{M2+} = 0.2 \pm 0.1$, $N_C/N_{M5+} = 0.80 \pm 0.03$, and $N_C/N_{M6+} = 2.2 \pm 0.1$, respectively. It seems plausible to assume that M-stars with spectral type later than M5 are on the TP-AGB. The N_C/N_M ratio in Local Group galaxies is discussed in [39].

3.6.3 Effective Temperatures of AGB Stars

Observationally, a notable dichotomy exists in effective temperatures and infrared colors between M-type and carbon stars. Carbon stars are systematically cooler and redder than M-type stars. HR-diagrams in the near infrared (e.g., K vs. $J-K$), provided by extensive surveys like DENIS and 2MASS, have clearly shown that carbon stars deviate from the M-type part of the AGB, populating a plume toward redder colors and lower $T_{\rm eff}$. This feature is closely related to marked differences in molecular opacities pertaining to the atmospheres of AGB stars with different C/O ratio. In summary, the abrupt change in the dominant opacity source as C/O grows from below to above unity, i.e., at the transition from oxygen-rich to carbon-rich configurations, causes a remarkable cooling of the HR TP-AGB tracks for carbon stars. This point, explored by [91] and more recently by [65], is illustrated in Figure 3.12 for a sample of Galactic giants.

The observed data indicate two major facts: (i) the almost complete segregation in effective temperature between M-type and carbon stars; and (ii) the relatively low C/O values (< 2) measured in carbon stars. A comparison is performed with two sets of TP-AGB synthetic models [65], which differ only in the adopted prescriptions for low-temperature opacities κ (for $\log T < 3.7$). They are assumed to be either fixed for solar-scaled abundances of C and O (hence C/O ≈ 0.48, bottom panel), or variable according to the actual C/O ratio, which controls the molecular concentrations involving C and O atoms (top panel). All models are followed from the first TP to the end of the AGB. The main result is that the models with variable opacities succeed in reproducing the observed range of effective temperatures of carbon stars, while this is not the case for fixed solar-scaled abundance opacities.

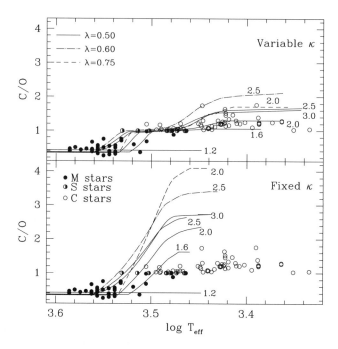

Fig. 3.12. Effective temperatures as a function of the C/O ratio in Galactic giants. Observed data for Galactic M, S, and C stars (circles) are compared to predictions of synthetic TP-AGB models (lines; see [65]) of different masses (indicated in $M_\odot$ near the corresponding curves), assuming different dredge-up efficiencies λ, and adopting either "chemically fixed" opacities (bottom panel), or "chemically variable" opacities (top panel)

3.6.4 Abundances in Planetary Nebulae

According to a commonly recognized scenario, AGB stars are the direct stellar progenitors of PNe, so that the nebular abundances are expected to be the record of the evolution of the photospheric composition during the stellar lifetimes. It follows that the abundances in PNe can be used to improve our understanding of stellar nucleosynthesis and convective mixing, as well as to constrain the basic model parameters in synthetic AGB evolution (i.e., dredge-up parameters, mass loss law, and initial conditions).

Three distinct groups of PNe progenitors are expected. The first group consists of stars with initial masses too low (e.g., $M_i < 1.2\,M_\odot$ for the case of the LMC) to experience the third dredge-up. These PNe should exhibit the same stellar abundances as changed—with respect to the main sequence values—by the occurrence of the first dredge-up and RGB extra-mixing process only. Specifically, a modest increase in He and N abundances is expected. The second group consists of stars with initial masses high enough to experience the third dredge-up but too low for HBB to occur on the TP-AGB (e.g., typically $1.2\,\mathrm{M}_\odot \lesssim M_\mathrm{i} \lesssim 3.5\,\mathrm{M}_\odot$). Prior to this phase, further chemical changes should have occurred via the first dredge-up, and possibly the second dredge-up for the massive stars of this group. An increase in He, C, and N abundances is usually predicted for the resulting PNe. The third group consists of stars massive enough to have HBB on the TP-AGB, in addition to the earlier first and second dredge-up processes (with $M_\mathrm{i} \gtrsim 3.5\text{–}4.0\,\mathrm{M}_\odot$). A more marked increment of He and N and a decrease of C are expected to characterize the nebular abundances. These theoretical groups should explain the empirical classification based on measured PNe abundances, e.g., type I and II PNe first introduced by [79].

If the lifetime of the PNe phase is t_PN, and since there is essentially no nucleosynthesis in the post-AGB and PNe phase itself, we can make the very approximate assumption that the space-average abundance pattern observed in PNe should represent the time-average abundance pattern of the ejecta in AGB stars over the last t_PN years of AGB evolution.

A complication is that during the running of the synthetic AGB evolution code, based on the Monte Carlo simulation method, one a priori does not know when a star has less than t_PN years of AGB evolution left. However, one can make a good estimate of the remaining AGB lifetime by calculating at each time step whether $M_\mathrm{env}/\dot{M} < t_\mathrm{PN}$. If so, one can then keep track of the amount of mass lost (the mass in the shell of a PN is another measurable quantity) and the average abundances in the ejecta.

There are a few additional comments to be made. First, not all AGB stars may actually turn into PNe, or be recognized as such. Low-mass stars may produce very low surface-brightness PNe that have not been discovered yet. In fact, the post-AGB and the central star of the PNe may evolve so slowly that the ejecta are dispersed in the interstellar medium before the central star is hot enough to ionize the ejecta. The procedure outlined above merely indicates that *if* a PN is observed, the abundance pattern should be the one calculated in this way. Another complication is from the observational side. The nebulae are strongly shaped by hydrodynamics, and so the abundance patterns, as ejected at the time of AGB evolution, may have been lost when these structures were formed. On average, the abundances should not have changed, but with a spectroscopic observation the slit has to be placed over the nebulae, and this may introduce a bias. The second effect is that gaseous material may condense into grains. This depends on the atom, and so specific elements may be depleted in the gas phase.

This scheme was applied by [33, 38, 66] to the abundances of PNe in the LMC and the Galaxy, respectively. The three groups of stars mentioned above could clearly be identified, and the overall agreement between observations and the synthetic model calculations was satisfactory. An example of these synthetic calculations of PNe abundances is presented in Figure 3.13, showing a comparison between predictions and observations of PNe in the LMC.

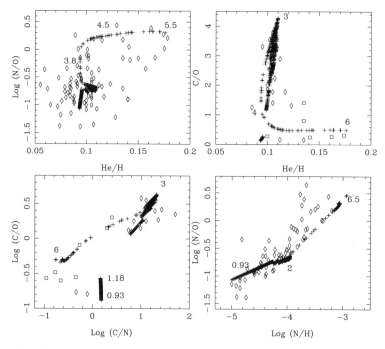

Fig. 3.13. Observed and predicted abundance ratios of PNe in the LMC. Model predictions (+ signs) and some initial stellar masses are indicated. In the C/O–He/H and C/O–C/N diagram the observed type I PNe (defined as having N/O > 0.5) are indicated by diamonds, the other observed PNe by squares. The density of the theoretical points reflect the IMF and SFR history. From [33]

3.6.5 Pulsating Stars

Historically, variable AGB stars have been divided into three classes: Mira variables, semiregular (or SR) variables, and irregular variables (see Chapters 1 and 2). The distinction between the classes is in the amplitude of the light variation in the V-band and the periodicity of the light variation.

Since stars are unstable to pulsation only in a small region of the Herzsprung–Russell diagram (the instability strip), and since Miras and SR variables are known to obey period–luminosity relations, variable stars can in principle provide constraints that can aid in an understanding of the evolution of AGB stars in general. Problems are that (i) the location of the instability strip in the T_{eff}-M_{bol} plane is not known a priori, (ii) the fact that stars can pulsate in different modes, or possibly even switch modes (e.g., [23, 119]), and (iii) that the so-called pulsation constant that theoretically relates pulsation period to stellar mass and radius is not accurately known.

Despite these difficulties, the only attempt so far to incorporate observations of variable stars in the LMC in synthetic evolution models was by [34]. There, details are given of the recipes used and the observational data available at the time. The conclusion was that long-period variables dominantly pulsate in the fundamental mode.

Since then, in particular, the observational data set has increased dramatically through the micro-lensing projects, MACHO, EROS, OGLE. Most impressively, [119] has used MACHO data to identify five distinct sequences of variable stars in the LMC that show that Miras pulsate in the fundamental mode, and SRs predominantly pulsate in the first, second or third overtone; see Chapter 2. Two of the five sequences identified do not correspond to single AGB star evolution. Similar work using EROS data has been done by [18].

3.6.6 Synthetic Colors

An additional constraint that can be used is the colors of stars. For a star with a certain luminosity, effective temperature, and mass-loss rate, at a certain distance, one can calculate the spectral energy distribution using a stellar atmosphere model and a dust radiative transfer program, and calculate the colors in the different photometric bands. To incorporate this in a synthetic evolution program would be very time-consuming; attempts have been made, however (e.g., [58]). A semianalytical approach was adopted by [32] and presented in their Appendix A. The specific question addressed there was whether the observed luminosity function of AGB stars in the LMC is biased because the stars with high mass-loss rates are so red and faint in the optical that they would have escaped detection using classical search methods. Alternatively, the approach was used to estimate which fraction of stars escaped detection with the IRAS satellite, which operated at 12 and 25 μm at the detection limit of the expected fluxes for mass-losing AGB stars. This aspect of obscured AGB stars in the Magellanic Clouds has gained considerable attention recently, both from the analysis of individual objects [97], and searches for, and studies of, a sample of objects using the ISO satellite (e.g., [31, 96, 104]). A major step forward will be provided by the 2MASS and DENIS near-infrared surveys [3, 22], which will detect virtually the entire AGB population in the direction of the Magellanic Clouds [17, 102].

One of the most surprising results of the IRAS satellite mission was the discovery of carbon stars with 60 μm excess emission [114]. It was shown by [121] that also some M- and S-stars show such an excess. The conclusion was that there must be a shell of very cold dust surrounding the star. Later on, this picture was confirmed with the detection of a thin shell of CO around some of these stars [62, 73, 74, 75] and more detailed studies of the extended dust emission [40, 51, 52]. The interpretation is that there was a brief phase of very high mass loss. After this phase, mass loss dropped, and the shell coasted outward. Recently, mass loss has resumed, as evidenced by the detection of CO close to the central star in some cases. This modulation of the mass loss, and the time scales involved, suggest a possible link with a modulation of the mass loss related to the change in luminosity over a TPC. Early attempts to determine the mass-loss rates in the various phases of the modulation, and to calculate the changes of the IRAS fluxes at 12, 25, and 60 μm as a function of time, are given by [15, 36, 115]. More recent studies that use the luminosity variations of full stellar evolutionary models, coupled to hydrodynamical calculations of the stellar wind, are given by [93]. They suggest that the observed thin shells are the result of interaction of winds with different velocity and density. Note, however, that for the less-studied objects with a 60 μm excess the alternative interpretation that this is simply due to Galactic cirrus cannot be excluded.

3.7 Conclusions and Outlook

This chapter describes the basic ingredients of a synthetic AGB evolution model. The advantages over full stellar evolutionary models are the computational speed and the fact that the dependence of parameters like the mass-loss law or dredge-up efficiency can be easily changed, and their influence on observable quantities investigated. In fact, by using synthetic models to constrain these parameters, important feedback can be given to the stellar evolution community.

The largest uncertainties for both the full and synthetic evolutionary models refer to the description of the mass loss and the third dredge-up process. Both are crucial in understanding the detailed evolution on the AGB and the formation of S-type and carbon stars. Advances can be made on several levels. First, by detailed modeling of the properties of individual stars. For example, it will be computationally possible in the near future to combine the output of a stellar evolution model $[L(t), T_{\mathrm{eff}}(t), \dot{M}(t)]$ with a hydrodynamical model (see Chapter 6) to make detailed predictions of the density structure around specific stars for specific mass-loss laws, and compare these to the density structure derived from interferometric observations.

Up to now, synthetic models have been used to fit data in our Galaxy, SMC, and LMC only. High-quality data on AGB populations in other galaxies have become available as well as information on the overall SFRs in these galaxies from detailed modeling of color–magnitude diagrams from HST observations (see Chapter 8). Since these galaxies cover a range of metallicity and SFR histories, fitting the luminosity function and other properties of C- and M-stars in these galaxies might give additional information on mass-loss laws and dredge-up efficiencies.

Finally, a new perspective on the role of pulsation in AGB evolution has been brought up by the microlensing surveys, as detailed in Section 3.6.5. That pulsation and mass loss are connected is beyond doubt, but a fully self-consistent picture is missing. Questions are whether AGB stars are pulsationally unstable during their entire evolution, and whether mode changes occur, possibly related to the phase in the thermal pulse cycle. The large number of red giant variable stars discovered in the MCs may allow these questions to be addressed in a statistical way.

Acknowledgments: P.M. thanks Léo Girardi for helpful advice during the preparation of the manuscript.

References

1. Arndt, T. U., Fleischer, A. J., and Sedlmayr, E. *A&A*, 327, 614, 1997.
2. Bedijn, P. J. *A&A*, 205, 105, 1988.
3. Beichman, C. A., Chester, T. J., Cutri, R., et al. *PASP*, 110, 367, 1998.
4. Bessell, M. S., Brett, J. M., Scholz, M., et al. In Nomoto, K., editor, *IAU Colloq. 108: Atmospheric Diagnostics of Stellar Evolution*, page 187. Springer-Verlag: Berlin, New York, 1988.
5. Blanco, V. M. and McCarthy, M. F. *AJ*, 88, 1442, 1983.
6. Blöcker, T. and Schönberner, D. *A&A*, 244, L43, 1991.
7. Bloecker, T. *A&A*, 297, 727, 1995.
8. Boothroyd, A. I. and Sackann, I. J. *ApJ*, 328, 632, 1988.
9. Boothroyd, A. I. and Sackmann, I. J. *ApJ*, 328, 641, 1988.
10. Boothroyd, A. I. and Sackmann, I. J. *ApJ*, 328, 653, 1988.
11. Boothroyd, A. I. and Sackmann, I. J. *ApJ*, 328, 671, 1988.
12. Boothroyd, A. I. and Sackmann, I. J. *ApJ*, 393, L21, 1992.
13. Bowen, G. H. *ApJ*, 329, 299, 1988.
14. Bryan, G. L., Volk, K., and Kwok, S. *ApJ*, 365, 301, 1990.
15. Chan, S. J. and Kwok, S. *JRASC*, 82, 287, 1988.
16. Chiappini, C., Matteucci, F., and Gratton, R. *ApJ*, 477, 765, 1997.
17. Cioni, M. R., Loup, C., Habing, H. J., et al. *A&AS*, 144, 235, 2000.
18. Cioni, M.-R., Marquette, J.-B., Loup, C., et al. *A&A*, 377, 945, 2001.
19. Costa, E. and Frogel, J. A. *AJ*, 112, 2607, 1996.
20. D'Antona, F. and Mazzitelli, I. *ApJ*, 470, 1093, 1996.
21. Dominguez, I., Chieffi, A., Limongi, M., and Straniero, O. *ApJ*, 524, 226, 1999.
22. Epchtein, N., Deul, E., Derriere, S., and et al. *A&A*, 349, 236, 1999.
23. Feast, M. In Le Bertre, T., Lébre, A., and Waelkens, C., editors, *IAU Symp. 191: Asymptotic Giant Branch Stars*, page 109. ASP: San Francisco, 1999.
24. Fleischer, A. J., Gauger, A., and Sedlmayr, E. *A&A*, 266, 321, 1992.
25. Frantsman, J. *A&A*, 319, 511, 1997.
26. Frost, C. A., Cannon, R. C., Lattanzio, J. C., Wood, P. R., and Forestini, M. *A&A*, 332, L17, 1998.
27. Frost, C. A. and Lattanzio, J. C. *ApJ*, 473, 383, 1996.
28. Gingold, R. A. *ApJ*, 193, 177, 1974.
29. Girardi, L. and Bertelli, G. *MNRAS*, 300, 533, 1998.
30. Girardi, L., Bressan, A., Bertelli, G., and Chiosi, C. *A&AS*, 141, 371, 2000.
31. Groenewegen, M. A. T. and Blommaert, J. A. D. L. *A&A*, 332, 25, 1998.
32. Groenewegen, M. A. T. and de Jong, T. *A&A*, 267, 410, 1993.
33. Groenewegen, M. A. T. and de Jong, T. *A&A*, 282, 127, 1994.
34. Groenewegen, M. A. T. and de Jong, T. *A&A*, 288, 782, 1994.
35. Groenewegen, M. A. T. and de Jong, T. *A&A*, 283, 463, 1994.
36. Groenewegen, M. A. T. and de Jong, T. *A&A*, 282, 115, 1994.
37. Groenewegen, M. A. T., Smith, C. H., Wood, P. R., Omont, A., and Fujiyoshi, T. *ApJ*, 449, L119, 1995.

38. Groenewegen, M. A. T., van den Hoek, L. B., and de Jong, T. *A&A*, 293, 381, 1995.
39. Groenewegen, M. In Le Bertre, T., Lébre, A., and Waelkens, C., editors, *IAU Symp. 191: Asymptotic Giant Branch Stars*, page 535. ASP: San Francisco, 1999.
40. Hashimoto, O., Izumiura, H., Kester, D. J. M., and Bontekoe, T. R. *A&A*, 329, 213, 1998.
41. Havazelet, D. and Barkat, Z. *ApJ*, 233, 589, 1979.
42. Herwig, F. In Noels, A., Fraiponi-Caro, D., Gabriel, M., Grevesse, N., and Demarque, P., editors, *32nd Liège Int. Astrophys. Coll.: Stellar Evolution: What should be done*, page 441. Institut d'astrophysique de Liège, 1997.
43. Herwig, F. *A&A*, 360, 952, 2000.
44. Herwig, F., Blöcker, T., Schönberner, D., and El Eid, M. In Truran, J. and Nomoto, K., editors, *IAU Symp. 187: Cosmic Chemical Evolution*, page 42. Kluwer Academic Publishers: Dordrecht, 1997.
45. Holtzman, J. A., Gallagher, J. S., Cole, A. A., and et al. *AJ*, 118, 2262, 1999.
46. Iben, I. *ApJ*, 217, 788, 1977.
47. Iben, I. *ApJ*, 246, 278, 1981.
48. Iben, I. and Laughlin, G. *ApJ*, 341, 312, 1989.
49. Iben, I. and Renzini, A. *ARA&A*, 21, 271, 1983.
50. Iben, I. and Truran, J. W. *ApJ*, 220, 980, 1978.
51. Izumiura, H., Hashimoto, O., Kawara, K., Yamamura, I., and Waters, L. B. F. M. *A&A*, 315, L221, 1996.
52. Izumiura, H., Waters, L. B. F. M., de Jong, T., et al. *A&A*, 323, 449, 1997.
53. Jeffries, R. D. *MNRAS*, 288, 585, 1997.
54. Kippenhahn, R. *A&A*, 102, 293, 1981.
55. Knapp, G. R. and Morris, M. *ApJ*, 292, 640, 1985.
56. Koester, D. and Reimers, D. *A&A*, 313, 810, 1996.
57. Kudritzki, R. P., Pauldrach, A., and Puls, J. *A&A*, 173, 293, 1987.
58. Lançon, A. In Le Bertre, T., Lébre, A., and Waelkens, C., editors, *IAU Symp. 191: Asymptotic Giant Branch Stars*, page 579. ASP: San Francisco, 1999.
59. Lattanzio, J. C. *ApJ*, 311, 708, 1986.
60. Lattanzio, J. C. *ApJ*, 344, L25, 1989.
61. Lattanzio, J. C. *Publ. of the Astr. Soc. of Australia*, 10, 120, 1992.
62. Lindqvist, M., Olofsson, H., Lucas, R., et al. *A&A*, 351, L1, 1999.
63. Marigo, P. *A&A*, 340, 463, 1998.
64. Marigo, P. *A&A*, 370, 194, 2001.
65. Marigo, P. *A&A*, 387, 507, 2002.
66. Marigo, P., Bressan, A., and Chiosi, C. *A&A*, 313, 545, 1996.
67. Marigo, P., Bressan, A., and Chiosi, C. *A&A*, 331, 564, 1998.
68. Marigo, P., Girardi, L., and Bressan, A. *A&A*, 344, 123, 1999.
69. Marigo, P., Girardi, L., and Chiosi, C. *A&A*, 316, L1, 1996.
70. Marigo, P., Girardi, L., Weiss, A., and Groenewegen, M. *A&A*, 351, 161, 1999.
71. Matteucci, F. *Fund. Cosmic Phys.*, 17, 283, 1996.
72. Mowlavi, N. *A&A*, 344, 617, 1999.
73. Olofsson, H., Bergman, P., Eriksson, K., and Gustafsson, B. *A&A*, 311, 587, 1996.
74. Olofsson, H., Bergman, P., Lucas, R., et al. *A&A*, 353, 583, 2000.
75. Olofsson, H., Carlstrom, U., Eriksson, K., Gustafsson, B., and Willson, L. A. *A&A*, 230, L13, 1990.

76. Paczynski, B. *Acta Astronomica*, 20, 47, 1970.
77. Paczynski, B. *ApJ*, 202, 558, 1975.
78. Pagel, B. E. J. and Tautvaisiene, G. *MNRAS*, 299, 535, 1998.
79. Peimbert, M. In Terzian, Y., editor, *IAU Symp. 76: Planetary Nebulae*, page 215. D. Reidel Publishing Co.: Dordrecht, 1978.
80. Pols, O. R., Schroder, K.-P., Hurley, J. R., Tout, C. A., and Eggleton, P. P. *MNRAS*, 298, 525, 1998.
81. Rebeirot, E., Azzopardi, M., and Westerlund, B. E. *A&AS*, 97, 603, 1993.
82. Refsdal, S. and Weigert, A. *A&A*, 6, 426, 1970.
83. Reid, I. N. *AJ*, 111, 2000, 1996.
84. Reimers, D. *Sociètè Royale des Sciences de Liège, Memoires*, 8, 369, 1975.
85. Renzini, A. and Buzzoni, A. In Chiosi, C. and Renzini, A., editors, *Spectral Evolution of Galaxies*, page 195. D. Reidel Publishing Co.: Dordrecht, 1986.
86. Renzini, A. and Fusi Pecci, F. *ARA&A*, 26, 199, 1988.
87. Renzini, A. and Voli, M. *A&A*, 94, 175, 1981.
88. Rocha-Pinto, H. J., Maciel, W. J., Scalo, J., and Flynn, C. *A&A*, 358, 850, 2000.
89. Rocha-Pinto, H. J., Scalo, J., Maciel, W. J., and Flynn, C. *A&A*, 358, 869, 2000.
90. Scalo, J. M., Despain, K. H., and Ulrich, R. K. *ApJ*, 196, 805, 1975.
91. Scalo, J. M. and Ulrich, R. K. *ApJ*, 200, 682, 1975.
92. Schaller, G., Schaerer, D., Meynet, G., and Maeder, A. *A&AS*, 96, 269, 1992.
93. Steffen, M., Szczerba, R., and Schoenberner, D. *A&A*, 337, 149, 1998.
94. Straniero, O., Chieffi, A., Limongi, M., et al. *ApJ*, 478, 332, 1997.
95. Tinsley, B. M. *Fundamentals of Cosmic Physics*, 5, 287, 1980.
96. Trams, N. R., van Loon, J. T., Waters, L., and et al. *A&A*, 346, 843, 1999.
97. Trams, N. R., van Loon, J. T., Zijlstra, A. A., and et al. *A&A*, 344, L17, 1999.
98. Travaglio, C., Galli, D., Gallino, R., et al. *ApJ*, 521, 691, 1999.
99. Tuchman, Y., Glasner, A., and Barkat, Z. *ApJ*, 268, 356, 1983.
100. van den Hoek, L. B. PhD thesis, Univ. of Amsterdam, 1997.
101. van den Hoek, L. B. and Groenewegen, M. A. T. *A&AS*, 123, 305, 1997.
102. van Dyk, S. D., Cutri, R. M., Skrutskie, M. F., and Egan, M. In *American Astronomical Society Meeting*, volume 193, page O809. American Astronomical Society, 1998.
103. van Loon, J. T. *A&A*, 354, 125, 2000.
104. van Loon, J. T., Groenewegen, M. A. T., de Koter, A., and et al. *A&A*, 351, 559, 1999.
105. van Loon, J. T., Zijlstra, A. A., Whitelock, P. A., and et al. *A&A*, 329, 169, 1998.
106. Vassiliadis, E. and Wood, P. R. *ApJ*, 413, 641, 1993.
107. Wagenhuber, J. PhD thesis, Techn. Univ. München, 1996.
108. Wagenhuber, J. and Groenewegen, M. A. T. *A&A*, 340, 183, 1998.
109. Weidemann, V. *A&A*, 188, 74, 1987.
110. Weidemann, V. In Rood, R. T. and Renzini, A., editors, *Advances in Stellar Evolution*, page 169. Cambridge University Press: Cambridge, 1997.
111. Weidemann, V. *A&A*, 363, 647, 2000.
112. Weidemann, V. and Koester, D. *A&A*, 121, 77, 1983.
113. Westerlund, B. E., Azzopardi, M., Breysacher, J., and Rebeirot, E. *A&A*, 303, 107, 1995.

114. Willems, F. J. and de Jong, T. *ApJ*, 309, L39, 1986.
115. Willems, F. J. and de Jong, T. *A&A*, 196, 173, 1988.
116. Wood, M. A. *ApJ*, 386, 539, 1992.
117. Wood, P. R. In Iben, I., J. and Renzini, A., editors, *Physical Processes in Red Giants*, page 135. D. Reidel Publishing Co.: Dordrecht, 1981.
118. Wood, P. R. In Cacciari, C. and Clementini, G., editors, *Confrontation Between Stellar Pulsation and Evolution*, page 355. ASP: San Francisco, 1990.
119. Wood, P. R., Alcock, C., Allsman, R. A., and et al. In Le Bertre, T., Lébre, A., and Waelkens, C., editors, *IAU Symp. 191: Asymptotic Giant Branch Stars*, page 151. ASP: San Francisco, 1999.
120. Wood, P. R. and Zarro, D. M. *ApJ*, 247, 247, 1981.
121. Zijlstra, A. A., Loup, C., Waters, L. B. F. M., and de Jong, T. *A&A*, 265, L5, 1992.

4 Atmospheres of AGB Stars

Bengt Gustafsson and Susanne Höfner

Department of Astronomy and Space Physics, Uppsala University

4.1 Introduction

The use of the term *atmosphere* for the outer layers of a star is not obvious. The atmosphere of a terrestrial planet is easily defined as the gaseous envelope extending from the crust (or ocean) out into free space. For stars, the lower boundary is less easily defined. For AGB stars, the outer boundary is not well defined either: How far out into the stellar wind or circumstellar envelope (CSE) should we apply the term atmosphere? A possibly stricter definition, applied by many practitioners in this field, is to define the atmosphere as the stellar region visible from the outside. The atmosphere would then be the region where most of the electromagnetic emission that is observed in the stellar spectrum originates.[1]

Clearly, these distinctions are not very sharp. One could perhaps draw a border between the atmosphere and the wind by defining the wind as the outer stellar region where the outflow velocity is higher than the escape velocity. All such definitions are, however, rather ad hoc, and confront us with one of the main points in the present chapter. We shall namely argue that a close coupling between the stellar interior, the atmosphere, and the wind is present in real stars, and should also characterize our attempts to understand and to model them. This will lead us in the direction of so-called unified models, where the different regions are given a uniform and consistent treatment. We shall see that this is natural. A proper understanding of the winds must be rooted in the physics of the stellar interior. A proper analysis of a stellar spectrum, e.g., for abundance determinations, must be founded on a model of the spectrum-forming regions, for which, however, knowledge of the transfer of mass, momentum, and energy to that region and from that region must be acquired. This couples the understanding of stellar atmospheres to

---

[1] The qualification "most of" is used here, since, e.g., the dust emission in the infrared and molecular millimeter lines are often sooner ascribed to the wind or CSE than to the atmosphere, except for when this term is used in a very general sense. Also note the qualification "electromagnetic": In principle, the central regions of a red giant star could be visible to neutrino telescopes. Even the word "originates" is ambiguous, since some of the radiation could be formed at considerable depths and scattered to the stellar surface.

that of stellar interiors and envelopes, and makes the spectroscopy of AGB stars intricate, and physically interesting.

In general, one could characterize a stellar atmosphere as a transition region between the relatively simple stellar interior and the complexity of interstellar matter. The stellar interior is simple, since equilibrium conditions prevail there in several respects. Local thermodynamical equilibrium (LTE) describes the condition of matter quite well (i.e., we may characterize all local properties of the gas by just two thermodynamical variables like temperature and pressure, and the elemental composition). Hydrostatic equilibrium is a good approximation for calculating pressures. Mass flows may be neglected for understanding the global properties of the interior if convection is treated schematically, and spherical symmetry is thought to be a reasonably good approximation, at least for the inner regions where the gravitation fields are strong. The radiation fields are very close to isotropic and are almost Planckian, with the local temperature as the characteristic parameter; and the resulting energy flow may be very well described in the diffusion approximation. This simplicity also makes it possible to describe the star as a function of a small number of parameters, essentially only its initial mass, its age, and its elemental composition. Possibly, its angular momentum must be added for a more complete description, but even so, the number of basic parameters of the system is appealingly small, reflecting the basic simplicity of stars as physical objects.

The interstellar medium, on the other hand, is an extremely complex system in which it is not possible to describe the physical status of the matter and its structural and dynamical configurations, as well as the radiation fields, with a restricted number of parameters; instead, a full and detailed account of the history of the system is required for such a description.

The stellar atmosphere is the transition region between these two systems, and its complexity grows the higher up in the atmosphere one looks. As is illustrated by the plethora of phenomena observed in the solar atmosphere, or by the intricate structures of AGB winds or planetary nebulae, the increased complexity in the state of the gas (with considerable deviations from LTE, and time-dependent nonequilibrium chemistry determining the composition of the wind) is matched by a strong increase in the complexity of the geometrical and dynamical structures in the atmosphere and the wind. This structure formation occurs in a region through which a strong energy flux is carried, pushing it far out of equilibrium. This aspect makes stellar atmospheres good examples of structure-forming physical systems, resembling the dissipative structures suggested and explored by Prigogine [145].

The situation just described is characteristic of most or all stellar atmospheres. What in particular characterizes the atmospheres of AGB stars, as well as the atmospheres of other red giants and supergiants? The low gravities of these stars make the usual assumption of spherical symmetry less realistic. Their atmospheres are unstable against convection in the deeper layers, due to hydrogen dissociation and hydrogen absorption. This instability continues inwards for most of the distance to the stellar center. It is also thought to lead to large-scale inhomogeneities ("giant granular cells") across the stellar surface. The stars are also unstable against pulsations, and these generate shock fronts that greatly affect the atmospheric structures. More dramatic phenomena, in particular the thermal pulses or helium shell flashes, occur in the stellar interior and expose the outer stellar regions to strong shocks. The radiation fields are highly complex, due to the very strong effects of molecular absorption. When dust forms, it has a great impact on the radiation field but even more on the momentum of the gas, due to the momentum transfer that takes place when dust grains absorb or scatter radiation and then drag the gas along with them via friction. The interaction between these different phenomena—convection, pulsations, radiation, molecular and dust formation and absorption, and the acceleration of the stellar wind—is intricate. In the present chapter we shall discuss these different aspects in some detail.

One may study stellar atmospheres with different objectives. One is because the atmospheres need to be understood, at least to some extent, in order for us to interpret stellar spectra in terms of basic parameters like stellar radius, mass, age, or chemical composition. For AGB stars, studying the composition of the atmosphere is of key interest, not the least since the stars themselves are producing heavy elements in their centers that are mixed out into the atmospheres. For some of these elements (perhaps carbon, probably nitrogen and fluorine, and most certainly many of the heavy elements beyond iron, the s-elements) these stars are main contributors in Galactic nucleosynthesis. The other objective is the study of the atmospheres as such, the arena of a complex interplay between physical phenomena of different character.

These two different objectives may lead to rather different approaches when the atmospheres are to be modeled. In the application-driven case, one may be willing to base some of the analysis on semiempirical recipes, with free parameters to be fitted to observations. However, one may also need very detailed modeling, e.g., of spectral regions of particular interest. This may require very extensive databases, in particular of atomic and molecular data to be used in the spectrum calculations. In the physics-driven case, the emphasis on theoretical self-consistency is usually stronger, while the need for completeness in basic data may be relaxed: The interest is often more in understanding certain physical mechanisms and the interplay between them than in achieving great realism in the predicted spectrum. However, the general experience from research in this area during the last decades is that the interplay between the different physical phenomena is so intricate that one cannot proceed very safely with only "toy models," even if one does not aim at detailed spectrum predictions. For example, a highly realistic treatment of molecular absorption is needed in order to obtain even a qualitatively correct picture of the mass loss of a pulsating AGB star. The development thus goes toward a detailed modeling of these atmospheres; all simplified toy models (which are certainly needed for understanding processes in a complex reality) will have to be based on, or checked versus, these detailed models. In the present chapter, we shall therefore—while sticking to basic physical principles and without elaborating on all the painstaking details—put the main emphasis on pointing the way toward considerably more realistic model atmospheres for AGB stars than those existing today and, in fact, argue that this road should be accessible even to beginning graduate students today. However, we shall not refrain from exploiting a number of very simplified models in order to illustrate some basic physical principles.

4.2 Observations

4.2.1 New Observational Opportunities

4.2.1.1 The IR Breakthrough

AGB stars are bright and numerous enough to enable detailed observational exploration of many stars. Most of their light is emitted in the infrared (IR) spectral region. Therefore, the quantity of observational data has increased drastically since the IRAS and ISO satellites opened the way for spectrophotometric studies in wavelength regions inaccessible from the ground. This has made it possible for the first time to obtain continuous flux spectra extending from the visual to the far infrared for many stars to compare with stellar models, though still at only moderate spectral resolution. Stellar spectral lines and bands from water vapor, carbon dioxide, silicon monoxide, and other molecules, impossible or difficult to study from the ground, have been explored systematically. Emission features from atoms in the far infrared, as well as from various dust components and polyaromatic hydrocarbons (PAHs), can now be studied as well. Examples of such results are presented in Figure 4.1, which shows that although the stellar fluxes in the near IR might be interpreted schematically as coming from a blackbody with a temperature of a few thousand degrees, this is certainly not the case for the entire spectrum. The stellar atmosphere is obviously a composite system, with physically different domains contributing to different spectral regions.

4.2.1.2 High-Resolution Spectroscopy

High-resolution spectrometers, the Fourier transform spectrometers, and more recently the cryogenic echelle spectrometers [74] can now record IR spectra of thousands of AGB stars at resolutions $\lambda/\Delta\lambda$ of about 10^5. The advent of such instruments has provided a wealth of information and opened up new avenues for the study of the structure, dynamics, and chemical composition of AGB star atmospheres. This is illustrated in Figures 4.2 through 4.4. One key factor is the fact that the spectra in the visual and near infrared

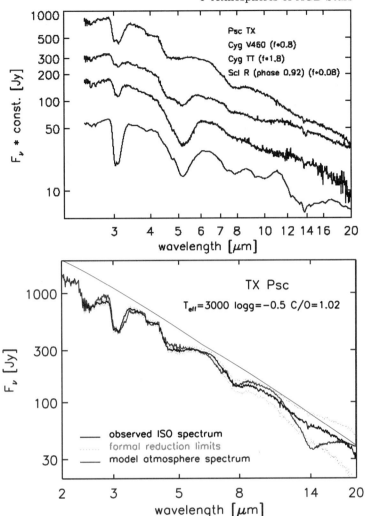

Fig. 4.1. Upper panel: Spectra of the four carbon stars TX Psc, V460 Cyg, TT Cyg, and R Scl, obtained with the Infrared Space Observatory (ISO) Short Wavelength Spectrometer (SWS). The spectral intensities have been scaled for better separation between the stars (scaling factors given in the legend). The molecular bands at 2.5 μm are attributed to CO and HCN; at 3 μm to HCN and C_2H_2; at 3.8 μm to CS and HCN; at 4.3–6 μm to CO and C_3; and at 7–9 μm to CS, HCN and C_3.
Lower panel: Observed spectrum for TX Psc (thick line), with upper and lower limits (dotted lines), is compared with the spectrum of a model atmosphere with representative parameters (thin continuous line). It is seen that the predicted deep feature around 14 μm due to HCN and C_2H_2 does not appear in the stellar spectrum (cf. the discussion in Sections 4.6.4 and 4.8.3). From [100]

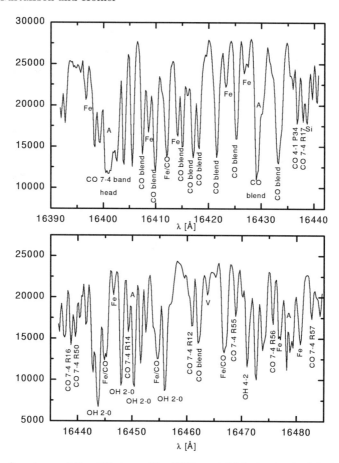

Fig. 4.2. Spectrum of the M6 star g Her (SRb variable) obtained with the Fourier Transform Spectrometer at Kitt Peak National Observatory. Several spectral features are marked. The CO features belong to the second-overtone vibration-rotation bands. "A" denotes a telluric feature. The units on the y-axis are arbitrary. From [121]

are so crowded that no safe continuum, or line-free, regions may be traced. It is also difficult to find any well-defined spectral lines free of blends from other lines. Beyond 2 μm, however, this problem is alleviated for many stars. Also, the vibration–rotation bands of molecules such as CO offer sequences of well-defined spectral lines of different strengths in the infrared that can be used to systematically explore the structure and dynamics at different depths in the atmospheres. Of particular interest are the possibilities of obtaining sequences of high-resolution spectra for variables: An example is the work on Miras by Hinkle and collaborators (see [123] and references therein). High-resolution spectra in infrared wavelength regions blocked by the terres-

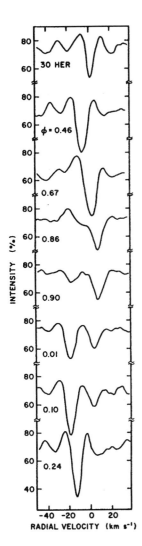

Fig. 4.3. Average CO second-overtone line profiles as a function of phase for the S-type Mira star χ Cyg, obtained with the FTS spectrometer at Kitt Peak. The velocity scale is heliocentric, and the intensity of the continuum is 100%. The spectrum of the low-amplitude variable 30 Her (M6 III) is shown on top for comparison. From [73]

trial atmosphere are still lacking, however. The EXES spectrometer, planned for the Stratospheric Observatory for Infrared Astronomy, will open up the spectral region from 5.5 to 28.5 µm to high-resolution studies of individual spectral lines for the first time [153].

4.2.1.3 Photometry, Polarimetry

A basic database of considerable importance for the interpretation of AGB star atmospheres consists of the broad-band fluxes and colors in the classical $UBVRIJHKLM$ system, and different versions thereof. The value of these data is, however, reduced by the fact that some of the filter widths were

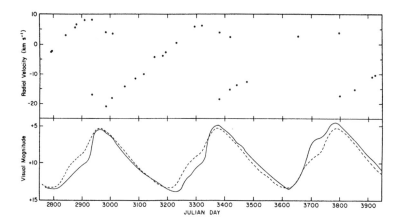

Fig. 4.4. Upper panel: Mean CO second-overtone velocity as a function of time (JD number − 2,440,000) for χ Cyg.
Lower panel: Visual (solid) and average (dotted) light curve for the star as a function of time. Note that the line splitting seen in the upper panel occurs at the light maxima. From [73]

chosen to be so wide that the actual band definition is highly affected by the transmission profile of the local atmosphere, and by some uncertainties in the absolute calibration (from magnitudes to physical fluxes) of the system. Also, the variability of almost all AGB stars makes it important to have observations in different bands made at the same time. For a number of stars there exist photometric data, light curves, or even flux distributions at different epochs (see, e.g., [34] and references given therein). More systematic data of this kind for stars of different variable type are, however, needed.

One interesting opportunity, which has not yet been explored systematically, is the use of polarimetry and spectropolarimetry to investigate inhomogeneities in the distribution of dust in the wind around the star or, say, strong TiO absorption in patches across the stellar surface. The point is that an uneven distribution of molecular patches would lead to a net polarization of perhaps a few percent, while an even distribution would not. For early polarimetric studies see [31, 43, 97]. Recently, polarization measurements and spectrophotometry have been combined to study the environment of a carbon star [152].

4.2.1.4 Interferometry

Many AGB stars are so big or relatively nearby that their diameters may be measured using phase or speckle interferometry, or lunar occultation methods. These data are of great significance for establishing a reliable effective-temperature scale for the AGB stars, as will be discussed below. For several stars, the diameters have been measured at different wavelengths, e.g., inside and outside strong molecular bands, and found to vary with wavelength. Studying this dependence of the diameter on wavelength is an important task, to check the predicted extension of the atmosphere by comparison to models (see [64] and references therein).

With the increasing resolution of new interferometers like the ESO Very Large Telescope Interferometer, we shall also obtain resolved images of a number of AGB surfaces in the near future. Such images are as yet available only for a few stars like the red supergiant α Ori (see, e.g., [200] and references therein), and the AGB stars Mira Ceti [104] and R Leo [78]. For α Ori, the varying surface seems to show asymmetries and possibly one or a few giant granules. Mira Ceti also shows an asymmetric surface pattern, but it is not clear how much of the asymmetry should be ascribed to a CSE. R Leo, though resolved, shows no clear signs of asymmetric patterns. The interferometric information available for the apparently biggest star in the sky except for the Sun, the nearby Mira R Dor [14], suggests an asymmetric appearance of the photosphere. Observations of this type are of great significance, not the least for giving information on convective inhomogeneities and other dynamic modes and surface structures of these stars. When combined with spectroscopy and polarimetry into spatially resolved spectropolarimetry of AGB surfaces, we shall have adequate observational means to match the progress in 3D hydrodynamic modeling that is presently being made for these stars. This will mark major progress in the future.

4.2.2 Observing Velocities and Magnetic Fields

Other important observations of AGB stars are related to their atmospheric velocity fields. In standard spectroscopic analyses, a microturbulence and a macroturbulence parameter may be estimated. These contain information on the velocity fields, although the classical interpretation of microturbulence as a measure of small-scale motions is oversimplified. Typical values of the microturbulence parameters for non-Mira AGB stars are between 2 and 3 $\mathrm{km\,s^{-1}}$, and somewhat more for the macroturbulence, as measured in high-resolution spectra in the infrared [67, 117, 183]. For Mira stars, the shock waves cause dramatically changing velocity fields through the atmospheres, which may broaden the spectral lines and even split them into several components. The variations in line profiles and line splitting as a function of phase give important information about how shocks progress up through the atmosphere (see, e.g., [2] and references therein). Other important observations of velocity fields in the upper atmospheres (or inner envelopes) are obtained through SiO masers (see, e.g., [37], and Chapter 7).

The study of masers also offers evidence for the existence and dynamical importance of magnetic fields in AGB stars. Very long baseline interferometry (VLBI) was used to map the magnetic field from SiO emission around the

Mira variable TX Cam in all Stokes parameters and traced a linear polarization suggesting fields on the order of 10 Gauss [105]. The spatial structure and polarization of the OH maser emission of R Crt suggest that magnetic fields caused the observed asymmetries [176], while the circular polarization of OH masers of four AGB stars gives evidence for turbulent magnetic fields carried by the outflows [138]. Additional indirect possible evidence for magnetic fields affecting AGB winds is the existence of semiperiodic concentric multiple arcs in the stellar envelopes, which could be the result of magnetic activity cycles [173].

The Zeeman splitting or broadening of spectral lines scales quadratically with their wavelengths, while the Doppler broadening scales linearly. Therefore, the development of high-resolution spectroscopy and spectropolarimetry in the infrared allows us to hope that photospheric magnetic fields of AGB stars could be observed in the near future, in spite of the large contribution to line broadening by turbulence and more systematic velocity fields.

4.2.3 Observing Chromospheres

AGB stars, like most late-type stars, show ultraviolet emission in the strong Mg II resonance lines at 280 nm, as well as in Ca II lines and many Fe II lines and C II lines. This suggests the presence of hot layers in the upper atmospheres, so-called chromospheres. These layers of red-giant stars, and of other stars, are not predicted by standard theoretical models; instead, their existence (as implied by observations) has long been regarded as a riddle. At least for low-mass stars, the chromospheric activity seems to increase as the star proceeds up along the giant branch [41], and this might be ascribed to increasingly efficient shock heating of the chromospheric layers either by waves generated by convective motions, or by pulsations. One should note, however, the demonstration by Carlsson and Stein [28] that the appearance of a solar nonmagnetic chromosphere may result from wave motions and radiative transfer alone, without any increase of a mean gas temperature. It is not known to what extent the chromospheric lines of AGB stars may be similarly explained.

The chromospheric lines show interesting differences when different stars are compared, and also vary significantly with phase for Miras (see, e.g., [130]). The Mg II emission is often overlaid by circumstellar and interstellar absorption lines, and varies considerably from time to time and from star to star (see [30] and references therein). Attempts have been made [93, 102] to match observations of the Mg II and CO lines as well as the HCN bands for the carbon star TX Psc using semiempirical models. These suggest that the layers above the stellar photosphere include an inhomogeneous, high-temperature region relatively close to hydrostatic equilibrium but with considerable turbulence as well as a deep but much more extended cool region with outflowing gas and dust (see also [29]). The existence of two such coexisting states of the upper atmosphere, one seemingly hot and one cool, is not astonishing in itself; a similar situation seems to be characteristic of hotter stars.

Emission lines are also found in the far IR. Thus, forbidden metal-line emission in the far IR from neutral metals ([Fe I], [S I]) in carbon star spectra, and more ionized lines ([Fe II], [Si II]) in M star spectra have been found [6]. They ascribe these to "rather dense and warm regions," which, however, are much cooler than the chromospheric regions and perhaps more related to the enigmatic, relatively stationary molecular layer traced by Tsuji [184] from CO line observations. Recently, an upper molecular layer for K giants and early M giants has been traced as well, by observing H_2O bands [185]. The physical structure and reason for these layers are not yet understood.

The chromospheres of AGB stars may also be studied at radio wavelengths as radio emission. Thus, the authors of [109] found an excess emission (relative to what is expected from the photospheres) for four AGB stars, which they tentatively ascribed to chromospheres extending out to several (optical) stellar radii.

A number of attempts have been made to model the observations of chromospheric lines of AGB stars (see, e.g., [128, 129]), but the complex geometry and velocity fields have prevented any more definitive models. Thus, detailed understanding of the properties and the origin of the complex upper layers of AGB star atmospheres is still missing. Further study of these regions is probably of key significance for understanding the acceleration of AGB winds and mass loss.

4.3 Physics and Characteristic Conditions

Radiation and its interaction with matter is a dominating phenomenon in stellar atmospheres in general. It plays a key role in both the energy and momentum transfer through the atmosphere, and thus strongly affects the atmospheric structure. Moreover, radiation leaking out of the atmosphere provides the single diagnostic tool for the study of the star. This makes it imperative to understand the formation of the radiation spectrum.

The atmospheres of AGB stars are strongly affected by time-dependent dynamical processes as well. Convection carries part of the energy flux in the deep photospheric layers and is supposed to produce large convection cells that are visible higher up, and possibly cause considerable deviations from spherical symmetry. Furthermore, the convective motions and pulsations of the star may trigger shock waves that propagate outwards through the atmosphere and change its structure both on a local and global scale. In the inner parts, the passing shocks cause a more or less periodic modulation of the structure. In the upper layers, the dissipation of mechanical energy leads to a levitation, i.e., a density enhancement of several orders of magnitude at certain heights. These effects influence the formation of molecules and dust grains and, consequently, the optical properties of the matter as well as the mass loss from the stars.

In order to explore the interaction of radiation with the atmospheric gas and the propagation of motions through the atmosphere, it is important to gain some insight into the physical conditions in the gas. We shall start here by making rough estimates of characteristic temperatures, densities, and velocities in the atmosphere.

4.3.1 Characteristic Temperatures

It is not trivial to determine the temperatures of the flux-forming regions in an AGB star. For a star like the Sun, the flux distribution may look much like a Planck function. This is essentially due to the limited variation of the continuous absorption coefficient with wavelength, which means that the observer sees roughly the same depth of the atmosphere, with a rather well-defined temperature, in the light at different wavelengths. The continuous absorption in the Sun is dominated by H^- bound–free and free–free absorption. This is also the case for AGB stars, but their flux distributions are very far from Planckian, as illustrated in Figure 4.12. Why is that? First, the AGB star flux is much more affected by spectral lines, in particular from molecular bands, all over the spectrum. Second, the flux maximum is pushed out toward the H^- minimum, and the free–free absorption that dominates at longer wavelengths is smaller relative to the bound–free absorption, which is effective at shorter wavelengths. Third, in the far infrared part of the flux distribution, emission from dust in the outer atmosphere or the envelope may dominate and thus introduce a much broader flux distribution than that from the photosphere.

If we neglect such extra emission in the infrared, we may estimate the typical temperature by attempting to fit a Planck function to the remaining flux in the visual and near infrared. Typical temperatures resulting from such a procedure for AGB stars lie in the interval between 2000 and 3500 K.

The procedure for determining the effective temperatures for the stars is somewhat better defined. The effective temperature is defined by the relation

$$\sigma_R T_{\text{eff}}^4 = \frac{L}{4\pi R_*^2}, \qquad (4.1)$$

where L is the luminosity, R_* the stellar radius, and σ_R the Stefan–Boltzmann constant. Obviously, instead of trying to fit the flux distribution we just assume that the star radiates as much energy per unit time as a blackbody with a certain temperature, and adopt this temperature as representative for the emitting gas. This procedure is simple but entails a number of problematic points. One seems to be that the luminosity and the radius must be known, which suggests that the distance to the star must also be determined. However, this is not so, since what is actually needed in (4.1) is L/R_*^2, which is equal to $16\pi f/\theta^2$, where f is the wavelength-integrated stellar flux measured above the Earth's atmosphere, and θ is the observed angular diameter of the star in radians. The flux f must be measured or estimated across the spectrum, which may be difficult, since observations at all wavelengths, particularly in the ultraviolet or in certain regions in the infrared, are rarely available. The flux f must also be corrected for circumstellar and interstellar extinction. Other difficult points are that R_* and θ are not well-defined but wavelength-dependent, because the absorption in the atmosphere varies with wavelength; and that the atmosphere is so extended that light seen at different wavelengths comes from layers at different distances from the stellar center. A third point of doubt may be that even if the effective temperature is well defined as such, T_{eff} may not be a good characteristic temperature of the atmospheric layers or even of the radiation emitted, since the star is certainly not radiating like a blackbody. Assumptions of similar character are, however, made even when much more refined methods are used to estimate temperatures. Thus, when model atmospheres are used to estimate the temperature structure in the surface layers, they are usually based on the assumption that the gas locally emits radiation like a blackbody. This is one aspect of the assumption of local thermodynamical equilibrium (LTE). LTE essentially means that thermodynamic equilibrium is assumed to be attained locally, with one local temperature characterizing the statistics of the microscopic motions in the gas, as well as the excitation, ionization, and dissociation of atoms and molecules, and the local emission of radiation.

One may ask whether LTE is a necessary condition for estimating, or even ascribing, any temperature to the stellar atmosphere. This is not the case. A temperature can usually be defined even if the distributions of atoms and molecules in different excitation and association states are far from those characterizing the thermal equilibrium, and even if the locally emitted radiation field is far from Planckian. As will be demonstrated below, this is because the elastic collisions in the gas, even in these rarefied media, are so frequent that the distribution of velocities of electrons, atoms, and molecules in most cases follows the Maxwell distribution. Thus one kinetic temperature may be derived for each region in the stellar atmosphere. This requires detailed modeling of the state of the atmosphere and of the radiation field.

In summary, when estimating temperatures of stellar atmospheres one may use measured angular diameters complemented with information from model atmospheres, or directly compare the observed flux distributions, as sketched above, to those from models. Alternatively, spectral lines are used to determine temperatures from excitation or dissociation conditions. Again, models play an important role in relating the observed spectral line strengths to the temperatures of the spectrum-forming layers.

Effective temperatures, when derived as indicated above, as well as excitation temperatures deduced from the line strengths of different spectral lines, are often found to be in the interval between 2000 and 3500 K for

AGB stars. The rough agreement between the different estimates of these characteristic temperatures does not mean that consistency is achieved in all determinations. Instead, the task of making accurate and consistent temperature determinations is still a major problem in the analysis of AGB stars, as will be further discussed in Section 4.9.1. Here, we shall just adopt a temperature in the middle of our interval, 3000 K, as a characteristic temperature of our objects.

4.3.2 Characteristic Densities and Scale Heights

We shall now try to estimate characteristic densities of the atmospheres of AGB stars. In order to do so, we shall first estimate the length Δ of a column of absorbing gas that has an optical depth of unity for radiation in the spectral continuum. The monochromatic optical depth τ_ν is defined according to

$$d\tau_\nu = \chi_\nu \, ds, \qquad (4.2)$$

where s is the geometrical length and χ_ν is the extinction coefficient per unit length.

To estimate χ_ν we note that at its maximum around 850 nm, the cross section for the dominating H^- bound–free absorption is $\sigma \approx 4 \times 10^{-17}$ cm^2 per H^- ion. Assuming LTE, the number density n_{H^-} may be obtained from the Saha equation

$$\frac{n_H}{n_{H^-}} = \left(\frac{2\pi m_e}{h^2}\right)^{3/2} (kT)^{5/2} p_e^{-1} \frac{2u(H)}{u(H^-)} \exp\left(\frac{-D_0}{kT}\right), \qquad (4.3)$$

where n_H is the number density of hydrogen atoms, p_e the electron pressure, $u(H)$ and $u(H^-)$ the respective partition functions, m_e the electron mass, and $D_0 = 0.75$ eV the ionization potential of the negative hydrogen ion. In cgs units this is

$$\frac{n_{H^-}}{n_H} = 0.75 \, p_e \, T^{-5/2} \, 10^{3830/T} \, . \qquad (4.4)$$

When estimating the pressures, we first note that in a gas of solar composition at stellar atmospheric pressures, most of the gas pressure is due to atoms of H, and most of the hydrogen is in atomic and not molecular form. Secondly, most of the electrons are contributed by the three elements Na, Al, and Ca, which are singly ionized to a considerable degree. Thus, we may write

$$p_{\text{gas}} \approx n_H \, kT \qquad (4.5)$$

and

$$p_e = 10^{-5} \, p_{\text{gas}}, \qquad (4.6)$$

since there is about one nucleus of Na, Al, or Ca per 10^5 hydrogen nuclei.[2]

Combining (4.2) through (4.6), we have

$$d\tau_{\text{cont}} \approx 2 \times 10^{-6}\, p_{\text{gas}}^2\, T^{-7/2}\, 10^{3830/T}\, ds, \qquad (4.7)$$

where τ_{cont} denotes the optical depth in the spectral continuum.

Integrating (4.2) to unit optical depth, we obtain a depth Δ of the photosphere, at our characteristic temperature of 3000 K:

$$\Delta \approx 4 \times 10^{16}\, p_{\text{gas}}^{-2}\, \text{cm}. \qquad (4.8)$$

As already mentioned above, there are some observations of AGB stars that are related to the extension of their atmospheres, namely angular-diameter measurements at different wavelengths. For instance, Quirrenbach et al. [148] used a long-baseline interferometer to obtain observations of 15 red giant stars in a strong TiO absorption band at 712 nm and in a band with less absorption at 754 nm. While the warmer giants (of spectral types K5 III and M0 III) have identical diameters in the two bands, the cooler stars (M3 to M5, III and II) appear about 10% larger in the 712 nm band. Quirrenbach et al. suggest that these data could be taken as an extension measure of the atmospheres. In the TiO band, one observes the outer layers of the stellar photosphere, while in the 754 nm band one looks much deeper into the photosphere. Converting the visibility measurements of the interferometer to real extension measures is, however, not simple. In particular, they depend on the assumed angular distribution of intensity across the stellar surface (the "center-to-limb variation") (cf. [89]). Let us anyhow adopt their results for estimating Δ. A typical value of the radius of an AGB star is 200 $R_\odot$ (cf., e.g., [45]), 10% of which is 1.4×10^{12} cm. With this value of Δ, (4.8) yields a pressure of about 1.7×10^{-4} atm. Using the gas law

$$p_{\text{gas}} = \frac{\rho k T}{m_H \mu}, \qquad (4.9)$$

where m_H is the atomic mass unit and $\mu \approx 1.4$ is the mean molecular weight for a slightly ionized atomic gas of solar composition, we obtain a density of $\rho \approx 10^{-9}\, \text{g cm}^{-3}$.

One may make a more theoretical estimate of the extent of the photosphere by assuming hydrostatic equilibrium, i.e.,

$$\nabla p = \rho \mathbf{g}, \qquad (4.10)$$

---

[2] True enough, these statements are based on some knowledge of characteristic pressures in the gas. At lower pressures, Mg and Fe may become significant electron donors; at higher pressures, Al and Ca remain neutral; and at considerably higher pressures, hydrogen molecules dominate. However, the characteristics given above hold true for a rather wide range of pressures at temperatures around 3000 K.

where **g** is the gravitational acceleration. Assuming spherical symmetry, we may write

$$g = GM/r^2, \quad (4.11)$$

where r is the distance from the center of the star, and M the stellar mass. Equating p with p_{gas}, i.e., disregarding radiation pressure and "turbulent pressure" (see Section 4.7.1 below), we obtain

$$\frac{\partial p_{\text{gas}}}{\partial r} = -\frac{\rho G M}{r^2}. \quad (4.12)$$

Applying (4.9), we get

$$\frac{\partial p_{\text{gas}}}{\partial r} = -\frac{Gm_{\text{H}}}{k}\frac{p_{\text{gas}} M \mu}{r^2 T}. \quad (4.13)$$

Neglecting the relatively slow variation of temperature and mean molecular weight in the atmosphere (as compared with the variation of pressure), we obtain the solution

$$p_{\text{gas}} = p_0 \exp\left[-\frac{Gm_{\text{H}}}{k}\frac{M\mu}{R_*^2 T}(r - r_0)\right], \quad (4.14)$$

where p_0 is the pressure at radius r_0. Thus, the gas pressure changes by a factor of $e \approx 2.72$ in the atmosphere with a characteristic scale height of

$$H = \frac{k}{Gm_{\text{H}}}\frac{R_*^2 T}{M\mu}. \quad (4.15)$$

Assuming a radius of $200\,R_\odot$ and a mass of $2\,M_\odot$, and again assuming $T = 3000$ K, we obtain $H = 1.5 \times 10^{11}$ cm. The optical depth in the continuum scales as p_{gas}^2 [see (4.7)]. Thus, if we define the photosphere to be the region extending from optical depth 10^{-4} to optical depth unity, then it has a height of approximately $4 \times 10/(2.72 \times 2) \times H = 1.1 \times 10^{12}$ cm. This is in good agreement with the Δ value estimated above from direct observations.

One should note, however, that there are clear observational and theoretical indications that many AGB stars, and in particular the Mira variables, may have more extended atmospheres. Speckle interferometry done by Hofmann et al. [77] in different wavelength bands seems to show that R Cas has an extended atmosphere of up to 30% of the total stellar radius in certain phases.[3] However, this does not mean that the pressure and density estimates given above will be overestimates for such extended atmospheres. As will be demonstrated in Section 4.7.2, the shocks in pulsating stars may push material outward and increase the densities considerably in the outer layers, while low temperatures and the resulting H_2 formation and low ionization may prevent H^- ions from forming, thus making the layers transparent in the continuum.

---

[3] Note, however, that the authors' warnings regarding the correct interpretation of their data in terms of stellar radii must be taken seriously.

4.3.3 Characteristic Velocities

A relevant quantity for the characteristic velocities in the atmospheres of AGB stars is the (adiabatic) sound speed

$$c_{\mathrm{s}} = \sqrt{\frac{\gamma kT}{m_{\mathrm{H}}\mu}}, \tag{4.16}$$

where $\gamma = C_{\mathrm{p}}/C_{\mathrm{v}}$. Putting in realistic values such as $\gamma \approx 5/3$, $\mu \approx 1.4$, and $T = 3000$ K, we obtain $c_{\mathrm{s}} = 5.4$ km s^{-1}. From this we see that the observed micro- and macroturbulent velocities in non-Miras (see Section 4.2 above) tend to be subsonic, which is to be expected in view of the high energies needed to drive supersonic velocities. However, the progression of the shock waves through the atmospheres of Miras, as seen in line splitting and line profile variations in their spectra, is no doubt faster than sound waves. For example, in a recent study based on IR CO and OH lines, an amplitude of 34 km s^{-1} in the pulsational velocities is derived [165].

4.4 Microscopic State of Matter

The physical conditions in the stellar gas are determined by a great number of processes in which atoms, molecules, and electrons interact with radiation and with one another in collision processes. Of key significance for the understanding of the stellar atmospheres and their spectra is the extent to which these processes are fully determined by local conditions (e.g., when temperature and pressure determine the rate of collision processes) and to what extent nonlocal effects play a role due to the fact that radiation from other regions in the atmosphere affects the conditions. Another important issue is the extent to which the distribution of atoms and molecules in different microscopic states depends on the history of the gas, i.e., whether the time dependence of the processes populating and depopulating these states must be considered. Here, we shall address these questions by discussing the excitation of atoms and molecules, the molecular equilibrium, and the formation of dust. We begin, however, with some considerations of the microscopic collisions.

4.4.1 Elastic Collisions

The atoms in the gas collide, with each other and with electrons, from time to time. These collisions may be elastic or inelastic. If elastic collisions dominate, they will tend to establish a Maxwellian distribution of particle speeds v,

$$f(v)\,\mathrm{d}v = \left(\frac{m}{2\pi kT}\right)^{3/2} \exp\left(\frac{-mv^2}{2kT}\right) 4\pi v^2\,\mathrm{d}v. \tag{4.17}$$

The mean speed $\langle v \rangle$ of the particles can be derived from (4.17),

$$\langle v \rangle = \left(\frac{8kT}{\pi m}\right)^{1/2} = 8 \left(\frac{T}{3000 \text{ K}}\right)^{1/2} \text{ km s}^{-1}, \qquad (4.18)$$

where the mass of the hydrogen atom has been used for m. The characteristic kinetic energy kT is 0.26 eV for $T = 3000$ K. The exponential in the distribution in (4.17) makes it improbable that a particle has a much greater energy than this; for example, one finds that for every particle with an energy of 2.6 ± 0.1 eV there are about 2×10^3 particles with an energy of 0.26 ± 0.1 eV. Therefore, collisions that excite atoms or molecules to high energies, or ionize them, are comparatively rare.

Which type of collisions dominates, elastic or inelastic? Hydrogen and helium are by far the most abundant atoms, and the overwhelming number of them are in the ground state. Since an energy of 10 eV is needed to excite neutral hydrogen atoms to their $n = 2$ levels, such collisional excitations should be very rare in the cool gas of AGB star atmospheres. Almost all atomic collisions will be elastic, as will the collisions between electrons and atoms. Thus the distributions of the particle speeds will be very close to Maxwellian.

For gas in the outer atmospheric layers of the (cooler) AGB stars, a considerable fraction of the hydrogen gas may be in molecular form, and the H_2 molecules have a great number of low-lying energy states that will be excited by collisions. However, these states are deexcited by collisions equally often. The probability that the states are deexcited by radiation, leaking out of the atmosphere and thus providing a sink in the velocity distribution at certain velocities, is small. This is because such radiative transitions are very weak, due to the symmetry principles of the homonuclear H_2 molecule. We thus conclude that for gas in the AGB star atmospheres, the Maxwell distribution should be a very good approximation, and its characteristic parameter T may be defined and taken as *the* temperature of the gas.

4.4.2 Excitation of Atoms and Molecules

The situation concerning the excitation equilibrium of atoms and molecules is different from that of their velocity distribution. While collisions control the velocities, radiation may well play a key role in establishing the populations of the discrete atomic states. The radiative rate R_{ij} (i.e., the number of transitions per particle per second) for a transition from a lower state i to an upper state j due to an absorption process in a particular atom or molecule may be written as

$$R_{ij} = 4\pi \int_0^\infty \alpha_{ij}(\nu) \frac{J_\nu}{h\nu} \, d\nu, \qquad (4.19)$$

where $\alpha_{ij}(\nu)$ is the absorption cross section for the transition $i \to j$ and $J_\nu = \frac{1}{4\pi} \int I_\nu d\Omega$ is the radiative mean intensity. We denote the cross section of the transition integrated over frequency by $f_{ij}\pi e^2/(mc)$, where f_{ij} is the oscillator strength. Then the radiative rate is

$$R_{ij} = \frac{4\pi^2 e^2 f_{ij}}{mc} \frac{\bar{J}_0}{h\nu_0}, \qquad (4.20)$$

where $\bar{J}_0$ represents an average of J_ν across the transition (spectral line), an average that is in fact defined by (4.20). In thermodynamic equilibrium, $\bar{J}_0$ is equal to the Planck function, and we may use that as an estimate, i.e.,

$$\bar{J}_0 \sim B_\nu(T_{\rm rad}) = \frac{2h\nu^3}{c^2} \left[\exp\left(\frac{h\nu}{kT_{\rm rad}}\right) - 1\right]^{-1}, \qquad (4.21)$$

where $T_{\rm rad}$ is the temperature of the radiation.

The radiative rate R_{ij} should now be compared with the corresponding collision rate C_{ij}. The number of transitions per particle and second will be

$$C_{ij} = n_{\rm c} \int_{v_0}^{\infty} \sigma_{ij}(v)\, vf(v)\, {\rm d}v, \qquad (4.22)$$

where $n_{\rm c}$ is the number density of particles colliding with the atoms or molecules under study and thus providing the excitations, and σ_{ij} is the velocity-dependent cross section for an exciting collision from state i to state j. The velocity v_0 corresponds to the minimum energy for the process to take place, so that if E_i and E_j are the excitation energies of the corresponding atomic or molecular states, then $mv_0^2/2 = E_0 = E_j - E_i$ for $E_j > E_i$, and $C_{ij} = 0$ for $E_j < E_i$.

One often introduces the energy-dependent cross section normalized to the Bohr-radius cross section,

$$Q_{ij}(E) = \sigma_{ij}(v)/(\pi a_0^2). \qquad (4.23)$$

This cross section and its energy dependence are often unknown, and must be estimated or guessed. Adopting a mean value $\overline{Q}_{ij}$ across the relevant energy interval, one may write (4.22) as

$$C_{ij} = n_{\rm c}\pi a_0^2 \left(\frac{8k}{m\pi}\right)^{1/2} T^{1/2} \exp\left(-\frac{E_0}{kT}\right) \frac{E_0}{kT} \overline{Q}_{ij}. \qquad (4.24)$$

The exponential in the Maxwell distribution in (4.17) gives the cross section near the threshold a heavy weight in the mean $\overline{Q}_{ij}$.

One may estimate C_{ij} from the cross section for the inverse process, collisional deexcitation, if the latter has been measured as a lifetime of the upper state. In thermodynamic equilibrium, the number of transitions $i \to j$ of any particular kind must be balanced by the same number of converse transitions $j \to i$ ("detailed balancing"). The number of transitions per unit volume will be $n_i^* C_{ij}$ and $n_j^* C_{ji}$, respectively, where n_i^* and n_j^* are the number densities of the respective states in thermodynamic equilibrium. Since the Boltzmann distribution for the populations of states is valid in thermodynamic equilibrium, we have

$$\frac{n_j^*}{n_i^*} = \frac{g_j}{g_i} \frac{\exp(-E_j/kT)}{\exp(-E_i/kT)}, \qquad (4.25)$$

where g_j and g_i are the statistical weights and E_j and E_i the excitation energies of the corresponding states.

From $n_i^* C_{ij} = n_j^* C_{ji}$, we find that

$$C_{ij} = \frac{g_j}{g_i} \frac{\exp(-E_j/kT_{\rm kin})}{\exp(-E_i/kT_{\rm kin})} C_{ji}. \qquad (4.26)$$

For given particle densities, the collisions must occur as in thermodynamic equilibrium; therefore, this expression is valid even if the gas is out of LTE.

The above expressions for R_{ij} and C_{ij} may in principle be modified to include transitions to the continuum, i.e., photoionizations and collisional ionizations. Similarly, rates of recombination processes, which lead to photon emission or add energy to colliding particles, may easily be found (cf. [131]).

The radiative and collisional rates together determine the populations of different atomic and molecular states. The ratio of radiative to collisional rates is decisive for the role of nonlocal radiative processes on the state of the gas. Now, using our expressions for R_{ij} and C_{ij}, let us estimate the rates for some characteristic transitions in atmospheres of AGB stars.

4.4.3 Three Examples of Important Microscopic Transitions

Our first example is *the formation of negative hydrogen ions*. The photodissociation of the H^- ion is the dominating continuous opacity source in the optical spectra of AGB star atmospheres. One may ask whether this dissociation is counterbalanced efficiently by collisions, so that the ion is as abundant as it would be in thermodynamic equilibrium. The dominant reactions by which H^- is destroyed and formed in cool stellar atmospheres are *associative detachment* ($H^- + H \longleftrightarrow H_2 + e^-$) and *photoionization* ($H^- + h\nu \longleftrightarrow H + e^-$). Molecular hydrogen is mainly provided by collision processes ($3H \longleftrightarrow H_2 + H$).

Lambert and Pagel [118] and Gebbie and Thomas [57] studied these reactions in late-type stellar atmospheres. The destruction by associative detachment has a rate of about $10^{-9}\,{\rm cm}^{-3}\,{\rm s}^{-1}$ per hydrogen atom and ion. Multiplying that by the number density of hydrogen atoms that corresponds to our characteristic temperature and density, one obtains a detachment rate per H^- ion of $4 \times 10^6\,{\rm s}^{-1}$, while the photoionization rate is about $2 \times 10^6\,{\rm s}^{-1}$ [118]. Thus, radiative processes are about as important as collisions, and we conclude that the effects of departures from LTE in the formation of H^- may be of some significance in the photospheres of AGB stars.

Next, let us turn to the ionization and excitation states of the atoms. The processes populating and depopulating atomic states should be taken into account for several reasons, not least for the determination of abundances. Atoms also contribute opacity, thus affecting the radiative energy transfer and the structures of the atmospheres. Moreover, it is important to understand the ionization of the most abundant of the easily ionized elements, i.e., those with ionization energies below about 7 eV. These elements also provide the electrons that associate with hydrogen atoms to form H^-, and are thus significant for the continuous opacity.

As a particular example, we shall take *the excitation and ionization of lithium atoms*. Lithium is so rare that it does not contribute many electrons to the gas, but its structure is not so different from that of electron donors like K and Na. Lithium is very important for AGB stars, since some of these stars show very strong Li lines, indicating that interesting mixing processes bring material processed in nuclear reactions up to the stellar surface. The formation of the line used in these studies, the $2s - 2p$ 670.8 nm resonance doublet, has therefore been explored in some detail.

The radiative rate for the excitation of the $2p$ state may be estimated from (4.20). Estimating $\bar{J}_0$ with the Planck function $B_{\nu_0}(T)$ at our characteristic temperature, we obtain $R_{(2s-2p)} = 8 \times 10^4$ s^{-1}. To estimate the collision cross section $\overline{Q}_{ij}$, we use the threshold value for σ_{ij} [33] and thus obtain $C_{(2s-2p)} \approx 7$ s^{-1} from (4.24). We see that for the photospheres of typical AGB stars, $R_{(2s-2p)} \gg C_{(2s-2p)}$. Obviously, radiative transitions dominate for the inner photospheres; therefore, they are even more pronounced at the much lower electron densities of the outer layers.

One might argue that the radiative rate R_{ij} was estimated with the Planck function calculated at 3000 K, whereas the estimate of $\bar{J}_0$ should be reduced, since the 670.8 nm line is strong and thus formed far out in the atmosphere, where the temperature is lower. True enough, the continuum radiation is considerably blocked by molecular absorption, in particular from TiO (cf. [171]); it should be less than the isotropic Planckian, because even if the radiation would penetrate from the deep photosphere into the upper layers, the incoming radiation is much less. These phenomena probably reduce the true $\bar{J}_0$ by a factor of about 50%, but this would not affect our conclusion that $R_{ij} \gg C_{ij}$ for this transition in AGB stars.

Using very similar arguments, one may also explore the processes that ionize lithium in atmospheres of AGB stars. The most important ionization channels will be ionization from the $2p$ state.[4] The transitions from this state to the continuum are again very much dominated by radiation processes: $R_{ik} \gg C_{ik}$. The threshold for this process corresponds to photons with wavelengths shorter than about 350 nm. Since the opacity is high in the ultraviolet

---

[4] Photoionization from the $2s$ ground state is less significant, since it requires photon energies corresponding to wavelengths shorter than 230 nm, at which the stellar fluxes are small.

due to heavy blocking of the spectrum by numerous molecular and atomic lines, the radiative UV field will reflect the local temperature fairly well. Some lithium atoms will be excited to higher bound states, again by radiation, and may be photoionized from these states by radiation of longer wavelengths, from greater atmospheric depths. In all, we find that the radiation field is of key significance for the ionization balance of lithium. This is further illustrated by Kiselman and Plez [106], who have made detailed calculations of Li in a model atmosphere of an oxygen-rich AGB star. These authors find an "overionization" relative to the LTE case that extends from relatively great depths around $\tau_{\rm cont}(500\,{\rm nm}) \approx 0.1$ and reduces the amount of Li I around $\tau_{\rm cont} \approx 0.001$ by a factor of two relative to the LTE value.[5]

The results for lithium are at least partially and qualitatively representative for what may be expected as regards other elements that contribute electrons, such as Na and K, and to some extent Al and Ca, or even Mg and Fe. Even if radiative processes dominate the excitation equilibria for most levels of these atoms, the degree to which this leads to population numbers very different from the Boltzmann–Saha distribution will depend on the radiation field. If the radiation field is close to Planckian and isotropic, with the local kinetic temperature as the characteristic parameter, one does not expect very large departures from LTE. However, populations far from the Boltzmann–Saha distribution may be expected if the field deviates from the local $B_\nu(T_{\rm kin})$. This may occur because the atmosphere is so transparent that deeper layers contribute to radiation in the outward direction, or the radiation coming from regions further out is significantly weaker, or if the source function is dominated by photons scattered in the lines and has little connection to the local gas temperature.

As a final example we shall now explore another significant constituent in the atmospheres of AGB stars, namely, the CO molecule and in particular *the formation of CO vibration–rotation lines*. CO contributes the impressive vibration–rotation bands in the infrared: the fundamental ($\Delta v = 1$) band around 4.6 μm, the first overtone ($\Delta v = 2$) around 2.3 μm, and the second overtone around 1.6 μm. These are important diagnostics for the atmospheric structure, velocity fields, and chemical abundances (including isotopic ratios) in AGB stars. Moreover, the bands are important as opacity sources in determining the atmospheric structure. If the transitions are coupled to the local gas through collisional processes, radiation from these transitions is able to cool the surface layers of red giants in general (cf. [61, 91]).

Each of the vibrational states of CO is split into a great number of rotational states with rather small separations between them. From (4.24) it is easy to see that the small energy difference E_0 between the states—along with

---

[5] The effects of this heavy reduction in the number density of Li atoms on the strength of the 670.8 nm line are compensated for by the circumstance that the line source function S_l ($= B_{\nu_0}$ in the LTE case) is reduced due to photon losses in the line.

the small radiative transition probabilities for the rotational lines—makes collisions dominate the transition rates, even in the tenuous outer atmospheres of AGB stars. Thus, the internal distribution of different rotational states within a vibrational state is expected to be close to the Boltzmann distribution. This is also verified by closer calculations [12].

What is the situation for the vibrational transitions? Again assuming $\bar{J}_0 = B_\nu(T)$ with T equal to our characteristic temperature 3000 K, we estimate typical radiation rates from (4.20) and obtain $R_{\Delta v=1} \approx 17 \text{ s}^{-1}$. We then use (4.26) and the lifetime in the $v = 1$ state from [12] to obtain $C_{ij} \approx 10^4 \text{ s}^{-1}$ at our characteristic pressure. Thus, the collision rates are much greater than the radiation rates in the inner atmospheres. Further out, radiative processes dominate and departures from LTE occur. This is illustrated by the calculations for Arcturus by [12], which suggests that serious departures from the Boltzmann distribution occur in the population of the $v = 1$ state when the density is about three orders of magnitude below our characteristic density.[6]

4.4.4 The Statistical Equilibrium

Summing up our collection of examples, we find that radiative processes are important, or even dominant, in populating most atomic states. This is also true for the different molecular electronic states and, in the outer atmospheres, the molecular vibrational states. The key issue is then whether the radiation field is isotropic and Planckian with the local kinetic temperature, or to what extent it departs from these LTE conditions. The further out in the atmosphere, and the more extended and inhomogeneous the atmosphere is, the more severe these departures may be. There are empirical ways to check for effects on the properties of the gas.[7] But empirical studies must be combined with theoretical calculations. In practice, important cases must be clarified with detailed calculations of the populations n_i for each molecular state i of the gas. In the general case we may write

$$\frac{dn_i}{dt} = -\nabla \cdot (n_i \mathbf{u}) + \sum_{j \neq i}(n_j P_{ji} - n_i P_{ij}). \tag{4.27}$$

Here, the first term on the right-hand side represents the net number of particles streaming into the unit volume due to the macroscopic velocity field $\mathbf{u}$, while P_{ij} denotes the total rate of transitions from any state i to any state j, $i \neq j$. Both radiation and collision-induced transitions contribute to P_{ij}, so that

---

[6] For the first overtone lines, the conditions are similar, the lower gf values for these transitions being balanced by the higher J_ν values in the radiative rates.

[7] For example, the CO molecule offers pairs of lines from different bands but with similar strengths and wavelengths. These lines can be compared and rather model-independent conclusions can be reached concerning the excitation conditions of the molecule; cf. [187].

$$P_{ij} = R_{ij} + C_{ij}. \tag{4.28}$$

In order to obtain a complete picture of the microscopic state of the gas in the stellar atmosphere, we obviously have to solve a huge set of coupled equations, one equation for each atomic and molecular state of significance, including the continuum. The R_{ij} terms also couple the radiation field to the state equations, which makes the problem very large. In the general case known as "the model atmosphere problem," (4.27) should be solved in combination with the equations describing the radiation field and the transport of mass, momentum, and energy.

In practice, one may often simplify the solution by neglecting the time derivative on the left-hand side of (4.27), as well as the divergence term on the right. As we have seen above, typical rates R_{ij} and C_{ij} for atoms may be in the interval 10–10^5 s^{-1}. Typical velocities in a stellar atmosphere may be on the order of 10 km s^{-1}, while the scale height for a drastic change of any number density or velocity (outside shocks) may be on the order of a pressure scale height $H \approx 10^{11}$ cm. From this we estimate the first term on the right-hand side of (4.27) to be on the order of $n_i \times 10^6$ cm s$^{-1}/10^{11}$ cm. In order for this term to become significant, the rates R_{ij} or C_{ij} have to be as low as 10^{-5} s^{-1}, or the region of characteristic change has to be compressed to about 10^5 cm = 1 km. Finally, in order for the time derivative on the left-hand side to be significant, there must be rapid variations in the rates. This might occur due to a rapidly varying radiation field or due to local heating by waves or shocks. Could such variations be generated by pulsations or convective motions? Such motions may be assumed to occur with a speed on the order of the speed of sound. Thus, one could expect changes in the rate terms on the right-hand side in a typical time of $H/c_s \approx 2 \times 10^5$ s. The rate is again at least 6 orders of magnitude below the rates on the right-hand side, implying that except for very drastic phenomena occurring on typical time scales of a few seconds, one may disregard the derivative.

We thus arrive at

$$\sum_{j \neq i}(n_j R_{ji} + n_j C_{ji} - n_i R_{ij} - n_i C_{ij}) = 0. \tag{4.29}$$

These are the *equations of statistical equilibrium*. They are possible to solve numerically for AGB stars in many important cases. Even though the collision cross sections in particular are often only poorly known, this may not be decisive for the problem in view of the fact that collisions are perhaps not so important anyway. Also, quantum-mechanical calculations of cross sections, both for inelastic collisions between atoms and electrons, and between atoms of hydrogen and other elements, are presently underway.

As yet, statistical-equilibrium calculations have hardly been applied to AGB stars. There are only a few exceptions, like the study of lithium [106], of titanium in giants and dwarfs [69], and of hydrogen and helium in cool giants [128]. However, due to important progress during the last two decades, adequate methods and computer codes do exist (cf. [5, 26, 27, 87, 161, 167] and references therein). Two major steps are particularly important: the invention of the approximate lambda iteration (ALI) method for radiative transfer—which goes back to early work by Cannon, was developed by Scharmer [160], and was further developed in many applications by many researchers—and the idea of grouping the states and transitions of complex atoms and molecules statistically into multigroup transitions, first tried by Anderson [5] and later developed and used extensively by many. There is no reason for a careful scientist to abstain from the pleasure of using these methods and programs. Until now, however, almost all work on AGB stars has been based on the sometimes gravely erroneous and in many cases dubious LTE approximation. This is not least the case when model atmospheres have been constructed. The reason for this neglect is the tremendous number of microscopic states, and transitions between states, that are actually significant for the atmospheric structure. This great complexity still constitutes an obstacle for making realistic models, an obstacle that, however, may now be dealt with considering the methodological progress just mentioned.

4.4.5 The Chemical Equilibrium

The number of different molecular species that affect the atmosphere and spectrum of an AGB star is considerable. In general, calculating the number densities of different kinds of molecules is a necessary and rather complex task, see also Chapter 5. For a diatomic molecule AB one may write expressions that are analogous to the Saha equation (4.3):

$$n(\mathrm{AB}) = n(\mathrm{A}) \times n(\mathrm{B}) \,/\, K(\mathrm{AB}). \qquad (4.30)$$

In LTE, K is a function only of the temperature, the partition functions of the molecule AB and the atoms A and B, and the dissociation energy D_0 of the molecule.[8] This equation is easily generalized to the polyatomic case.

In what follows we shall assume that LTE gives a good description of the molecular equilibrium. This is not necessarily true for AGB stars. One problem may be the radiative dissociation of certain molecules, which may be more significant than corresponding collisional processes. The hydrides with their fairly low dissociation energies around 4–5 eV may be particularly problematic. Moreover, the time scales of the dissociation, for example in a shock, may be much shorter than the corresponding association time scales in a tenuous plasma.

The abundances of the different atoms and molecules are linked together, since each type of atom usually appears in several different molecules. The resulting system of nonlinear equations for the molecules and atoms involved may be solved numerically, a matter that is possible but not trivial and is

---

[8] The quantity K, or a related quantity defined by the ratio of partial pressures instead of number densities, is often called the "dissociation equilibrium constant," strictly speaking, an inadequate term.

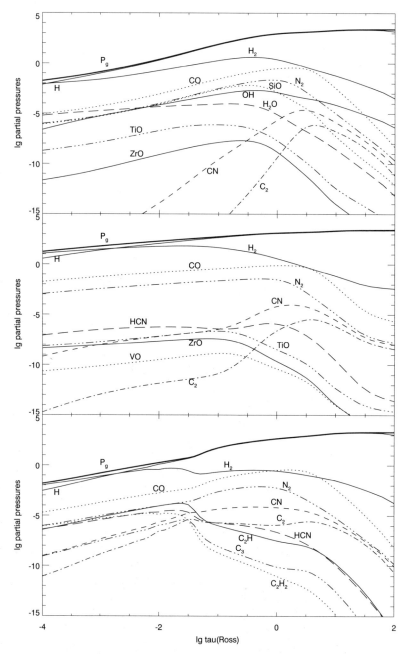

Fig. 4.5. Selected molecular partial pressures as a function of depth for three models with $T_{\text{eff}} = 3000$ K, $1\,M_\odot$, $\log g = 0.0$. The **top** panel represents an M giant (solar composition), the **middle** panel an S star ($\epsilon_C/\epsilon_O = 0.98$), and the **bottom** panel a carbon star ($\epsilon_C/\epsilon_O = 1.2$)

often accomplished by using a multidimensional Newton–Raphson technique. Examples of results from such calculations were given by Tsuji [177]. Molecular equilibria for three AGB model atmospheres with different chemical compositions are given as illustrations in Figure 4.5.

From simple arguments based on the dissociation energies of the molecules, one may understand a number of significant properties of the resulting solutions.

A basic fact that affects the structure and spectra of AGB stars profoundly is the high dissociation energy (11.1 eV) of carbon monoxide. Thus, except in the hottest deep atmospheric layers, CO is very efficient in consuming all of the carbon in M stars (which have a higher oxygen than carbon abundance, $\epsilon_C \leq \epsilon_O$, as measured by number of atoms) and conversely all of the oxygen in carbon stars, or C stars (which have $\epsilon_C \geq \epsilon_O$). The dichotomy was studied by Russell [156], who credited this explanation of carbon-star spectra to R.H. Curtiss (who realized this "years ago," according to Russell) and C.D. Shane. The phenomenon makes oxygen-rich molecules like TiO and H_2O abundant and prominent in M-star spectra, while making C-star spectra appear very different because carbon-rich molecules like C_2, CN, HCN, and C_2H_2 are dominant.

A consequence of the importance of CO is that the abundances of oxygen-rich molecules in M stars are sooner dependent on the abundance difference $\epsilon_O - \epsilon_C$ than on the oxygen abundance itself. Of these oxygen-rich molecules, OH ($D_0 = 4.2$ eV) and SiO ($D_0 = 8.3$ eV) are the most abundant, together with H_2O for the cooler stars. The amount of free oxygen is, however, much greater than that of these molecules, except for the very coolest atmospheric layers. Similarly, the $\epsilon_C - \epsilon_O$ difference sets the availability of carbon for carbon molecules in the C stars. The most abundant carbon molecules, after CO and free carbon atoms, are CH, which is abundant due to the high number density of carbon and hydrogen atoms (in spite of $D_0 = 3.5$ eV), C_2 ($D_0 = 6.1$ eV), CN ($D_0 = 7.9$ eV), CS ($D_0 = 7.9$ eV), C_2H ($D_0 = 6.4$ eV for $C_2H \to C_2 + H$), HCN ($D_0 = 4.9$ eV for HCN $\to$ CN + H) at relatively small $\epsilon_C - \epsilon_O > 0$, and C_2H_2 and C_3 at greater $\epsilon_C - \epsilon_O$ values.

In this connection one should also note the possibility of dissociation of CO by UV emission, e.g., from chromospheres. Even if the CO molecule still determines whether a carbon or an oxygen chemistry develops, the radiative dissociation may free up carbon atoms in stars with $\epsilon_C \leq \epsilon_O$, and thus offer possibilities for the formation of some carbon-rich dust in oxygen-rich atmospheres (cf. [13]).

Disregarding hydrogen and helium, the significance of CO and the remaining free oxygen or carbon atoms as the dominating species is contested only in the very coolest atmospheric layers by H_2O and CO_2 in M stars and by CH_4 (for high-gravity objects) and other polyatomic molecules, as well as carbon dust, in cool carbon stars.

The temperature at which the polyatomic molecules become important is pressure-dependent. The number density of, e.g., H_2O in a warm M star is proportional to the number density of H atoms and OH molecules, and the latter is in turn proportional to the number density of H atoms and to $\epsilon_O - \epsilon_C$. Thus, the partial pressure of water vapor is roughly proportional to p_{gas}^3, while that of OH scales as p_{gas}^2. Therefore, the significance of H_2O relative to OH depends on the surface gravity of the star, and is greater for dwarfs than for giants. However, the variation with pressure may change depending on the situation. For instance, almost all of the available oxygen in a cool star may be in the form of H_2O, the partial pressure of which will then be directly proportional to the gas pressure. Equation (4.30) then implies that the partial pressure of OH in this case will be independent of p_{gas}. At higher densities, in which most of the hydrogen is in molecular form, the partial pressure of hydrogen atoms will be proportional to $p_{gas}^{1/2}$. In this case, applying (4.30) to the OH–H–H_2O balance, we see that the OH partial pressure will also be proportional to $p_{gas}^{1/2}$.

When the carbon abundance gets close to that of oxygen, so that almost all oxygen is bound in CO, very pronounced spectral effects occur. This corresponds observationally to the spectral sequence M–MS–S (a sequence that continues for even higher carbon abundances to spectral type C). As the carbon abundance increases in the sequence, the TiO and H_2O bands get much weaker, while the most stable oxides like YO, LaO, and ZrO remain relatively less depleted than those of less-stable oxides like TiO. Thus, some of the increase we observe in, e.g., YO lines relative to TiO lines when moving from M to S stars is due to changes in the molecular equilibrium, and not solely to changes in the abundances of atoms (cf. [159]). A similar situation occurs for carbides in proceeding from spectral type CS to C. In general, the details of the molecular equilibrium of S-type stars depend in a rather complex way on a number of molecules, and the abundances of their constituent atoms.

While the abundances of the different carbon- and oxygen-carrying molecules depend not only on the availability of carbon and oxygen but also on the other atoms in the molecules as well, the situation is different for different species. An important group consists of the molecules TiO (dissociation energy 6.9 eV), YO (7.4 eV), ZrO (7.9 eV), and LaO (8.2 eV) in M and S star spectra. The dissociation energies of these elements are so high, in fact significantly higher than the corresponding ionization energies of the metals, that the molecules form immediately when the ions recombine. The quantity of free neutral atoms is therefore small; for cool gas the molecules totally dominate. At the transition temperature, around 3000 K at typical pressures, the situation may be dependent on the possible overionization of metal atoms due to hot nonlocal radiation, thus diminishing the amount of molecules that form (cf. [69]). TiO remains the dominating form of titanium for the very cool stars, making the partial pressure of TiO directly proportional to the gas pressure, the Ti abundance, and the abundance difference

$\epsilon_O - \epsilon_C$. In contrast, as the temperature decreases to values below 3000 K, YO and LaO rapidly form the dioxides YO_2 and LaO_2, so the partial pressures of the monoxides depend on the temperature and gravity in a nontrivial way. ZrO is an intermediate case.

Let us, as one more example, look at the nitrogen-rich molecules. Most of the nitrogen in both oxygen- and carbon-rich gas is in the form of N_2, due to its high dissociation energy (9.8 eV). This dipole inactive molecule is, however, not a strong opacity contributor. From (4.30) we find that the partial pressure of free nitrogen atoms is proportional to $p_{gas}^{1/2}$. Applying the same equation to the N–|O−C|–CN balance, we find that the partial pressure of CN is proportional to $p_{gas}^{3/2}$. This also makes the abundance of CN proportional to $\epsilon_N^{1/2} |\epsilon_O - \epsilon_C|$, which means that CN spectral lines will be more sensitive measures of the difference between the carbon and oxygen abundances than of the nitrogen abundance. The square-root dependence on the N abundance is also present for other nitrogen molecules such as NH and HCN. This makes it somewhat difficult to measure accurate nitrogen abundances in AGB stars.

An important aspect of the molecular equilibria in dynamic atmospheres, in particular in the outer layers with low densities, is the time scale for molecule formation. To what extent does the molecular equilibrium depend not only on the local conditions and radiation fields but also on the history of the gas? One possible situation, which may be of relevance for Mira stars, is the following: If the CO molecules are dissociated in a shock wave, the formation of CO may have a characteristic time scale of hours, or more, in the outer atmospheres, which is comparable to typical hydrodynamical time scales. [This has been studied in the solar case; cf. [11, 187].]

4.4.6 Dust Formation

In many cases, the outer layers of the atmospheres of AGB stars are so cool and dense that not only molecules but also dust grains may form. There are different types of observational evidence for the presence of grains in AGB stars, ranging from the general shape of the spectral energy distribution (e.g., the existence of a secondary maximum at longer wavelengths than the maximum of the photospheric emission) to the actual identification of emission features caused by dust; see Chapter 7. The effects of dust grains on the structure and dynamics of the stellar atmosphere and on the observable properties of the star are discussed in Sections 4.7.3 and 4.8 and Chapter 6. Here, we are concerned with the microphysics of dust formation and the properties of the dust grains. We give a simple kinetic description of grain formation, aimed at deriving typical time scales; a more detailed discussion of the processes, including chemical aspects, can be found in Chapter 5.

4.4.6.1 The Kinetic Picture

Let us consider how dust grains form from the gas. With decreasing temperature, the composition of the gas will shift to more complex molecules, so the relative abundance of polyatomic species relative to atoms and diatomic molecules will increase. At any given time there is a certain probability that a few of the bigger molecules will continue to grow by addition of more atoms or small molecules, thus forming clusters. At first, this is a purely random process with a certain statistical probability that depends on the thermodynamical conditions of the gas; but at some point, at a so-called critical cluster size, it becomes energetically favorable to add more atoms to the cluster. Any cluster larger than this critical size is a seed nucleus for a dust grain. It will have the tendency to grow as long as there is a sufficient supply of condensible material in the gas. When it has reached a certain size, i.e., the number of atoms at which the properties of the cluster start to be like those of bulk solid material, the cluster can be called a dust grain. The details of this first step of grain formation, called nucleation, are quite intricate, and a thorough description is beyond the scope of this section (see also Chapters 5 and 6). To simplify the following discussion we will assume that these condensation nuclei are present and that their formation rate is an "external input" that can be derived from chemical considerations.

To follow the further evolution of a dust grain we consider the grain and its surrounding gas as a closed system with a given constant temperature below the condensation limit. Statistically, atoms evaporate from the surface of the grain at a certain rate. On the other hand, atoms and molecules of the gas hit the surface and have a certain probability to stick to it. As long as the rate of addition of atoms (condensation) is larger than the rate of atoms evaporating from the surface, the grain will grow; otherwise, it will shrink and eventually evaporate completely.

The rate describing the addition of atoms from the gas phase to the surface of the grain is a rather complex quantity, combining grain properties and quantities that describe the conditions of the gas phase. For simplicity, we assume that the dust grain is not moving (not drifting) systematically relative to the gas, and that random thermal motions of the grain are negligible due to its large mass. Then the flux of gas particles of mass m hitting the surface of the grain is

$$\frac{1}{4}n\langle v \rangle = n\sqrt{\frac{kT_{\text{gas}}}{2\pi m}}, \qquad (4.31)$$

where n is the number density of these particles and $\langle v \rangle$ is their mean speed as defined by (4.18). [Note that the temperature T_{gas} of the gas may differ from that of the dust grain.] In addition to these kinetic quantities that determine the flux of atoms hitting the grain surface, there is a probability that a hit will actually cause the incoming atom to stick to the surface, i.e., that a chemical reaction will occur that binds it to the grain.

We assume that the grains consist of only one type of chemical element or molecule. This basic building block is called a monomer. We further assume that grains grow only by one type of reaction involving this monomer (homogeneous grain growth). The rate of addition per surface area can then be formulated as $\mathcal{R}_{\rm cond} = \alpha_1 \, n_1 \langle v \rangle / 4$, where n_1 denotes the number density of monomers in the gas phase, and α_1 is the sticking coefficient, or the probability for a reaction that binds the incoming monomer to the grain surface.

The evaporation rate per surface area, $\mathcal{R}_{\rm evap}$, mainly depends on the energy of the individual atoms in the grain, and therefore increases with the grain temperature. In the discussion below we are interested in grain growth, in which $\mathcal{R}_{\rm evap}/\mathcal{R}_{\rm cond} \ll 1$. Therefore, we will not discuss the evaporation rate in detail (cf. Chapters 5 and 6).

Subtracting the evaporation rate from the addition rate, we get the net growth rate of the grains,

$$\frac{dN}{dt} = A_N \left(\mathcal{R}_{\rm cond} - \mathcal{R}_{\rm evap} \right) = A_N \alpha_1 \, n_1 \frac{1}{4} \langle v \rangle \left(1 - \frac{\mathcal{R}_{\rm evap}}{\mathcal{R}_{\rm cond}} \right), \tag{4.32}$$

where $A_N = 4\pi a_{\rm gr}^2$ is the surface area of a grain with radius $a_{\rm gr}$ (assuming the grains to be spherical), and N is the number of monomers contained in the grain.

Considering the closed box of gas, this net growth rate will eventually go toward zero as the abundance of the condensible material in the gas phase decreases due to the condensation process and the value of n_1 drops. The growth rate becomes identical to zero when a dynamical equilibrium has been reached, where the same number of atoms are evaporating from the surface of the grain per second as are added. For the given temperature, total gas density, and elemental abundances, this equilibrium defines a fraction of material in the condensed state, or an equilibrium degree of condensation.

In general, the grains are often built up of more than one type of basic building block, and they grow by many different chemical reactions (heterogeneous grain growth). In that case, proper sums over all growth and evaporation processes must be taken in calculating the net growth rate, but the basic principle is the same, and a generalized definition of equilibrium condensation can be deduced from the condition of the net growth rate being zero (detailed balance for the different elements). Further modifications of this simple picture are necessary if the grains drift systematically relative to the gas, because then the distribution of the gas particles hitting the grain is not isotropic. For high drift velocities, the incoming atoms and molecules may even have enough kinetic energy to cause a partial destruction of the grain. An additional complication is that the temperature of the grains may differ from that of the surrounding gas; this happens if the interaction with the radiation field is the dominant energy exchange process and the grains reach their radiative equilibrium temperature, which depends on their optical properties.

4.4.6.2 Nonequilibrium Grain Growth

In discussing the equilibrium degree of condensation we have tacitly assumed that the grains have enough time to grow to this asymptotic limit, or in other words, that the external conditions like temperature and density change on much longer time scales than those required for grain growth. Simple estimates, however, demonstrate that this is probably not the case in the atmospheres of AGB stars.

Even a grain that is considerably smaller than the grains found in the interstellar medium, say a carbon grain with a radius of 10^{-6} cm, contains on the order of 10^8 atoms. This means that it takes the same order of magnitude of individual chemical reactions to form this grain. To get an idea of the typical time scales we make a simple kinetic estimate based on the growth rate given above: We take (4.32) as a starting point and define a growth rate

$$\mathcal{R}_{\text{growth}} = A_1 \alpha_1 n_1 \frac{1}{4} \langle v \rangle \left(1 - \frac{\mathcal{R}_{\text{evap}}}{\mathcal{R}_{\text{cond}}}\right) \tag{4.33}$$

normalized to a unit surface area $A_1 = 4\pi a_1^2$, where a_1 is the hypothetical radius of a monomer, i.e., the space it takes up in the solid material, defined as $a_1^3 = a_{\text{gr}}^3/N$. By definition, $\mathcal{R}_{\text{growth}}$ does not depend explicitly on the size of the grain but only on the number density of the monomers in the gas phase and on the gas temperature, as well as the material properties of the solid. Using the relation

$$A_N = 4\pi a_{\text{gr}}^2 = 4\pi (a_1 N^{1/3})^2, \tag{4.34}$$

which links the grain surface to the number of monomers contained in the grain, we can rewrite (4.32) in the following way:

$$\frac{dN}{dt} = \mathcal{R}_{\text{growth}} \frac{A_N}{A_1} = \mathcal{R}_{\text{growth}} N^{2/3}. \tag{4.35}$$

Introducing a normalized grain radius

$$\hat{a}_{\text{gr}} = \frac{a_{\text{gr}}}{a_1} = N^{1/3} \tag{4.36}$$

and substituting it into the equation above, we obtain

$$\frac{d\hat{a}_{\text{gr}}}{dt} = \frac{1}{3} \mathcal{R}_{\text{growth}}. \tag{4.37}$$

In a closed system, the right-hand side of this equation depends implicitly on the grain size: As the grain grows, n_1 changes because the condensible material is depleted in the gas phase. If we restrict our discussion to the phases of grain growth before the depletion becomes significant (low degree of condensation) and assume a constant temperature, then the growth rate $\mathcal{R}_{\text{growth}}$ is constant, and we can integrate the equation. Assuming furthermore that the initial grain size is negligible compared to the final size, we obtain

$$\hat{a}_{\rm gr} = \frac{1}{3}\int_0^{t_{\rm growth}} \mathcal{R}_{\rm growth} dt = \frac{1}{3}\mathcal{R}_{\rm growth}\, t_{\rm growth}. \tag{4.38}$$

Therefore, the growth time scale of the grains can be expressed as

$$t_{\rm growth} = 3\frac{a_{\rm gr}}{a_1}\mathcal{R}_{\rm growth}^{-1}. \tag{4.39}$$

To get an idea of the numerical value, let us consider typical conditions in the dust-formation zone. Grain formation occurs at temperatures of about 1000 K (cf. Chapter 5), at a distance from the photosphere where the total gas density is typically several orders of magnitude lower than in the photosphere, $10^{-14}\,{\rm g\,cm^{-3}}$ or less. This value for the gas density is taken from dynamical models, but it may be compared with the following order-of-magnitude estimate: Assuming a stationary, spherically symmetric outflow with a mass-loss rate $\dot{M}$ and velocity $v(r)$, the conservation of mass requires $\dot{M} = 4\pi r^2 \rho v$ at each radius r. Thus, for a given value of the mass-loss rate and a specified velocity profile $v(r)$ we can calculate the corresponding density

$$\rho(r) = \frac{\dot{M}}{4\pi r^2 v(r)}. \tag{4.40}$$

For simplicity, we assume a spatially constant flow velocity $v = 10\,{\rm km\,s^{-1}}$ and a typical mass-loss rate of $\dot{M} = 10^{-6}\,{\rm M_\odot\,yr^{-1}}$. Dust formation usually starts at around $2\,R_*$, which translates into $r = 4\times 10^{13}$ cm, assuming a stellar radius of about $300\,{\rm R_\odot}$. With these numbers we obtain a density of about $3\times 10^{-15}\,{\rm g\,cm^{-3}}$, which is a little lower than the value given above. However, since the flow velocity in the dust-formation zone is not constant but will increase outwards, our simple extrapolation from the wind region will tend to underestimate the real value; i.e., the density profile in the inner layers will be steeper than in the zone where the outflow has reached its final velocity.

We now specifically assume that the atmosphere is C-rich and that the grains consist of carbon. To estimate the number density of monomers in the gas we use an elemental carbon abundance $\tilde{\epsilon}_{\rm C}$ that is reduced by the amount of carbon locked up in CO, $\tilde{\epsilon}_{\rm C} = (\epsilon_{\rm C}/\epsilon_{\rm O} - 1)\epsilon_{\rm O}$. Assuming $\epsilon_{\rm O} = 10^{-3.18} = 6.607\times 10^{-4}$ and a C/O ratio of 1.4, we obtain a number density of $n_1 = 1.13\times 10^6\,{\rm cm^{-3}}$. The monomer radius can be calculated from the bulk density $\rho_{\rm cond}$ of the condensed material, and the mass of the monomer m_1 according to

$$a_1 = \left(\frac{3 m_1}{4\pi \rho_{\rm cond}}\right)^{1/3}. \tag{4.41}$$

Using $\rho_{\rm cond} = 2.25\,{\rm g\,cm^{-3}}$ and the mass $m_1 = 2.01\times 10^{-23}$ g of a carbon atom, we obtain a value of $a_1 = 1.29\times 10^{-8}$ cm. Typical grain sizes are 10^{-6} to 10^{-5} cm, and we adopt $a_{\rm gr}/a_1 = 5\times 10^2$. For simplicity we choose a sticking coefficient $\alpha_1 = 1$ and assume that the evaporation rate is negligible compared to the condensation rate (high supersaturation), $\mathcal{R}_{\rm evap}/\mathcal{R}_{\rm cond} \ll 1$. Inserting numbers into the expression for the growth time scale, we obtain

$t_\text{growth} = 2 \times 10^7$ s. This time scale should be compared to the typical time scales that are characteristic of the changes in the thermodynamical quantities controlling grain growth.

A very simple estimate for the dilution of the gas in the stellar wind is the following: Let us assume a stationary outflow at a constant velocity v. In this case, the density will behave as $\rho \propto 1/r^2$, where r denotes the distance from the center of the star. During the growth time, the material will travel over a distance $\Delta r = v\, t_\text{growth}$, and the density will change by a factor of

$$\frac{\rho}{\rho_0} = \frac{1}{\left(1 + \frac{\Delta r}{r_0}\right)^2}, \qquad (4.42)$$

with $r_0 = r(t=0)$ and $\rho_0 = \rho(t=0)$. Assuming a typical wind velocity of $v = 10\,\text{km}\,\text{s}^{-1}$, we obtain $\Delta r = 2 \times 10^{13}$ cm, which is on the order of magnitude of the stellar radius. Since the dust formation usually starts at about $2\,R_*$ the ratio $\Delta r/r_0$ is typically $1/2$, leading to a decrease of the density during t_growth by a factor of about 2.

In reality, however, the velocity in the dust-formation region will not be constant but increase outwards, resulting in a steeper density gradient. In any case, the density will decrease considerably during the estimated growth time, meaning that the growth rate will actually decrease significantly even before the depletion of the condensible material sets in; this again will increase the growth time of the grains, resulting in an even stronger decrease of the density during grain growth, etc. In many cases, this runaway feedback process will actually cause the grain growth to stop at some point due to the effects of expansion (i.e., decreasing gas density) and not due to depletion, via condensation, of the condensible material in the gas phase.

Another important time scale is the pulsation period of the star. The thermodynamic conditions in a given layer will change significantly during the pulsation cycle, both through the global changes in luminosity and effective temperature with a typical time scale of half a period between extremes, and through the effects of passing shock waves on a small fraction of the period. As typical pulsation periods are on the order of a year, the relevant time scales are on the order of 10^6 to 10^7 seconds, which is comparable to or smaller than t_growth. Therefore, grain growth will usually proceed far from equilibrium, and it is necessary to use a time-dependent description of grain growth to determine the degree of condensation and other relevant properties of the dust.

4.5 The Radiation Field

4.5.1 The Transfer of Radiation

Radiation is a key constituent of stellar atmospheres. We have already found that the microscopic state of the atmospheric gas is, to a great extent, determined by the radiation field. Radiation also provides the main transport mechanism for energy, and adds momentum to the gas and dust to such an extent that it may drive stellar mass loss. Moreover, for us as observers it provides all of the observational information we have on the stars. For these reasons, a detailed quantitative understanding of radiation in stellar atmospheres is necessary.

We shall consider the interaction between radiation and the atmospheric gas. The radiation flow may be calculated from the equation of radiative transfer (i.e., the Boltzmann equation for the radiation field)

$$\left(\frac{1}{c}\frac{\partial}{\partial t} + \frac{\partial}{\partial s}\right) I(\mathbf{r}, n, \nu, t) = \eta(\mathbf{r}, n, \nu, t) - \chi(\mathbf{r}, n, \nu, t)\, I(\mathbf{r}, n, \nu, t)\,, \quad (4.43)$$

where I is the specific intensity, η the emissivity, and χ the extinction coefficient (for a discussion and derivation, see Mihalas [131]). The first term on the left-hand side may be neglected for the AGB stars, since the characteristic velocities affecting the state of the gas are on the order of the speed of sound, which is five orders of magnitude below the speed of light c. One exception might be concentrated eruptions with sudden great releases of radiation, as in solar flares; detailed modeling of such phenomena might require consideration of the first term.

In the time-independent and spherically symmetric case, we have

$$\left(\mu\frac{\partial}{\partial r} + \frac{1-\mu^2}{r}\frac{\partial}{\partial \mu}\right) I(r,\mu,\nu) = \eta(r,\mu,\nu) - \chi(r,\nu)\, I(r,\mu,\nu)\,, \quad (4.44)$$

where $\mu = \cos\theta$, θ is the angle between the ray (direction) under consideration and the direction from the stellar center to the point where the equation is applied, and η and χ are complicated functions of the local characteristics of the gas. If η contains a scattering part and the scattering may be assumed to be isotropic (which is often reasonable), η includes the mean intensity J_ν. For example, for spectral lines formed in scattering, we have

$$\eta(r,\nu) = n_i\, \sigma(r) \int_0^\infty R(\nu',\nu) J(r,\nu')\, d\nu' + \cdots\,, \quad (4.45)$$

where n_i is the number density of scattering atoms, $\sigma(r)$ is a scattering coefficient, $R(\nu',\nu)$ is the *redistribution function* that describes the probability for radiation of frequency ν' to be redistributed to frequency ν, and "$+\cdots$" indicates further emission terms. Note that (4.45) assumes R to be independent of angle, which may be an oversimplification in certain cases. For spectral lines, it also depends on the microscopic line broadening processes (radiation damping, collisions, thermal broadening) as well as macroscopic motions in the gas through the Doppler effect.

Physically, both the emissivity and the extinction coefficient couple indirectly to the radiation field through the populations of different atomic and molecular levels n_i ($i = 1, \ldots$), which are dependent on the great number of radiative rates R_{ij}. Therefore, the system of rate equations, i.e., (4.29) in the time-independent case, must be solved simultaneously with the equation of radiative transfer.

4.5.2 Continuous Absorption and Scattering

What are the main contributing species to η and χ; that is, what particles affect the spectrum and the thermal conditions in the atmospheres?

The continuous absorption is dominated by H^- bound–free and free–free absorption throughout the spectrum (except in the ultraviolet and violet spectral ranges, where photoionization of heavier elements like C, Si, and Fe may be significant). For hotter stars like the Sun, the wavelength dependence of the H^- opacity is not very pronounced over the spectral region that contains the dominating stellar flux. For cooler stars, however, the minimum in the absorption coefficient around 1.6 μm, at the photodissociation threshold, is a significant feature. Its effects are accentuated by the fact that relative to hotter stars, the stellar flux maximum is shifted out into the infrared, so that the H^- absorption minimum is clearly visible in the flux spectrum as a peak. Another significant continuous opacity mechanism in AGB stars is free–free absorption by H_2^-, in which an electron passing a hydrogen molecule temporarily induces a dipole moment in the molecule that interacts with the radiation field. This H_2^- absorption fills in the H^- minimum, in particular in the upper cool layers. In the deepest atmospheric layers ($\tau_{\text{cont}} > 10$), where the temperature is high enough for H I to be excited to the $n = 2$ and $n = 3$ levels, H I absorption takes over from H^-. Also, the radiative flux in the UV and the blue is shifted to the Balmer and Paschen continua, thus increasing the significance of H I and the metal continua relative to H^- there.

In addition to continuous absorption, Rayleigh scattering by H atoms and H_2 molecules is significant. It is easy to estimate the scattering coefficient due to Rayleigh scattering (cf. equation 4–128 in [131]) at our characteristic temperature and pressure, and compare it with the H^- absorption coefficient (cf. Section 4.3.2). One finds that H^- absorption is almost of the same order of magnitude as H I scattering. While H^- absorption dominates at greater depths, the much lower electron pressure further out in the atmosphere proportionally diminishes the H^- abundance, so that H scattering begins to take over. Because it varies with λ^{-4}, scattering is more significant at shorter wavelengths and less so at longer wavelengths. Since the cross section for H_2 Rayleigh scattering is of the same order of magnitude (per particle) as for H I scattering, H_2 scattering dominates the continuous extinction in the very surface layers of cool AGB atmospheres.

In representing the scattering contribution to η in (4.44) and (4.45) we note that the redistribution in wavelength is insignificant with respect to the continuous spectrum. Assuming isotropic scattering, we may write

$$\eta(r,\nu) = [n_{\mathrm{HI}}\,\sigma_{\mathrm{HI}}(r,\nu) + n_{\mathrm{H_2}}\,\sigma_{\mathrm{H_2}}(r,\nu)]\,J(r,\nu) + \cdots \;. \qquad (4.46)$$

Scattering terms must also be added to $\chi(r,\nu)$, along with other sources of extinction such as continuous absorption due to H^- and $\mathrm{H\,I}$ bound–free and free–free absorption, continuous absorption by metals (in particular in the ultraviolet and violet–blue), or spectral line absorption from atoms or molecules, which will now be discussed.

4.5.3 Line Absorption

The significance of line absorption for AGB star atmospheres is obvious from their spectra; hardly any other types of stars show such rich spectra, in which the continuum may be very difficult to trace. Tens of thousands of spectral features can be identified in high-resolution spectra, and a great fraction of all features visible are blends of several spectral lines each. In detailed synthetic spectrum calculations, one often has more than 100 spectral lines contributing significantly to the absorption at a single wavelength point. It is easy to understand that compiling vast tables of spectral line data is of key significance for the understanding and modeling of AGB spectra and atmospheres.

We shall now comment on the main sources of line opacity and on the available data, as illustrated in the figures.

Atomic lines, in particular from Fe I, are of major significance for gases hotter than 4500 K, and also significant at cooler temperatures for the wavelength interval below 450 nm. Present compilations such as the VALD database [111, 143] contain experimental data from various sources, and calculated data for millions of transitions from Kurucz [112, 113, 114, 115]. The calculated data often have errors in gf values greater than a factor of 2, but hopefully these errors cancel rather well as regards the effects on the blocking. There are indications that there are still missing lines in these compilations that would add a "veil" of weak absorption to the stars in the violet and ultraviolet. Such "unknown opacity" has been traced for the Sun (cf. [15, 61, 86]) and for red giants [60]. However, since AGB stars have so little flux in these short-wavelength regions, the uncertainties in metal-line absorption are probably not very significant for the model structures.

The effects of molecular lines on the atmospheres of AGB stars and cool stars in general are much more significant than those of atomic lines. Several factors determine the significance of a molecular opacity source in a stellar atmosphere: (1) the abundance of the molecule, as discussed in Section 4.4.5; (2) the wavelength-integrated absorption coefficient of the molecule at the relevant temperature, reflecting the number of lines in the absorption system and its transition probabilities; and (3) the wavelength of the bulk of the absorption, relative to the wavelength maximum of the emitted flux in the layers of the photosphere where the molecule can form.

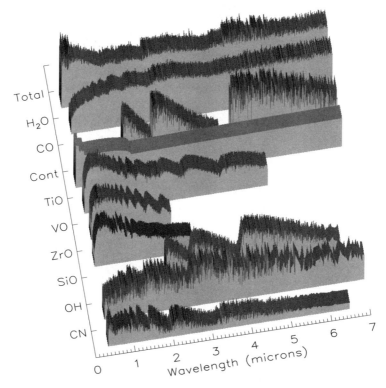

Fig. 4.6. Contributors to the opacity (mass absorption coefficient, logarithmic scale) in the upper layers of a model with $T_{\rm eff} = 3000$ K, $1\,{\rm M}_\odot$, $\log g = 0.0$, and solar composition. The continuous opacity is mainly due to H^- (bound–free and free–free), except at the shortest wavelengths, where bound–free metal absorption is dominant

It must, however, be noted that the significance of a particular molecule or molecular band is not merely the result of the strength of the transitions or the value of the absorption coefficient. A quite strong saturated band may have little significance for the atmosphere, except in the uppermost layers that are optically thin even to the radiation in the band. On the other hand, a band with spectral lines on the linear part of the curve of growth may have great effects, both on the emergent spectrum and on the atmospheric structure. These effects are discussed in Section 4.6.2.

Figures 4.6 and 4.7 show the contributions of different molecules to the absorption coefficient in one oxygen-rich and one carbon-rich case. The most important molecular opacity sources are CO, CN, C_2, TiO, and H_2O, as well as HCN, C_2H_2, and C_3. It is interesting to note that although CO is of significance for the uppermost layers in G and K stars (see Section 4.6.2), it is of minor significance for M stars compared to TiO, in spite of its

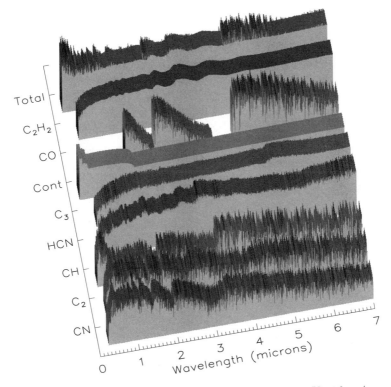

Fig. 4.7. Same as Figure 4.6, but for a model with $T_{\rm eff} = 2800$ K and $\epsilon_C/\epsilon_O = 1.1$. Note that the main opacity contributors are different

much greater abundance. This is because the important CO bands are the vibration–rotation (VR) bands in the infrared, and though quite strong, they are concentrated in a few limited wavelength regions. The CO electronic transitions are located in the ultraviolet and are masked by other strong opacity sources in a wavelength region of little significance for the energy transfer in AGB stars. On the other hand, the numerous electronic bands of TiO are in the visual and near infrared, and their integrated absorption coefficient per molecule is much higher than that of the CO VR bands. The TiO bands therefore block out a considerable fraction of the total flux. For lower temperatures, H_2O is even more important than TiO for the temperature structure, in particular for higher gravities.

CN, C_2, HCN, C_2H_2, and C_3 are important opacity sources primarily in carbon stars. The most significant CN feature is the extended but rather sparse CN red electronic system in the visual and near infrared. C_2 is represented by several electronic systems in the visual and near infrared. For the diatomic molecules, the data needed for stellar atmosphere modeling (line lists with wavelengths, excitation energies, transition probabilities, and possibly identifications) are relatively complete and of reasonable quality in many cases. The line lists are often based on experimental determinations of wavelengths and transition probabilities or lifetimes, though theoretical or semiempirical methods are used to extend the available laboratory data. For the polyatomic molecules, with their hundreds of millions of contributing VR lines in the infrared, the data are still far from definite, but significant improvements have taken place, in particular due to extensive ab initio calculations (see [63, 70, 99] and references therein; also [101, 139, 155]).

For the S stars, other molecules such as ZrO, VO, LaO, YO, and FeH become important opacity sources. The present data are still unsatisfactory for several of these.

Finally, it should be noted that there are many more molecular bands that are of great value as diagnostics of AGB stars, even if they hardly affect the total radiative energy transfer of the stars.

4.5.4 Dust Opacity

Regarding opacities, dust formation influences the stellar atmosphere in two ways. From the viewpoint of the gas phase chemistry, dust formation results in a depletion of certain elements in the gas, which influences the molecular composition of the gas and consequently the corresponding opacity. On the other hand, the dust grains themselves have a rather high mass absorption coefficient, which often more than compensates for the reduced gas opacity.

In general, the total opacity of an ensemble of spherical dust grains can be formulated as

$$\chi(\lambda) = \int_0^\infty a_{gr}^2 \pi \, Q_{ext}(a_{gr}, \lambda) \, n(a_{gr}) \, da_{gr} , \qquad (4.47)$$

where $n(a_{gr}) \, da_{gr}$ is the number density of grains in the grain radius interval between a_{gr} and $a_{gr} + da_{gr}$, and Q_{ext} is the extinction efficiency, or ratio of the extinction cross section to the geometrical cross section of the grain. In general, the size distribution function of the dust grains must be known. However, if the size of the particles is small compared to the wavelength of the radiation, the quantity $Q'_{ext} = Q_{ext}/a_{gr}$ becomes independent of the grain radius a_{gr}, and the dependence of the opacity on wavelength and grain size can be separated into two independent factors. The dust opacity for particles that are small compared to the wavelength can therefore be rewritten as

$$\chi(\lambda) = Q'_{ext}(\lambda) \, \pi \int_0^\infty a_{gr}^3 \, n(a_{gr}) \, da_{gr} , \qquad (4.48)$$

where the integral is proportional to the degree of condensation, i.e., the total amount of material condensed into grains per unit volume of the atmosphere, which is a direct result of the description of grain formation.

The photospheric spectral energy distribution of AGB stars has its maximum around wavelengths of 1 μm, with a sharp decline of the stellar flux toward shorter wavelengths. Both observations and theoretical arguments indicate that the typical grain sizes are much smaller than 1 μm. Thus, the simplifying assumption that the particles are small compared to the wavelength is valid for a major fraction of the spectrum, and the opacities necessary for calculating the structure of the atmosphere and stellar wind can be computed using (4.48) if the degree of condensation is known. The quantity Q'_{ext} is defined by the optical properties of the grain material and can, for example, be calculated from the complex refractive index. The complex refractive index of the material as a function of wavelength can be determined by laboratory measurements, e.g., on bulk samples. Often, however, if grains are used instead of bulk samples, the measured quantity $Q'_{ext} = Q_{ext}/a_{gr}$ is directly specified.

4.6 The Modelling of AGB Star Atmospheres

indexatmosphere, static model We shall now change perspective from the microscopic properties of gas and radiation in AGB stellar atmospheres to the macroscopic properties and overall structure of the atmospheres. We shall begin by writing some very general equations and later reduce them to the static case, where the atmosphere is assumed to be structured in homogeneous layers or shells, and mass motions are neglected in several respects. After presenting results and models of atmospheres calculated under those assumptions, we shall turn to more dynamical (and realistic) situations.

4.6.1 Basic Equations

In general, one may write a conservation equation for quantities like mass, momentum, and energy as follows:

$$\frac{\partial}{\partial t}(density\ of\ quantity) + \mathrm{div}\,(flux\ of\ quantity) = sources - sinks\ . \quad (4.49)$$

The first term describes the time variation of a quantity at a point, the second one the transfer of the quantity across the point due to mass motions (or a moving coordinate system). On the right-hand side there are terms measuring the creation or destruction of the quantity per unit time. For mass under conditions found in stellar atmospheres, the terms on the right-hand side are zero. For momentum, those terms correspond to forces, and for energy they correspond to energy absorbed or emitted, primarily due to radiation.

From mass conservation, we have

$$\frac{\partial \rho}{\partial t} + \nabla \cdot (\rho \mathbf{u}) = 0 \quad (4.50)$$

for the mass density ρ and velocity $\mathbf{u}$.

The momentum equation, also known as the equation of motion or the Euler equation, is

$$\rho \frac{\partial \mathbf{u}}{\partial t} + \rho(\mathbf{u} \cdot \nabla)\mathbf{u} = -\nabla p_{\text{gas}} + \rho \mathbf{g} + \frac{1}{c} \int_0^\infty \chi_\nu \mathbf{F}_\nu \, d\nu, \qquad (4.51)$$

where p_{gas} is the gas pressure and $\mathbf{g}$ is the gravitational acceleration, and the last term on the right is the force exerted by the radiation ("radiation pressure").[9] The net radiative energy flux is calculated by integrating over all directions $\mathbf{n}$:

$$\mathbf{F}_\nu = \oint_\Omega I_\nu(\mathbf{n}) \mathbf{n} \, d\omega, \qquad (4.52)$$

where $I_\nu(\mathbf{n})$ is the specific intensity. Because of the low densities in AGB atmospheres, viscous forces were neglected in (4.51).

The energy equation is

$$\frac{\partial}{\partial t}\left(\frac{1}{2}\rho u^2 + \rho e\right) + \nabla \cdot \left[\left(\frac{1}{2}\rho u^2 + \rho e\right)\mathbf{u}\right] = \qquad (4.53)$$

$$-\nabla \cdot (\mathbf{u} p_{\text{gas}}) + \rho \mathbf{g} \cdot \mathbf{u} + \mathbf{u} \cdot \frac{1}{c} \int_0^\infty \chi_\nu \mathbf{F}_\nu d\nu + 4\pi \int_0^\infty \chi_\nu (J_\nu - S_\nu) d\nu,$$

where e is the internal energy of the gas per gram, $S_\nu = \eta_\nu/\chi_\nu$ is the source function, and the energy losses due to viscosity are neglected. The four terms on the right-hand side represent the work done against pressure, the work done by gravity, the work done by radiation forces, and the difference between radiative heating by absorption and cooling by emission.

These equations, with their relevant initial and boundary conditions, must be solved together with the equation of radiative transfer, (4.43), to obtain I_ν, J_ν, and $\mathbf{F}_\nu$. In order to calculate χ_ν, η_ν, and S_ν one must also simultaneously solve the system of rate equations, (4.27). Altogether, this is a formidable numerical problem that has not yet been solved for any realistic case.

In this situation, approximations must be made. These go in different directions for different types of stars and different applications. One possibility is to set all time derivatives and velocities $\mathbf{u}$ in the equations equal to zero, yielding the static, or radiative, case. Other possibilities are to assume spherical symmetry, or to keep the time dependence and velocities but simplify the treatment of radiation, e.g., by assuming LTE and calculating only crudely the integrals representing radiative force in (4.51) and radiative heating and cooling in (4.53). Even with these simplifications, this is a very difficult task due to the nonlinear terms in the hydrodynamic equations. Further simplifications must be made, and will be discussed below in Section 4.7. Before that, however, we shall discuss the static case in some detail.

---

[9] Note that the left-hand side of (4.51) has been modified relative to the generic form in (4.49) by using (4.50).

4.6.2 Static Models

4.6.2.1 Further Approximations

In the static case it is very natural to assume that the matter in the atmosphere is stratified in homogeneous, spherically symmetric layers, since the gravitational acceleration **g** is directed toward the stellar center (for a nonrotating star). In this case, the Euler equation (4.51) reduces to the hydrostatic equilibrium equation (4.10), except that it also includes the integral representing the radiation force:

$$\frac{\partial p_{\text{gas}}}{\partial r} = -G\rho \frac{M}{r^2} + \frac{1}{c}\int_0^\infty \chi_\nu F_\nu \, d\nu \ . \qquad (4.54)$$

In the energy equation only the last integral is nonzero, and it is thus reduced to the condition of *radiative equilibrium* :

$$\int_0^\infty \chi_\nu (J_\nu - S_\nu) \, d\nu = 0. \qquad (4.55)$$

In spite of these very considerable simplifications, even the static approximation is quite difficult to apply to AGB stars because of the enormous number of radiative transitions between all of the atomic and molecular states that affect the energy transfer. The population numbers should be calculated from the statistical equilibrium equations (4.29), which must be solved together with the transfer equation (4.44). Numerically, solutions to complex problems of this character are within reach today; see, e.g., the methods developed by Anderson [5], Hubeny and Lanz [87], and [70] and references therein), where the wealth of transitions is treated statistically. It still remains, however, to apply them to AGB stars. Not only are there numerical problems in doing so, but an additional difficulty is that many of the cross sections for radiative and especially for collisional transitions are still quite uncertain. Still, in view of the possible significance of departures from LTE (e.g., for lines from TiO and other oxides in M stars, and for electronic transitions of the carbon-rich molecules in carbon stars), it is due time to proceed further toward non-LTE modeling of AGB star atmospheres and spectra.

Lacking results of such studies, we shall turn our attention to yet much more simplified models for which the level populations are calculated using the Boltzmann–Saha equation. From these populations we calculate $\chi_\nu = \kappa_\nu + \sigma_\nu$, where κ_ν is the true absorption coefficient (including line absorption) and σ_ν is the continuous scattering coefficient. The source function S_ν is assumed to be composed of one thermal contribution, adopted from Kirchhoff's law, and one scattering contribution, cf. (4.46):

$$S_\nu = \frac{\kappa_\nu}{\kappa_\nu + \sigma_\nu} B_\nu + \frac{\sigma_\nu}{\kappa_\nu + \sigma_\nu} J_\nu \ . \qquad (4.56)$$

These are the approximations of local thermodynamical equilibrium.

Inserting (4.56) into (4.55), one obtains

$$\int_0^\infty \kappa_\nu B_\nu(T)\,\mathrm{d}\nu = \int_0^\infty \kappa_\nu J_\nu\,\mathrm{d}\nu\,, \qquad (4.57)$$

which is sometimes called the *Strömgren equation*.

Extensive grids of models of AGB star atmospheres have been calculated with these approximations and applied, e.g., to abundance analyses. The computer codes used to calculate those grids often allow a stationary convective flux, which is estimated in the so-called mixing-length approximation. This requires an extra term to be added to (4.57), representing the convective flux derivative.[10] The effects of this convective flux are generally very small for AGB star atmospheres, reflecting the fact that the convective flux in this approximation is proportional to the very low gas density, and primarily becomes important in the hydrogen ionization zone at very great atmospheric depths. In reality, the mixing-length approximation not only contains unphysical free parameters, but is also physically incorrect in describing convection as a phenomenon resembling molecular diffusion. For more realistic treatments of convection, see Section 4.7.2.

4.6.2.2 Heating and Cooling

Before going into results of detailed numerical solutions to the system of equations mentioned above, we shall study absorption at different wavelengths more schematically. Our attention will focus mainly on the effects of line absorption on the atmospheric structure (usually called "blanketing"). Since the atmospheres of AGB stars may be geometrically extended, e.g., due to increased opacity, radiative forces, or turbulence and shocks, we shall also illustrate how the extension of the atmospheres affects their temperature structures.

Spectral lines obviously block radiation that would otherwise be transmitted outwards. However, a certain amount of energy is released in the stellar center by nuclear reactions, and at least in the time-independent case it must be radiated away from the atmosphere. This occurs between the blocking lines in the spectrum, which means that the corresponding flux-forming layers radiate more intensively. Generally, this is accomplished by means of an increase in the temperature of these layers. We can simply estimate this temperature increase by saying that if the atmosphere is not blocked by spectral lines, then the flux radiated by the atmosphere is approximately $\pi B_\nu(T)$, where T is the characteristic temperature of the flux-forming layers. The total emitted flux is then $\sigma_R T^4$. With the same degree of approximation, if a fraction β_b of the flux is blocked by spectral lines, then a temperature increase of ΔT is required to produce the same total flux as in the unblocked

---

[10] In practice, $\int F_\nu \mathrm{d}\nu + F_{\mathrm{conv}} = \sigma_R T_{\mathrm{eff}}{}^4$ is often used instead, in particular in the deeper layers of the models.

case, i.e.,
$$\sigma_R(T+\Delta T)^4(1-\beta_\mathrm{b}) \approx \sigma_R T^4 \, . \tag{4.58}$$

Thus we obtain
$$\frac{\Delta T}{T} \approx \frac{1}{4}\frac{\beta_\mathrm{b}}{(1-\beta_\mathrm{b})} \, . \tag{4.59}$$

It is clear from this equation that if 40% of the flux is blocked by spectral lines, then we can expect a temperature increase of about 500 K in the continuum-forming layers for a star with a characteristic temperature of about 3000 K. Although it is quite approximate, it should be noted that this estimate is relevant independently of the detailed formation mechanism of the spectral lines. Thus, even if the lines form far from LTE, for instance if they are formed totally in scattering processes, the layers forming the continuous flux will increase in temperature as long as they are dominated by processes in LTE or are at least closely coupled to the local temperature.

What are the direct local effects of the interaction between the gas and the radiation in the spectral lines? Let us assume first that line absorption occurs suddenly at rather great heights in the atmosphere, e.g., originating from a molecule like H_2O or C_2H_2 that dissociates further inwards because of the higher temperatures there. This molecule will pick up the radiation field from below, which is hot in the sense that it was generated at greater depths and has been relatively unaffected by absorption as it passes through the layers where it formed to the upper regions where it is absorbed.

Assuming radiative equilibrium and LTE, we may use (4.57) to explore what happens if we add a line opacity l_ν to the continuous opacity $(\kappa_\nu)_\mathrm{cont}$. Let us consider the upper atmosphere. Obviously, the mean intensity has a major contribution from layers below, since the temperature there is higher in most cases and the downward-directed radiation weaker. At the surface, this incoming radiation is negligible, and for an atmosphere where line absorption is neglected we may write

$$J_\nu(\tau) \approx \frac{1}{2} B_\nu(T_\mathrm{eff}) \cdot \left(\frac{R_\mathrm{i}}{R_\mathrm{o}}\right)^2 \tag{4.60}$$

by taking the effective temperature as a characteristic temperature of the continuum-forming regions. The last factor on the right-hand side accounts for the fact that the atmospheres of AGB stars are geometrically extended. This means that the radiation field at radius R_o in the outer atmosphere is diluted by a factor of $(R_\mathrm{i}/R_\mathrm{o})^2$, where R_i denotes the stellar radius at which the continuum is formed. Inserting this into (4.57), we get

$$\int_0^\infty (\kappa_\nu)_\mathrm{cont} \left[B_\nu(T) - \frac{1}{2} B_\nu(T_\mathrm{eff}) \cdot \left(\frac{R_\mathrm{i}}{R_\mathrm{o}}\right)^2 \right] d\nu = 0 \, . \tag{4.61}$$

We now add the opacity l_ν from spectral lines in the upper part of the atmosphere, and compare the new temperature $T'(r)$ to the $T(r)$ we obtained with only continuous opacity. Assuming that the continuous opacity is practically grey (i.e., wavelength-independent: $(\kappa_\nu)_{\text{cont}} = \kappa_{\text{cont}}$) and that the extra opacity in the layer we are exploring is so weak that the mean intensity is not affected, we find that

$$\Delta T = \frac{1}{4\sigma_R T^3} \int_0^\infty \frac{l_\nu}{\kappa_{\text{cont}}} \left[\frac{1}{2} B_\nu(T_{\text{eff}}) \cdot \left(\frac{R_i}{R_o}\right)^2 - B_\nu(T') \right] d\nu. \quad (4.62)$$

In the infrared part of the spectrum, on the long-wavelength side of the maximum of the Planck function, the temperature sensitivity of B_ν is rather small, so the integrand is negative and the extra opacity in the IR cools the upper layers. In the blue part, on the short-wavelength side of the local Planck maximum, the integrand may be positive. Physically, this means that in spite of the fact that the radiation field is diluted and contains only outward-directed rays, the radiation from below is so much more intense than the local cooling emission that it heats the surface layers.

By putting in some characteristic numbers, it is easy to verify that the effects on the thermal structure by cooling IR line opacity, or by heating opacity in the visual, may be very considerable, typically several hundred degrees. In the infrared, the cooling effects of CO VR bands on G- and K-type stars, and of H_2O VR bands on M stars, is of this order of magnitude. For M stars, the TiO bands in the red and near-IR are far enough to the blue of the Planck maximum to cause efficient heating, at least if they form in true absorption [98, 110]. As has already been pointed out, the opacity in (4.57) is only the true absorption; the effects of scattering canceled in the derivation of (4.57) when the transfer equation was integrated across the spectrum. Therefore, if, for example, the TiO electronic transitions are affected by non-LTE in the sense that they form mainly in scattering processes (as suggested by [75]), the effects of these transitions will be much less than one would estimate from (4.57).

Now, if the line absorption l_ν is so large that the mean intensity in (4.60) is a severe overestimate, then the integrand in (4.62) will be less, i.e., more negative. In this case, too, opacities in the blue may be cooling. A well-known example is the effect of metal lines on the atmospheres of solar-type stars and K giants.

In general terms, most strong molecular bands in AGB stars originate rather high up in the photospheres, due to the relatively small dissociation energies of molecules like TiO, H_2O, HCN, and C_2H_2. This makes the estimate in (4.62) relatively realistic. Exceptions are the CO, CN, and C_2 bands, which are formed throughout much of the atmosphere.

We may also use (4.61) to estimate the effects on the temperature structure from the dilution of the radiation field due to the extension of the atmosphere as such. If we assume κ_ν to be grey, such that it can be brought outside the integral, we may easily integrate and get

$$T \approx 0.84\, T_{\text{eff}} \sqrt{\frac{R_{\text{i}}}{R_{\text{o}}}}\,. \tag{4.63}$$

Obviously, an expansion of the atmosphere by 50% ($R_{\text{i}}/R_{\text{o}} \approx 1/1.5$, which is not uncommon for AGB stars) leads to a cooling of a grey atmosphere by typically 500 K for a star with $T_{\text{eff}} = 3000$ K. One may also see from (4.61) that the amount of cooling due to the extension depends on where in the stellar spectrum the dominating opacity sources occur. If new molecules form due to extension cooling, then the interplay between extension, temperature, and new opacity sources may be complex. One may even find several solutions to the static model-atmosphere problem, with quite different surface temperatures.

What would cause such an extra expansion? One reason may be the very formation of polyatomic molecules, and their strong and widespread absorption across the spectrum. If, for example, carbon-rich atmospheres cool below $T_{\text{eff}} = 3000$ K, then HCN and C_2H_2 molecules form and increase the overall opacity; this in turn decreases the gas pressures. One can see this from (4.2) and (4.10), which, together with the boundary condition $p \to 0$ when $\tau \to 0$, lead to

$$\frac{dp}{d\tau} = \frac{\rho g}{\chi}\,. \tag{4.64}$$

This shows that an increase in χ will inevitably lead to lower pressures at corresponding τ values. Like the continuous H^- opacity, the abundances of polyatomic molecules are also pressure-sensitive. The estimate of the extension of the atmosphere in (4.8) was based on the H^- opacity, but a similar pressure-dependence of the extension is also relevant if the polyatomic molecules provide the dominant opacity. Thus, [48] found that the extension of a carbon-rich model atmosphere with $T_{\text{eff}} = 2800$ K increased by more than a factor of 5 when absorption from polyatomic molecules (in particular HCN) was added, and [144] found similar but smaller effects for M giants as a result of H_2O formation.

Radiative forces might also affect the structure and expand the atmosphere very considerably through their action on molecules, i.e., due to the increase of χ_ν in the integral in (4.54). This is in particular the case if the molecule forms rather abruptly and rather far out in the atmosphere. Then, the flux F_ν may still be strong and not much reduced by absorption further in, so that it contributes weight to the integral. Such is the case, e.g., for TiO in the earlier M stars, as well as for the polyatomic molecules in cooler stars. The drastic effects of TiO are also seen in Figure 4.8. The photon momentum absorbed by both molecules and dust might even contribute so strongly to the last term in the equation that the star is brought out of hydrostatic equilibrium (see, e.g., [103, 144]).[11] The expansion leads to cooling

---

[11] If this happens, the radiation field from the star below will be attenuated in proportion to $(R_i/r)^2$ for a given element in the gas phase; but the gravity will

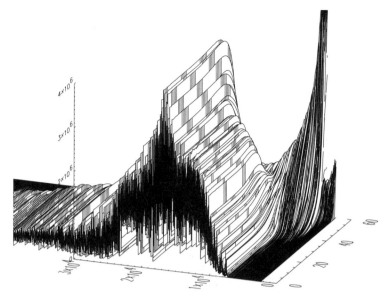

Fig. 4.8. Radiative flux as a function of wavelength (in Å, right to left) and depth point "inwards" from the model surface (0 to 60), for a model with $T_{\rm eff} = 3000$ K, $1\,{\rm M}_\odot$, $\log g = 0.0$, and solar composition. Note the heavy blocking (by TiO) in the upper atmosphere, which dramatically shifts the flux from the visual and near infrared to longer wavelengths. From [62]

and more formation of molecules and dust, and this may increase the value of the integral further.

Other causes for the expansion of atmospheres may be turbulence and pulsations. These phenomena and their consequences will be discussed later.

4.6.3 Grids of Static Model Atmospheres

Even the restricted task of calculating static model atmospheres in LTE for AGB stars is a cumbersome undertaking if one aims at a fair representation of the radiative energy transfer and an accurate calculated or synthetic spectrum. In fact, this application of models to spectrum calculations, including calculation of lower-resolution spectral energy distributions and colors, is a driver of that research. The aim is then to compare the calculated spectra to observations in order to determine stellar fundamental parameters such as abundances, radii, masses, or luminosities, for example to study nucleosynthesis and stellar or galactic evolution. As we shall see, the calculated spectra may fit the observations quite well. This does not mean, however, that the

---

also decrease correspondingly, so that the force terms directed outwards will together stay positive.

static LTE models are accurate representations of the stars, or even that the fundamental parameters derived from comparing models to observations are correct within the formal errors of the observations.

Within the particular framework set by the questionable assumptions of static LTE conditions, there are basically two major problems left for these calculations. The first is to acquire atomic and molecular data of satisfactory quality and completeness, in particular for calculating absorption coefficients. The progress in this respect has been considerable during the two last decades of last century, although further work is needed. The second problem is to devise algorithms that represent this absorption in a reasonable way, without requiring the time-consuming calculation of a fully resolved spectrum in evaluating the integrals in (4.54) and (4.55). It is easy to estimate how many points are needed to resolve an AGB spectrum. The typical width of a spectral line may be estimated from the Doppler shift of radiation from an individual atom or molecule. A characteristic thermal velocity is given by (4.18). Take, for example, HCN, which contributes very many lines to carbon-star spectra. At 3000 K we obtain $\langle v \rangle_{\rm HCN} = 1.5$ km s^{-1}, which corresponds to $\delta\nu/\nu = 5 \times 10^{-6}$. Even if the spectral lines are broadened additionally by more collective gas motions ("micro- and macroturbulence"), which may well contribute velocities of about 5 km s^{-1}, we see that the number of frequency points needed to resolve the full spectrum is several hundred thousand. The computing time for solving the radiative transfer part of the problem, which dominates the total computing work, will generally scale in proportion to the number of points used to represent the full spectrum.

In practice, in particular when many models are needed, it is still both possible and necessary to diminish the number of points considerably below that required for a fully resolved spectrum. This may be done using statistical methods. One is to divide the spectrum into spectral intervals and transform the opacity within each interval into opacity distribution functions (ODFs). Then the flux, the mean intensity, and the opacity within each interval may be represented by a small number of quadrature points in calculating the integrals of (4.54) and (4.55) (see [61, 116]). Altogether, typically fewer than 500 frequency points may be sufficient. A major drawback with this method is that it assumes a strong correlation between the absorption at different atmospheric depths, in the sense that strong absorption at one depth should correspond to strong absorption at another depth ("the ODF approximation"). For F, G, and K stars this approximation has been found to be valid (cf. [61]), but for AGB stars it is certainly dangerous, since these stars are characterized by different wavelength dependences of line opacities at different atmospheric depths; absorption from TiO, for example, falls across transparent and less-transparent wavelength regions in deeper layers, where metal lines dominate the line opacity. Similarly, in upper cool atmospheric layers, the polyatomic molecules add absorption in spectral regions different from those of the diatomic absorption that dominates in the deeper layers of cooler stars. Severe errors resulted from using the standard ODF method for carbon-star model atmospheres as shown by [46].

Instead, another method is to be preferred, the so-called opacity sampling (OS) method [46, 96, 172]. Here, the quadrature points are merely distributed along the spectrum, more or less randomly, but not in such numbers that the spectrum is resolved. Typically, a few thousand points are needed to get a model accurate to within about 50 K in temperature at each depth. One problem, however, is that the error in the model decreases only slowly with the number n_ν of frequency points, as for typical Monte Carlo methods with $n_\nu^{-1/2}$. Therefore, it is relatively expensive to improve the accuracy considerably. Moreover, rather few spectral lines with particularly high opacity may be decisive for the structure, and these lines must be represented, but not overrepresented, in the sampling. To catch such effects, hybrid methods combining the three discussed here—detailed spectrum calculations, the ODF method, and the OS method—have been suggested but not used extensively. Today, most of the models and grids for AGB stars are constructed by using a straight OS method. This method also has the advantage of being relatively easy to generalize to statistical equilibrium ("non-LTE cases") and even dynamical situations.

Pioneering grids of static LTE models for AGB stars were calculated already in the 1970s by Tsuji [178] and Johnson et al. [94] for M stars, and by Querci et al. [146, 147] and Johnson [92] for C stars. The representation of the molecular opacities in these models was rather primitive. Later and gradually more refined grids were those in [23], for M stars in [144], and [103, 117] for C stars. Presently, at least two independent groups are announcing extensive grids of spherically symmetric, static OS models: Hauschildt, Allard, et al.[70] and a group with Edvardsson, Eriksson, Gustafsson, Jørgensen, and Plez using the MARCS code. Some models used as illustrations for the present chapter are taken from the latter group (see Figures 4.9–4.13).

The studies mentioned in the previous paragraph clearly demonstrate the very strong effects of molecular line opacities on the temperature structures of the models (see Figures 4.11 and 4.12). This makes the models very sensitive to the chemical abundances. In particular, the CNO abundances are crucial for carbon stars; but also, e.g., the Ti abundance for M stars and even some heavy metal abundances for S stars are of great importance. This makes it important to base quantitative analyses of spectra of these stars on model atmospheres with adequate chemical compositions. Therefore, the chemical analysis of these stars must be an iterative procedure between spectrum calculation, comparison with observed spectra, calculation of new models and spectra, and so on.

4.6.4 Spectrum and Flux Comparisons Based on Static Models

The calculated spectral energy distributions (SEDs) from static model atmospheres have been compared to observations in a number of studies during

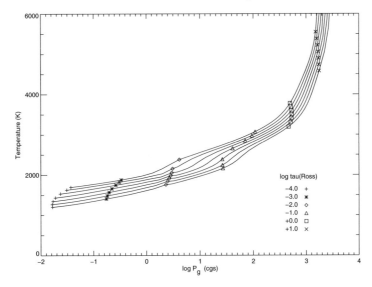

Fig. 4.9. A sequence of model structures for carbon stars ($\log g = 0.0$, $1\,M_\odot$, $\epsilon_C/\epsilon_O = 1.1$) with different T_{eff} values (2600 K to 3200 K in steps of 100 K). Based on MARCS models. Symbols indicate Rosseland optical depths from $\tau = 10^{-4}$ to $\tau = 10$ in steps of $\Delta \log \tau = 1$

the last decade. For the most recent models, and for stars that are not very cool and not varying at large amplitudes, these comparisons show very good, though not perfect, agreement.

Models discussed in [144] reproduce the SEDs of α Tau (K5 III) and μ Gem (M3 III) well, but have problems with the near-IR spectral region of g Her (M6 III). More recently, much better agreement has been obtained with new TiO line lists ([4], [125]). The near-IR narrow-band colors of M giants with $T \leq 3100$ K are now reproduced quite well with static models (see Figure 4.13 [4]). For Mira stars, there were clear discrepancies.

Broad-band colors, bolometric corrections, and temperature calibrations have also been calculated by [18] using models by [144] and more recent models by Plez for M giants. They found a generally excellent agreement for stars hotter than 3000 K. Exceptions were the $B - V$ colors, which tend to be systematically bluer than the models by about 0.2 magnitudes. Missing model opacities in the blue-visual spectral region are a probable explanation.

Static models can also be applied successfully to IR broad-band colors and their value as luminosity and abundance criteria for long-period variables (LPVs) and other luminous red stars [3], even though, for the LPVs, a number of significant discrepancies between models and observations were found.

Rather good agreement was found between calculated and observed SEDs in the IR for carbon stars [117], although the fits were very sensitive to the adopted effective temperature and the atmospheric extension, not the least

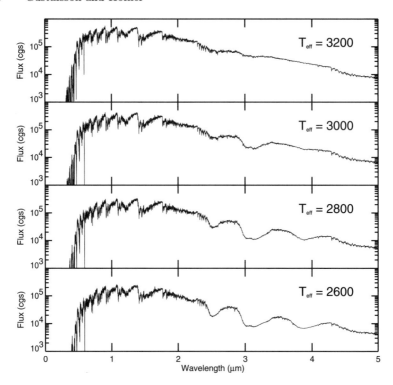

Fig. 4.10. Fluxes for selected models from Figure 4.9. Note the changes in the features due to HCN and C_2H_2 at 2.5 μm, 3 μm, and 3.8 μm

due to the dramatic effects of polyatomic molecules. More recently, it has been shown that computed spectra based on static models agree with observed ones in the 0.7–2.5 μm region [127], as well as with ISO SWS spectra in the 2.4–10 μm region [100]; see Figure 4.1. A flux excess is found in the observations beyond 10 μm and is tentatively ascribed to warm dusty clumps, obscuring about 10% of the photosphere. Most of the stellar fluxes are well matched, even for the Mira variable CU Cru, with some discrepancies in certain wavelength regions. The fits are, however, dependent on the assumed effective temperatures, the carbon and oxygen abundances, and the equilibrium constant (K in 4.30) for the C_3 molecule, which is still uncertain. A good agreement is also found at higher spectral resolution; see, e.g., the work on the ISO spectra of R Dor around 3 μm [157].

The positive outcome of the comparison between models and observations of individual galactic stars makes systematic studies, for example of the populations of AGB stars in galaxies in the Local Group, a natural next step. However, it is also important to continue detailed studies of stars in the solar neighborhood, combining flux measurements and spectra at various resolutions as well as other observations, and doing fully consistent comparisons

$$H = \frac{R_*^2}{GM}\left(\frac{kT}{m_H\mu} + \beta_t v_t^2\right). \quad (4.66)$$

Inserting $T = 3000$ K, $\mu = 1.4$, and $\beta_t = 0.5$, and setting v_t to the speed of sound $c_s \approx 5$ km s^{-1}, one finds that for a typical star with $GM/R_*^2 \approx 1$ cm s^{-2}, the atmospheric extension increases by about 50%. It is also easy to see that the modified model structure approximately mimics that of a model in which the turbulent pressure is not included in the calculation but whose surface gravity g is modified to

$$g_{\text{mod}} \approx g\left(1 + 168\beta_t v_t^2/T_{\text{eff}}\right)^{-1}, \quad (4.67)$$

where v_t is expressed in km s^{-1}. It should be noted that although (4.66) and (4.67) give reasonable estimates of the extension effects due to turbulence, systematic motions like pulsations and outgoing shock waves may give very different and more severe effects (cf. Section 4.7.3).

4.7.2 Convection

In the deep layers of the atmospheres of AGB stars, the temperature will be high enough to excite and begin ionizing hydrogen atoms. The excitation leads to a strong increase of the continuous opacity in the ultraviolet and visual wavelength regions. This opacity increase requires a steeper temperature gradient in order to transport outwards the energy flux produced in the core of the star.[12] The ionization of hydrogen, on the other hand, affects the adiabatic temperature gradient: If a parcel of ionized gas were to be moved upwards adiabatically, it would cool, but the recombination of hydrogen would then release energy, and the temperature of the parcel would not decrease as much as it would have for a perfect gas. These two effects contribute to a situation in which the radiative temperature gradient is steeper than the adiabatic temperature gradient, so that the Schwarzschild stability criterion is no longer fulfilled.

In order to calculate the convective flux, the hydrodynamical equations, (4.50), (4.51), and (4.53), must be solved. This is a totally nontrivial problem, even with computers of very high performance. The basic reason for this is the existence of nonlinear terms in the equations, which couple motions and inhomogeneities of different scales in a very complicated way. In addition, the equations must be solved together with the radiative transfer problem. This is because the character of the motions and the inhomogeneities in the stellar atmospheres is very much dependent on the interplay between the dynamics of the gas and its cooling and heating by radiation. Since the convection zone penetrates into layers close to optical depths of around unity, neither the

---

[12] This may easily be seen from the diffusion equation, or equation of heat transfer, which is applicable to the energy transfer in the stellar interior, and in which the inverse opacity plays the role of the diffusion constant, or conductivity.

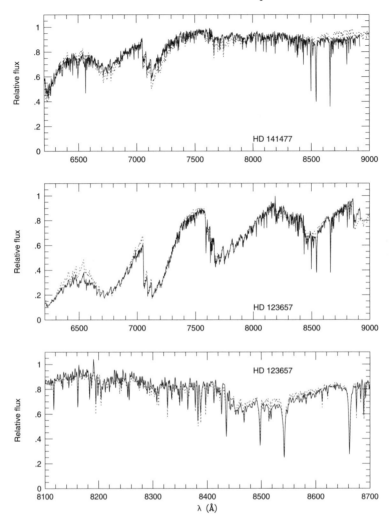

Fig. 4.13. Spectra of models (dotted lines) compared with observations (solid lines) of the two giants HD 141477 (M0 III) and HD 123857 (M5 III), from [168]. The models have $T_{\rm eff} = 3800$ K and 3500 K, and $\log g = 1.5$ and 0.9, respectively, and solar composition. The figure is based on work by [4]

where the parameter β_t is of order 0.5–1, depending on the angular distribution of the motions. How does this pressure affect the atmospheric structure? Following [61], it is easy to estimate the effects by adding p_t to the pressure term in (4.10) to obtain a modified expression for the pressure scale height H; cf. (4.15). Again disregarding variations in T, μ, and also in β_t and v_t as functions of depth, we obtain

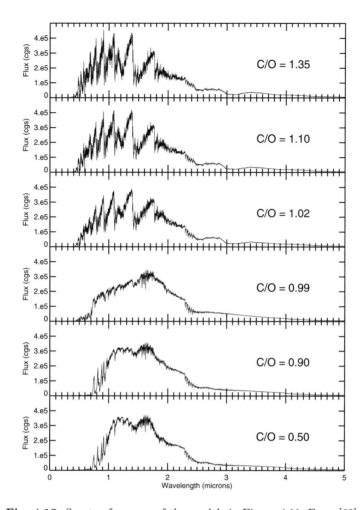

Fig. 4.12. Spectra for some of the models in Figure 4.11. From [62]

a spherically symmetric hydrostatic stratification, discussing both the basic physics of dynamical processes and giving examples of state-of-the-art models.

4.7.1 Turbulent Pressure

As we noted in Section 4.2, there is empirical spectroscopic evidence for motions in the AGB star atmospheres, due to turbulence or other flows. The typical velocities are a few kilometers per second. Small-scale motions lead to a turbulent pressure that may be written as

$$p_\mathrm{t} = \beta_\mathrm{t}\, \rho\, v_\mathrm{t}^2 \;, \tag{4.65}$$

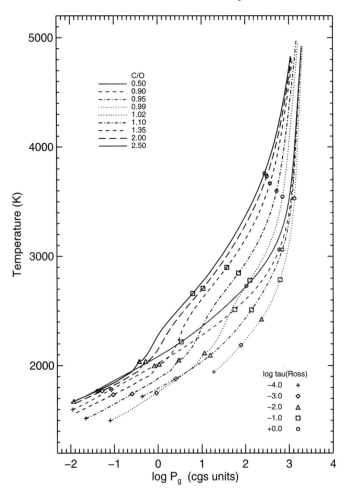

Fig. 4.11. Model structures, all with $T_{\text{eff}} = 3000$ K, $1\,M_\odot$ and $\log g = 0.0$ for different ϵ_C/ϵ_O ratios. Note the drastic changes as C $\to$ O (corresponding to the sequence M $\to$ S $\to$ C stars). From [62]

and analyses with static models, in order to learn more about the possibilities and shortcomings of those models.

4.7 Dynamics

In the preceding section we have assumed that the structure of the atmosphere in terms of gas density, etc., as a function of depth, is determined by hydrostatic equilibrium. Now we shall take a closer look at this assumption, or rather at the processes that cause deviations from this simple picture of optically thin nor the optically thick (diffusion) approximation is satisfactory for treating the radiative energy exchange. Radiative transfer should then be calculated allowing for spatial temperature–density inhomogeneities in three dimensions and for velocity fields with Doppler shifts of spectral lines.

In spite of the complexity, remarkable progress has been made in this field of research for solar-type stars, in particular due to the 3D simulations of Nordlund and collaborators (cf. [9, 10, 40, 133, 134, 175]). In the first simulations, the gas was assumed to be anelastic ($\nabla \cdot \mathbf{u} = 0$), and the quantities were represented by Fourier series in the horizontal and by cubic splines in the vertical direction, with periodic boundary conditions on the vertical boundaries of a box. For the solar case the resolution was about 90 km, and the simulated box was about 3000 km × 3000 km, enough to contain a few granules. In the latter simulations (see [174] for a description of the methods), the anelastic approximation has been relaxed, the box size has been doubled in both directions, and the spatial resolution has been improved; also, 3D radiative transfer has been included and the opacity represented by a four-bin ODF.

The characteristics of the simulations by Nordlund, Stein, and collaborators are not dependent on adjustable parameters. The simulations show remarkable agreement with observations in a number of respects. Images of solar granulation, with their intensity and size distributions, are well reproduced; observed asymmetries and bisectors for spectral lines of different strengths, elements, and excitation are also well reproduced, as are differences between stars with different fundamental parameters. The observed frequencies of solar acoustic waves (p modes) are also reproduced when the standard 1D solar models are modified with results from 3D models. How much of this progress can be applied to studies of AGB stars?

The general picture of convection in solar-type stars that emerges from these simulations shows strong narrow downdrafts surrounded by hot gas, which rises slowly in wide regions and approaches the surface. Due to the pressure above the granules, the flow is diverted horizontally and a fountain-like topology results. If the pressure is insufficient to push enough mass out horizontally, the density builds up over the granule until the pressure is raised sufficiently to expel it. The excess pressure also decelerates the upflow and thus decreases the energy flow to the surface, in particular near the center of the granule, which then cools. Therefore, the growing granules show an edge-brightened appearance. The horizontal flows encounter surrounding expanding granules and push against them. The granules with the largest pressures driving their expansion win this competition and grow at the expense of others, which have their growth limited and may even shrink. In regions where the ascent velocities drop, the density increases and the velocity may shift sign. This instability accelerates. Small-scale narrow features develop with cool spots or lanes that divide the granules into fragments (cf. the discussion in [174]).

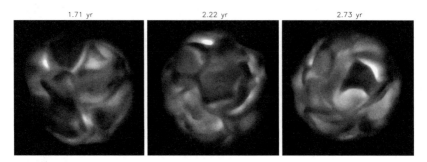

Fig. 4.14. A sequence of surface-intensity snapshots from a 3D simulation of a red supergiant with parameters $T_{\rm eff} = 3300$ K, $M = 5\,{\rm M}_\odot$, and $R = 650\,{\rm R}_\odot$ by B. Freytag (unpublished). The sequence covers one year

It is not easy to generalize these findings for solar-type stars to red giants. From simple scaling arguments Schwarzschild [166] suggested, however, that the granules of red giants, and of AGB stars in particular, may have sizes that are considerable as compared to the stars themselves. Recently, Freytag [51] has performed 3D hydrodynamical simulations for the red supergiant Betelgeuse, representing the full star by a Cartesian grid. Three-dimensional nonlocal radiative transfer is found to be significant and is included, though as yet with only grey opacities. Impressive movies (cf. Figure 4.14 and http://www.astro.uu.se/~bf/) show giant granules that cover considerable fractions of the stellar surface and give the model star a fascinating structure. Obviously, Freytag's simulations support the conjecture that convection inhomogeneities on AGB stars may be very large, even larger than those estimated by Schwarzschild. More systematic exploration of these structures, using further 3D simulations in combination with interferometric studies of red giant surfaces and high-resolution spectra for studying velocity fields in AGB star atmospheres, is of very great interest.

4.7.3 Effects of Pulsation

Basic textbooks on hydrodynamics tell us that an object that is embedded in a compressible medium and is periodically changing its volume creates sound waves (i.e., periodic compression and expansion at a fixed location) in the surrounding medium. As the spherical waves travel away from the pulsating source at their center, their amplitude will usually decrease due to energy conservation, since a growing volume and therefore a growing mass is affected by each wave (assuming a medium with a constant density).

As a first approximation, we can consider a radially pulsating star (or rather its interior up to and including the layers that are causing the pulsation) as such a volume-changing object, and the layers above the pulsation zone and the stellar atmosphere as the surrounding medium. There is, however, a significant difference compared to the typical textbook example outlined above: The gas surrounding the pulsating source of waves is not homogeneous but characterized by a strong radial density and temperature gradient. Therefore, waves traveling outwards will encounter less and less mass per volume and (as the density gradient gets steeper and steeper) eventually even a decreasing mass in the total volume affected by the wave at a given instant. Thus, the ratio of the kinetic energy in the wave compared to internal energy of the gas will increase, and the momentum transferred to the gas by the propagating wave may become important for the structure of the outer layers of the atmosphere.

Furthermore, the wave itself is usually neither adiabatic nor isothermal,[13] since it radiates energy at a finite but nonnegligible rate. The actual value of this rate and the question of whether non-LTE radiative processes are relevant in this context has been a subject of intensive debate for more than a decade (e.g., [189, 190, 195]).

4.7.3.1 Shock Waves and Levitation

Let us consider the propagating waves and their mechanical effects on the atmospheric structure in more detail. The slow (i.e., subsonic) periodic motions induced by the pulsating interior of the star result in sound waves that travel outwards. Due to nonlinear effects, waves with finite amplitude generally have the tendency to steepen, since the parts of the wave corresponding to more compressed (hotter) gas will propagate faster than the thinner regions in front of them (i.e., the maximum of the wave gradually catches up with the front). The ultimate result of this process is the development of a shock wave, with a sudden change of thermodynamical conditions in the shock front on a very short length scale (typically on the order of the mean free path), followed by a slow relaxation on a spatial scale that is orders of magnitude larger.

The strongly decreasing temperature and density in the stellar atmosphere enhance this steepening of the waves, as illustrated in Figure 4.15. The fact that the wave gradually runs out of material as described above will lead to an increase in wave amplitude; the local sound velocity decreases outwards, which tends to turn any motion eventually into a supersonic phenomenon. Usually, the waves created by the pulsation will have developed into strong shocks deep in the photospheric layers. The waves change from almost adiabatic to radiative shocks, losing part of their energy by radiation from the hot compressed gas, as they reach optically thin layers (which counteracts further steepening to some extent).

To derive characteristic properties of the pulsating atmosphere, we consider the motions of mass shells, taking only two forces into account at first:

---

[13] This is meant in the local sense of the gas temperature being equal to the local radiative equilibrium temperature, and not in the sense of the atmosphere as a whole being isothermal.

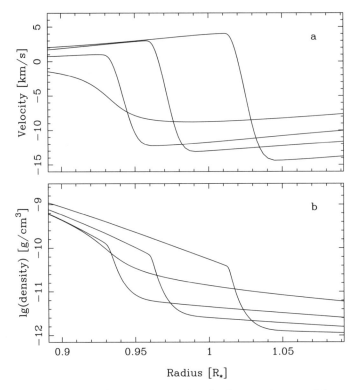

Fig. 4.15. Steepening and propagation of a shock wave in a model atmosphere: (a) velocity and (b) density as a function of the radius (in units of the photospheric radius of the corresponding hydrostatic model)

pressure gradients and gravitation (other effects will be discussed later). For gas in a spherical shell under the influence of gravity, the equation of motion (4.51) can be written as

$$\rho \frac{\partial u}{\partial t} + \rho (u \cdot \nabla) u = -\nabla p_{\text{gas}} - \frac{GM\rho}{r^2} \ , \qquad (4.68)$$

where M denotes the stellar mass (up to the location of the shell), and r the distance from the center of the star.

For the purpose of describing the movement of a gas shell that is hit by shock waves at more or less regular intervals, the pressure gradient can be regarded, in a first-order approximation, as a short impulse (delta function) around the time t_0 when the shock wave is passing through the gas shell, and as negligible during the rest of the time.[14] This impulse can be specified as a starting velocity of the shell $u_0 = u(t_0)$. Using these assumptions, the

---

[14] This reflects the fact that, due to the mechanical energy input, the average pressure gradient in a pulsating atmosphere is smaller than for the hydrostatic

equation of motion is reduced to

$$\frac{du}{dt} = -\frac{GM}{r^2} \tag{4.69}$$

for $t \neq t_0$, where du/dt now represents the derivative in the comoving frame. This equation can be integrated, yielding the velocity of the mass shell as a function of time, and then integrated a second time to result in an (implicit) description for the radius of the mass shell as a function of time (see, e.g., [191]). However, the resulting expression is rather complicated, and for the purpose of deriving some order-of-magnitude estimates, we reduce the problem further by assuming a constant gravitational acceleration $g = GM/r^2 \approx GM/r_0^2 = g_0$, where $r_0 = r(t_0)$; i.e., we assume that the change of the radius r is small compared to its starting value r_0. In this case, the integration of the equation of motion is trivial, and we obtain

$$u(t) = -g_0(t - t_0) + u_0, \tag{4.70}$$

$$r(t) = -\frac{g_0}{2}(t - t_0)^2 + u_0(t - t_0) + r_0, \tag{4.71}$$

which is the well-known ballistic case of motion. If we require the motion of the gas shell to be periodic in the sense that it returns to its initial location r_0 after a time $P = t_1 - t_0$, we find that the initial velocity u_0 and the period P are linked by $u_0 = g_0 P/2$, or $P = 2 u_0/g_0$. The maximum distance from the center $r_{\max} = u_0 P/4 + r_0$ is reached at time $t = t_0 + P/2$ during the ballistic motion, and we obtain

$$\frac{r_{\max} - r_0}{r_0} = \frac{u_0^2}{2 g_0 r_0} = \left(\frac{u_0}{u_{\text{esc}}}\right)^2, \tag{4.72}$$

where $u_{\text{esc}} = \sqrt{2GM/r_0}$ is the escape velocity at the starting point r_0. Note that our initial assumption of constant gravitational acceleration corresponds to assuming that $(r_{\max} - r_0)/r_0 \ll 1$, or that the initial velocity u_0 is small compared to the local escape velocity.

Since the velocity is smallest around $r_{\max}$, the mass shell will spend a significant part of the period close to this point. The times at which a certain value of r is reached during the ballistic trajectory can be expressed as

$$t_r = \frac{P}{2}\left(1 \pm \sqrt{1 - \frac{r - r_0}{r_{\max} - r_0}}\right) + t_0. \tag{4.73}$$

Therefore, the time spent above a given r is equal to the difference between these two values, or, expressed as a fraction of the period,

$$\frac{\Delta t_r}{P} = \sqrt{1 - \frac{r - r_0}{r_{\max} - r_0}}. \tag{4.74}$$

---

case in which the gas pressure (together with the radiative pressure) balances the gravitational force.

Thus, the gas stays in the top quarter of its trajectory for half of the period P, and in the upper half for more than 70% of the time.

To decide whether the approximation of constant gravitational acceleration as used above is reasonable in AGB stars, we derive an alternative expression for the relative dynamical extension in the following way: The kinetic energy transferred to the gas by the shock wave is converted into potential energy as the mass shell moves outwards and is decelerated by gravity. At some point the velocity of the gas reaches zero before the shell starts to fall back toward the center. We set the initial kinetic energy of the shell equal to the difference in potential energy at the starting point and at the maximum distance from the center reached by the shell, obtaining[15]

$$u_0^2 = -2\,GM\left(\frac{1}{r_{\max}} - \frac{1}{r_0}\right). \qquad (4.75)$$

Therefore, we obtain

$$\frac{r_0}{r_{\max}} = 1 - \left(\frac{u_0}{u_{\text{esc}}}\right)^2, \quad \text{or} \quad \frac{r_{\max} - r_0}{r_{\max}} = \left(\frac{u_0}{u_{\text{esc}}}\right)^2, \qquad (4.76)$$

which differs from the case of constant gravitational acceleration $g = g_0$ discussed above. Note that this new expression is valid even when the change in r is not small compared to r_0, provided that the gas shell is still bound by gravity, i.e., as long as $u_0^2/u_{\text{esc}}^2 < 1$. [In the limit of small initial velocity, i.e., $r_{\max} \approx r_0$, the two expressions approach each other.]

If we consider a star with one solar mass and a radius of about $300\,R_\odot$, the local escape velocity at that point is on the order of $35\,\text{km}\,\text{s}^{-1}$. An estimate for u_0 can be derived, e.g., from line doubling of CO lines (see Section 4.8.3 and Figure 4.4): The splitting between infalling and outflowing material is typically $30\,\text{km}\,\text{s}^{-1}$, corresponding to about $15\,\text{km}\,\text{s}^{-1}$ for the outflowing material relative to the center of mass in our simple picture where $u_0 = -u\,(t = t_0 + P)$. Therefore, u_0/u_{esc} may be as large as 0.5, meaning that the constant-g approximation may be problematic. Assuming $u_0/u_{\text{esc}} = 1/2$, we obtain $r_{\max}/r_0 = 4/3$ from (4.76), while the limit of constant g results in a value of $5/4$.

So far, we have assumed that the only relevant forces acting on the mass shell are gravity and the gas-pressure gradient. Another obvious candidate is the radiation pressure, which may contribute considerably to the forces acting on the atmosphere and in particular on the outer layers in cool, luminous objects. Since the radiation pressure is directed outward it will counteract gravity to some extent, modifying the ballistic motion described above. If the luminosity is constant, the radiative flux will vary proportionally to r^{-2}, i.e., in the same way as the gravitational force. The radiation pressure can

---

[15] The same expression results from integrating (4.69), using $du/dt = u\,du/dr$ and setting $r = r_{\max}$ afterwards.

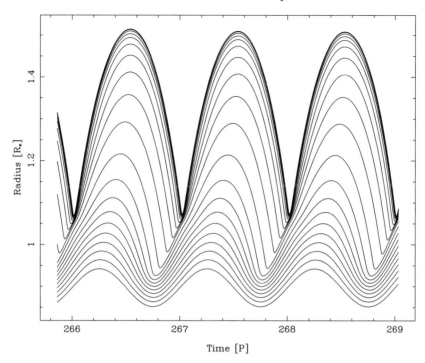

Fig. 4.16. Position of selected mass shells as a function of time for a dynamical model atmosphere [83] with the following parameters: $T_{\text{eff}} = 2800\,\text{K}$, $L = 7000\,\text{L}_\odot$, $M = 1\,\text{M}_\odot$, $\epsilon_C/\epsilon_O = 1.4$, $P = 390\,\text{d}$, and piston velocity amplitude $2\,\text{km/s}$. (Time in periods, radius in units of the photospheric radius $R_* = 355\,\text{R}_\odot$ of the hydrostatic initial model.)

therefore be treated formally as a reduced mass (reduced escape velocity), assuming that the opacity does not change too drastically. In large-amplitude variables, however, the temporal variation of the luminosity during the cycle is considerable, resulting in a variation of the radiation pressure. In general, this will lead to an asymmetry in the force acting on the gas during expansion and infall; the size of the effect depends on the phase lag between the variable luminosity and the moment t_0 at which the gas is hit by the shock wave. Therefore, in contrast to the purely ballistic case, the trajectories will not be symmetric. Furthermore, the opacity will not only change during the motion but will show similar "nonsymmetric effects" because it depends on the changing temperature and density of the gas, which are not simple functions of the radial distance r but depend on processes like radiative cooling after the passage of a shock. Figure 4.16 shows trajectories resulting from a detailed hydrodynamical simulation of an atmosphere of a pulsating star.

In fact, only the gravitational force can be treated properly in the simple local picture of a moving mass shell as outlined above, while the forces due to gas-pressure gradients and radiation pressure are intrinsically nonlocal. To determine the gas-pressure gradient, we need a simultaneous description of all mass shells in order to construct a snapshot of the global atmospheric structure at any time. The thermodynamical quantities of the gas that determine, e.g., the opacity (and thus the radiation pressure) result from complex interactions between dynamics and the radiation field, the latter again being a highly nonlocal phenomenon.

Consistent time-dependent models of pulsating atmospheres therefore require the simultaneous solution of hydrodynamics and radiative transfer with realistic microphysical data. This complex problem can only be solved numerically, using state-of-the-art computers and efficient methods. Before we start a description of existing models, however, we want to discuss another important ingredient: dust and its influence on the structure and dynamics of the atmosphere.

4.7.3.2 Dust Formation and the Connection Between the Atmosphere and the Stellar Wind

From the point of view of atmospheric structure and dynamics, it is tempting to regard dust as just another source of opacity. However, this is problematic for the following reasons: First, dust grains often interact with radiation much more efficiently than molecules and atoms, meaning that the dust opacity may actually dominate the total opacity and even give rise to a radiation pressure that is sufficient to overcome gravity in the outer layers of the extended atmosphere. Secondly, as we have shown in Section 4.4.6, the time scales for grain growth are of the same order of magnitude as the dynamical and thermodynamical variations in the atmosphere. Consequently, the formation and growth of dust grains will typically proceed far from equilibrium; the dust properties (including the dust opacity) are not merely a function of the local thermodynamical conditions. The time-dependent growth of the grains must be treated simultaneously with dynamics and radiative transfer; see also Chapter 6.

One condition necessary for dust formation is a low temperature, making the condensate stable against evaporation. However, this is not a sufficient condition in a dynamical atmosphere, where the time available for grain growth may be severely limited, e.g., due to the gas being reheated periodically, or due to the gradual dilution of the condensible material caused by decreasing density in a stellar wind. The main factor controlling the efficiency of dust formation (i.e., the growth time scale of the grains) is the number densities of condensible species in the gas phase, which translates into a rate of adding atoms and molecules to a grain by collisions (cf. Section 4.4.6).

Therefore, two basic conditions must be fulfilled for dust formation in AGB star atmospheres: low temperature combined with high density. In terms of atmospheric structure, these two conditions seem contradictory: Low temperatures can be reached only at a certain distance above the photosphere, where densities will be orders of magnitude lower than in the photosphere. In a hydrostatic atmosphere, the gas density basically decreases exponentially with the distance from the photosphere. In the atmosphere of a pulsating star, however, the propagating shock waves will cause a levitation of the outer atmospheric layers, creating a temporary reservoir of relatively dense gas at a certain distance from the photosphere and thus increasing the efficiency of dust formation.

Assuming that dust formation can occur in a given atmosphere, how much dust is necessary to drive a stellar wind, or at least influence significantly the structure and dynamics of the atmosphere? To obtain an estimate, we assume that gas and dust are efficiently coupled dynamically, i.e., that most of the momentum gained by the dust grains through radiative acceleration is immediately transferred to the gas. We demand that the radiative acceleration be large enough to overcome gravity:

$$\langle\chi\rangle \frac{L}{c\,4\pi\,r^2} > \frac{GM}{r^2}\rho \quad\Longrightarrow\quad \frac{\langle\chi\rangle}{\rho} > 4\pi\,c\,G\,\frac{M}{L}\,, \qquad (4.77)$$

where L and M are the luminosity and the mass of the star, respectively, r is the distance from the center and $\langle\chi\rangle$ the flux-mean dust opacity defined by

$$\langle\chi\rangle = \frac{\int \chi(\lambda) F_\lambda\,d\lambda}{\int F_\lambda\,d\lambda}. \qquad (4.78)$$

Next we use the small-particle limit for the dust opacity given in (4.48), and reformulate this expression in terms of the bulk density $\rho_{\rm cond}$ of the condensed material (Section 4.4.6) and the mass density $\rho_{\rm d}$ of the dust in the atmosphere. This yields

$$\frac{\rho_{\rm d}}{\rho}\frac{3}{4}\frac{\langle Q'_{\rm ext}\rangle}{\rho_{\rm cond}} > 4\pi\,c\,G\,\frac{M}{L} \quad\Longrightarrow\quad \frac{\rho_{\rm d}}{\rho} > \frac{16\pi\,c\,G}{3}\frac{\rho_{\rm cond}}{\langle Q'_{\rm ext}\rangle}\frac{M}{L}\,, \qquad (4.79)$$

where $\langle Q'_{\rm ext}\rangle$ is the flux-averaged mean of $Q'_{\rm ext}$.

To consider a quantitative example, we choose typical values for carbon grains, i.e., $\rho_{\rm cond} = 2\,{\rm g\,cm^{-3}}$ and $\langle Q'_{\rm ext}\rangle = 5\times 10^3\,{\rm cm^{-1}}$, as well as a stellar mass $M=1\,{\rm M_\odot}$ and a luminosity $L=5\times 10^3\,{\rm L_\odot}$. This leads to $\rho_{\rm d}/\rho > 1.4\times 10^{-3}$ for the dust-to-gas mass ratio, which is the same order of magnitude as the abundance by mass of the excess carbon in a carbon-rich atmosphere. Therefore, a considerable fraction of the carbon that is not bound in CO will typically have to be condensed into grains in order to drive a stellar wind by radiation pressure on dust.

We should, however, mention in this context that our simple criterion will tend to overestimate the amount of dust that is necessary to drive an outflow. Apart from the radiation pressure on dust, there are other forces that counteract gravity: the thermal pressure of the gas, the radiation pressure on molecules, and in pulsating atmospheres the gas that may actually be moving outwards due to the influence of the shock waves at the point where dust formation occurs. Therefore, a stellar wind may be present even in a situation where the dust-to-gas ratio is smaller than the value derived above, or, in other words, where the ratio of the radiative force acting on the grains to the gravitational force is somewhat smaller than unity, as demonstrated by several of the detailed numerical models discussed below.

Since mass loss has a significant influence on the observable properties and the evolution of AGB stars, the causes and consequences of stellar winds are discussed in detail in Section 4.8. However, in the present context we stress that wind characteristics like outflow velocities and mass-loss rates will depend sensitively on the conditions in the zone where the flow originates. Therefore, a realistic modeling of the atmospheric region is crucial for our understanding of mass loss and evolution on the asymptotic giant branch.

4.7.4 Model Atmospheres for Pulsating Stars

Compared to the construction of static model atmospheres discussed in Section 4.6, the dynamical modeling of AGB star atmospheres is a relatively young field of research. Probably the most important reason is that the computational effort is much larger: For each set of stellar parameters, instead of one hydrostatic structure, a consistent solution to the hydrodynamic equations and the radiative transfer problem is required at every instant in time. The computation must span a considerable time interval, first to avoid transient effects, e.g., caused by starting the calculation from a hydrostatic model, and second because even in the "final" model, some time-dependent phenomena may span several pulsation periods.

To keep the computational effort at a level that can be handled reasonably with present computers, the problem of computing dynamic model atmospheres is divided into two steps: First, the structure of the atmosphere is calculated by solving the equations of hydrodynamics together with a simplified (often grey) treatment of radiative transfer. This directly yields the wind characteristics (mass-loss rates, outflow velocities, dust properties, etc.). In a second step, snapshots of the time-dependent structures are used as input for detailed radiative transfer calculations, to compute the required observable properties (synthetic spectra, colors, light curves, visibilities, line profiles, etc.).

Presently, most models do not contain the driving zone of the stellar pulsation. This zone is usually simulated by a variable inner boundary of the model below the stellar photosphere. The prescriptions for the motions of the gas and the variation of the bolometric luminosity are either specified as simple periodic functions or derived from pulsation calculations. This introduces new parameters (period and amplitude of the variations) in addition to the usual stellar parameters (mass, luminosity, effective temperature, elemental abundances) that determine a hydrostatic model.

The time-dependent modeling of atmospheres and winds of AGB stars started in the late 1970s. Wood [196] investigated the possibilities of creating a slow, massive stellar wind by stellar pulsation with or without the influence of radiation pressure on dust. Instead of including an energy equation for the gas, he used the limit of an adiabatic flow or assumed that the temperature is equal to a prescribed radius-dependent value. In the first case, the mass-loss rates were much too high, but in the second case a combination of pulsation and radiation pressure on dust produced reasonable wind characteristics. In his models, dust was featured only as a parameterized additional opacity, and the pulsation was simulated by a pressure variation at the inner boundary of the model.

Almost a decade later, Bowen [21] published his landmark paper on mass loss in AGB stars. In contrast to Wood's pioneering work, he used a semianalytic treatment of radiative transfer to determine the radiative-equilibrium temperature structure. He explored the limits of adiabatic flows and isothermal shocks (in the latter case, assuming that the gas temperature is equal to the radiative-equilibrium temperature everywhere at any given time) and confirmed Wood's results. In addition, Bowen studied models with an explicit prescription for radiative cooling behind shocks, based on non-LTE assumptions. Since the adopted cooling rate is relatively slow in some cases, this led to situations in which the gas could not return to the equilibrium temperature before the next shock arrived. In certain models this resulted in a hot zone above the stellar photosphere referred to as the "calorisphere," which opened up the possibility for a thermally driven wind. Bowen also used a parameterized opacity (depending on temperature only) to describe the dust component. The stellar pulsation was simulated by a sinusoidal motion of the inner boundary ("piston model"). Bowen performed a systematic study of the dependence of mass loss on the fundamental stellar parameters and the pulsation parameters of his models.

Fleischer et al. [49] presented models for C-rich stars based on a modeling method similar to Bowen's (assuming the isothermal limit for shocks) but including a consistent, time-dependent description of the dust-formation process (moment method [54, 55].[16] They demonstrated that the interaction of dynamics and dust formation can result in a shell-like structure of the CSE. The strong coupling between thermodynamical conditions, dust formation, and dynamics may even lead to a multi- or nonperiodic behavior of the CSE, since the dust component is governed by its own time scales. Fleischer et al. [50] and Höfner et al. [82] described in detail how (more or less periodic) instabilities can develop in purely dust-driven winds, without shocks created by the pulsation of the star (dust-induced κ-mechanism). If this situation occurs in the atmosphere of a pulsating star, the interference

---

[16] Due to the much more complicated nature of dust formation in an O-rich environment, similar models for M stars are lagging behind. First results have been presented by [90].

of the dust-formation cycle with the periodically passing shocks may cause complicated time-dependent phenomena.

Höfner and Dorfi [81] used a radiation-hydrodynamical description of pulsating atmospheres of AGB stars, including time-dependent dust formation (for C stars; treated with the moment method). The equations of hydrodynamics were solved together with the zeroth and first moment equations of the radiative transfer equation (which describe the energy and momentum balance of the radiation field). The authors studied the temporal variations of the atmosphere and CSE, such as multiperiodicity; see Figure 4.17. They also explored the dependence of mass-loss rates and wind velocities on fundamental stellar parameters.

The various models described above have two things in common: They were based on a simple grey treatment of radiative transfer (or even an explicit prescription of the temperature as a function of depth in the early models), and they were mainly used to investigate the conditions for, and consequences of, mass loss by stellar winds. Apart from a few qualitative exploratory studies, the prediction of observable properties based on time-dependent dynamical models started only fairly recently (see below). Basic assumptions like grey radiative transfer and simple variable boundary conditions to simulate pulsation have allowed a qualitative insight into basic processes in dynamical AGB star atmospheres. However, they do not lead to realistic quantitative models that can be compared directly to observations.

When applied to a cool line-blanketed atmosphere, grey radiative transfer causes a noticeably incorrect representation of the crucial energy- and momentum-exchange processes between matter and the radiation field. This results in quantitatively incorrect temperature–density structures and therefore in synthetic spectra and other calculated observables that may be very far from the observed ones. During the last decade, various ways of working around this problem have been applied, e.g., using the density structure from the dynamical calculation and recalculating the temperature structure when computing the detailed radiative transfer (e.g., [19]; see below). The only satisfactory, physically consistent solution to this problem is the improvement of fundamental assumptions on which the models are based.

The main reason why grey radiative transfer was and is still used in many cases is a computational one: Calculating a time-dependent model corresponds roughly to the computational effort of solving the "standard" atmosphere problem typically 10^4 times for one given set of stellar parameters. This means that at present—even with fast computers—it is impractical if not impossible to include detailed radiative transfer with (tens of) thousands of frequency points in a time-dependent hydrodynamic model, as done in hydrostatic models. However, studies by Helling and Jørgensen [71] that systematically reduce the number of frequencies in hydrostatic models have demonstrated that as few as 20–50 frequency points may be sufficient to produce structures that are close enough to results of more detailed models to

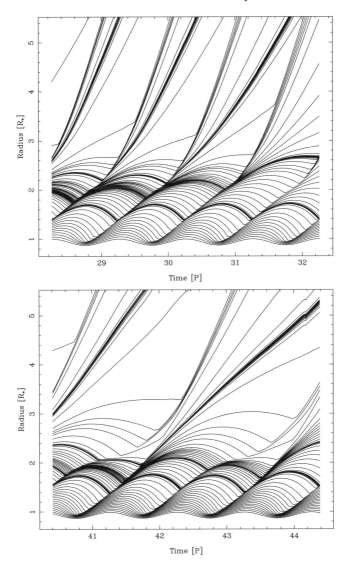

Fig. 4.17. Positions of selected mass shells as a function of time for models P1 (**top**) and P2 (**bottom**) of [81]. Model P1 is a typical single-periodic model. During each piston cycle a new dust layer is formed, triggered by enhanced density behind the shock waves. Below about $2\,R_*$ the dust-free atmosphere is periodically passed by strong shocks (marked by sharp bends in the lines). The formation of dust layers and their subsequent acceleration due to radiation pressure (indicated by the steepening of the lines) takes place between 2 and $3\,R_*$. In model P2 (which differs from P1 by a lower ϵ_C/ϵ_O value and thus less efficient dust formation), we encounter a simple form of multiperiodicity: A dust layer is formed every second piston period; see [81] for details. (Time in pulsation periods, radius in units of the stellar radius R_* of the corresponding hydrostatic initial model.)

give reasonably realistic synthetic spectra. Höfner et al. [80, 83] recently presented a new generation of models combining time-dependent hydrodynamics with frequency-dependent radiative transfer and a detailed description of dust formation. In these models, the frequency-averaged opacities that are required to close the system of conservation equations all come from a solution of the frequency-dependent radiative transfer equation that uses opacity sampling at about 50 frequency points. Prototypes of these models compare well to detailed standard models in the hydrostatic limit and show good agreement with various observational results (e.g., [8, 126]).

Current dynamic model atmospheres represent a transition from qualitative modeling of physical processes to realistic quantitative predictions and interpretations of observations. In this context we must not only worry about the internal consistency of the atmosphere model but also about how to simulate the effects of pulsation on this part of the star. Combined detailed models of the pulsation zone and stellar atmosphere are certainly an important goal. First steps in this direction have been taken by deducing variable boundary conditions for detailed atmosphere models from simpler dynamic models that describe the pulsating interior of the star (e.g., [79]). A major part of this problem, however, seems to lie in constructing realistic self-consistent models of the pulsating stellar interior (see Chapter 2).

4.7.5 Comparison of Models and Observations

In principle, and just like hydrostatic model atmospheres, dynamic models can be used to calculate any kind of observable properties by solving detailed radiative transfer for a given snapshot of the spatial structure. Densities, temperatures, etc., are taken from the dynamical calculation and used to compute the quantities like detailed opacities that are required to solve the radiative transfer equation. In this section, we will concentrate on the synthetic observable properties that are of particular relevance for dynamical atmospheres, i.e., features that are used to study effects of pulsation and dynamical extension as well as the temporal variability of the atmosphere. In addition, a detailed description of observable properties related to stellar winds is given in Section 4.8.3.

Early exploratory studies investigating the influence of atmospheric extension on observable properties of M-type Mira stars were performed by Bessell et al. [17] and Scholz and Takeda [164]. Using parameterized density structures similar to the models of [196], they obtained infrared spectra and a phase and wavelength dependence of monochromatic radii in reasonable agreement with observations.

Based on the pulsation models of [197], Bessell et al. [19] investigated the photospheric structure and observable properties of Mira variables. Using only the density structures resulting from the dynamical models, they recalculated the gas temperatures with a more realistic treatment of radiative transfer. The general appearance and the variability of the resulting synthetic spectra that were used to derive monochromatic radii are in good agreement with observations. The velocity fields lead to a complex time-dependent behavior of absorption line profiles and influence the integrated line strengths [163]. Certain features not only show a variation with phase but also differ at a given phase from cycle to cycle, reflecting a nonperiodic behavior of the atmospheric layers in which they originate. Similar behavior is seen in some observed line profiles.

Alvarez and Plez [4] investigated the relative phase shifts of light curves of M-type Mira stars observed in different narrow band colors. They demonstrated that this phenomenon could not be explained by a series of detailed hydrostatic model atmospheres with varying stellar parameters, although these models give good results for somewhat hotter and less variable stars. Using a dynamic model [84] they were able to reproduce the phase shifts and interpret them as modulations of the structure by propagating shocks. Another example of observations that could not be reproduced with hydrostatic models but at least qualitatively understood with dynamic models is the behavior of the SiO rotation–vibration bands around 4 μm studied by [7]. While hydrostatic model atmospheres predict overly strong features for very cool and extended objects, the experimental dynamical models used in this paper are in principle able to explain the whole observed range of equivalent widths and their variations. Similar problems with the strength and variation of water features around 3 μm have been largely resolved by using prototypes of a new generation of dynamical models [8].

The near- and mid-infrared spectra obtained with ISO have recently opened the possibility to study simultaneously the molecular and dust features originating in different layers of extended atmospheres and stellar winds. A detailed interpretation of these spectra will probably be a significant step toward understanding the mass-loss mechanisms in AGB stars. Observational progress has inspired the development of semiempirical and self-consistent models for the atmosphere and inner wind region. Since it is difficult to separate the discussion of "atmospheric" and "wind" features in the spectra, we give a common overview in Section 4.8.3.

4.8 Mass Loss

Mass loss is a significant phenomenon in stellar evolution. Most stars have a stellar wind, i.e., an outflow of gas from the stellar surface; but mass-loss rates and outflow velocities depend strongly on the stellar parameters. Mass loss is particularly significant in the AGB stage. Only a few types of stars share the fate of AGB stars, which can lose mass at such high rates that mass loss, rather than nuclear burning, may eventually become the process that determines the maximum luminosity, the time scale on the giant branch, and the star's further evolution.

4.8.1 Atmosphere and Circumstellar Envelope

To simplify the discussion of physical processes, modeling, and observational results, the atmosphere of the star and its surrounding material can be roughly divided into three zones.

The innermost layers of the extended stellar atmosphere (zone A) show a steeply (i.e., exponentially) decreasing density and temperature profile similar to a hydrostatic atmosphere as a background structure that is locally modified by propagating shock waves due to stellar pulsation. Here, the variations in thermodynamical conditions in a given layer are basically periodic, dominated by the more-or-less ballistic movement of the gas as discussed in Section 4.7.3.

The other extreme is the circumstellar envelope (CSE; zone C), which can be defined as the asymptotic regime where the stellar wind has reached its final outflow velocity. Ignoring small-scale structures due to variations in the outflow on time scales of up to a few pulsation periods (caused, e.g., by nonperiodic dust formation), as well as changes of the stellar wind on evolutionary time scales, the overall picture of the CSE resembles a stationary outflow with an r^{-2} density distribution. However, observations seem to indicate that this widespread "standard model" of the circumstellar envelope may be oversimplified. A detailed discussion of zone C is given in Chapter 7.

Between these two extremes of zone A ("atmosphere") and zone C ("CSE") we find a transition region with a complex, time-dependent structure (zone B). Here, the conditions for the mass loss are set and the wind is accelerated. Due to recent developments in observational techniques in the infrared, in particular the successful mission of the Infrared Space Observatory (ISO), along with the progress of theoretical models, a better picture of this zone is emerging.

Before we begin describing various possible driving mechanisms for winds of AGB stars, we express the warning that the simple "ABC scenario" above should be regarded only as a mind map for the discussion of mass loss, not a suggestion for semiempirical modeling. While a reasonable separation of zone C from B is feasible, a clear-cut division between A and B seems difficult. A physically motivated border could be the (variable!) "point of no return" where the moving gas is accelerated beyond the local escape velocity. However, while this is a possible concept for theoretical models, it is difficult to apply to observations.

4.8.2 Driving Mechanisms

A key problem in understanding the mass loss of AGB stars is the quantitative description of the driving mechanisms from first principles. Let us start with some basic facts: A stellar wind requires a source of momentum, i.e., an outward-directed force that can overcome gravity. This statement might seem trivial, but many discussions and a lot of computational effort have been devoted to clarifying which are the (dominant) mechanisms responsible for mass loss in different types of stars, and for AGB stars in particular; see also Chapter 6.

According to the nature of the dominating outward-directed force, stellar winds can be divided into two groups. *Thermal winds* are driven by the (gradient of) the gas pressure; this requires a high pressure in the wind-formation zone, which may be created, e.g., by dissipation of sound waves or by radiative heating of the gas. *Radiation-driven winds* receive their momentum directly from the radiation field of the star by absorbing the outward-directed stellar radiation and reemitting photons in all directions; this creates a net force (radiation pressure) directed away from the star. A further possible way of accelerating gas away from the stellar surface is through waves: sound waves, shocks, magneto-hydrodynamic waves, etc. This is a rather inefficient mechanism for directly driving a wind by momentum input. Nevertheless, waves may contribute considerably to the creation of thermal winds (heating, increasing the gas pressure) or even radiation-driven outflows (by enhancing the density of the outer atmospheric layers so that relevant absorbing species like molecules or dust grains can form more efficiently). A detailed physical description of different driving mechanisms for stellar winds can be found in the textbook [120].

4.8.2.1 Characteristics of AGB Star Winds

Compared to winds of other types of stars, winds of AGB stars are slow, with typical terminal velocities of 10–20 km s^{-1}. Since observations indicate that dust is a common ingredient of these outflows, it has been a long-standing suggestion that the mass loss is driven by radiation pressure on dust grains. One observational argument for the importance of the radiation pressure for mass loss in AGB stars is based on the quantity $\beta = (v_\infty \dot{M})/(L/c)$, i.e., the ratio of the momentum carried away by the wind relative to that of the radiation field (where v_∞ is the terminal velocity of the wind, $\dot{M}$ the mass loss rate, L the stellar luminosity, and c the speed of light). For an optically thin wind accelerated by radiation pressure on dust, β must be less than 1 (which seems to be the case, e.g., for most of the stars investigated by [108] and [124]). However, in the case of optically thick winds (in the sense of the flux-averaged optical depth being larger than 1), the value of 1 is not an upper limit for β, due to the fact that more than one interaction between dust grains and photons can take place on their way out [52, 132].

While the terminal velocities of AGB star winds are on the lower end of the scale of stellar wind velocities, the mass-loss rates are high, typically 10^{-7} to 10^{-4} M$_\odot$ yr^{-1}.[17] These two facts together imply high gas densities in the outflow, indicating that the wind originates in a dense environment.

---

[17] It may be instructive to compare with the solar wind, with final speeds of typically 500 km s^{-1} and mass-loss rates of typically 10^{-14} M$_\odot$ yr^{-1}.

However, density–temperature profiles inferred from hydrostatic model atmospheres result in far too low densities at temperatures where dust grains might condense. Therefore, most mass-loss scenarios are based on a two-step process: One mechanism increases the scale height of the atmosphere by lifting the gas to a certain distance from the photosphere, thus creating a reservoir of matter in the region where a second mechanism (e.g., radiation pressure on newly formed dust grains) can set in, giving the gas the necessary momentum to escape from the star.

Since large-amplitude pulsations are a common phenomenon among AGB stars, one natural mechanism for lifting the gas is through the shock waves that are propagating in the stellar atmosphere. In fact, observations indicate the presence of both shocks and a correlation between mass loss and pulsation characteristics. However, a number of alternative processes have been suggested to increase the scale height of the atmosphere. Furthermore, there have been extensive discussions on whether radiation pressure on dust (or molecules) is the actual driving force of the wind. Alternative scenarios in which the dust forms as a by-product of the outflow have also been studied (see below).

4.8.2.2 Pulsation-Enhanced Dust-Driven Winds

The most generally accepted mass-loss scenario at present combines (1) the dynamical extension of the stellar atmosphere by shock waves due to pulsation as a levitation mechanism with (2) radiation pressure on dust grains, possibly in combination with radiation pressure on molecules, as the source of momentum for the wind. Models of this type have attained the highest level of self-consistency, with few free parameters and the possibility of predicting the dependence of mass-loss rates on stellar parameters.

The "atmospheric part" of this scenario has been discussed in detail in Section 4.7.3 and will only be summarized here. The interior pulsation of the star leads to large-scale motions of the outer layers, resulting in waves that steepen as they propagate through the density gradient of the atmosphere and develop into strong radiating shocks. As the shock front passes, the gas is compressed, heated, and pushed outwards. Eventually, the temperature in the shocked gas drops again by radiative cooling, resulting in a relatively cool, dense outward-moving shell. If too little or no dust forms, this motion will finally come to a standstill and be reversed; that is, the gas starts falling back to its original position until it is hit by the next propagating shock.

If conditions are favorable through a part of this ballistic trajectory, i.e., if densities are high enough at sufficiently cool temperatures, then dust grains may form in the gas. Temperature mainly acts as a "threshold" for the condensation process, but the prevailing densities determine the time scales for grain growth, or the efficiency of the dust formation, since the condensation process must compete with the changes in thermodynamical conditions in a given layer (unlike a static atmosphere, where in principle an infinite time is available for condensation).

Since the solid particles (in particular amorphous carbon grains) tend to have higher mass absorption coefficients than the gas from which they form, they feel a higher radiation pressure. They start to move outwards, dragging the gas along via momentum transfer by gas–grain collisions. During this acceleration phase, the further growth of grains proceeds in an environment in which falling temperature (which favors condensation) competes with decreasing gas density (which slows down grain growth); eventually, the condensation process may come to a halt, possibly long before the available condensible material is exhausted (cf. Section 4.4.6). The final degree of condensation, together with the stellar parameters, determines the terminal velocity of the wind.

During the past two decades, detailed time-dependent numerical calculations have been performed independently by several groups. In Section 4.7.4 we have described the physical input of the models. The results concerning mass loss can be summarized in the following way. Stellar pulsation alone cannot produce realistic mass-loss rates; a combination of stellar pulsation (shock waves) and radiative driving (dust) is necessary to do this. The treatment of thermal relaxation behind shocks (finite rate or immediate cooling) has a considerable influence on the atmospheric structure and the mass-loss rate (e.g., [21, 196]). The interaction of time-dependent formation, growth, and evaporation of dust grains with the dynamics of the atmosphere may lead to a shell-like structure of the CSE and instabilities in the outflow (dust-induced κ-mechanism), depending on the stellar parameters (e.g., [49, 50, 82]). The dependence of the mass-loss rate on stellar parameters indicates a strong increase of mass loss during the evolution along the asymptotic giant branch (e.g., [22, 81]).

In summary, the existing models of dust-driven winds originating from pulsating AGB stars predict mass-loss rates and wind velocities comparable to values derived from observations, indicating that this is a likely mass-loss scenario. Nevertheless, we will discuss some alternative mechanisms that have been investigated during the past 20 years.

4.8.2.3 Alternative Scenarios

Historically, the first wind models were actually stationary, i.e., describing a time-independent flow pattern, and could not take into account the effects of propagating radiating shock waves in a proper way. Therefore, the formation of an extended atmosphere seemed to be a major problem. Most of the ideas for levitation mechanisms other than pulsation-induced shock waves date back to these times and center on small-amplitude waves—acoustic or magneto-hydrodynamic—created in the photospheric layers, e.g., by convection.

Hartmann and MacGregor [68] proposed a model of stellar winds driven by Alfvén waves, using a linearized stationary theory based on small-amplitude Alfvén waves with wavelengths much smaller than the density scale height. The time-averaged momentum input can be expressed as a wave pressure gradient in the equation of motion. In contrast to a thermal wind, where the gas-pressure gradient may be due to heating by the dissipation of waves, the waves in this model directly transfer momentum to the gas. This scenario has two major difficulties: (1) the unclear origin and strength of the waves, represented by free parameters in the models, and (2) resulting wind velocities that are too large. The latter problem was overcome (or rather deferred) by introducing a dissipation length (free parameter) for the Alfvén waves of order one stellar radius.

Pijpers and Hearn [142] suggested sound waves as a driving mechanism for stellar winds. Formally, their model is similar to the Alfvén-wave-driven outflows mentioned above, just replacing one type of small-amplitude, short-wavelength waves with a different one. One advantage of this model compared to the one by Hartmann and MacGregor is that the resulting wind speeds are lower (due to the fact that the sound speed is significantly smaller than the Alfvén speed near the star). However, the energy input by the waves at the inner boundary remains a free parameter. The wind is accelerated slowly over a large distance, and dust formation is not required to drive the mass loss but might occur as a by-product of the outflow [141].

Gail and Sedlmayr [53] have presented self-consistent stationary models of dust-driven winds around carbon-rich stars, obtained by simultaneously solving the hydrodynamical equations, an approximate treatment of radiative transfer in extended stellar atmospheres, and the equations describing dust formation and growth. A large grid of such self-consistent stationary models has been calculated by [39] to investigate the dependence of dust-driven mass loss on basic stellar parameters (mass, effective temperature, luminosity) and on the chemical composition (C/O ratio). They demonstrate that it is possible to drive a slow massive stellar wind by radiation pressure on dust alone, without pulsation-induced shock waves or other levitating processes. However, unfortunately, this mechanism works only for very luminous cool stars (with high C/O ratios) that are known to have large-amplitude pulsations.

In this context we would like to mention that a stationary picture of the atmosphere and wind acceleration zone—or even the term "extended scale height"—may be seriously misleading. It implies that a reservoir of dense material (e.g., a cool layer of gas where dust can form) must be present at all times. In contrast, time-dependent dynamical models of dust-driven winds indicate that it may suffice to have a layer of dense cool gas present for a small part of the pulsation cycle, just long enough to allow for the condensation of a certain amount of dust. Probably it is much easier to create this short-lived "jump-off point" for the stellar wind than to supply a stationary reservoir of dense cool gas constantly. In that sense, a time-dependent levitation mechanism based on large-amplitude shock waves may be more efficient.

However, as mentioned before, one problem associated with modeling shock waves in atmospheres of AGB stars is to obtain realistic radiative cooling rates. Many models assume LTE cooling, which is very efficient and leads to dense cool layers close to the stellar photosphere, where dust grains can form and drive the stellar wind. While LTE cooling is certainly a good approximation in the dense inner layers of the atmosphere, Willson and Bowen [190] argue that this assumption overestimates cooling behind shocks in the critical outer zones where the wind is accelerated.[18] This may have a strong effect not only on the detailed structure of the atmosphere and the resulting spectra but also on the mass-loss mechanism itself. If the cooling between the passage of successive shocks through a given layer is slow enough, the star might have a warm extended atmosphere (or "calorisphere," as Willson et al. call it) instead of a cool one. This could inhibit dust formation close to the star, making a different driving mechanism necessary. Bowen [21] presented models in which the outflow is actually driven by the high gas pressure in the calorisphere, i.e., a thermal wind instead of a radiation-driven outflow.

A problem that is often overlooked in the discussion of driving mechanisms is to explain the observed high dust-to-gas ratios. In models that use sources of momentum other than the radiation pressure on dust to initiate the wind, it is often assumed that the grains will form somewhere in the outflow, as soon as the temperature drops below a certain value. Unfortunately, while a low temperature is a necessary condition for dust formation, it is not sufficient. Even simple kinetic arguments show that grain growth is a relatively slow process that proceeds far from equilibrium under the typical conditions in the atmospheres of AGB stars (see Section 4.4.6). The gas density is a critical factor for the growth rates and the final degree of condensation. Therefore, even if the temperature eventually becomes low enough in the wind, it may not be possible to produce enough dust simply as a by-product of the outflow, because the gas becomes diluted too fast to allow significant condensation.

4.8.3 Observables for the Inner Wind Regions

Detailed observations of the inner wind region, i.e., the zone from the stellar photosphere out to about 10–20 stellar radii, are crucial for our understanding of mass loss. Especially important in this context are monitoring programs that cover at least one pulsation cycle of a star. The time-dependent spatial structures of gas and dust densities, the temperatures, and velocity fields must be studied systematically in order to settle the question of mass-loss

---

[18] Presently the discussion centers on the actual value of the critical gas density below which the cooling must be described by non-LTE processes, and e.g., [195] arrives at much lower values than [190].

mechanisms, and to permit the construction of self-consistent models that produce reliable predictions for synthetic spectra and mass-loss rates from first principles.

4.8.3.1 General Considerations

The visual and infrared spectra of AGB stars provide a wealth of information in terms of atomic lines, molecular bands (electronic, vibrational, and rotational transitions covering a wide range of energies), and dust features. It is, however, not easy to select suitable observables for the outer atmosphere and inner wind region (in particular with the restrictions imposed on ground-based observation by atmospheric extinction in critical infrared regions). Some general criteria can be summarized in the following way:

- When probing the spatial density and temperature structure, one should consider in particular those spectral features that are incompatible with a photospheric origin (in the classical hydrostatic sense of the word), e.g., those caused by molecules that are expected to have negligible abundances at the hot temperatures prevailing in the inner atmosphere or at the low densities characteristic of the cool outer layers of hydrostatic atmospheres. The mere presence of such features can be a telltale sign of atmospheric levitation by dynamical effects (shocks) or a stellar wind; a detailed analysis may lead to quantitative predictions of densities and temperatures.
- A more direct picture of the kinematics of the outer atmosphere and inner wind regions requires high-resolution spectroscopy in order to relate observed variable line profiles to the actual velocity fields via comparison with synthetic lines from dynamical models. Here, in contrast to the first point, it is actually an advantage if the abundances of the respective molecules are not very sensitive to densities and temperatures, so that the line strengths and shapes reflect primarily the Doppler shifts caused by the flow velocities.

In both cases, a simultaneous monitoring of several features originating at different distances from the stellar photosphere is important to reconstruct "snapshots" of the overall structure of the extended atmosphere, and to trace propagating shocks and outflowing matter.

The last decade has seen considerable progress in observational techniques, as regards high spatial resolution (interferometry), high spectral resolution (variability of line profiles; kinematics), and facilities providing access to wavelength regions that are not observable from the ground (e.g., ISO, HST). The results have provided valuable constraints for models and probable mass-loss mechanisms. Due to the present rapid developments in this field, the following discussion is only a short overview that is far from complete. The aim is to give representative examples; the reader is referred to recent reviews for a more detailed picture.

4.8.3.2 Spectroscopic Data and Photometric Colors

For obvious reasons, the most common type of observational data (in particular when it comes to time series for individual objects) are photometric colors and magnitudes. These are of great importance, not least due to the fact that a high accuracy in absolute calibrations may be obtained. However, the very nature of the data makes a theoretical interpretation difficult. First of all, a number of different physical phenomena are "folded" into a single observational quantity (combination of features originating in different layers of the atmosphere, depending on the global thermodynamical structure, velocity fields, molecular abundance variations, etc.). Secondly, some of the basic data (such as opacities or chemical data) entering into the synthetic colors derived from models are uncertain. Both effects combine to make, e.g., the prediction of absolute values of colors difficult. However, features like relative phase lags between light curves taken in different filters may give valuable hints about the propagation of shock waves through the atmosphere (local variations of the thermodynamical quantities due to passing shocks) or the formation of new dust shells (redistribution of light to other spectral regions). In summary, at the present stage of qualitative or at best semiquantitative modeling, the use of photometric data is a tricky diagnostic tool. The full value of existing data and ongoing monitoring programs will become obvious only when the predictive power of self-consistent dynamical models has reached a reliable quantitative level, which may take some years yet. Spectroscopic data are more immediately applicable.

A set of spectroscopic data that deserves special discussion in the context of the extended atmosphere and inner wind region consists of the low- and mid-resolution spectra obtained with the ISO Short Wavelength Spectrometer (SWS). This unique instrument covered a wavelength range of about 2.5–40 µm, i.e., including spectral regions that are not at all accessible from the ground, and allowed for a simultaneous measurement of molecular vibration–rotation bands originating in different layers as well as measurement of dust features.

In the atmospheres of O-rich AGB stars, the molecules CO, SiO, OH, and H_2O are expected to be abundant; and they are indeed detected in the ISO-SWS spectra. However, the discovery of features ascribed to CO_2 and SO_2 [158, 199] was more surprising, because these dioxides (and in particular SO_2) were not expected to be very abundant. Their presence can be interpreted as an indication of high densities in the cool outer layers of the extended atmospheres.[19] Even for the abundant molecules (e.g., SiO, H_2O) the strengths and shapes of features often deviate from synthetic spectra based

---

[19] Note, however, that the quantitative deduction of gas densities is difficult, since, e.g., the validity of chemical equilibrium seems to be highly questionable in this context; nevertheless, local densities and temperatures derived from semiempirical fitting of the features are in qualitative agreement with current dynamical models (e.g., [198]).

on (hydrostatic) photospheric models, and, not surprisingly, the discrepancies increase with decreasing effective temperature and increasing pulsation amplitude of the stars. These deviations led Tsuji et al. [186] to propose a *warm molecular envelope*, i.e., a dense, relatively warm layer of gas above the stellar photosphere that could account for additional contributions to the purely photospheric features. In terms of dynamical models it is, at least qualitatively, simple to explain the presence of such a layer; however, a detailed quantitative modeling of the molecular spectra is a nontrivial problem still under investigation. [For example, [8] have reported recent progress for features of H_2O based on dynamical models.]

The ISO-SWS spectra of C-rich AGB stars clearly show a number of features of diatomic and polyatomic molecules, i.e., CO, CN, C_2, CS, C_3, C_2H_2, and HCN. Investigations of the near- and mid-IR spectra of hot, not strongly variable C-stars found good agreement with hydrostatic model atmospheres for wavelengths up to about $10\,\mu m$ [100, 126]. At longer wavelengths, even these only mildly variable stars show significant differences compared to the models, which the authors attribute to nonphotospheric contributions to the spectrum. Of particular interest in this context is the combined feature of C_2H_2 and HCN around $14\,\mu m$, which is clearly visible in somewhat cooler stars like R Scl but is very weak or not detectable in hot stars like TX Psc (see Figure 4.1). Hydrostatic models, however, predict a strong absorption feature at these wavelengths, even for the hot C-stars. While it is not unusual for a feature seen in an observed spectrum to be missing in the synthetic counterpart, e.g., due to incomplete opacity data, the opposite case, as encountered here, is somewhat alarming (in particular since this feature is seen in other stars, and is due to molecular transitions similar to those other features in the same spectra that are fitted well; see [100] for details). Several possible explanations for this phenomenon have been discussed, but a preliminary study with dynamical models indicates that the $14\,\mu m$ feature can be filled in by molecular emission contributions from the extended atmosphere [85, 126]. In general, the discrepancies between (static) model atmospheres and ISO spectra increase with decreasing effective temperature and increasing pulsation amplitude of the stars. This is similar to the O-rich case.

High-resolution spectroscopy of molecular vibration–rotation lines in the infrared region seems to be a very promising tool for studying the kinematics of atmospheres and wind acceleration regions. Among various molecules that have been used to trace velocity fields, the CO molecule deserves special attention. Due to its high bond energy, it forms deep inside the atmosphere and therefore allows us to probe many different layers, is stable against dissociation, is very abundant, and exists in both M and C stars. Several decades of ground-based monitoring programs in the near IR (e.g., [73, 122], have led to the following general results. Second-overtone lines around $1.6\,\mu m$ originating in layers with temperatures between about 2200 and 4000 K nicely trace the pulsations of the stars. Both M- and C-type Mira stars show characteristic S-shaped, discontinuous radial velocity curves with line doubling occurring around the luminosity maximum (see Figure 4.4). The amplitude of these curves (difference between maximum and minimum velocity) is about 30 km s^{-1} with only small deviations for individual stars. The line doubling (splitting of individual lines into two components with different Doppler shifts) is interpreted as a shock wave passing through the line-formation zone with an abrupt change in velocity at the shock front. Depending on pulsation phase, both infalling and outward-moving material (in the star's frame of reference) is seen.

First-overtone lines at about 2.3 µm can be divided into two groups, according to their temporal behavior: High-excitation lines behave similarly to the second overtone lines, showing periodic S-shaped velocity curves with line doubling, while low-excitation lines (tracing layers with about 800 K) usually show only moderate velocity variations around the center-of-mass velocity of the star with no clear periodicity. Line doubling seldom occurs. Fundamental lines around 4.6 µm are presumably produced in the stellar wind region, showing outflowing material at all phases. Synthetic line profile variations of fundamental and first-overtone lines based on dynamical models [56, 192, 194] are in qualitative agreement with observations.

The second-overtone velocity curves of semiregular variables (SRVs) show significantly smaller amplitudes that distinguish them clearly from Mira variables, and no typical shape of the velocity variation has been found [76, 121].

4.8.3.3 Spatially Resolved Observations

To decide whether radiation pressure on dust is the dominant driving force for the stellar wind or whether the grains form merely as a by-product of the outflow, it is important to know the distance of the dust shell's inner radius from the stellar photosphere. Danchi et al. [32] have used the Infrared Spatial Interferometer (ISI) at 11 µm to measure the inner radii of dust shells for different types of late-type stars. They conclude that Mira variables as well as semiregular and irregular variables form dust very close to the star, i.e., at a few stellar radii, which gives strong support to the pulsation-enhanced dust-driven wind model. Ridgway and Keady [154] obtained similar results with interferometric and lunar occultation measurements for the dust-enshrouded IRC +10216 at shorter wavelengths. Winters et al. [193] have presented synthetic brightness distributions and visibilities of C-rich LPVs with heavy mass loss, based on self-consistent time-dependent dynamical models. They find temporal variations in accordance with observations and interpret variable structures seen in observed visibility functions as an onion-skin-like structure of the circumstellar dust shell.

Apart from the various types of thermal radiation discussed above, maser emission has turned out to be a useful tool for studying stellar winds. A discussion of these observations and their implications can be found in Chapter 7, but some relatively recent results should be mentioned here. In contrast to OH and H_2O masers, which trace layers further away from the star, SiO maser emission originates quite close to the star, i.e., within a few stellar radii, probably inside the dust-formation radius. The SiO emission comes from small patches arranged in a shell around the star [38]. Long-term monitoring programs allow us to trace the evolution of these patterns (interpreted as movements of individual gas elements) and deduce time-dependent and spatially resolved information about the kinematics of this region. It is interesting to note that both outflow and infall are observed, in accordance with the scenario in which the emission is formed inside of the wind acceleration region. The variation of the SiO maser flux density for Miras is similar to that of the optical flux, but with a phase lag of 0.2 periods. Due to the more complicated processes that generate the maser emission, the interpretation of the observations is possibly not as straightforward as for the molecular absorption line profiles mentioned above; but the results of recent modeling attempts look promising [88]. H_2O masers form at somewhat larger distances from the star, presumably within the dust formation and wind acceleration region (e.g., [47, 149]). In contrast to SiO masers, the correlation of the intensities with the optical light curve is weak or nonexistent for Miras and semiregulars.

4.9 Abundances and Other Fundamental Parameters

The determination of elemental abundances is a main objective in the study of AGB star atmospheres. One reason is that such determinations offer insight into the role played by these stars in the nucleosynthesis in galaxies. It is thought that stars in these stages are important sources of Li, C, N, F, and a great number of elements beyond the iron peak, the so-called s-elements. Moreover, the study of abundances in the atmospheres of AGB stars gives insight into their internal structure, mixing processes, and evolution in general. Finally, these stars are among the brightest stars in galaxies and may be used for determining the general abundance characteristics of these systems.

However, the determination of abundances in these stars belongs to the most difficult tasks in stellar abundance work, and therefore rather few reliable results (i.e., results with well-known errors) have been obtained as yet. For a review that is not yet totally out of date—its relevance more than a decade later is in itself a characteristic of the unsatisfactory situation—see [59]. The main reason for the difficulty is the crowdedness of the spectra, which makes it almost mandatory to synthesize the relevant spectral regions from model atmospheres. However, as has been illustrated above, these models are still far from realistic. In practice, almost all abundance work has until recently been based on static model atmospheres. Also, the spectrum calculations, even within narrow wavelength bands, require a great set of atomic and molecular data, because hundreds of spectral lines may affect the strength of each abundance criterion. This, in turn, puts heavy demands on the quality of the relevant physical data that often cannot be met by published empirical or theoretical results from atomic and molecular physics, even if the situation in this respect has improved during the last decade.

The situation is somewhat better in the infrared wavelength region and in particular beyond 1.5 μm, at least for stars with $T_{\text{eff}} > 3000$ K, where the polyatomic molecular absorption has not yet become a major characteristic feature that depresses the continuum throughout the spectrum. For those stars, it is possible to find "continuum windows" that in principle allow the measurement of equivalent widths. An example of work that takes advantage of this opportunity is the extensive study of carbon-star abundances by Lambert et al. [117], where Fourier transform spectrometer (FTS) data at high resolution were used in the wavelength interval 1.6–4 μm. That study concentrated on the CNO elements, which simplified matters, since several molecular systems with those elements offer many different spectral bands and lines as criteria. The high-resolution Fourier transform spectrometers, and more recently the cryogenic echelles [74], have in general widened the possibilities of this field of science dramatically.

In Section 4.9.2 we shall make some more general comments and review some of the results in abundance determinations for AGB stars. We refer the reader who is interested in details and in particular the interpretation of results in terms of stellar evolution and nucleosynthesis to other chapters in this book, as well as to the current literature in the field.

4.9.1 Effective Temperatures and Surface Gravities

In order to be able to interpret observed spectra in terms of abundances, fundamental parameters like the stellar luminosity L, radius R_*, and mass M must be known or guessed. From these one may calculate the effective temperature T_{eff} and the surface gravity g [see (4.1) and (4.11)]. However, since the radius is not well-defined for such stars with extended atmospheres but relates to a particularly defined optical depth, T_{eff} and g are also ambiguous unless this optical depth is specified.[20] In practice, the surface gravity is not ideal for use as a fundamental parameter, but should be substituted by the radius and the mass. This is because the stellar spectrum and atmospheric structure are in principle functions of both these parameters, and it may in fact be possible to determine both, and not just a combination of the two in terms of M/R_*^2 observationally (see [162] and references therein).

The overall metallicity and in particular the abundances of the most important electron donors, which affect the dominating continuous opacity H^-, should also be known or determined from the spectra. This is not always simple, since these elements are represented by rather few clean spectral lines

---

[20] A reasonable and frequent choice that is often used in practice to define stellar radii is to take the radius at which the optical depth in the Rosseland mean opacity is unity.

in AGB spectra, which are overcrowded by molecular bands. In addition, the abundances of the elements that contribute the dominating line opacities (i.e., the CNO elements as well as some metals such as Ti with significant oxides) must be determined in the process, because they affect the atmospheric structure and, directly and indirectly, the spectrum. It is clear that the procedure to be used for abundance determination is an iterative one, along with first guesses of relevant parameters, model and spectrum calculations, comparisons to the observed spectra, modifications of abundances, and other parameters.

In principle, L, R_*, and M might also be determined from the spectra and adjusted in the iterative process. This is, however, both difficult and very uncertain in view of all the uncertainties involved in present models of stellar atmospheres and spectra. However, additional astronomical information is available and is most often used. The stellar angular diameter θ may be known from interferometric measurements or from time-resolved observations of occultations of the star by the Moon, either for the program star or for stars of very similar spectral types. This, when combined with observations of the apparent bolometric flux f, can be used for estimating an effective temperature T_{eff}. For abundance analyses this is the most important fundamental parameter in view of the strong temperature dependence of the abundance criteria used. We have (cf. Section 3.1)

$$f = \sigma_R T_{\text{eff}}^4 \theta^2 / 4. \qquad (4.80)$$

Here, both T_{eff} and θ must be consistent with the definition of the radius. In practice, it is not quite simple to determine θ from the measurements, since assumptions must be made in converting the observations of interferometric data (e.g., "visibility amplitudes") to angular diameters. The conversion depends on what is assumed as regards the spectral energy distribution and the intensity distribution across the stellar disk. The correction from angular diameters measured in particular filters—in which the atmospheres may be more or less transparent—needs some model-atmosphere calibration. One must also calibrate the measurements for the effects of the intensity variation across the stellar disk.

Generally, measurements are presented as angular diameters for an assumed uniform stellar disk. Available model atmospheres suggest that these diameters should be corrected by a factor between 1.02 to 1.04; but depending on the extension of the atmosphere and the effects on the atmosphere due to stellar pulsations, this correction may be even larger. The bolometric flux f (which should in principle be measured above the Earth's atmosphere) is usually estimated from broad-band colors and data from infrared space telescopes such as the Infrared Space Observatory, supplemented by model-atmosphere fluxes. Angular diameters and effective temperatures based on interferometric measurements for altogether more than 100 AGB stars, both late M giants and C giants, have recently been published (e.g.,

[42, 44, 45, 64, 140, 150, 151, 188]). Once a set of angular diameters has been obtained it can be used to establish a relation between the colors and the effective temperature scale. Recently, the effective temperatures have been published of 390 carbon-rich stars, based on a calibration of photometric and spectrophotometric data [16].

An important method for estimating effective temperatures is the infrared flux method invented by Blackwell and Shallis [20]. In this method, the flux f_i in a particular band in the infrared, e.g., the L band at 3 μm, is measured observationally relative to the total integrated flux f, which is a direct measure of the effective temperature. Since the Planck function, and thus hopefully the stellar flux, is only weakly dependent on the temperature at wavelengths much longer than that of the Planck maximum, the ratio f_i/f may be only marginally dependent on the detailed model structure and reflect mainly the effective temperature. Therefore, this ratio may also be calibrated with simple model atmospheres and used for determining T_{eff}. The models can also be used to correct the chosen IR band for the effects of molecular absorption. The method was used already on M and C giants by Tsuji [179, 180, 181], who found a satisfactory agreement with angular diameter measurements. More recent studies with much larger samples of angular-diameter stars verify this finding, even if systematic discrepancies seem to exist for the cooler stars (cf. [16]).

In addition to the effective temperature, the gravitational acceleration $g = GM/R_*^2$ is of significance for the abundance analysis. This parameter sets the characteristic pressures in the atmospheres, at least as long as the atmospheres are close to hydrostatic equilibrium. Spectroscopic means for determining pressures are available, since the molecular equilibria are pressure-dependent (cf. Section 4.4.5) and the relative abundances of different molecules (such as H_2O relative to TiO) depend on the pressure. Other features that may be used for estimating pressures in stellar atmospheres are the (forbidden and thus weak) quadrupole lines of H_2 in the infrared [25, 72, 95, 117, 182]. However, spectroscopic methods for estimating the gravity are still very uncertain. This is partially due to the high temperature sensitivity of the molecular abundances, requiring not only good temperature scales but also good knowledge about the thermal structure of the stellar atmospheres.

If both the distance D to the star as well as the interstellar and circumstellar reddening are known, then the luminosity L may be obtained from the apparent bolometric magnitude, and one may derive the surface gravity from a guess of the mass and the relation

$$g = 4\pi\sigma_R G \frac{MT_{\mathrm{eff}}^4}{L} . \qquad (4.81)$$

This method obviously contains a number of uncertainties, the most important for galactic field stars being the uncertainty in distance. Only a handful of AGB stars in our Galaxy have parallaxes with an accuracy better than 30%

(cf. [107]), and the distance comes in quadratically (through L) in (4.81). For stars in clusters, as well as in nearby galaxies such as the Magellanic Clouds, the situation is better, since the system distance may be known. For cluster stars the cluster age also may give important bounds as regards the stellar mass.

In general, however, we conclude that the uncertainty regarding the stellar surface gravity is a severe obstacle in abundance analyses of AGB stars. The gravity is hardly known to an accuracy better than a factor of two for any single star. One should also note that even if M, T_{eff}, and L were well known, the extension of the atmosphere (e.g., as caused by dynamical effects) may considerably change atmospheric pressures to values far from those of static models with g from (4.81). From this point of view, as long as model atmospheres are not describing the dynamical phenomena properly, spectroscopic gravity criteria such as H_2 lines should probably be preferred.

4.9.2 Abundance Determinations

Spectroscopic analyses of stellar elemental abundances are generally based on measurements of spectral line strengths. Here, it is important to find quite weak lines,[21] since their strengths scale linearly with the abundance and are not directly dependent on complex and uncertain processes such as dynamical phenomena and collisional line broadening. However, if one does not have access to high-resolution spectra beyond 1.5 μm, such weak features are almost always blended in the spectra of AGB stars. Moreover, in the visual and near infrared it is very difficult to find spectral regions that are so free from line absorption that continua can be traced as is required for measuring equivalent widths.

In studying crowded spectral regions, one should always use synthetic spectra, i.e., spectra calculated for different assumed abundances of the element under study, and compare these with the observed spectra. However, one must then remember that even if the final fit resulting from such a comparison between theory and observation may seem very reassuring, the absence of proper atomic and molecular data (e.g., for faint absorption from a wealth of weak transitions that may overlay the spectral region under study) could well lead to very severe systematic errors in the resulting abundances.

In spite of these difficulties, chemical abundances for many elements have been derived for AGB stars during the last two decades. Here, just a few examples will be given (for more historical remarks, see [59]).

Smith and Lambert [169, 170] studied 21 M, MS, and S stars spectroscopically, using highly resolved FTS spectra of CO, OH, and NH vibration–rotation lines as well as lines from the CN electronic red system to get the CNO element abundances. In addition to deriving $^{12}C/^{13}C$ ratios, they also

---

[21] In practice, lines should have equivalent widths such that $\log(W_\lambda/\lambda) \leq -5$.

obtained abundances for a number of metals (Ti, Fe, Ni), including important s-elements (Sr, Y, Zr, Ba, Nd), from spectra in the near infrared. Rb abundances were determined from the Rb I 780 nm line for a similar sample of stars by Lambert et al. [119]. They used the ratio of Rb/Sr to estimate the neutron density at the time of s-processing, which was found to be consistent with the operation of the s-process during the interpulse intervals in low-mass AGB stars. More recently, Abia and Wallerstein [1] have published heavy-element abundances (from Sr to Eu, including Rb and Tc) relative to Fe for 7 SC stars and 4 S stars. Harris and Lambert [65] and Harris et al. [67] determined oxygen isotopic ratios for M and MS stars. Isotopic ratios for a number of isotopes (Mg, Si, Cl, Ti, Zr) have been determined from molecular bands by various groups, producing results that are of interest for attempts to trace the origin or more precise conditions for the s-process. For instance, [119] found the absence of ^{96}ZrO bands at 693 nm in MS and S giants to be consistent with their finding from Rb/Zr ratios that the observed stars do not have massive cores and are thus of low mass. For a more general discussion of the s-process operation in AGB stars as traced from observed elemental abundances, see Busso [24].

Lambert et al. [117] obtained abundances of C, N, O, and metals, as well as ^{12}C/^{13}C ratios for about 30 carbon stars, and Harris et al. [66] obtained oxygen isotope abundances for these stars. These authors used high-resolution FTS spectra in the 1.6 and 2.2 μm regions. Molecular lines from CO, C_2, and CN were used, as well as a small number of metal lines. The results of Lambert et al. are in fair agreement with later independent studies by Ohnaka et al. [137], with agreements often within 0.1 dex (about 25%), which is gratifying. However, the important C/O ratios differ systematically between the two studies, with larger values being obtained by Ohnaka et al. This may be partly due to a systematic difference in effective temperatures, and partly due to differences in the model atmospheres. Another example of diverging results, illustrating the difficulties in the analysis, are the carbon isotope ratios obtained by Lambert et al. from IR FTS spectra as compared with those by Ohnaka and Tsuji [135] using spectra in the near infrared. The latter ^{12}C/^{13}C ratios are systematically lower, a fact that [35] ascribe to methodological problems in the crowded near infrared (cf. [36, 136]). In general, the results discussed above support the view that carbon stars are formed mainly as a result of the addition of ^{12}C to the stellar atmospheric gas (due to the mixing of material that has undergone helium burning to the atmospheric layers). Carbon stars also tend to have overall metal abundances somewhat below solar and—more unexpectedly—rather low nitrogen abundances.

In a survey of 112 red giants in the Magellanic Clouds, Smith et al. [171] detected the Li 671 nm resonance line in 35 stars, all on the asymptotic giant branch. In spite of great difficulties due to blocking by TiO in this spectral region, the authors attempted abundance analyses and found abundances ranging from 1 to 4 on the logarithmic scale where the number of hydrogen atoms is 12. The authors found their result to be consistent with the hypothesis that Li is produced in hot bottom convective envelope burning; see Chapter 2.

As model atmospheres improve, studies at lower resolution may also give reasonable abundance estimates. One example of this is the work by [127], who obtain reasonable C/O ratios and ^{12}C/^{13}C ratios for some carbon stars from spectra at a resolution $R = \lambda/\delta\lambda$ of about 1100.

The examples above give some glimpses of a difficult but very interesting field of research in which theories of stellar evolution, nuclear burning, models of envelope mixing, and stellar atmospheres are all more or less uncertain but gradually becoming constrained empirically by observations of rapidly increasing volume and quality. This development will now accelerate rapidly as the next generation of high-resolution infrared spectrometers are introduced, e.g., the CRIRES spectrometer at the ESO Very Large Telescope. With this instrument, it will be possible to study spectroscopically individual AGB stars in the Magellanic Clouds and other Local Group galaxies at a resolution of $R \approx 10^5$. No doubt, this will lead to many more detailed and reliable abundance results.

4.10 Conclusions and Recommendations

It should be clear from the present chapter and other chapters in this book that the study of AGB star atmospheres is of great significance for the understanding not only of the stars themselves but also of their rôle in stellar evolution, in the evolution of galaxies, and in the formation of the chemical elements. The study of the atmospheres provides almost all empirical evidence and tests our understanding of what happens in the interior. Since that understanding is still preliminary, observations of the atmospheres offer central verification and inspiration for new ideas. And since the atmosphere forms the boundary layer from which the stellar wind is accelerated, studies of atmospheres are of profound significance for the understanding of winds and CSEs. Finally, observations and the interpretation of these observations give direct information on properties of the stars that can be exploited in a wide range of astronomical and astrophysical studies, far outside the field of stellar physics per se.

The study of atmospheres of AGB stars, in which observational and theoretical efforts must be closely linked, is a difficult and intricate activity with many problems. However, it should also be clear that a number of interesting new possibilities are presently opening up due to very important fundamental progress. This includes much improved physical data, impressive progress in numerical hydrodynamics and not the least the very remarkable development in observational techniques, which includes interferometric imaging of AGB star surfaces and high-resolution spectroscopy in the infrared for many stars in different environments and different dynamical phases.

It will be a great challenge to use all these new possibilities in an optimal way. A very large number of interesting research projects can be formulated. Which will be the most important and prolific ones for the present decade?

This is certainly a difficult question of taste. The present authors would, however, suggest the following list of urgent projects:

- Explore the properties of AGB star convection, using high-resolution spectra in the infrared to show line asymmetries due to convective motions, and simultaneous interferometric studies of the stellar surface. Make detailed 3D simulations for comparison with observations. Explore the phenomena that set the scale of the convection patterns.
- Explore, observationally as well as theoretically in 3D simulations, the coupling between convection and pulsations in AGB stars.
- Study the dust-formation process and the coupling between dust, pulsations, and mass loss, with more detailed kinetic models for the interaction between dust and gas.
- Investigate the warm molecular layer in the upper photospheric layers, both in terms of spatial distribution across the star, as well as its depth structure and its dynamics.
- Scrutinize systematically the consistency, or lack of consistency, in abundance determinations for AGB stars in using different criteria, such as CH and C_2 lines and NH, CN, and HCN lines. Also, study carefully departures from LTE, observationally as well as theoretically, in molecular and ionization equilibria as well as in the excitation and formation of spectral lines.
- Determine atmospheric abundances for AGB stars in different stellar systems with different overall metallicity. The Local Group offers many objects for such a study, within the reach of cryogenic echelles.

About two decades ago, one of us published a review on contemporary and future research on red-giant atmospheres, in which a cartoon was presented (Gustafsson [58]). In that picture the classical theory, depicted as a Greek temple, rested on four hills (the classical basic assumptions of hydrostatic equilibrium, homogeneous stratification, LTE, and the mixing-length theory of convection), eroded by new observations. One could envisage a "New Theory" in which these assumptions had been relaxed, and this dream was being sketched by an ambitious artist who appeared in the cartoon. He appears again in Figure 4.18, but now with many companions drawing drastically different pictures of an AGB star from different perspectives.[22] Nevertheless, despite the differences, many scientists are now making coherent and persistent efforts to construct a much improved and much more solid theory. Maybe, after another 20 years, a unified theory of AGB stars that encompasses everything from their centers to their outer winds will have resulted

---

[22] The reader interested in art may easily trace the trend from modernism to postmodernism during these two decades.

Fig. 4.18. Different aspects in studies of AGB star atmospheres. A postmodernistic view

from these efforts. How different will this theory be from the attempts we have presented here?

On the way to this unified theory, we shall certainly see full non-LTE models, models with nonequilibrium time-dependent chemistry and dust formation, models with 3D hydrodynamics and detailed radiative transfer, and models with magnetic fields. On the observational side, the excellent possibilities presently opening up will support these models, or disprove them. As usual in astrophysics, it is conceivable or even probable that a number of quite new and very significant phenomena will be discovered.

Acknowledgments: A number of colleagues and collaborators have contributed very significantly with comments and discussion. We thank Anja Andersen, Bernd Freytag, Uffe Gråe Jørgensen, Rita Loidl, Bertrand Plez, Nils Ryde, Christer Sandin, and Rurik Wahlin. Particular thanks are due to Kjell Eriksson and Nikolai Piskunov, who also produced many of the figures, and Michelle Mizuno-Wiedner, who read the manuscript very carefully and made numerous comments. Finally, we wish to thank the editors for their remarks and patience.

References

1. Abia, C. and Wallerstein, G. *MNRAS*, 293, 89+, 1998.
2. Alvarez, R., Jorissen, A., Plez, B., et al. *A&A*, 379, 305, 2001.
3. Alvarez, R., Lançon, A., Plez, B., and Wood, P. R. *A&A*, 353, 322, 2000.
4. Alvarez, R. and Plez, B. *A&A*, 330, 1109, 1998.
5. Anderson, L. S. *ApJ*, 339, 558, 1989.
6. Aoki, W., Tsuji, T., and Ohnaka, K. *A&A*, 350, 945, 1999.
7. Aringer, B., Höfner, S., Wiedemann, G., et al. *A&A*, 342, 799, 1999.
8. Aringer, B., Jørgensen, U. G., Kerschbaum, F., Hron, J., and Höfner, S. In Aerts, C., Bedding, T., and Christensen-Dalsgaard, J., editors, *IAU Colloq. 185: Radial and Nonradial Pulsations as Probes of Stellar Physics*, page 538. ASP: San Franscisco, 2002.
9. Asplund, M., Nordlund, Å., Trampedach, R., Allende Prieto, C., and Stein, R. F. *A&A*, 359, 729, 2000.
10. Asplund, M., Nordlund, Å., Trampedach, R., and Stein, R. F. *A&A*, 346, L17, 1999.
11. Avrett, E., Hoeflich, P., Uitenbroek, H., and Ulmschneider, P. In Pallivicini, R. and Dupree, A., editors, *Ninth Cambridge Workshop on Cool Stars, Stellar Systems, and the Sun*, page 105. ASP: San Francisco, 1996.
12. Ayres, T. R. and Wiedemann, G. R. *ApJ*, 338, 1033, 1989.
13. Beck, H. K. B., Gail, H.-P., Henkel, R., and Sedlmayr, E. *A&A*, 265, 626, 1992.
14. Bedding, T. R., Zijlstra, A. A., von der Luhe, O., et al. *MNRAS*, 286, 957, 1997.
15. Bell, R. A., Balachandran, S. C., and Bautista, M. *ApJ*, 546, L65, 2001.
16. Bergeat, J., Knapik, A., and Rutily, B. *A&A*, 369, 178, 2001.
17. Bessell, M. S., Brett, J. M., Wood, P. R., and Scholz, M. *A&A*, 213, 209, 1989.
18. Bessell, M. S., Castelli, F., and Plez, B. *A&A*, 333, 231, 1998.
19. Bessell, M. S., Scholz, M., and Wood, P. R. *A&A*, 307, 481, 1996.
20. Blackwell, D. E. and Shallis, M. J. *MNRAS*, 180, 177, 1977.
21. Bowen, G. H. *ApJ*, 329, 299, 1988.
22. Bowen, G. H. and Willson, L. A. *ApJ*, 375, L53, 1991.
23. Brown, J. A., Johnson, H. R., Cutright, L. C., Alexander, D. R., and Sharp, C. M. *ApJS*, 71, 623, 1989.
24. Busso, M., Gallino, R., Lambert, D. L., Travaglio, C., and Smith, V. V. *ApJ*, 557, 802, 2001.
25. Carbon, D. F., Augason, G. C., and Johnson, H. R. *BAAS*, 18, 987, 1986.
26. Carlsson, M. *Uppsala Astronomical Observatory Reports*, 33, 1986.
27. Carlsson, M. In *7th Cambridge Workshop on Cool Stars, Stellar Systems, and the Sun*, page 499. ASP: San Francisco, 1992.
28. Carlsson, M. and Stein, R. F. *ApJ*, 440, L29, 1995.
29. Carpenter, K. G., Robinson, R. D., and Johnson, H. R. *American Astronomical Society Meeting*, 29, 1230, 1997.
30. Carpenter, K. G., Robinson, R. D., Johnson, H. R., et al. *ApJ*, 486, 457, 1997.
31. Chen, P. C. *A study of the polarization of red variable stars*. PhD thesis, Case Western Reserve University, Cleveland, 1979.
32. Danchi, W. C., Bester, M., Degiacomi, C. G., Greenhill, L. J., and Townes, C. H. *AJ*, 107, 1469, 1994.

33. de La Reza, R. and Querci, F. *A&A*, 67, 7, 1978.
34. de Laverny, P., Geoffray, H., Jorda, L., and Kopp, M. *A&AS*, 122, 415, 1997.
35. de Laverny, P. and Gustafsson, B. *A&A*, 332, 661, 1998.
36. de Laverny, P. and Gustafsson, B. *A&A*, 346, 520, 1999.
37. Diamond, P. J. and Kemball, A. J. In Le Bertre, T., Lebre, A., and Waelkens, C., editors, *IAU Symp. 191: Asymptotic Giant Branch Stars*, page 195. ASP: San Francisco, 1999.
38. Diamond, P. J., Kemball, A. J., Junor, W., et al. *ApJ*, 430, L61, 1994.
39. Dominik, C., Gail, H. P., Sedlmayr, E., and Winters, J. M. *A&A*, 240, 365, 1990.
40. Dravins, D. and Nordlund, A. *A&A*, 228, 184, 1990.
41. Dupree, A. K., Hartmann, L., and Smith, G. H. In *6th Cambridge workshop on Cool Stars, Stellar Systems, and the Sun*, page 408. ASP: San Francisco, 1990.
42. Dyck, H. M., Benson, J. A., van Belle, G. T., and Ridgway, S. T. *AJ*, 111, 1705, 1996.
43. Dyck, H. M. and Jennings, M. C. *AJ*, 76, 431, 1971.
44. Dyck, H. M., van Belle, G. T., and Benson, J. A. *AJ*, 112, 294, 1996.
45. Dyck, H. M., van Belle, G. T., and Thompson, R. R. *AJ*, 116, 981, 1998.
46. Ekberg, U., Eriksson, K., and Gustafsson, B. *A&A*, 167, 304, 1986.
47. Engels, D., Winnberg, A., Brand, J., and Walmsley, C. M. In Le Bertre, T., Lebre, A., and Waelkens, C., editors, *IAU Symp. 191: Asymptotic Giant Branch Stars*, page 373. ASP: San Francisco, 1999.
48. Eriksson, K., Gustafsson, B., Jørgensen, U. G., and Nordlund, A. *A&A*, 132, 37, 1984.
49. Fleischer, A. J., Gauger, A., and Sedlmayr, E. *A&A*, 266, 321, 1992.
50. Fleischer, A. J., Gauger, A., and Sedlmayr, E. *A&A*, 297, 543, 1995.
51. Freytag, B. In *11th Cambridge Workshop on Cool Stars, Stellar Systems and the Sun*, page 785. ASP: San Francisco, 2001.
52. Gail, H.-P. and Sedlmayr, E. *A&A*, 161, 201, 1986.
53. Gail, H. P. and Sedlmayr, E. *A&A*, 171, 197, 1987.
54. Gail, H.-P. and Sedlmayr, E. *A&A*, 206, 153, 1988.
55. Gauger, A., Sedlmayr, E., and Gail, H.-P. *A&A*, 235, 345, 1990.
56. Gauger, A., Winters, J. M., Fleischer, A., and Keady, J. J. In Kaper, L. and Fullerton, A., editors, *Cyclical Variability in Stellar Winds*, page 309. Springer-Verlag: Berlin, New York, 1998.
57. Gebbie, K. B. and Thomas, R. N. *ApJ*, 161, 229, 1970.
58. Gustafsson, B. In Iben, I. and Renzini, A., editors, *Physical Processes in Red Giants*, page 25. D. Reidel Publishing Co.: Dordrecht, 1981.
59. Gustafsson, B. *ARA&A*, 27, 701, 1989.
60. Gustafsson, B. and Bell, R. A. *A&A*, 74, 313, 1979.
61. Gustafsson, B., Bell, R. A., Eriksson, K., and Nordlund, A. *A&A*, 42, 407, 1975.
62. Gustafsson, B., Edvardsson, B., Eriksson, K., et al. In Hubeny, I., Mihalas, D., and Werner, K., editors, *Workshop on Stellar Atmosphere Modeling*. ASP: San Francisco, in press.
63. Gustafsson, B. and Jørgensen, U. G. *A&A Rev.*, 6, 19, 1994.
64. Haniff, C. A., Scholz, M., and Tuthill, P. G. *MNRAS*, 276, 640, 1995.
65. Harris, M. J. and Lambert, D. L. *ApJ*, 285, 674, 1984.

66. Harris, M. J., Lambert, D. L., Hinkle, K. H., Gustafsson, B., and Eriksson, K. *ApJ*, 316, 294, 1987.
67. Harris, M. J., Lambert, D. L., and Smith, V. V. *ApJ*, 299, 375, 1985.
68. Hartmann, L. and MacGregor, K. B. *ApJ*, 242, 260, 1980.
69. Hauschildt, P. H., Allard, F., Alexander, D. R., and Baron, E. *ApJ*, 488, 428, 1997.
70. Hauschildt, P. H., Allard, F., Barman, T., et al. In Woodward, C., Bicay, M., and Shull, J., editors, *Tetons 4: Galactic Structure, Stars and the Interstellar Medium*. ASP: San Francisco, 2002.
71. Helling, C. and Jørgensen, U. G. *A&A*, 337, 477, 1998.
72. Hinkle, K. H., Aringer, B., Lebzelter, T., Martin, C. L., and Ridgway, S. T. *A&A*, 363, 1065, 2000.
73. Hinkle, K. H., Hall, D. N. B., and Ridgway, S. T. *ApJ*, 252, 697, 1982.
74. Hinkle, K. H., Joyce, R. R., Sharp, N., and Valenti, J. A. In Iye, M. and Moorwood, A., editors, *Optical and IR Telescope Instrumentation and Detectors*, volume 4008, page 720. SPIE, 2000.
75. Hinkle, K. H. and Lambert, D. L. *MNRAS*, 170, 447, 1975.
76. Hinkle, K. H., Lebzelter, T., and Scharlach, W. W. G. *AJ*, 114, 2686, 1997.
77. Hofmann, K.-H., Balega, Y., Scholz, M., and Weigelt, G. *A&A*, 353, 1016, 2000.
78. Hofmann, K.-H., Balega, Y., Scholz, M., and Weigelt, G. *A&A*, 376, 518, 2001.
79. Hofmann, K.-H., Scholz, M., and Wood, P. R. *A&A*, 339, 846, 1998.
80. Höfner, S. *A&A*, 346, L9, 1999.
81. Höfner, S. and Dorfi, E. A. *A&A*, 319, 648, 1997.
82. Höfner, S., Feuchtinger, M. U., and Dorfi, E. A. *A&A*, 297, 815, 1995.
83. Höfner, S., Gautschy-Loidl, R., Aringer, B., and Jørgensen, U. G. *A&A*, 399, 589, 2003.
84. Höfner, S., Jørgensen, U. G., Loidl, R., and Aringer, B. *A&A*, 340, 497, 1998.
85. Höfner, S., Loidl, R., Aringer, B., Jørgensen, U. G., and Hron, J. In Salama, A., Kessler, M., Leech, K., and Schulz, B., editors, *ISO beyond the peaks: The 2nd ISO workshop on analytical spectroscopy*, page 299. ESA Special Production No. 456, 2000.
86. Holweger, H. *A&A*, 4, 11, 1970.
87. Hubeny, I. and Lanz, T. *ApJ*, 439, 875, 1995.
88. Humphreys, E. M. L., Gray, M. D., Yates, J. A., et al. *A&A*, 386, 256, 2002.
89. Jacob, A. P., Bedding, T. R., Robertson, J. G., and Scholz, M. *MNRAS*, 312, 733, 2000.
90. Jeong, K. S., Winters, J. M., and Sedlmayr, E. In Le Bertre, T., Lebre, A., and Waelkens, C., editors, *IAU Symp. 191: Asymptotic Giant Branch Stars*, page 233. ASP: San Francisco, 1999.
91. Johnson, H. R. *ApJ*, 180, 81, 1973.
92. Johnson, H. R. *ApJ*, 260, 254, 1982.
93. Johnson, H. R. *A&A*, 249, 455, 1991.
94. Johnson, H. R., Bernat, A. P., and Krupp, B. M. *ApJS*, 42, 501, 1980.
95. Johnson, H. R., Goebel, J. H., Goorvitch, D., and Ridgway, S. T. *ApJ*, 270, L63, 1983.
96. Johnson, H. R. and Krupp, B. M. *ApJ*, 206, 201, 1976.
97. Johnson, J. J. and Jones, T. J. *AJ*, 101, 1735, 1991.
98. Jørgensen, U. G. *A&A*, 284, 179, 1994.

99. Jørgensen, U. G. In van Dishoeck, E., editor, *IAU Symp. 178: Molecules in Astrophysics: Probes & Processes*, page 441. Kluwer Academic Publishers: Dordrecht, 1996.
100. Jørgensen, U. G., Hron, J., and Loidl, R. *A&A*, 356, 253, 2000.
101. Jørgensen, U. G., Jensen, P., Sørensen, G. O., and Aringer, B. *A&A*, 372, 249, 2001.
102. Jørgensen, U. G. and Johnson, H. R. *A&A*, 244, 462, 1991.
103. Jørgensen, U. G. and Johnson, H. R. *A&A*, 265, 168, 1992.
104. Karovska, M. In Le Bertre, T., Lebre, A., and Waelkens, C., editors, *IAU Symp. 191: Asymptotic Giant Branch Stars*, page 139. ASP: San Francisco, 1999.
105. Kemball, A. J. and Diamond, P. J. *ApJ*, 481, L111, 1997.
106. Kiselman, D. and Plez, B. *Memorie della Societa Astronomica Italiana*, 66, 429, 1995.
107. Knapik, A., Bergeat, J., and Rutily, B. *A&A*, 334, 545, 1998.
108. Knapp, G. R. *ApJ*, 311, 731, 1986.
109. Knapp, G. R., Bowers, P. F., Young, K., and Phillips, T. G. *ApJ*, 455, 293, 1995.
110. Krupp, B. M., Collins, J. G., and Johnson, H. R. *ApJ*, 219, 963, 1978.
111. Kupka, F., Piskunov, N., Ryabchikova, T. A., Stempels, H. C., and Weiss, W. W. *A&AS*, 138, 119, 1999.
112. Kurucz, R. SYNTHE Spectrum Synthesis Programs and Line Data. Kurucz CD-ROM No. 16, 1993.
113. Kurucz, R. Atomic Data for Ca, Sc, Ti, V, and Cr. Kurucz CD-ROM No. 20, 1994.
114. Kurucz, R. Atomic Data for Fe and Ni. Kurucz CD-ROM No. 22, 1994.
115. Kurucz, R. Atomic Data for Mn and Co. Kurucz CD-ROM No. 21, 1994.
116. Kurucz, R. L. *ApJS*, 40, 1, 1979.
117. Lambert, D. L., Gustafsson, B., Eriksson, K., and Hinkle, K. H. *ApJS*, 62, 373, 1986.
118. Lambert, D. L. and Pagel, B. E. J. *MNRAS*, 141, 299, 1968.
119. Lambert, D. L., Smith, V. V., Busso, M., Gallino, R., and Straniero, O. *ApJ*, 450, 302, 1995.
120. Lamers, H. J. G. L. M. and Cassinelli, J. P. *Introduction to Stellar Winds*. Cambridge University Press: Cambridge, 1999.
121. Lebzelter, T. *A&A*, 351, 644, 1999.
122. Lebzelter, T. and Hinkle, K. H. In Aerts, C., Bedding, T., and Christensen-Dalsgaard, J., editors, *IAU Colloq. 185: Radial and Nonradial Pulsations as Probes of Stellar Physics*, page 556. ASP: San Francisco, 2002.
123. Lebzelter, T., Hinkle, K. H., and Hron, J. *A&A*, 341, 224, 1999.
124. Lefevre, J. In Delache, Ph., Laloe, S., Magnan, C., and Tran Thanh Van, J., editors, *Modeling the Stellar Environment: How and Why?*, page 133. Editions Frontières: Gif sur Yvette, 1989.
125. Loidl, R., Aringer, B., Hron, J., et al. In Cox, P. and Kessler, M., editors, *The Universe as Seen by ISO*, page 365. ESA Special Production 427, 1999.
126. Loidl, R., Hron, J., Jørgensen, U. G., and Höfner, S. In Salama, A., Kessler, M., Leech, K., and Schulz, B., editors, *ISO beyond the peaks: The 2nd ISO workshop on analytical spectroscopy*, page 315. ESA Special Production 456, 2000.

127. Loidl, R., Lançon, A., and Jørgensen, U. G. *A&A*, 371, 1065, 2001.
128. Luttermoser, D. G. and Johnson, H. R. *ApJ*, 388, 579, 1992.
129. Luttermoser, D. G., Johnson, H. R., Avrett, E. H., and Loeser, R. *ApJ*, 345, 543, 1989.
130. Luttermoser, D. G. and Mahar, S. In Donahue, R. and Bookbinder, J., editors, *Tenth Cambridge Workshop on Cool Stars, Stellar Systems, and the Sun*, page 1613. ASP: San Francisco, 1998.
131. Mihalas, D. *Stellar Atmospheres*. W. H. Freeman and Co.: San Francisco, 2nd edition, 1978.
132. Netzer, N. and Elitzur, M. *ApJ*, 410, 701, 1993.
133. Nordlund, A. *A&A*, 107, 1, 1982.
134. Nordlund, A. and Dravins, D. *A&A*, 228, 155, 1990.
135. Ohnaka, K. and Tsuji, T. *A&A*, 310, 933, 1996.
136. Ohnaka, K. and Tsuji, T. *A&A*, 335, 1018, 1998.
137. Ohnaka, K., Tsuji, T., and Aoki, W. *A&A*, 353, 528, 2000.
138. Palen, S. *ApJ*, 547, L57, 2001.
139. Partridge, H. and Schwenke, D. W. *J. Chem. Phys.*, 106, 4618, 1997.
140. Perrin, G., Coude Du Foresto, V., Ridgway, S. T., et al. *A&A*, 331, 619, 1998.
141. Pijpers, F. P. and Habing, H. J. *A&A*, 215, 334, 1989.
142. Pijpers, F. P. and Hearn, A. G. *A&A*, 209, 198, 1989.
143. Piskunov, N. E., Kupka, F., Ryabchikova, T. A., Weiss, W. W., and Jeffery, C. S. *A&AS*, 112, 525, 1995.
144. Plez, B., Brett, J. M., and Nordlund, A. *A&A*, 256, 551, 1992.
145. Prigogine, I. *From being to becoming. Time and complexity in the physical sciences*. W. H. Freeman and Co.: New York, 1980.
146. Querci, F. and Querci, M. *A&A*, 39, 113, 1975.
147. Querci, F., Querci, M., and Tsuji, T. *A&A*, 31, 265, 1974.
148. Quirrenbach, A., Mozurkewich, D., Armstrong, J. T., Buscher, D. F., and Hummel, C. A. *ApJ*, 406, 215, 1993.
149. Richards, A. M. S., Cohen, R. J., Bains, I., and Yates, J. A. In Le Bertre, T., Lebre, A., and Waelkens, C., editors, *IAU Symp. 191: Asymptotic Giant Branch Stars*, page 315. ASP: San Francisco, 1999.
150. Richichi, A., Ragland, S., and Fabbroni, L. *A&A*, 330, 578, 1998.
151. Richichi, A., Ragland, S., Stecklum, B., and Leinert, C. *A&A*, 338, 527, 1998.
152. Richichi, A., Stecklum, B., Herbst, T. M., Lagage, P.-O., and Thamm, E. *A&A*, 334, 585, 1998.
153. Richter, M. J., Lacy, J. H., Jaffe, D. T., Greathouse, T. K., and Hemenway, M. K. In Melugin, R. and Roeser, H.-P., editors, *Airborne Telescope Systems*, volume 4014, page 54. SPIE, 2000.
154. Ridgway, S. and Keady, J. J. *ApJ*, 326, 843, 1988.
155. Rothman, L. S., Gamache, R. R., Tipping, R. H., et al. *J. Quant. Spec. Radiat. Transf.*, 48, 469, 1992.
156. Russell, H. N. *ApJ*, 79, 317, 1934.
157. Ryde, N. and Eriksson, K. *A&A*, 386, 874, 2002.
158. Ryde, N., Eriksson, K., and Gustafsson, B. *A&A*, 341, 579, 1999.
159. Scalo, J. M. and Ross, J. E. *A&A*, 48, 219, 1976.
160. Scharmer, G. B. *ApJ*, 249, 720, 1981.
161. Scharmer, G. B. and Carlsson, M. *Journal of Computational Physics*, 59, 56, 1985.

162. Scholz, M. *A&A*, 145, 251, 1985.
163. Scholz, M. *A&A*, 253, 203, 1992.
164. Scholz, M. and Takeda, Y. *A&A*, 186, 200, 1987.
165. Scholz, M. and Wood, P. R. *A&A*, 362, 1065, 2000.
166. Schwarzschild, M. *ApJ*, 195, 137, 1975.
167. Schweitzer, A., Hauschildt, P. H., and Baron, E. *ApJ*, 541, 1004, 2000.
168. Serote Roos, M., Boisson, C., and Joly, M. *A&AS*, 117, 93, 1996.
169. Smith, V. V. and Lambert, D. L. *ApJ*, 294, 326, 1985.
170. Smith, V. V. and Lambert, D. L. *ApJ*, 311, 843, 1986.
171. Smith, V. V., Plez, B., Lambert, D. L., and Lubowich, D. A. *ApJ*, 441, 735, 1995.
172. Sneden, C., Johnson, H. R., and Krupp, B. M. *ApJ*, 204, 281, 1976.
173. Soker, N. *ApJ*, 570, 369, 2002.
174. Stein, R. F. and Nordlund, A. *ApJ*, 499, 914, 1998.
175. Stein, R. F. and Nordlund, A. In *SOHO-9 Workshop "Helioseismic Diagnostics of Solar Convection and Activity"*, 1999.
176. Szymczak, M., Cohen, R. J., and Richards, A. M. S. *MNRAS*, 304, 877, 1999.
177. Tsuji, T. *A&A*, 23, 411, 1973.
178. Tsuji, T. *A&A*, 62, 29, 1978.
179. Tsuji, T. *J. Astroph. Astron.*, 2, 213, 1981.
180. Tsuji, T. *J. Astroph. Astron.*, 2, 95, 1981.
181. Tsuji, T. *A&A*, 99, 48, 1981.
182. Tsuji, T. *A&A*, 122, 314, 1983.
183. Tsuji, T. *A&A*, 156, 8, 1986.
184. Tsuji, T. *A&A*, 197, 185, 1988.
185. Tsuji, T. *A&A*, 376, L1, 2001.
186. Tsuji, T., Ohnaka, K., Aoki, W., and Yamamura, I. *A&A*, 320, L1, 1997.
187. Uitenbroek, H. *ApJ*, 536, 481, 2000.
188. van Belle, G. T., Lane, B. F., Thompson, R. R., et al. *AJ*, 117, 521, 1999.
189. Willson, L. A. *ARA&A*, 38, 573, 2000.
190. Willson, L. A. and Bowen, G. H. In Kaper, L. and Fullerton, A., editors, *Cyclical Variability in Stellar Winds*, page 294. Springer-Verlag: Berlin, New York, 1998.
191. Willson, L. A. and Hill, S. J. *ApJ*, 228, 854, 1979.
192. Windsteig, W., Höfner, S., Aringer, B., and Dorfi, E. A. In Kaper, L. and Fullerton, A., editors, *Cyclical Variability in Stellar Winds*, page 308. Springer-Verlag: Berlin, New York, 1998.
193. Winters, J. M., Fleischer, A. J., Gauger, A., and Sedlmayr, E. *A&A*, 302, 483, 1995.
194. Winters, J. M., Keady, J. J., Gauger, A., and Sada, P. V. *A&A*, 359, 651, 2000.
195. Woitke, P. In Kaper, L. and Fullerton, A., editors, *Cyclical Variability in Stellar Winds*, page 278. Springer-Verlag: Berlin, New York, 1998.
196. Wood, P. R. *ApJ*, 227, 220, 1979.
197. Wood, P. R. In Mennessier, M. and Omont, A., editors, *From Miras to Planetary Nebulae: Which Path for Stellar Evolution?*, page 67. Editions Frontières: Gif sur Yvette, 1990.
198. Yamamura, I. and de Jong, T. In Salama, A., Kessler, M., Leech, K., and Schulz, B., editors, *ISO beyond the peaks: The 2nd ISO workshop on analytical spectroscopy*, page 155. ESA Special Production No. 456, 2000.

199. Yamamura, I., de Jong, T., Onaka, T., Cami, J., and Waters, L. B. F. M. *A&A*, 341, L9, 1999.
200. Young, J. S., Baldwin, J. E., Boysen, R. C., et al. *MNRAS*, 315, 635, 2000.

5 Molecule and Dust Grain Formation

Tom J. Millar

Department of Physics, UMIST

5.1 Introduction

AGB stars provide a fascinating observational laboratory for investigating a range of chemical processes important in many applications of molecular astrophysics. Figure 5.1 outlines the basic chemical structure of an AGB circumstellar envelope (CSE) and identifies the processes that dominate in each region. These include chemistry in local thermodynamic equilibrium (LTE), which occurs at high density and temperature close to the stellar photosphere; the effects of shock waves driven by stellar pulsations, which expose molecules formed in LTE to new physical conditions and which can form "nonequilibrium" molecules; the nucleation of solid particles; the interaction of stellar photons with the dust particles and the subsequent interaction between the gas and the dust as the solid matter is ejected to large distances by radiation pressure; the growth of dust grains; and finally, the interaction of stellar material with the interstellar radiation field.

In this chapter we shall discuss all of these applications for the cases of O-rich and C-rich AGB stars under the assumption that these objects can be adequately modeled by assuming spherical symmetry. Complications introduced by effects such as asymmetric mass loss, clumps, and disks are discussed briefly in Section 5.5. For the case of spherically symmetric, constant-velocity mass loss at a rate $\dot{M}$, we can write the density of H_2 molecules per unit volume, $n(r)$, and the outward radial column density of H_2, $N(r)$, as

$$n(r) = X\dot{M}/8\pi r^2 \mu m_H v_\infty, \qquad (5.1)$$

$$N(r) = X\dot{M}/8\pi r \mu m_H v_\infty, \qquad (5.2)$$

where X is the mass fraction of hydrogen, and v_∞ the expansion velocity of the gas. We can rewrite these equations as ($X = 0.7$)

$$n(r) = 7.1 \times 10^4 \left[\frac{\dot{M}}{10^{-5}}\right] \left[\frac{10^{16}}{r}\right]^2 \left[\frac{15}{v_\infty}\right] \text{ cm}^{-3}, \qquad (5.3)$$

$$N(r) = 7.1 \times 10^{20} \left[\frac{\dot{M}}{10^{-5}}\right] \left[\frac{10^{16}}{r}\right] \left[\frac{15}{v_\infty}\right] \text{ cm}^{-2}, \qquad (5.4)$$

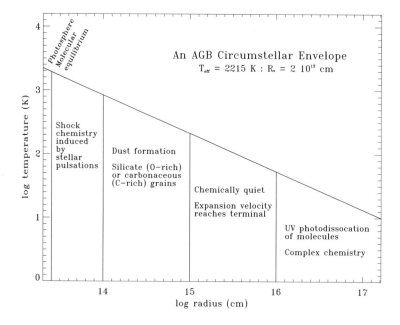

Fig. 5.1. Schematic chemical structure in an AGB envelope [77]

where $\dot{M}$ is measured in units of $M_\odot$ yr^{-1}, r in units of cm, and v_∞ in km s^{-1}. In the outer CSE the temperature is thought to follow a power-law behavior

$$T(r) \propto r^{-\alpha}, \tag{5.5}$$

with $\alpha \approx 0.6$–0.7, although in the inner CSE the effects of shock waves cause this behavior to break down. The extinction at 100 nm from a radial point r to infinity can be written as

$$A_{\rm UV}(r) = 7.6 \left[\frac{\psi}{0.01}\right] \left[\frac{\dot{M}}{10^{-5}}\right] \left[\frac{\chi_{\rm r}}{10}\right] \left[\frac{10^{16}}{r}\right] \left[\frac{15}{v_\infty}\right] \text{ mag}, \tag{5.6}$$

where $\chi_{\rm r}$ is the ratio of extinctions at 100 nm and V, and ψ is the dust-to-gas ratio by mass. The quantity $\chi_{\rm r}$ is difficult to determine, while ψ probably varies from object to object, but may have a typical value of 0.003. These parameters determine the chemistry in the outer CSEs of AGB stars, in particular the competition between reactive collisions in the gas, which determines the growth of chemical complexity, and photodissociation, which works to reduce complexity. Although photons ultimately win, there is, as we shall see, a region in the outer CSE in which reactive radicals produced by photons are able to build up a range of exotic molecules.

Molecular line observations, and particularly interferometric studies, of AGB CSEs have greatly increased our knowledge of the column densities and spatial distributions of molecules and provide ever more stringent tests for theory; see Chapter 7. In addition, observation of PAH emission bands in post-AGB objects provides a much-needed observational link to the chemistry of dust grain nucleation, while the identification of presolar dust grains, specifically dust grains formed in AGB stars, in primitive meteorites has focused renewed attention on specific nucleation routes and on the role of carbides as condensation nuclei. The identification of these dust grains relies on measurements of isotopic ratios in elements such as C, N, O, and Si. For example, these have led to the identification of six different classes of silicon carbide presolar dust grains. Some 90% of these fall into the mainstream type, which shows evidence of the CNO cycle, partial He-burning, and s-processing, thus revealing their origin in low-mass ($1–3\,M_\odot$) C-rich AGB stars. AGB stars, and indeed other processes such as novae and supernove, leave imprints of specific nucleosynthesis events on the dust grains that they condense. In AGB stars, the unique isotopic ratios arise in the dredge-up process, the first of which enriches the surface in ^{13}C and ^{14}N; see Chapter 2. Stars with masses greater than about $3.5\,M_\odot$ have a second dredge-up of CNO-processed material and have ^{12}C/^{13}C ratios of 20–30 and ^{14}N/^{15}N ratios of 600–1600. Thermal instabilities in the He-burning shell lead to a third dredge-up, which progressively enriches ^{12}C, such that C/O > 1 and ^{12}C/^{13}C $\approx$ 30–200. The C, N, Si, and Mg-Al isotopic ratios of SiC dust grains of type Z have been measured by [54]. These have ^{12}C/^{13}C ratios of $\approx$ 11–120 and ^{14}N/^{15}N ratios of $\approx$ 1100–19000. These dust grains, which form about 1% of presolar SiC dust grains recovered from meteorites, most likely form in AGB stars with masses less than $2.3\,M_\odot$. In these stars, cool bottom burning, which mixes CNO-processed material from a nonburning zone to the surface subsequent to the first dredge-up, lowers the ^{12}C/^{13}C ratio.

Other examples of presolar dust grains extracted from meteorites include diamond, which has a complex formation history [2], graphite [1], titanium carbide [9], corundum [62], silicon nitride [55], and hibonite, $CaAl_{12}O_{19}$, [27]. . The identification of presolar dust grains from O-rich stars has proven very difficult because of the fact that calcium–aluminum inclusions were also formed in the solar system and because the chemical techniques conventionally used to isolate presolar dust grains destroy presolar silicates, although a corundum (Al_2O_3) dust grain greatly enriched in ^{17}O and ^{26}Mg has been extracted from the Orgueil meteorite [63]. In this dust grain ^{16}O/^{17}O is 1028 compared to a solar system value of 2610, while the ^{26}Mg excess, which derives from the radioactive decay of ^{26}Al, corresponds to a ^{26}Al/^{27}Al ratio about 20 times larger than that in solar system solids. Even larger enrichments have beeeen discovered in a $3\,\mu m$ corundum dust grain in the Bishnapur meteorite [62]. Such enrichments, particularly those in ^{17}O, are consistent with formation in O-rich AGB stars. To date, around 120 presolar oxide dust grains have been found. The majority are corundum, although a small number composed of spinel ($MgAl_2O_4$), hibonite ($CaAl_{12}O_{19}$), and titanium dioxide (TiO_2) have been identified.

In addition to the chemical composition, there has also been information derived on structures. Graphite spherules in the Murchison meteorite have been studied [10] with the following results:

- Many of the spherules have a composite structure, sometimes with a core of nanocrystalline carbon at the center surrounded by well-graphitized carbon, at other times randomly embedded in the graphite. In the former case, the nanocrystals appear to be composed of small sheets, "graphene," resembling polycyclic aromatic hydrocarbons (PAHs).
- One-third of the spherules contained internal crystals of refractory carbides, mostly titanium carbides, with some being centrally condensed and others randomly dispersed.
- SiC was found in only one spherule, indicating that the formation of the carbonaceous spherules occurred before the formation of SiC.

Any reasonable theory of dust nucleation and growth in carbon stars must be able to account for these findings.

5.2 Chemical Processes for Molecule and Dust Formation

In this section, we shall discuss the processes responsible for molecule and dust formation in AGB stars. Any chemical processes must take place on a faster time scale than the expansion of the gas to be important. The expansion time of the gas, t_e, is given by

$$t_e = \left[v \frac{1}{n(r)} \frac{dn(r)}{dr}\right]^{-1} \approx 3 \times 10^9 \left[\frac{r}{10^{16}}\right] \left[\frac{15}{v_\infty}\right] \text{ s.} \quad (5.7)$$

Close to the star, where densities and temperatures are large, H_2 is the most important reactant, with a typical rate coefficient for reaction of around 10^{-11} cm^3 s^{-1} and an associated reaction time-scale, t_r

$$t_r = [10^{-11} n(r)]^{-1} \approx 10^6 \left[\frac{r}{10^{16}}\right]^2 \left[\frac{v_\infty}{15}\right] \left[\frac{10^{-5}}{\dot{M}}\right] \text{ s.} \quad (5.8)$$

Close to the star, e.g., at $r = 10^{13}$ cm, t_r is much smaller than t_e, and chemistry is important. Most reactions involving H_2 possess activation energy barriers, so that they become negligible as material flows away from the star, and material is expected to retain a "frozen-in" composition. This picture can be perturbed by shock waves, which can heat, compress, and trap the gas. At large radial distances, $\approx 10^{17}$ cm, the interstellar UV radiation field penetrates the CSE and destroys molecules on the time scale $t_r \approx 10^{10}$ s, also much shorter than the expansion time.

The evolution of objects beyond the AGB phase, to form protoplanetary nebulae and planetary nebulae (PNe), brings additional physical complications, including accounting for nonsphericity, the evolving internal radiation field of the central object, the interaction of the fast stellar wind with the slow wind in the AGB phase, the presence of shock waves, clumpy structures, and so on; see Chapter 10. As a result, only the most simplistic models following chemical evolution beyond the AGB have been developed [98]. A review of the chemistry associated with this phase, which will not be discussed further here, has been given by [59].

5.2.1 LTE Chemistry

At the high densities and temperatures appropriate to the photosphere of an AGB star, molecular abundances are determined by local thermodynamic equilibrium (LTE). In LTE the Gibbs free energy of the system is at a minimum. The free energy of the system is given by

$$G = \sum f_i x_i, \tag{5.9}$$

where x_i is the number of moles of species i, and f_i its chemical potential:

$$f_i = \left(\frac{G}{RT}\right)_i + \ln P + \ln\left(\frac{x_i}{\bar{x}}\right), \tag{5.10}$$

where $(G/RT)_i$ is the Gibbs free energy of species i, P the total pressure of the system, $\bar{x} = \Sigma x_i$, and R the gas constant. The free energy of i is a temperature-dependent quantity and is tabulated in the NIST-JANAF tables, for example.

The abundances x_i are found numerically by the condition that they minimize G. They are calculated most readily by the technique of *steepest descents*. Given initial elemental abundances and a trial solution, the method is rapid. Table 5.1 shows the results of such calculations for the case of an O-rich AGB star, the M-type Mira R Cas with C/O = 0.75 and $T = 2215$ K, and the C-rich Mira, IRC+10216 with C/O = 1.5 and $T = 2300$ K. In both cases, the pressure was chosen to be 1.033×10^{-3} atm, corresponding to a total hydrogen density of 3×10^{15} cm^{-3}.

Minimization of free energy preferentially forms molecules with strong bonds, i.e., high dissociation energies. The CO molecule has a very large bond energy and is formed in abundance. Table 5.1 shows that the LTE abundances of the objects differ dramatically. In both cases, the most abundant molecules are H$_2$ and CO, the latter of which takes up essentially all of the available O (in the C-rich star) or C (in the O-rich star). This dichotomy is well represented by Figures 5.2 and 5.3, which show the partial pressures of CS and SO as a function of the C/O ratio and temperature. Note that at C/O ≈ 1, the abundances change by around 5 orders of magnitude, due to the fact that

Table 5.1. The atoms and top 20 molecules produced in LTE calculations for O-rich and C-rich stars with parameters as discussed in the text. $f(X)$ is the fractional abundance of species X relative to H_2. $[a(b) = a \times 10^b]$ [77]

	O-rich Species	$f(X)$	C-rich Species	$f(X)$
	H	2.1(−1)		3.5(−1)
	C	4.1(−13)		2.3(−6)
	N	5.8(−9)		1.4(−8)
	O	1.3(−6)		3.2(−12)
	Si	4.0(−8)		6.3(−5)
	S	2.6(−5)		3.4(−6)
	P	4.7(−7)		6.2(−7)
	Cl	5.2(−8)		8.4(−8)
1	H_2	1	H_2	1
2	CO	1.1(−3)	CO	1.6(−3)
3	H_2O	2.9(−4)	C_2H_2	2.2(−4)
4	N_2	1.3(−4)	C_2H	1.1(−4)
5	SiO	6.9(−5)	N_2	9.5(−5)
6	OH	9.0(−6)	HCN	8.5(−5)
7	SH	7.7(−6)	CS	2.3(−5)
8	H_2S	7.2(−7)	SiS	9.8(−6)
9	HCl	3.4(−7)	C_3H	9.5(−6)
10	SiS	2.0(−7)	CN	1.6(−6)
11	HF	1.7(−7)	SH	7.0(−7)
12	TiO	1.6(−7)	SiH	6.7(−7)
13	PO	9.7(−8)	SiC_2	3.7(−7)
14	NP	8.2(−8)	HCl	3.4(−7)
15	CO_2	6.3(−8)	CH_3	2.6(−7)
16	SO	4.0(−8)	CH	1.6(−7)
17	MgH	3.9(−8)	C_2	6.2(−8)
18	AlH	3.2(−8)	NP	5.5(−8)
19	AlOH	1.5(−8)	SiO	4.8(−8)
20	CrH	1.5(−8)	H_2S	4.4(−8)

the amounts of free O and C, not taken up by CO, alter rapidly. Figure 5.4 shows how the LTE abundances of a number of molecules vary as a function of temperature, in the range 2000–3000 K, in the C-rich case [77].

5.2.2 The Effects of Periodic Shock Waves

AGB stars undergo pulsations with a typical period of several hundred days. These pulsations give rise to compression waves, which propagate through the atmosphere, steepening into shock waves as they progress; see Chapter 4. Shock waves that emerge into the photosphere can lose energy radiatively or through doing work on the gas. When the density is high, radiative cooling

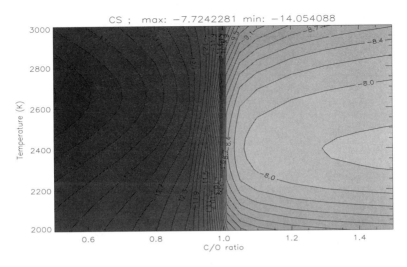

Fig. 5.2. The CS fractional abundance in LTE as a function of temperature and C/O ratio. The maximum and minimum fractional abundances are also given [77]

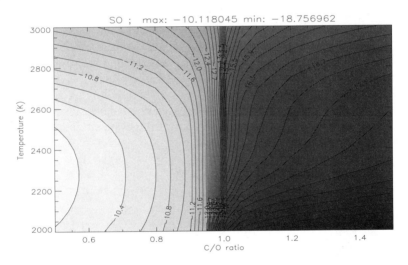

Fig. 5.3. The SO fractional abundance in LTE as a function of temperature and C/O ratio. The maximum and minimum fractional abundances are also given [77]

is efficient, and the shocks can be treated as isothermal [15]. If the density is low, the radiative cooling time is long, and the shock loses energy by adiabatic expansion. The evolution of shocks in AGB stars has been studied in [15], which modeled the response of a spherical model atmosphere to periodic driving at its inner boundary. Shocks are initially weak and irregular, but they are capable of moving the mass upwards, changing the density structure of

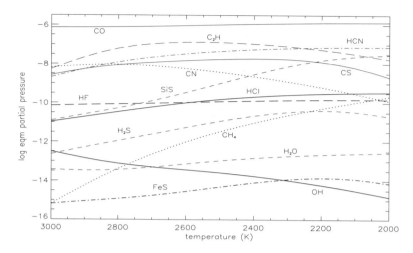

Fig. 5.4. LTE partial pressures calculated as a function of temperature for a C/O ratio of 1.5 [77]

the star. After some cycles, a strong shock forms at some distance from the photosphere and is even more efficient at moving the mass upwards, reducing the density gradient and thereby producing an extended atmosphere. A particular parcel of gas thus receives an outward impulse and follows a roughly ballistic trajectory under the effect of gravity before falling back toward the stellar surface. If the gas parcel experiences a second shock before it returns to its initial position, it may attain a net outward movement and may initiate a mass-loss process.

The pulsation causes a shock with velocity v_s to give rise to a density profile [26]

$$\rho(r) = \rho(r_s) \exp\left(-\frac{r_s(1-\gamma^2)}{0.4 H_0(r_s)}\left[1 - \left(\frac{r}{r_s}\right)^{-0.4}\right]\right), \quad (5.11)$$

where r_s is the radius at which the shock forms, and H_0 the scale height, given by

$$H_0 = 2c_s^2 r_s / v_{\mathrm{esc}}^2, \quad (5.12)$$

where c_s is the local sound speed, v_{esc} the escape velocity, and $\gamma = v_s/v_{\mathrm{esc}}$. As the shock propagates outward, its velocity decreases as [97]

$$v_s(r) = v_s(r_s)\left(\frac{r}{r_s}\right)^{-1/2}. \quad (5.13)$$

Table 5.2 shows the densities and temperatures produced by a shock with $v_s = 20\,\mathrm{km\,s^{-1}}$ at 1.2 stellar radii (R_*). Figure 5.5 shows a typical run of gas

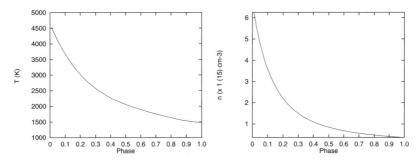

Fig. 5.5. Temperature and density as a function of phase following a $20\,\mathrm{km\,s^{-1}}$ shock at a distance of $1.2\,R_*$ [95]

Table 5.2. Preshock (labeled 0) and postshock (labeled 1) densities $(\mathrm{cm^{-3}})$ and temperatures as a function of radial distance in units of R_*. $[a(b) = a \times 10^b]$ [20]

Distance	v_s	n_0	T_0	n_1	T_1
1.2	20.0	3.7(14)	2062	2.0(15)	19722
1.4	18.5	1.4(14)	1879	7.3(14)	17016
1.7	16.8	4.2(13)	1673	2.3(14)	14118
1.9	15.9	2.3(13)	1565	1.2(14)	12646
2.4	14.1	6.6(12)	1360	3.4(13)	10176
3.0	12.6	2.2(12)	1190	1.2(13)	8218

temperature and density at this radius as a function of phase for the excursion induced by a $20\,\mathrm{km\,s^{-1}}$ shock [95].

Once heated, the gas cools, initially through the dissociation of H_2 by atomic hydrogen, [32], on a length scale of

$$l_d = \frac{v_s}{k n_H n_j},\quad (5.14)$$

where k is the collisional dissociation rate coefficient, n_H is the abundance of atomic hydrogen per unit volume, and n_j is the ratio of the immediate postshock density to the preshock density (equal to 4 for an atomic gas, and 6 for a molecular gas). The dissociation length is typically on the order of 1–100 km.

The hydrodynamic cooling of the gas has been studied in [11], which gives expressions for the density and temperature as a function of phase in the pulsation. In this regime, the rate equation for a species i is

$$\frac{dn_i}{dt} = [F - D] - \frac{n_i}{v}\frac{dv}{dt},$$

where F and D are the formation and destruction rates $(\mathrm{cm^{-3}\,s^{-1}})$ of i, and v is the gas velocity.

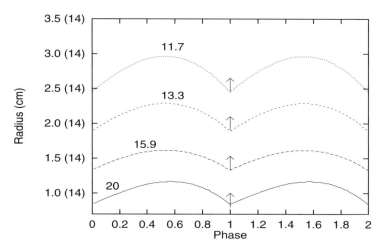

Fig. 5.6. Gas excursions induced by shocks of different strengths (in km s^{-1}). The shock velocity, initially 20 km s^{-1}, decreases as it moves out through the CSE, and the individual points are labeled by the shock velocity at that radius. The arrows show where the gas is shocked at the end of the first period [95]

This approach has been implemented for the chemistry of sulfur and silicon in the inner wind of IRC+10216 [95]. Here it was assumed that a gas parcel eventually fell back to its original position, thus modeling a quasi-stationary layer rather than a mass-loss process, and followed the chemistry for two pulsation periods in order to ensure that the behavior is in fact periodic. Figure 5.6 shows how a parcel of gas moves under the influence of different shock strengths.

The calculation starts from a chemical composition calculated using the LTE model described in the previous section. Since the densities are high and the effective temperature of the star low, only neutral–neutral processes, bimolecular and termolecular, need to be considered. The authors of [95] find that the parent molecules C_2H_2 and CO are unaffected by the shocks but that the abundances of other molecules can be significantly altered by the shock chemistry. For fast (20 km s^{-1}) shocks, the dynamical time is too small to allow any significant dissociation of H_2, but at lower velocities the abundance of atomic hydrogen can be larger than that of molecular hydrogen. For most molecules, the important region of change occurs in the hydrodynamic cooling region, where the phase is on the order of 0.02–0.6.

The model indicates that shock production close to the photosphere can be significantly more important than LTE chemistry. An example is SiO, which in LTE takes up around 10^{-4} of the available silicon in IRC+10216. In fast shocks, some CO is destroyed by collisional dissociation and by reaction with H atoms to release atomic oxygen to the gas. These atoms quickly react with H_2 to form OH and H_2O. Atomic silicon, which is the dominant form of this element according to the LTE calculations, reacts with OH to produce SiO. At $5\,R_*$, the authors of [95] estimate a SiO fractional abundance of 2.2×10^{-7}, close to that derived from IR observations, 8.0×10^{-7} [68]. Other molecules produced efficiently in the shock include SiS and CS. We note that the abundances of saturated molecules such as NH_3 and CH_4 are low in the shocked gas, in agreement with the nondetection of NH_3 through IR interferometry. NH_3 in IRC+10216 and the supergiant VY CMa forms at radii greater than $80\,R_*$ and $40\,R_*$, respectively [84].

The hydrocarbon chemistry in the postshock gas has been investigated in [20], where it is found that the major chemical process thatoccurs is the formation of PAH molecules. Such species, and carbon solids, are seen to form readily in terrestrial flames. In his model, he considers formation of PAHs with up to 7 aromatic rings. The first stage in the growth of PAHs is ring closure to form phenyl, C_6H_5. A number of routes are possible [33]:

1. 1-buyen-3-ynyl (C_4H_3U or n-C_4H_3) to phenyl,

$$C_4H_3U + C_2H_2 \longrightarrow C_6H_5\,,$$

where U means that the molecule has a vinyl radical located on a terminal carbon atom;

2. 1,3-butadienyl (C_4H_5U or n-C_4H_5) to benzene,

$$C_4H_5U + C_2H_2 \longrightarrow C_6H_6 + H\,,$$

3. the reaction of propargyl radicals,

$$C_3H_3 + C_3H_3 \longrightarrow C_6H_5 + H\,,$$

which has been measured in the laboratory; and

4. formation via benzyne (C_6H_4),

$$C_4H_3U + C_2H_2 \longrightarrow C_6H_4 + H\,,$$
$$C_6H_4 + H \longrightarrow C_6H_5\,.$$

Once the first ring is formed, the next stage is to close a second ring. The most efficient mechanism appears to be one proposed in [33], in which a ring with a radical site, i.e., a missing H atom, undergoes reaction with C_2H_2 followed by H abstraction and subsequent reaction with C_2H_2 to close the ring:

$$A_i(r) + C_2H_2 \longrightarrow A_i\dot{C}_2\dot{H} + H\,,$$
$$A_iC_2H + H \longrightarrow A_iC_2H^* + H_2\,,$$
$$A_iC_2H^* + C_2H_2 \longrightarrow A_{i+1}(r)\,.$$

Here the notation is that i is the number of rings in the molecule, * represents a radical with an unpaired electron located on an aromatic ring, and (r)

Table 5.3. Abundances, A_i (cm^{-3}) of PAHs containing i rings as a function of stellar radius. $[a(b) = a \times 10^b]$ [20]

r	A_1	A_2	A_3	A_4	A_5	A_6	A_7
1.7	2.7(5)	4.4(2)	2.1(-1)	1.4(1)	8.5(0)	7.4(0)	1.7(2)
1.9	2.0(1)	4.8(−1)	4.2(0)	4.7(0)	3.6(0)	4.2(0)	1.3(3)
2.4	1.6(−1)	2.9(−2)	1.8(0)	2.3(0)	2.0(0)	1.8(−4)	6.2(2)
3.0	1.8(−3)	6.4(−4)	2.7(−1)	2.7(−1)	3.0(−2)	5.3(−8)	2.1(2)

represents other aromatic radicals. Except for the first ring closure, reaction-rate coefficients for these are unknown. Cau [20] has also considered the formation of PAH dimers formed by reactions such as [25]

$$C_6H_6 + C_6H_5 \longrightarrow C_6H_5C_6H_5 + H.$$

For larger dimers, he has included the effect of van der Waals forces in enhancing the rate coefficient. The efficiency of PAH production can be investigated through the concept of *yield*, defined to be the number of carbon atoms in a particular class of species relative to the total number of carbon atoms in all hydrocarbons. At $1.4\,R_*$, benzene is the dominant ring molecule, but it falls off at larger radii as molecules with more and more rings are produced. Table 5.3 gives the abundances of species A_i, where i is the number of rings, for radii between 1.7 and $3.0\,R_*$.

The results show that the most abundant ring molecule is A_7, the terminal species in this analysis, indicating that growth to larger species should be extremely efficient in the postshock gas. At a particular radial distance, there is a point in the phase of the pulsation at which the addition of acetylene, which builds rings, is faster than H abstraction, which decreases complexity. Comparison with the temperature profile shows that this condition is met when the temperature reaches about 1600 K, similar to that found in flame experiments. This is significantly larger than the window for PAH formation identified in earlier studies [26, 33], which was around 900–1100 K or 700–900 K. These models found lower temperatures because they adopted much lower gas densities.

The pulsation model has also been applied to O-rich AGB stars [29]. The motivation to investigate such models is provided by the detection of IR emission bands from CO_2 by ISO [19, 78, 89] and by the observation of several carbon-bearing molecules, such as HCN, HNC, CN, and H_2CO, in the outer CSEs of these stars. As is clear from Table 5.1, these species are not expected to form in LTE chemistry. The pulsation model described above has also been applied [29] to the specific case of the M-type Mira IK Tau, which has a pulsation period of around 470 days and a C/O ratio of 0.75. They choose a shock velocity of $32\,\text{km}\,\text{s}^{-1}$ at $1\,R_*$, consistent with velocities derived from CO IR line profiles and calculate abundances at various radial distances out to $2.2\,R_*$, where the shock velocity has fallen to $21.6\,\text{km}\,\text{s}^{-1}$.

The calculations show that the abundance of CO_2 is greatly enhanced, by a factor of more than 1000, over the LTE value at radial distances greater than $2\,R_*$. The efficient formation of this molecule is related to the collisional dissociation of water by atomic hydrogen, followed by reaction of hydroxyl radicals with CO, which is important at small radii, and a three-body assisted association between CO and O atoms, which causes a large increase in the CO_2 at $2\,R_*$:

$$H_2O + H \longrightarrow OH + H + H,$$
$$CO + OH \longrightarrow CO_2 + H,$$
$$CO + O + M \longrightarrow CO_2 + M.$$

The rate coefficient for the last reaction is very uncertain, but even with a three-body value of $2 \times 10^{-32}\,cm^6\,s^{-1}$, the CO_2 fractional abundance reaches 10^{-6}. Two other species whose LTE abundances are greatly enhanced by the shock chemistry are CS and HCN, with the abundance of the former reaching 3×10^{-7} and the latter reaching 2×10^{-6} at $2.2\,R_*$. These values are in good agreement with those derived from millimeter wave observations of the outer CSE, and the abundances, radii, and temperatures derived from the ISO observations of CO_2 are well matched by the model predictions. A recent survey of SiO and HCN toward 30 AGB stars has been completed [13]. This finds that large fractional abundances of HCN must be present in the inner CSE in order to account for the strengths of the high-excitation lines they detect. Although the abundance is difficult to determine because of uncertainties surrounding the dust properties, a value of around 10^{-6} fits the observations reasonably well.

Further evidence that pulsations play an important role in determining the chemical composition of the inner CSE comes from the detection of vibrationally excited HCN (J=8–7 and the 3–2, ν_2=1) in the S star χ Cygni [30]. The latter transition lies 1050 K above the ground state and has a critical density of about $5 \times 10^{11}\,cm^{-3}$, which is already lower than the density of the inner CSE of χ Cygni at $2\,R_*$. It is likely that the vibrational state is pumped by IR photons at 14 µm, which constrains the emitting region to be less than $33\,R_*$.

Duari and Hatchell [30] applied the pulsation model to χ Cygni, adopting a C/O ratio of 0.95 and a shock velocity of $32\,km\,s^{-1}$ at $1\,R_*$, and found that the HCN fractional abundance rose from 1.6×10^{-9} in LTE to 1.8×10^{-4} at $1\,R_*$, falling to 4.8×10^{-6} at $2\,R_*$. In this object, and in O-rich stars, the formation of HCN (and CS) is aided by CN:

$$CN + H_2 \longrightarrow HCN + H,$$
$$CN + S \longrightarrow CS + N.$$

The dynamical models have thus proven very successful in explaining the origin and chemical composition of this *quasi-stationary layer* and will undoubtably prove very useful in explaining dust grain formation in AGB stars. At the present time, a comprehensive study of parameter space needs to be carried out. Warm (600 K) SO_2 has been detected in the ISO spectra of three AGB stars [100]. The analysis indicates that the SO_2 lies at a distance of a few R_* and locks up a significant fraction of sulphur. However, neither the LTE models (Table 5.1) nor the shock models predict much SO_2, although this latter point needs to be explored more fully. Yamamura et al. [100] find that in the M-type Mira T Cep the number of emitting molecules and the radius of the SO_2 layer decrease with phase over two pulsation periods. This is difficult to reconcile with the models described above, since the chemistry repeats itself each cycle. Yamamura et al. suggest that the variations are due to photodissociation of SO_2 by stellar UV photons, a process not yet included in the theoretical models.

More detailed hydrodynamical models of AGB atmospheres have been developed in [52] and [53], which have included a more accurate treatment of radiative transfer and molecular opacities; see Chapter 4. Höfner et al. [53] include grey radiative transfer, using Planck mean opacities, an improvement over the constant gas absorption coefficient introduced in [15], and include a time-dependent description of dust formation. They find that the mass-loss rates and dust grain condensation are significantly reduced by the more accurate treatment of opacity for C-rich stars, whereas the outflow velocities are less affected, probably implying that radiation pressure on the gas, rather than the dust grains, dominates the mass-loss process. The coupling of radiation to molecules is efficient only when the shock waves can lift the gas to cooler layers; hydrostatic models show no evidence of radiation-driven mass loss. Subsequently, [52] implemented a frequency-dependent radiative transfer, albeit at a very limited number of frequencies. Although dust formation was not included, the models show that densities in the outer atmosphere are enhanced compared to the grey radiative transfer models. Since dust grain and molecule formation are strong functions of density, it will be important to include chemistry into these models.

A similar approach has been taken in [49], which considered how this quasi-stationary warm layer could be produced in carbon-rich stars. The authors investigated time-dependent hydrodynamic models coupled with radiative transfer, dust dust grain nucleation, and growth, using the method of moments, and chemical equilibrium. They find that periodic shocks form and that these shocks levitate material, resulting in orders-of-magnitude increase in density over static models. Dust dust grain formation amplifies this process because of the extra momentum coupling the dust grains provide. Further, back-warming by emission from the dust also heats the layer. The result is a stratified atmosphere in which molecules group according to the physical conditions they experience. It should be noted that chemical equilibrium is assumed, whereas shock waves are intrinsically nonequilibrium processes; the incorporation of such processes into this model is difficult and computationally demanding and is still awaited.

5.2.3 Dust Grain Formation

Solid particles can condense out of a cooling gas if the partial pressure, P_p, of the condensible species is larger than the vapor pressure, P_v, of the species. In fact, it is usually the case that the partial pressure must be very much larger; that is, the supersaturation ratio $S = P_p/P_v$ must be much larger than 1. The condensation temperature, T_c, at which the species condenses is determined by equating P_p and P_v. Thus, for species M, $S = 1$ requires that

$$P_p = n_M k T_c = P_M e^{-T_M/T_c} = P_v, \qquad (5.15)$$

where P_M and T_M are constants different for different species. The condensation temperature, above which material cannot undergo the phase change from gas to solid, ranges from ≈ 500 K for magnetite (Fe_3O_4) to ≈ 1400 K for magnesium silicates to ≈ 1800 K for corundum (Al_2O_3) and titanium oxides.

Early work on dust grain formation in C-rich stars used classical nucleation theory. This theory describes the growth of a solid particle from its gas-phase monomers and identifies a critical cluster size above which dust grain growth by addition of momomers becomes energetically favorable. In general, such a theory cannot be applied to AGB stars for a number of reasons: The gas, particularly in O-rich atmospheres, does not contain monomers with the same composition as the dust grains; an equilibrium state may not be reached within a dynamical (expansion) time scale; and bulk properties are assumed to be applicable to very small clusters, that is, below the critical size. This is unlikely to be true and can lead to significant error in calculating the critical size and subsequent growth. A more accurate, although more difficult, approach is to include kinetic effects.

In Section 5.2.2, it was discussed how pulsational shock waves can aid in the formation of large PAH molecules. Once formed, these species can either coagulate to form larger particles, i.e., dust grains, or grow through the accretion of acetylene molecules. Coagulation and accretion are difficult processes to model, since one must follow the evolution of each individual dust grain. A more effective means of following dust grain growth is to study the evolution of the dust grain size distribution via the so-called method of moments [18]; see Chapter 6. This article showed that the first moment is related to the average number of carbon atoms in dust grains, and therefore dust grain size and mass, assuming spherical dust grains. The usual kinetic terms representing coagulation and accretion can be written in terms of the three lowest order moments. When accretion is neglected, the resulting dust grains can have an average diameter of several Å[18, 20]. When this process is included, dust grains can grow to several microns, although when they reach this size, almost all acetylene is removed from the gas. Thus, the detection of acetylene in AGB CSEs indicates that sticking of acetylene is not an efficient process; detailed calculations indicate that the dust grain properties in IRC+10216 can be reproduced if the sticking coefficient of acetylene is on the order of 0.01 [20].

The first models to combine realistic hydrodynamic models of AGB CSEs with the nucleation and growth of carbon dust grains were made in [31]. In stationary atmospheres, back-warming of the interior regions by IR emission from dust further out helps keep the gas temperature too high for PAH formation and subsequent dust grain growth. In pulsational models, however, the shock waves move material to regions of higher density but lower temperature further from the stellar photosphere where dust grains can nucleate and grow. As a result, radiation pressure accelerates the dust grains, whose movement induces a new shock to form and drive the mass loss more efficiently. These models thus predict that the dust should have a radial structure, notably the formation of shells. Whether or not dust shells form, every period depends on the C/O ratio, larger values producing shells at every pulsation. While this study is very interesting, particularly in its discovery of how dust formation itself can generate pulsations, a significant shortcoming is that it uses nucleation theory rather than a kinetic description of dust formation. We note here that dust shells have now been discovered in IRC+10216 [79, 80] and are discussed further in Section 5.5.2.

In O-rich stars the dominant dust grains are composed of amorphous silicates with evidence for a smaller population of crystalline silicates. Models for the formation of these complex heterogeneous grains are poorly developed as yet, so that the detailed conversion of gaseous species to grains solids is difficult to follow. There have been a number of equilibrium studies of grain condensation, e.g., [92], but grain formation in the expanding, cooling gas around an AGB star is inherently a nonequilibrium kinetic process.

After the very stable molecule CO, which does not take part in grain formation, the most abundant oxide is SiO, which does not nucleate until the temperature is less than 600 K. However, from observations it is known that dust nucleates at around 1000 K in massive M-stars undergoing substantial mass loss. Thus, it is the less abundant oxides such as TiO, AlO, and Al_2O_3 that are likely to be the first species to condense.

The formation of these first condensates, which can then act as seed nuclei for the growth of macroscopic grains, has been discussed in [34]. Titanium forms a number of simple oxides, including TiO, TiO_2, and Ti_2O_3, of which TiO has the largest bond energy and can therefore form and survive at high temperatures. TiO consumes almost all of the available Ti below 1600–2500 K, while TiO_2 is the dominant species for temperatures less than 1400 K, independent of pressure. Here, TiO_2 results from the reaction

$$TiO + H_2O \longrightarrow TiO_2 + H_2.$$

Solid TiO forms from reaction of TiO_2 with H_2,

$$\mathrm{TiO_2 + H_2 \longrightarrow TiO(s) + H_2O},$$

with other solid oxides forming in similar reactions,

$$\mathrm{2TiO_2 + H_2 \longrightarrow Ti_2O_3(s) + H_2O},$$
$$\mathrm{3TiO_2 + H_2 \longrightarrow Ti_3O_5(s) + H_2O},$$
$$\mathrm{4TiO_2 + H_2 \longrightarrow Ti_4O_7(s) + H_2O},$$

but it is likely that homogeneous nucleation dominates to produce solid $\mathrm{TiO_2}$ as the dominant particles, which then act as seed nuclei on which silicates condense [34].

Other high-temperature condensates involving Ti can condense. These include perovskite, $\mathrm{CaTiO_3}$, which is more stable than $\mathrm{TiO_2}$ above 1400 K. Hence it is possible that this might be the first titanium compound to condense. However, its formation requires a three-body process

$$\mathrm{TiO_2 + H_2O + Ca \longrightarrow CaTi_2O_3 + H_2},$$

which results in its abundance being lower than that of $\mathrm{TiO_2}$.

At around 1000 K, corundum ($\mathrm{Al_2O_3}$) and compounds such as spinel ($\mathrm{MgAl_2O_4}$) and diopside ($\mathrm{CaMgSi_2O_6}$) can condense. The most stable of these is corundum, and this is the most abundant Al-containing molecule below 1400 K [92]. However, the nucleation rate of $\mathrm{Al_2O_3}$ is low, since its supersaturation ratio is large only at high temperatures, above 1400 K, when in fact, its abundance is low [34]. Hence corundum is difficult to form in homogeneous nucleation and may not act as seed nuclei. We note that, as discussed in Section 5.1, around 90% of all presolar oxide grains found in meteorites are corundum that has condensed in AGB stars. At first glance, this appears to contradict the theoretical expectation. However, the laboratory experiments actually destroy most of the presolar oxide grains.

The condensation of silicates and iron grains has been discussed in [35], which notes that condensates can grow on seed nuclei once the condensed phase becomes thermodynamically stable. In particular, it is noted that for olivine, a mixed magnesium–iron silicate, stability is determined not by thermal decomposition but by reaction with $\mathrm{H_2}$. Such a process could well be important for other potential grain materials. In thermal equilibrium, simple silicon oxides, such as SiO and $\mathrm{SiO_2}$ (quartz), are not formed, since all of the available silicon is incorporated into silicates. However, in stellar outflows, incorporation is not complete, and gaseous SiO exists at 1000 K. Solid $\mathrm{SiO_2}$ is formed from the gas phase by

$$\mathrm{SiO + H_2O \longrightarrow SiO_2(s) + H_2}.$$

Quartz becomes stable at around the same temperature as the iron-poor silicates but above that of the iron-rich silicates, so that the detailed condensation sequence depends on whether iron is available to be incorporated into the condensing material. Solid SiO is stable only at very low temperatures and does not condense as a separate grain species. The situation is different for solid MgO, which condenses out at around 1000 K from the atomic magnesium not consumed by silicate formation

$$\mathrm{Mg} + \mathrm{H_2O} \longrightarrow \mathrm{MgO(s)} + \mathrm{H_2}\,.$$

Pure iron grains are also less stable than silicates, so that iron–magnesium silicates are the material out of which dust grains grow on seed nuclei such as TiO_2. Although iron can be incorporated into the silicate structure directly, it is also possible that condensation produces iron inclusions on the surface, which either eventually break off or are submerged by further grain growth.

Gail and Sedlmayr [35] have also discussed the growth of silicates by the addition of SiO, Mg, and Fe at the dust-grain surface followed by oxidation by H_2O. The processes involved are complex, with uncertain kinetics, and include diffusion of ions within the grain lattice, evaporation, sputtering, and annealing. Detailed modeling of a stellar wind shows that silicates form efficiently. In addition, iron grains and MgO solids form, but quartz growth is inhibited due to the fact that the measured sticking coefficient of SiO_2 at 1000 K is small. During the early stages of olivine growth the iron content of the silicate is low, and a near-equilibrium composition results. However, once significant Mg is incorporated into the dust, the iron content increases. The resulting grains thus consist of a nucleation center, most likely composed of TiO_2, surrounded by an iron-poor silicate and an iron-rich mantle.

5.2.4 Gas–Dust Interaction

The time scale for collision between a gas-phase species and dust grains is roughly

$$t_{\mathrm{gd}} = (\pi a_{\mathrm{gr}}^2 v_{\mathrm{dr}} n_{\mathrm{d}})^{-1}, \qquad (5.16)$$

where the grains are assumed to be spherical with radius a_{gr} and have a number density n_{d}, and v_{dr} is the relative gas–grain velocity. Comparison with the expansion time scale, r/v_∞, shows that the collisional time scale is shorter for distances less than $\approx 10^{15}$ cm. Thus, in principle, the chemical composition of the gas could be affected by this interaction, provided that the gas species (i) are bound to the surface, and (ii) find a reactive partner before thermal evaporation.

Although there has been little attempt to study the gas–grain interaction when both gas and dust are hot, and therefore little real understanding of the likely chemistry, the very short time scale before evaporation implies that hydrogen must be the dominant reaction partner for any species binding to the surface of a dust particle.

However, the observational evidence that grain surfaces contribute to molecule formation in the inner CSE is persuasive, if not quite conclusive as yet. The 10 μm spectrum of IRC+10216 was studied in [68], which concluded that silane, SiH$_4$, has a negligible abundance inside 40 R_*, consistent with the LTE calculations, but a fractional abundance of 2.2×10^{-7} beyond this. Similarly, they concluded that methane, CH$_4$, is produced at distances larger than 125 R_*. For ammonia, NH$_3$, they found that there is little present in the dust-formation zone at 5 R_*, but that it is present by 12 R_*. For CS, the situation is different. CS is observed to have a high abundance, 4×10^{-6}, at less than 12 R_*, a lower abundance for distances greater than about 20 R_*, and an even lower abundance in the external CSE, where interstellar photons penetrate. This behavior is also seen in interferometry maps of thermal SiO radio line emission [90]. They find that for three Mira variables, the SiO fractional abundance is centrally peaked and decreases as a power law in radius as the gas is accelerated. This decrease happens inside the photodissociation shell and is likely due to accretion onto dust grains. In order for sticking to be effective, the grains must be cold enough to retain the accreted species against thermal desorption. Since the evaporation time scale depends exponentially on the ratio of the binding energy of a species relative to the dust temperature, accretion occurs only for distances greater than some critical radius, which will be different for different species and smallest for those having the largest binding energies. For a binding energy for SiO equivalent to 29,500 K [66], binding occurs efficiently for $T_\mathrm{d} < 550$ K, or at a distance of around 4×10^{14} cm for IRC+10216, assuming a distance of 150 pc.

Similar results have been found from an interferometric study of SiS in IRC+10216 [14], which shows that its abundance is centrally peaked and decreases substantially out to a distance of $\approx 2 \times 10^{15}$ cm, assuming a distance of 100 pc. The circumstellar chemistry of silicon in IRC+10216, using SiS as a parent and taking into account its depletion onto dust as it flows out through the CSE, has been discussed [42]. As far as I am aware, this is the only chemical model that incorporates accretion onto dust.

Finally, one should point out that when the grains are cold enough, they will bind O atoms and H$_2$O molecules if the collision time is shorter than the expansion time. Since this condition is most likely to occur for high mass-loss rates and low expansion velocities, the possibility exists that gas-grain interactions in AGB CSEs can lead to the formation of ice mantles. Such mantles are indeed observed; efficient accretion of water could lead to little photoproduction of OH, so that extremely red objects may not have OH maser emission [66].

5.2.5 Photochemical Shells

The interstellar radiation starts to affect molecular abundances once the radial UV extinction to infinity falls below a value of 10 magnitudes or so, that is, for distances greater than 10^{16} cm for a mass-loss rate of around 5×10^{-5} M$_\odot$ yr^{-1} and an expansion velocity of 15 km s^{-1}. The radial distance at which photoprocesses become important is directly proportional to the ratio of mass-loss rate to expansion velocity, at least for grains that absorb rather than scatter UV radiation. Thus, objects with high mass-loss rates and low velocities have photoshells further removed from the central star than those with low mass-loss rates and high velocities.

The effect of dust is appreciable. Consider a molecule injected from the inner CSE with an initial fractional abundance f_0 and an unshielded photodissociation rate β_0 s^{-1}. In the absence of any other chemical effects, photodissociation will limit the radial extent of the molecule, with the abundance given by

$$f(r) = f_0 e^{-\beta_0 r/v_\infty}. \tag{5.17}$$

For $\beta_0 = 1.0 \times 10^{-10}$ s^{-1} and $v_\infty = 15$ km s^{-1}, this gives a length scale, $l_0 = v_\infty/\beta_0$, of 1.5×10^{16} cm. When dust is present, the photodissociation rate is modified and may be represented by, for the purposes of discussion,

$$\beta(r) = \beta_0 e^{-d/r}, \tag{5.18}$$

where d is a constant, different for different species. In this case, the new scale length becomes

$$l = l_0/E_2(d/r), \tag{5.19}$$

where $E_2(y)$ is the exponential integral of the second kind, and has a value less than 1. For a typical value of $d = 2 \times 10^{16}$ cm, and $r = l_0 = 1.5 \times 10^{16}$ cm, $E_2(d/r) = 0.1$; thus dust shielding increases the distance of the photodissociation scale length by an order of magnitude.

5.2.5.1 Self-Shielding

The overall chemical structure of the external CSE is also determined to a large extent by the efficiency of CO and H_2 in surviving photodissociation. The effects of both species are rather different. H_2 is a very reactive molecule that helps build molecular complexity, particularly through ion–molecule reactions, while its photodissociation product, H, is essentially unreactive. CO, on the other hand, is very unreactive, while its photoproducts, C and C$^+$, are highly reactive and important in carbon-chain growth in C-rich AGB stars. Both H_2 and CO are destroyed by line absorption and thus can self-shield. In addition, H_2 provides some additional shielding for CO.

The H/H_2 ratio in AGB stars has been investigated [39]. In stars with photospheric temperatures above 2500 K, atomic hydrogen is the dominant form. Once dust forms, the formation of H_2 by surface reactions becomes possible. For a grain surface rate of formation $\mathcal{R}$ in units of the interstellar value 3×10^{-17} cm^3 s^{-1}, the time scale for H_2 formation is less than the expansion time for

$$r < \alpha\, 10^{15}\, \mathcal{R} \left[\frac{\dot{M}}{10^{-5}}\right] \left[\frac{15}{v_\infty}\right]^2 \text{ cm},$$

where α is the ratio of atomic to molecular hydrogen. Molecular hydrogen will form only if the thermal evaporation time for H atoms is less than the expansion time, that is, if

$$\nu_0^{-1} e^{E_b/kT_d} < r/v_\infty,$$

where ν_0 is a typical vibrational frequency $\approx 10^{13}\,\mathrm{s}^{-1}$, E_b the binding energy, and T_d the grain temperature. For $r = 10^{15}$ cm and $v_\infty = 15\,\mathrm{km\,s}^{-1}$, this requires $E_b/kT_d < 50$. Since the dust grains are warm close to the star, we conclude that only sites that are able to chemically bind hydrogen atoms will be able to form H_2. Given the fact that grains are newly formed, we may expect that sufficient sites exist to convert all hydrogen to molecular form upon injection to the outer CSE, at $r > 10^{16}$ cm. However, objects in which $\dot{M}$ is low, or in which chromospheric radiation is important, could have a substantial atomic hydrogen component.

Glassgold and Huggins [39] also considered the H_2 to H transition region at the outer edge of the CSE, where interstellar photons are capable of dissociating H_2. They found that self-shielding of H_2 in the Lyman bands is extremely efficient, such that for an object similar to IRC+10216 the transition zone occurs near 10^{18} cm from the star, which is, as we shall see, much further out than the chemically active region. Thus, for the purposes of modeling the photochemistry in the outer CSE, one can assume that "all" hydrogen is in molecular form and remains so throughout the entire CSE.

The photodissociation of CO in AGB CSEs has been studied [75]. CO is destroyed by line absorption in some 30 bands over the range 91 to 112 nm. Because H_2 is so abundant, photodissociation of this molecule takes place in the radiative wings of the Lyman bands. As a result, one Lyman line can block a CO band up to several nanometers away. Fortunately, the effect on the total CO photodissociation rate is limited, because CO bands distant from H_2 bands contribute 50% of the total photodissociation rate [75]. These authors have modeled the CO–C–C$^+$ transition in some detail for a wide range of parameters appropriate to an AGB star. They find that for $\dot{M} = 10^{-5}\,M_\odot\,\mathrm{yr}^{-1}$ and $v_\infty = 15\,\mathrm{km\,s}^{-1}$, self-shielding causes the size of the CO distribution to be increased by more than a factor of 5 compared to that when only dust shielding is included. The radial distribution of CO can be written as [75]

$$f = f_0 e^{[-(r/r_{1/2})^a \ln 2]}, \tag{5.20}$$

where f_0 is the fractional abundance of CO injected from the photosphere, and a and $r_{1/2}$ are constants that depend on expansion velocity and mass loss rate, with $f/f_0 = \frac{1}{2}$ at $r = r_{1/2}$. Table 5.4 gives the values of a and $r_{1/2}$ for an expansion velocity of $15\,\mathrm{km\,s}^{-1}$ [75].

Table 5.4. Fit parameters for the CO radial distribution for an expansion velocity of 15 km s^{-1}. $[a(b) = a \times 10^b]$ [75]

M [$M_\odot$ yr^{-1}]	a	$r_{1/2}$ [cm]
1(−7)	1.74	1.85(16)
5(−7)	2.09	4.05(16)
1(−6)	2.24	5.95(16)
5(−6)	2.61	1.54(17)
1(−5)	2.79	2.35(17)
5(−5)	3.20	6.67(17)
1(−4)	3.39	1.07(18)

The radial distribution of ^{13}CO is also affected by shielding—nine of its bands overlap with ^{12}CO bands—but to a lesser extent than ^{12}CO. Thus, it is more easily photodissociated; however, the fractionation reaction between ^{13}C$^+$ and ^{12}CO forms ^{13}CO efficiently and offsets its increased photodissociation to a great extent [75].

5.2.6 Cosmic-Ray, Radioactive, and Photospheric Ionization

Cosmic ray particles can penetrate the CSE and provide a source of ionization in regions from which interstellar UV photons are excluded. In addition to ionization of hydrogen and helium, the particles also provide a low-level flux of UV line photons: the Prasad–Tarafdar mechanism. In this process, secondary electrons produced in the ionization of H$_2$ and He collisionally excite H$_2$, which then decays, emitting UV photons [43].

The cosmic-ray ionization of H$_2$ leads to the formation of H$_3^+$, which because of its low proton affinity quickly transfers its proton to neutrals such as CO, C$_2$H$_2$, and H$_2$O. Since H$_3^+$ has been detected in the interstellar medium [36], it is of interest to calculate the column density that might be present in an AGB CSE. Since it is destroyed by ion–molecule reactions and by dissociative recombination, its abundance per unit volume can be written

$$n_{\mathrm{H}_3^+}(r) = \frac{\zeta n_{\mathrm{H}_2}(r)}{\sum k_i n_i(r) + \alpha_e n_e(r)}, \quad (5.21)$$

where ζ is the cosmic-ray ionization rate, and $n_i(r)$ is the abundance at radial distance r of neutral species i, which reacts with H$_3^+$ with rate coefficient k_i, and α_e is the dissociative recombination rate coefficient. Deep in the CSE, where the IS UV field is negligible, the recombination rate is much smaller than the proton transfer rate, so that the expression becomes

$$n_{\mathrm{H}_3^+}(r) = \frac{\zeta}{\sum k_i f_i(r)}, \quad (5.22)$$

where $f_i(r)$ is the fractional abundance of i. The abundance of H_3^+ is thus fairly independent of the type of star, C-rich or O-rich, and since the denominator is almost constant throughout the CSE, the fractional abundance of H_3^+ varies as r^2, and its column density can be written

$$N_{H_3^+} = \frac{r_m \zeta}{\sum k_i f_i(r)}, \quad (5.23)$$

where r_m is the radius at which dissociative recombination rather than proton transfer dominates the loss of H_3^+. Noting that $k_i \approx 10^{-9} \, \mathrm{cm^3 \, s^{-1}}$ for all i, we obtain

$$N_{H_3^+} = 10^{12} \left[\frac{\zeta}{10^{-17}}\right] \left[\frac{r_m}{10^{17}}\right] \left[\frac{10^{-9}}{k}\right] \left[\frac{10^{-3}}{f}\right] \, \mathrm{cm^{-2}}, \quad (5.24)$$

at least an order of magnitude less than the column densities detected in the interstellar medium [36].

The influence of radioactive decay, particularly that of ^{26}Al, on the ionization structure of AGB stars has been considered [38]. Evidence for an AGB production of ^{26}Al comes from the detection of large amounts of extinct ^{26}Al in presolar grains, as discussed in Section 5.1; see also Chapter 2. If this radioactive element is produced in AGB stars, then its decay via β-decay produces a positron, which ionizes H_2 directly. The ejected electron can produce secondary ionizations. For live ^{26}Al trapped in dust in the CSE, positrons still escape the grain, since their mean free path is larger than the typical grain radius [38], so that they, too, provide ionization. The amount of ^{26}Al in AGB stars is difficult to determine, but the amount of extinct ^{26}Al found in meteorites can give a crude estimate. With such an analysis, [38] derived an ionization rate due to ^{26}Al decay of $3 \times 10^{-17} \, \mathrm{s^{-1}}$, similar to that caused by cosmic rays in interstellar clouds. However, as pointed out in this paper, the radioactive ionization could be more important than cosmic ray ionization in AGB stars because of the possibility that stellar magnetic fields exclude low-energy cosmic rays.

Proton transfer reactions of H_3^+ lead to the formation of the important (and observable) molecular ions HCO^+ and H_3O^+, the latter in O-rich CSEs only. The abundance of HCO^+ gives a rather direct measure of the ionization rate, since it is not produced by any photo-process and since it has a relatively low proton affinity, thereby simplifying its chemistry. HCO^+ has been detected in IRC+10216 [71] with an abundance indicating that the ionization rate is about $3 \times 10^{-18} \, \mathrm{s^{-1}}$, or an ^{26}Al/^{27}Al ratio of 2×10^{-3}, consistent with the values found in SiC grains in the Murchison meteorite.

Finally, we discuss the effects of chromospheric radiation, which can provide ionization of the inner CSE. In most chemical models of AGB stars this is neglected, but it has been considered in modeling α Orionis [40]. In this object, the chromosphere has a temperature of $\approx 8000 \, \mathrm{K}$ and emits photons that can ionize metals such as C, S, Si, and K. This star also emits strongly at Lyα, but this is a minor effect compared to continuum radiation. Because the dust-to-gas ratio in α Ori is low, by a factor of 10–20 relative to the ISM, IS UV photons penetrate relatively deeply into the core and dominate the chromospheric flux once r is greater than about $10\,R_*$.

5.3 Detailed Models: Carbon-Rich CSEs

In this section we review the basic chemical models for the outer CSEs of carbon-rich AGB stars.

Models of the photo-induced chemistry in the outer CSE depend on the adoption of initial conditions related to stellar physics (mass-loss rate, expansion velocity, dust-to-gas ratio, dust extinction law, temperature distribution) and chemistry (initial molecular abundances). In this section, we will concentrate on models developed for IRC+10216, the most heavily observed C-rich AGB star.

5.3.1 Hydrocarbons

The key parent molecule in carbon stars is acetylene, C_2H_2, which is injected to the outer CSE with a fractional abundance of about 10^{-5}. Acetylene can be both photodissociated and photoionized

$$C_2H_2 + h\nu \longrightarrow C_2H + H,$$
$$C_2H_2 + h\nu \longrightarrow C_2H_2^+ + e.$$

The ethynyl radical C_2H can be further photodissociated in a chain of reactions, which ultimately reduce one acetylene molecule to two carbon ions [41]

$$C_2H + h\nu \longrightarrow C_2 + H,$$
$$C_2 + h\nu \longrightarrow C + C,$$
$$C + h\nu \longrightarrow C^+ + e.$$

This photodissociation leads to shell distributions at increasing radii for the daughter products and could explain the column density and radial distribution of C_2H in IRC+10216. Figure 5.7 shows the typical radial distribution of these species in a model of IRC+10216 chemistry.

Acetylene and its photo-products are reactive species, particularly with one another. The overlap between the distributions of reactive daughters and (still) abundant parent molecule, together with the innate reactivity of these species, ensures efficient growth of hydrocarbons. For example, C_4H and C_4H_2 are formed from

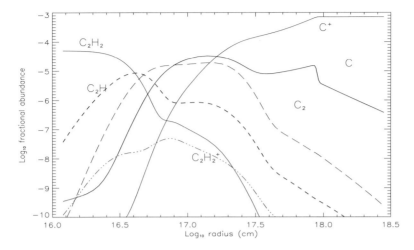

Fig. 5.7. The radial distributions of acetylene and its daughter species are shown. The formation of molecular shells of C_2H and C_2 is evident. The parameters used are representative of IRC+10216

$$C_2H + C_2H_2 \longrightarrow C_4H_2 + H,$$
$$C_2H + C_2H \longrightarrow C_4H + H,$$
$$C_2 + C_2H_2 \longrightarrow C_4H + H,$$
$$C_4H_2 + h\nu \longrightarrow C_4H + H.$$

In addition, C atoms are known to insert efficiently and thus build chain backbones one atom at a time. Similarly, C^+ insertion reactions, such as

$$C^+ + C_2H_2 \longrightarrow C_3H^+ + H,$$

followed by hydrogen addition and dissociative recombination, leads to observed species such as C_3H and C_3H_2. Carbon chains with relatively few H atoms attached are preferentially formed, despite the overwhelmingly large abundance of H_2, because of thermodynamic effects, most reactions between hydrocarbon ions and H_2 being endothermic.

In addition, since C_2H_2 has a large cross-section for photoioization to $C_2H_2^+$, this ion also plays a fundamental role in building hydrocarbons, for example

$$C_2H_2^+ + C_2H_2 \longrightarrow C_4H_3^+ + H,$$
$$C_2H_2^+ + C_2H_2 \longrightarrow C_4H_2^+ + H_2.$$

The chemistry of hydrocarbon molecules containing six or more carbon atoms is highly uncertain, due to a lack of laboratory information, in particular. While some information is available on the reaction of hydrocarbons

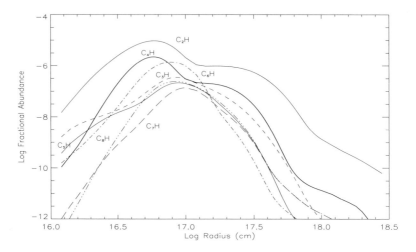

Fig. 5.8. The radial distributions of the carbon hydrides in a model for IRC+10216 (after [83])

with H atoms and H_2, there are no studies as yet on the synthetic routes involving chain growth, for example

$$C_2H + C_mH_2 \longrightarrow C_{m+2}H_2 + H,$$
$$C_2H + C_mH \longrightarrow C_{m+2}H + H.$$

The growth of hydrocarbon chains in IRC+10216 has been discussed [82] in the light of laboratory data on C atom insertion reactions and shows that appreciable abundances of molecules with up to 8 carbon atoms could be achieved with a reasonable extrapolation of experimental data. This work was extended to species with up to 23 carbon atoms [83]. For the intermediate sized chains, that is, up to about 10 carbon atoms, ion–molecule chemistry plays a significant role in the formation of C_nH. For the larger hydrocarbons, as well as the cumulenes and polyacetylenes, C_nH_2, routes involving negative ions play a more important role as the size of the chain increases, for example

$$H + C_{15}^- \longrightarrow C_{15}H + e.$$

Figure 5.8 shows the radial distributions of the hydrides.

5.3.2 Cyanopolyynes

The cyanopolyynes up to HC_9N have been detected in IRC+10216. Early attempts to reproduce the observed column densities concentrated on ion–molecule routes, such as

$$C_3H_3^+ + N \longrightarrow HC_3NH^+ + H,$$

followed by dissociative recombination, but the neutral–neutral reaction

$$CN + C_2H_2 \longrightarrow HC_3N + H$$

is far more efficient [57]. Here CN is the photodissociation product of parent HCN. Routes that involve parents and daughter species reacting directly are generally more efficient because of better radial overlap of their respective abundance distributions. Similar reactions,

$$CN + C_mH_2 \longrightarrow HC_{m+1}N + H,$$

can synthesize the larger cyanoployynes. Once again, while laboratory measurements are available for the reaction of $CN + C_2H_2$, there are no data available for the analogous reactions forming the larger cyanopolyynes. In addition to $CN + C_2H_2$, another parent–daughter reaction that can form HC_3N is that of $C_2H + HCN$. Reactions of C_2H can drive the formation of larger cyanopolyynes through a boot-strapping process [82]

$$HC_mN + C_2H \longrightarrow HC_{m+2}N + H,$$

although rates are again unknown. The work of [83], which considered cyanopolyynes up to $HC_{23}N$, indicates that with rates appropriate for the formation of HC_3N, the higher cyanoployynes are produced most efficiently by this bootstrap mechanism. Figure 5.9 shows the radial distribution of the fractional abundances of the cyanopolyyne species. Again, one notes the formation of molecular shells and that the radii at which the fractional abundances peak increase as the size of the molecule increases.

Table 5.5 compares observed and calculated radial column densities for IRC+10216 [82, 83]. In this comparison it is important to note that the observed column densities are often beam-averaged; see Chapter 7.

The agreement between theory and observation is quite encouraging, particularly for the cyanopolyynes, and, to a lesser extent, for the hydrides. For these species, the models overpredict the column densities by over an order of magnitude (C_7H, C_8H). The results of [83] also indicate that the formation of very large molecules in the outer CSE of IRC+10216 is fairly efficient; for example, the column density of the cyanopolyynes falls by a factor of about four as one progresses from one member of the chain to the next. The column density of $HC_{19}N$ is 4.5×10^{10} cm^{-2}. However, one should note that the abundances are somewhat sensitive to the adopted photodissociation rates, which for large species are unknown. They could be larger or smaller than those used in the calculation.

5.3.3 Negative Ions

The model calculations by Millar et al. [83] were the first to include negative ion chemistry in a systematic manner. The electron affinities of hydrocarbon species are quite large, so that association reactions such as

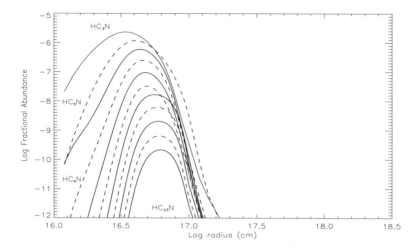

Fig. 5.9. The radial distributions of the cyanopolyynes, up to $HC_{23}N$, in a model for IRC+10216 (after [83])

$$C_mH_n + e \longrightarrow C_mH_n^- + h\nu$$

can be rapid. The negative ions so formed are of interest in their own right, since many have significant electric dipole moments and may be detectable via rotational emission. Further, some negative ions such as C_7^- have been implicated in the formation of the diffuse interstellar bands, so that C-rich AGB CSEs may prove an interesting laboratory in which to test this hypothesis. Finally, negative ions are very reactive and can help build chemical complexity, for example,

$$C_m^- + C_n \longrightarrow C_{m+n}^- + h\nu,$$
$$C_{m+n}^- + H \longrightarrow C_{m+n}H + e,$$
$$C_{m+n}H + e \longrightarrow C_{m+n}H^- + h\nu,$$
$$C_{m+n}H^- + H \longrightarrow C_{m+n}H_2 + e.$$

Because negative ions are formed by attachment of electrons to neutrals and since the fractional ionization is increasing with radius, the fractional abundance of anions remains large beyond 10^{17} cm (Figure 5.10). Table 5.6 gives calculated column densities for some of the more abundant negative ions. These column densities are substantial, with the column density of C_7^- around 0.1 that of C_7. It has been suggested that C_7^- is the carrier of certain of the diffuse interstellar bands. Although this now appears to be unlikely, we note that the radial column densities of many anions in IRC+10216 are larger, by about an order of magnitude, than those required to produce the interstellar bands. An interesting check on whether large anions (or other

Table 5.5. Comparison of observed and calculated column densities (cm^{-2}). MHB [83]; MH [82]. [$a(b) = a \times 10^b$]

Species	MHB	MH	Observed	Obs. ref.
C	1.0(16)	2.7(16)	1.1(16)	[69]
C_3	6.5(14)	4.7(14)	1(15)	[51]
C_5	7.5(14)	1.1(15)	1(14)	[8]
C_2H	5.7(15)	1.8(16)	3–5(15)	[5, 44, 46]
C_3H	1.4(14)	1.4(14)	3(13)	[46, 67]
C_4H	1.0(15)	5.5(15)	2–9(15)	[5, 28, 46, 67]
C_5H	8.7(13)	5.5(13)	2–50(13)	[46, 67]
C_6H	5.8(14)	4.5(14)	3–30(13)	[46, 67]
C_7H	4.5(13)	5.4(12)	1(12)	[46]
C_8H	1.1(14)	3.6(13)	5(12)	[46]
C_3H_2	2.1(13)	3.0(13)	2(13)	[67]
C_4H_2 [a]	2.9(15)	4.7(15)	3–20(12)	[21, 67]
C_3N	3.2(14)	2.2(14)	2–4(14)	[48, 67]
C_5N	1.4(14)	2.3(14)	3(12)	[48]
HC_3N	1.8(15)	2.8(15)	1–2(15)	[48, 67]
HC_5N	7.1(14)	1.2(15)	2–3(14)	[48, 67]
HC_7N	2.2(14)	2.6(14)	1(14)	[67]
HC_9N	5.8(13)	5.1(13)	3(13)	[67]
HCO^+	2.4(12)	—	3(12)	[86]
CH_3CN	3.4(12)	—	6(12)	[45]

[a] Observations refer only to the cumulene form

Table 5.6. Calculated column densities (cm^{-2}) for selected negative ions for selected species in IRC+10216 [83]. [$a(b) = a \times 10^b$]

Spec.	N	Spec.	N	Spec.	N	Spec.	N
C_7^-	1.4(13)	C_7H^-	1.2(13)	C_8^-	1.4(13)	C_8H^-	2.7(13)
C_9^-	6.6(12)	C_9H^-	7.9(12)	C_{10}^-	2.6(12)	$C_{10}H^-$	4.8(12)
C_3N^-	4.0(11)	C_5N^-	1.3(13)	C_7N^-	7.1(11)	C_9N^-	1.8(11)

hydrocarbon chains) are responsible for the bands would be to search for them in optical absorption observations through the outer shell of IRC+10216. Such a study has been carried out [70], but no features intrinsic to the CSE were found.

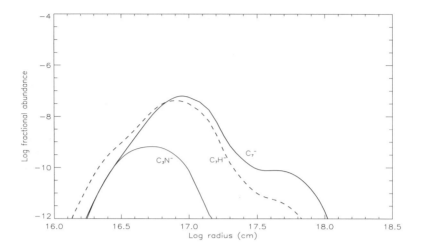

Fig. 5.10. The radial distributions of some negative ions in a model for IRC+10216 (after [83])

5.3.4 Silicon Chemistry

A number of silicon-bearing molecules have been detected in IRC+10216, including SiS, SiC, SiC_2, SiC_3, SiC_4, SiO, SiH_4, SiN, and SiCN. Table 5.7 gives observed column densities together with information on whether the molecule arises in the inner or outer part of the CSE.

Models of Si-bearing chemistry have been investigated [42, 57, 74]. Parent SiC_2 can directly produce SiC through photodissociation [57], but a significant problem is that the SiC_2 abundance is actually higher in the outer than in the inner CSE, indicating that some formation process must be operating. The photodissociation of parent molecules such as SiS and SiO ultimately forms Si^+, which can react with acetylene

$$Si^+ + C_2H_2 \longrightarrow SiC_2H^+ + H,$$

which is followed by dissociative recombination to produce SiC_2 in the outer CSE. The SiC_2H^+ ion can also react with acetylene

$$SiC_2H^+ + C_2H_2 \longrightarrow SiC_4H^+ + 2H,$$

which, following dissociative recombination, produces SiC_4, and possibly SiC_3. Photodissociation of SiC_4 can also produce SiC_3. Two alternative routes to SiC_4 involve SiC_2 [85]

$$SiC_2 + C_2H_2 \longrightarrow SiC_4 + H_2,$$
$$SiC_2 + C_2H \longrightarrow SiC_4 + H.$$

Table 5.7. Observed column densities (cm^{-2}) for silicon-bearing molecules in IRC+10216. [$a(b) = a \times 10^b$]

Species	N	Region	Ref.
SiO	1(15)	Inner	[86]
SiS	7(15)	Inner	[12]
SiH$_4$	2(15)	Inner	[68]
SiC$_2$	5(15)	Inner/Outer	[37]
SiC	5(13)	Outer	[22]
SiC$_3$	4(12)	Outer	[3]
SiC$_4$	5(12)	Outer	[85]
SiCN	2(12)	Outer	[47]
SiN	4(13)	Outer	[93]

The first of these reactions involves two parent molecules and would be so efficient that it would reduce the SiC$_2$ column density far below that observed. The column density of SiC$_3$ is predicted to be 2×10^{12} cm^{-2} [57], a factor of two less than observed, although this prediction is based on highly uncertain photodissociation rates and on parent abundances that have been revised since. Parent SiS may be directly involved in the formation of SiC$_2$ through the reactions [42]

$$SiS + C_2H_2^+ \longrightarrow SiC_2H^+ + HS,$$
$$SiS + C_2H_3^+ \longrightarrow SiC_2H^+ + H_2S,$$

followed by dissociative recombination.

The detection of SiH$_4$ [68] led to a reinvestigation of silicon chemistry [74]. The photodissociation of silane produces SiH$_2$ directly and provides reactive Si and Si$^+$ in the outer CSE. SiN can be formed via the fast reaction

$$Si^+ + NH_3 \longrightarrow SiNH_2^+ + H,$$

followed by dissociative recombination. The recently detected radical SiCN may be formed by a similar reaction involving parent HCN [47]

$$Si^+ + HCN \longrightarrow SiCNH^+ + h\nu.$$

Since this is a radiative association reaction, the rate coefficient is low, about 1000 less than that of Si$^+$ and C$_2$H$_2$, a fact that may explain the low ratio of SiCN/SiC$_2$ observed.

Despite the qualitative successes in understanding silicon chemistry, there are quantitative problems, in particular reproducing the large SiC$_2$ abundance in the outer CSE, where it is about an order of magnitude more abundant than in the inner CSE. Finally, it is worth pointing out that some information is available on the spatial distribution of SiS, which decreases in abundance appreciably as it flows out from the photosphere to about 2×10^{15} cm, before being photodissociated away around 3×10^{16} cm [14]. Similar results have been found in IR observations of SiS in IRC+10216 [16], which derived that the fractional abundance decreased from 4.3×10^{-5} a few stellar radii from the star to 4.3×10^{-6} at $12\,R_*$. The implication is that SiS depletes onto the grains. With the exception of the model described in [42], the gas–grain interaction has not been considered.

5.3.5 Sulphur Chemistry

The species CS, C_2S, and C_3S have all been detected in IRC+10216 [14, 23, 99], and a tentative detection of C_5S has been claimed by the latter authors. CS is now known to have a large abundance close to the photosphere [50, 68, 73], so that it and SiS provide the source of sulphur to the outer CSE. The recent study of vibrationally excited CS [50] indicates that the fractional abundance of CS is in the range $(3-7) \times 10^{-5}$ at around $14\,R_*$. This is slightly larger than suggested by the IR observations 4×10^{-6} at $12\,R_*$ [68] or by the interferometry 10^{-6} [73], the latter two observations also indicating that the CS abundance falls off to larger radii, probably due to accretion on to dust.

Photodissociation of CS and SiS leads to photochemical shells of S and S^+ that are very reactive, via both ion-neutral and neutral reactions, with the photoproducts of acetylene

$$S^+ + C_2H_2 \longrightarrow HC_2S^+ + H,$$
$$S + C_2H_2^+ \longrightarrow HC_2S^+ + H,$$

which lead to C_2S upon dissociative recombination, and

$$S + C_2H \longrightarrow C_2S + H.$$

This latter reaction has been studied theoretically by J.R. Flores (priv. comm.), who finds it to be exothermic with a rate coefficient that varies in the range $(3.3-2.6) \times 10^{-10}\,\mathrm{cm^3\,s^{-1}}$ over the temperature range 300–60 K, almost an order of magnitude greater than the value of $5 \times 10^{-11}\,\mathrm{cm^3\,s^{-1}}$ adopted in [83].

Larger organo-sulphur chain molecules can form through similar reactions with C_n chains ($n > 2$). Because the C_4 chain is more abundant than the C_3 chain, the major route to C_3S is

$$S^+ + C_4H \longrightarrow C_4S^+ + H,$$
$$S^+ + C_4H_2 \longrightarrow HC_4S^+ + H,$$
$$S + C_4H_2^+ \longrightarrow HC_4S^+ + H,$$

followed by dissociative recombination to C_3S and C_4S. The neutral reaction

Table 5.8. Observed and calculated column densities (cm^{-2}) for C_nS molecules in IRC+10216. The column labeled MBH refers to the model by [83], while the label New refers to [81], which assumes parent CS and the new rate coefficient for S + C_2H. The radius at which the fractional abundances peak in the "New" model is also given. $[a(b) = a \times 10^b]$

Species	MHB	New	r_p [cm]	Obs
C_2S	3.5(13)	3.8(14)	5.6(16)	1.5(14)
C_3S	6.8(12)	2.8(14)	4.8(16)	1.1(14)
C_4S	8.7(11)	2.3(12)	5.8(16)	–

$$CS + C_2H \longrightarrow C_3S + H$$

may also be effective in producing C_3S, although there is no measurement of its rate coefficient as yet.

These reactions were included in a model of IRC+10216 that assigned a parent sulphur abundance of 10^{-6} to H_2S [83]. The calculations have now been repeated [81] using CS as a parent with a fractional abundance of 4×10^{-6}. Table 5.8 lists, in column two, the calculated column densities of C_2S, C_3S, and C_4S from [83] and, in column three, from the new model [81], for comparison with observations. The influence of the reaction between CS and C_2H, which has an adopted rate coefficient of 5×10^{-11} cm^3 s^{-1}, and that of the new rate coerfficient for S + C_2H are clearly evident. The increases are an order of magnitude for C_2S and a factor of 40 for C_3S. The resulting increases give much better agreement with observation. The importance of S and CS reactions in building larger organo-sulphur species may be very significant, but unfortunately, there are no accurate estimates of rate coefficients [81].

5.4 Detailed Models: Oxygen-Rich CSEs

Oxygen-rich AGB CSEs have been widely observed in maser transitions of SiO, H_2O, and OH, but data are relatively scarce for thermal emission from molecules compared to the situation for carbon-rich objects. In part, this difference reflects the ability of carbon to form bonds, in part to the lack of a star equivalent in mass-loss rate and proximity to IRC+10216. Since CO self-shields against dissociating UV radiation, the dominant parent as far as chemistry is concerned is H_2O, whose photodissociation product OH is a very reactive radical. Since H_2O does not have a large photoionization channel, unlike C_2H_2, the chemistry in O-rich CSEs generally involves neutral–neutral reactions. This, together with the presence of abundant OH, ensures that diatomic oxides are produced efficiently.

In O-rich objects, the abundance of H_2O is large enough that self-shielding becomes important in determining its radial extent. The photodissociation of water has been modeled with a three-band approximation, and detailed cross-sections, photo-rates and shielding lengths are available [76]. The size of the OH shells as a function of several parameters, including mass-loss rate, has been investigated in [61], which found a good agreement between theory and observation.

The major puzzle in observations of O-rich CSEs is the presence of carbon atoms in many molecules, currently CN, CS, HCN, HNC, OCS, HCO^+, and H_2CO. Because the carbon tied up in CO is unavailable in the region in which the molecules are detected, much effort has gone into identifying the source of reactive carbon. Table 5.1 shows that after CO, the most abundant carbon-bearing molecule at LTE is CO_2, some four orders of magnitude less abundant. Unfortunately, CO_2 does not provide reactive carbon since it photodissociates to CO in the outer CSE. Provision of reactive carbon can be aided by pulsational shock chemistry, which produces CS and HCN [29] with fractional abundances of 3×10^{-7} and 2×10^{-6}, respectively, at $2.2\,R_*$, see Section 5.2.2. Another possibility is that the gas–grain interaction can produce CH_4. The influence of injected CH_4 on the outer CSE chemistry of R Dor, TX Cam, and IK Tau, including the chemistry of silicon, chlorine, and phosphorus, has been studied [96].

OH and its photoproduct, O, can form O_2 in the outer CSE

$$O + OH \longrightarrow O_2 + H.$$

In addition, although it is often assumed that CO is unreactive, it does take part in a number of chemical reactions. In C-rich CSEs, it can associate with $C_2H_2^+$ to form HC_3O^+, a possible precursor to C_3O, and with CH_3^+ to form CH_3CO^+, the precursor of ketene, CH_2CO. In O-rich CSEs, CO reacts with OH to form CO_2, although the amount produced is rather less than that synthesized in the pulsational shock. Finally, as in C-rich objects, CO reacts with H_3^+ to form HCO^+.

If methane is a parent molecule, although recent ISO searches make this unlikely, it can aid the production of species such as methanol, ketene, and formaldehyde

$$CH_3^+ + H_2O \longrightarrow CH_3OH_2^+ + h\nu,$$
$$CH_3OH_2^+ + e \longrightarrow CH_3OH + H,$$
$$CH_3^+ + CO \longrightarrow CH_3CO^+ + h\nu,$$
$$CH_3CO^+ + e \longrightarrow CH_2CO + H,$$
$$CH_2 + OH \longrightarrow H_2CO + H,$$
$$CH_3 + O \longrightarrow H_2CO + H.$$

Methanol and C_2H have been searched for in TX Cam, IRC+10011, and IK Tau without success [24], and the conclusion is that methane is not the source of reactive carbon.

Nitrogen chemistry is driven by the breakup of parent N_2, by either He^+ or photons. Some of the N and N^+ produced form nitrogen hydrides, including NH_3, and the interaction of the nitrogen and carbon chemistries produces HCN and HNC in the outer shell. In C-rich CSEs, the abundance of HCN is orders of magnitude greater than that of HNC, because the former is a parent species. In O-rich CSEs, both are chemical daughters and can cycle between one and another using $HCNH^+$ as an intermediary, so that their abundances are similar, as observed.

The photodissociation of HCN and HNC produces the radical CN, which is easily detected. However, the CN/HCN ratio is observed to be around an order of magnitude less in O-rich stars than in C-rich CSEs [6], despite the fact that it is the direct photoproduct of HCN in both cases. This suggests that an additional destruction mechanism, which works only in O-rich CSEs, is reducing the CN abundance. O_2 is known to have a fast reaction with CN, but its abundance, at least far out in the CSE, is never large enough to be significant. It is possible that some nonequilibrium process is operational, that is, that CN is destroyed not by photons but closer to the star. There is some indication that this could be true, because the rotational temperatures derived for CN are larger in O-rich than C-rich CSEs [6].

Sulphur is injected into the outer CSE as S, HS, H_2S, and SiS. S and HS are very reactive with O and OH, forming SO

$$S + OH \longrightarrow SO + H,$$
$$HS + O \longrightarrow SO + H,$$

which reacts with OH to form SO_2,

$$SO + OH \longrightarrow SO_2 + H.$$

The CS molecule can be formed by reaction between SO and HS with C atoms, while the ion SO^+ forms from the reaction of S^+ with OH. This ion is unreactive with H_2 and could be detectable in O-rich CSEs; it has been observed in interstellar clouds, although its low dipole moment and the low mass of CSEs will make detection difficult.

The chemistry of chlorine makes for an interesting possibility for detection of HCl in CSEs. LTE calculations (Table 5.1) show that HCl takes up all of the available chlorine. It is readily photodissociated, ultimately to Cl^+, in the external CSE. However, Cl^+ reacts rapidly with H_2, with successive reactions producing H_2Cl^+, which can reform HCl following dissociative recombination. This recycling is rapid when H_2 is abundant and therefore keeps the abundance of HCl surprisingly large in the outer CSE. As a result, it might be possible to detect HCl through its $J = 1-0$ rotational transition at 625.9 GHz. Other chlorine-bearing molecules are predicted to have very small abundances.

Finally, we discuss the ionization in O-rich CSEs, a topic considered in detail in [76]. Ionization is provided mainly by cosmic rays in the inner CSE and photons in the external parts of the CSE. Cosmic ray ionization of H_2 rapidly produces H_3^+, which proton transfers with abundant parent molecules to produce HCO^+, H_3O^+, and N_2H^+. Photodissociation of CO and H_2O lead to C^+ and OH. The interaction of C^+ with water and OH forms HCO^+

$$C^+ + H_2O \longrightarrow HCO^+ + H,$$
$$C^+ + OH \longrightarrow CO^+ + H,$$
$$CO^+ + H_2 \longrightarrow HCO^+ + H.$$

The major destruction for HCO^+ is proton transfer with H_2O

$$HCO^+ + H_2O \longrightarrow H_3O^+ + CO,$$

and dissociative recombination. The fractional abundances of HCO^+ and H_3O^+ can be as large as 10^{-7} [76].

5.5 Complications

The above discussion has been restricted to spherically symmetric steady mass loss at a constant rate. While these assumptions may hold for certain periods in the evolution of AGB stars, they do not hold for the entire duration of the mass-loss process. In this section, we discuss issues that complicate this simple picture.

5.5.1 Asymmetric Mass Loss

While stars are spherical, post-AGB objects usually show clear evidence of nonsphericity, particularly bipolar shapes suggesting a concentration of remnant AGB matter in a toroidal region. Some degree of CSE anisotropy presumably exists in the AGB phase, even if only toward the end, which determines the anisotropy seen in pre-PNe. High-resolution observations of molecular lines in AGB winds also show some indication of anisotropy (see the observational evidence discussed in Chapter 7). Anisotropy can be due to variation with direction of properties such as nonradial pulsations, density, temperature, UV photon flux, and mass-loss rate, and in some cases may be related to the presence of a binary companion. Anisotropy using an oblate density distribution for the material around the star was first considered in [64]. The adopted distribution was

$$n(r, \theta) = \frac{\dot{M}}{4\pi r^2 m_H \langle v \rangle} F(a, \theta),$$

where $n(r,\theta)$ is the number density of hydrogen nuclei at radial distance r and polar angle θ, $\langle v \rangle$ the average velocity of the wind, and $F(a,\theta)$ is given by

$$F(a,\theta) = \frac{a(1-a^2)^{1/2}}{\left(\arctan\left[(1-a^2)^{1/2}/a\right]\right)\left[(a^2-1)\sin^2\theta + 1\right]},$$

where $0 < a < 1$ (a spherically symmetric wind has $a=1$).

Jura [64] calculated the UV radiation field for this density distribution and used it to determine the distribution of CO in the the polar direction, i.e., along the symmetry axis ($\theta = 0$), and orthogonal to this, finding that parent molecules in the polar direction could be photodissociated closer to the star than in the case of a spherically symmetric outflow. The work of Jura was extended by Howe and Millar [58], who calculated the UV radiation field at an arbitrary outflow angle for Jura's oblate density distribution with azimuthal symmetry. This enabled them to compute photorates as a function of radial distance and polar angle and to produce maps of molecular distributions in the (r, θ) plane. They found that asymmetries in molecular distributions greater than those adopted for the underlying mass distribution, i.e., the H_2 distribution, were largest for molecules for which a number of intermediate chemical steps were necessary for their formation, with polar holes being formed. Given the simple mass and velocity distributions adopted, it is perhaps not too surprising that comparison with observations of AGB stars, in particular IRC+10216, is less than ideal [58].

5.5.2 Clumps and Shells

The CSEs of AGB stars are probably very clumpy; see also Chapter 7. For example, [94] imaged IRC+10216 at a resolution of 76 mas in the K′-band and detected five separate dust clumps within 0.21 arcsec, or a few stellar radii, of the star. More recent imaging [88] shows evidence for clump evolution as well as a bipolar structure, indicating asymmetric mass loss in the very central regions of the star. Striking evidence of clumpiness in the outer CSE comes from Figure 4 in [72], which plots the location of "clumps"—more accurately, peaks in molecular line emission—in IRC+10216. Although in broad terms, molecules such as CN and HNC do exhibit shell distributions, the shells contain considerable substructure.

A small number of carbon stars show evidence of a detached CO shell, with typical radii of $\approx 10^{17}$ cm and a shell thickness/radius ratio of less than 0.1. The shells show remarkable spherical symmetry, with the radius of that around TT Cyg, $\approx 2.7 \times 10^{17}$ cm, varying by less than 3 %. Radiative transfer models of these shells indicate that the CO emission is best fit by clumpy models [7]. As examples, the shell around S Scuti may have 900 clumps of radius around 2×10^{16} cm, while that around TT Cyg may have about 7300 clumps of radius 4×10^{15} cm [87]. It is notable that PNe contain many clumps of molecular gas [60].

If such shells are formed by a period of rapid mass loss, $\approx 10^{-5}\,M_\odot\,yr^{-1}$, then the duration of the event must have been only a few hundred years. In addition, the nature of the shells shows that the mass loss must have been very isotropic in both mass and velocity, the latter to within a few tenths of a km s^{-1} [87]. In this scenario, the clumps may be the result of Rayleigh–Taylor instabilities and therefore not indicative of clumps in standard CSEs. An alternative possibility is that the shells result from the episodic mass loss driven by pulsating dust formation, as proposed in [91]. The results of 48 model calculations from [31] were fitted in [4] to a simple equation that showed that for C-rich AGB stars with effective temperatures in the range 2500–3000 K, the mass-loss rate was extremely sensitive to stellar temperature ($T^{-8.3}$) and luminosity ($L^{1.5}$). The sensitivity to temperature is the result of mass loss being a dust-driven process for high mass-loss rates, and the sensitivity of dust-related chemistry, growth, and evaporation to temperature. Furthermore, since the process is dependent on radiation pressure, there is a minimum luminosity needed to get large mass-loss rates. For stars with initial masses less than about 1.1 M$_\odot$, this luminosity is never reached. For masses in the range 1.1–1.3 M$_\odot$, [91] found that pulsational shock waves, to which grain formation is intimately connected, as discussed in Sections 5.2.2 and 5.2.3, stimulate episodic dust formation, which drives high mass-loss events. The shells so produced agree well, in mass and thickness, with the detached shells observed.

Dust shells have also been detected recently around IRC+10216 in deep B- and V-band imaging [79, 80]. The observations indicate a multiple shell structure, with a shell–intershell density ratio of about 2–3. The shells are extremely thin, with thicknesses of about 0.5–3 arcsec, corresponding to a time scale of 20–120 yr for a distance to the star of 120 pc. The spacing is irregular, and there is a tendency for the shells to be thicker at greater radii, indicating a dispersion velocity of about 0.7 km s^{-1}. Mauron and Huggins have compared the dust distributions to those of molecular gas [72] and find extremely good correlations, indicating that the gas and dust clumps overlap. This implies that the gas–grain drift velocity is low. The effect of these shells on molecular line formation has not yet been considered, but Brown and Millar [17] have investigated their effects on the radial distributions of molecules. The presence of rings results in distributions that are more sharply peaked than in the smooth r^{-2} outflow (Figure 5.11). This is due to the additional extinction provided by the rings, which reduces the flux of interstellar photons penetrating deep into the shell, hence decreasing the rate at which daughter molecules are formed. Material within rings, through their density enhancement, also experiences more rapid collisions, which help synthesize molecules. In their study, Brown and Millar found that not all rings affected the chemistry. Too far in, and no photons reached them, too far out, and

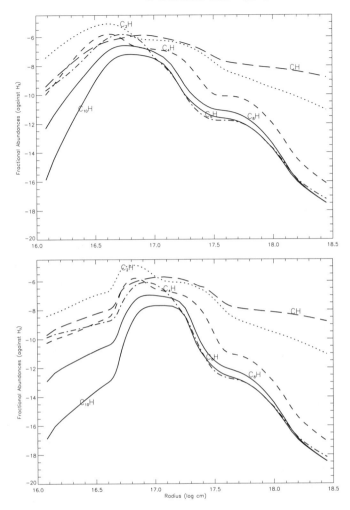

Fig. 5.11. Comparison of the effects of including enhanced density shells in the calculation of molecular abundances in the CSE of IRC+10216. The top panel shows the radial distribution of C_nH molecules in the usual $1/r^2$ density distribution, the bottom panel shows the distribution that results when several narrow shells of density enhanced by a factor of six are included [17]

the densities in the rings were too low to provide either much additional extinction or collisions at a rate faster than the expansion rate. Rings at the "photochemical" distance, equivalent to a visual extinction of a few, have the largest effect, as one might expect.

Chemical models of clumps in PN have been studied [56], but clumps have not yet been discussed in the context of AGB CSE chemistry. In PN, clumps are essential if the molecules are to be protected from the strong stellar UV field. In AGB CSEs, one needs photons to drive the chemistry, so a clumpy medium could, depending on the density enhancement, enable parents to travel further out from the central star before photodissociation.

5.5.3 Disks

A small number of late-type stars exhibit very narrow CO line profiles, with widths typically 1–2 km s^{-1}, indicative of material orbiting in disks rather than outflowing from the star [65]. It appears that most of these systems might be binaries in which gas (and dust) is trapped. The result of such trapping can be grain growth as well as cystallization. Jura and Kahane [65] have searched BM Gem, a C-rich star surrounded by O-rich dust, for a number of simple molecules in addition to CO. The upper limits for both HCN/CO and SiO/CO are less than those observed in C-rich and O-rich AGB stars, respectively, but can be explained by the presence of a disk. The HCN/CO ratio is depressed because HCN is photodissociated by the companion—CO is able to self-shield to some extent—while the SiO is incorporated into silicate grains in the disk, with strong evidence from IR observations that the silicate grains and the dust-to-gas ratio are larger than normal. SiO can also be photodissociated more readily than CO.

5.6 Conclusions and Outlook

The chemical composition of AGB CSEs is, as we have seen, determined by a range of processes. Fortunately, the properties of the CSEs are such that different processes dominate at different radial distances, so that to first order, we can make piecemeal models. These have allowed great insight in interpretating the many observations of dust and gas. As a result, we believe that we understand the main chemical processes that occur at different radii, although important information such as rate coefficients and photodissociation cross sections are still lacking. One should not underestimate the difficulty of obtaining these data. Neutral–neutral reactions, which dominate in the inner CSE and are important in the outer CSE, are difficult to study in the laboratory because each experiment must be designed to follow the abundance of a particular reactant or product. In contrast to the case of ion–neutral systems, there is no generic technology such as mass spectometry that makes monitoring abundances easy. Furthermore, such rate coefficients are needed over a wide temperature range, from 10–2000 K in many cases, with, at the highest temperatures, three-body, rather than binary, reactions dominating. Similarly, photodissociation cross-sections are much more useful than the rate coefficients, since the latter have an adopted radiation field already incorporated. Provision of the cross-sections allows one to model the radiative transfer with particular radiation fields, grain properties, and geometries. Although cross-sections are available for a number of species, there is still a need for data on radicals and larger molecules for which quantum calculations rather than experiment are likely to be necessary.

Despite this lack of data, we are in a position to make more holistic models of the chemistry, that is, a self-consistent study of particular objects from the photosphere to the interstellar medium. While this remains a formidable intellectual and computational challenge, the increasingly detailed observational results, and particularly those that will emanate from the new class of 8 m optical telescopes and large (sub)mm-wave interferometers, such as ALMA, will drive the subject in this direction. The future is bright.

Acknowledgments: Astrophysics at UMIST is supported by a grant from PPARC. I am grateful to P. Cau and A. Markwick for material used in this article.

References

1. Amari, S., Hoppe, P., Zinner, E., and Lewis, R. S. *Nature*, 365, 806, 1993.
2. Andersen, A. C., Jorgensen, U. G., Nicolaisen, F. M., Sorensen, P. G., and Glejbol, K. *A&A*, 330, 1080, 1998.
3. Apponi, A. J., McCarthy, M. C., Gottlieb, C. A., and Thaddeus, P. *ApJ*, 516, L103, 1999.
4. Arndt, T. U., Fleischer, A. J., and Sedlmayr, E. *A&A*, 327, 614, 1997.
5. Avery, L. W., Amano, T., Bell, M. B., et al. *ApJS*, 83, 363, 1992.
6. Bachiller, R., Fuente, A., Bujarrabal, V., et al. *A&A*, 319, 235, 1997.
7. Bergman, P., Carlstrom, U., and Olofsson, H. *A&A*, 268, 685, 1993.
8. Bernath, P. F., Hinkle, K. H., and Keady, J. J. *Science*, 244, 562, 1989.
9. Bernatowicz, T. J., Amari, S., Zinner, E. K., and Lewis, R. S. *ApJ*, 373, L73, 1991.
10. Bernatowicz, T. J., Cowsik, R., Gibbons, P. C., et al. *ApJ*, 472, 760, 1996.
11. Bertschinger, E. and Chevalier, R. A. *ApJ*, 299, 167, 1985.
12. Bieging, J. H. and Nguyen-Quang-Rieu. *ApJ*, 343, L25, 1989.
13. Bieging, J. H., Shaked, S., and Gensheimer, P. D. *ApJ*, 543, 897, 2000.
14. Bieging, J. H. and Tafalla, M. *AJ*, 105, 576, 1993.
15. Bowen, G. H. *ApJ*, 329, 299, 1988.
16. Boyle, R. J., Keady, J. J., Jennings, D. E., Hirsch, K. L., and Wiedemann, G. R. *ApJ*, 420, 863, 1994.
17. Brown, J. M. and Millar, T. J. *MNRAS*, 339, 1041, 2003.
18. Cadwell, B. J., Wang, H., Feigelson, E. D., and Frenklach, M. *ApJ*, 429, 285, 1994.
19. Cami, J., Yamamura, I., de Jong, T., et al. *A&A*, 360, 562, 2000.
20. Cau, P. *Dust formation in carbon-rich AGB stars: the case of IRC+10216*. PhD thesis, UMIST, 2001.
21. Cernicharo, J., Gottlieb, C. A., Guélin, M., et al. *ApJ*, 368, L43, 1991.
22. Cernicharo, J., Gottlieb, C. A., Guélin, M., Thaddeus, P., and Vrtilek, J. M. *ApJ*, 341, L25, 1989.
23. Cernicharo, J., Kahane, C., Guélin, M., and Hein, H. *A&A*, 181, L9, 1987.

24. Charnley, S. B. and Latter, W. B. *MNRAS*, 287, 538, 1997.
25. Cherchneff, I. In van Dishoeck, E., editor, *IAU Symp. 178: Molecules in Astrophysics: Probes & Processes*, volume 178, page 469. Kluwer Academic Publishing: Dordrecht, 1997.
26. Cherchneff, I., Barker, J. R., and Tielens, A. G. G. M. *ApJ*, 401, 269, 1992.
27. Choi, B., Wasserburg, G. J., and Huss, G. R. *ApJ*, 522, L133, 1999.
28. Dayal, A. and Bieging, J. H. *ApJ*, 407, L37, 1993.
29. Duari, D., Cherchneff, I., and Willacy, K. *A&A*, 341, L47, 1999.
30. Duari, D. and Hatchell, J. *A&A*, 358, L25, 2000.
31. Fleischer, A. J., Gauger, A., and Sedlmayr, E. *A&A*, 266, 321, 1992.
32. Fox, M. W. and Wood, P. R. *ApJ*, 297, 455, 1985.
33. Frenklach, M. and Feigelson, E. D. *ApJ*, 341, 372, 1989.
34. Gail, H.-P. and Sedlmayr, E. *Farad. Disc.*, 109, 303, 1998.
35. Gail, H.-P. and Sedlmayr, E. *A&A*, 347, 594, 1999.
36. Geballe, T. R. *Phil. Trans. R. Soc. Lond. A*, 358, 2503, 2000.
37. Gensheimer, P. D., Likkel, L., and Snyder, L. E. *ApJ*, 439, 445, 1995.
38. Glassgold, A. E. *ApJ*, 438, L111, 1995.
39. Glassgold, A. E. and Huggins, P. J. *MNRAS*, 203, 517, 1983.
40. Glassgold, A. E. and Huggins, P. J. *ApJ*, 306, 605, 1986.
41. Glassgold, A. E., Lucas, R., and Omont, A. *A&A*, 157, 35, 1986.
42. Glassgold, A. E. and Mamon, G. A. In Bohme, D. K., editor, *Chemistry and Spectroscopy of Interstellar Molecules*, page 261. University of Tokyo Press: Tokyo, 1992.
43. Gredel, R., Lepp, S., Dalgarno, A., and Herbst, E. *ApJ*, 347, 289, 1989.
44. Groesbeck, T. D., Phillips, T. G., and Blake, G. A. *ApJS*, 94, 147, 1994.
45. Guélin, M. and Cernicharo, J. *A&A*, 244, L21, 1991.
46. Guélin, M., Cernicharo, J., Travers, M. J., et al. *A&A*, 317, L1, 1997.
47. Guélin, M., Muller, S., Cernicharo, J., et al. *A&A*, 363, L9, 2000.
48. Guélin, M., Neininger, N., and Cernicharo, J. *A&A*, 335, L1, 1998.
49. Helling, C. and Winters, J. M. *A&A*, 366, 229, 2001.
50. Highberger, J. L., Apponi, A. J., Bieging, J. H., Ziurys, L. M., and Mangum, J. G. *ApJ*, 544, 881, 2000.
51. Hinkle, K. W., Keady, J. J., and Bernath, P. F. *Science*, 241, 1319, 1988.
52. Höfner, S. *A&A*, 346, L9, 1999.
53. Höfner, S., Jorgensen, U. G., Loidl, R., and Aringer, B. *A&A*, 340, 497, 1998.
54. Hoppe, P., Annen, P., Strebel, R., et al. *ApJ*, 487, L101, 1997.
55. Hoppe, P., Strebel, R., Eberhardt, P., Amari, S., and Lewis, R. S. In *Lunar and Planetary Institute Conference*, volume 25, page 563, 1994.
56. Howe, D. A., Hartquist, T. W., and Williams, D. A. *MNRAS*, 271, 811, 1994.
57. Howe, D. A. and Millar, T. J. *MNRAS*, 244, 444, 1990.
58. Howe, D. A. and Millar, T. J. *MNRAS*, 282, L21, 1996.
59. Howe, D. A. and Williams, D. A. In Hartquist, T. and Williams, D., editors, *The Molecular Astrophysics of Stars and Galaxies*, page 347. Clarendon Press: Oxford, 1998.
60. Huggins, P. J., Bachiller, R., Cox, P., and Forveille, T. *ApJ*, 401, L43, 1992.
61. Huggins, P. J. and Glassgold, A. E. *AJ*, 87, 1828, 1982.
62. Huss, G. R., Fahey, A. J., Gallino, R., and Wasserburg, G. J. *ApJ*, 430, L81, 1994.
63. Hutcheon, I. D., Huss, G. R., Fahey, A. J., and Wasserburg, G. J. *ApJ*, 425, L97, 1994.

64. Jura, M. *ApJ*, 275, 683, 1983.
65. Jura, M. and Kahane, C. *ApJ*, 521, 302, 1999.
66. Jura, M. and Morris, M. *ApJ*, 292, 487, 1985.
67. Kawaguchi, K., Kasai, Y., Ishikawa, S., and Kaifu, N. *PASJ*, 47, 853, 1995.
68. Keady, J. J. and Ridgway, S. T. *ApJ*, 406, 199, 1993.
69. Keene, J., Young, K., Phillips, T. G., Buettgenbach, T. H., and Carlstrom, J. E. *ApJ*, 415, L131, 1993.
70. Kendall, T. R., Mauron, N., McCombie, J., and Sarre, P. *A&A*, 387, 624, 2002.
71. Lucas, R. and Guélin, M. In Watt, G. and Webster, A., editors, *Submillimeter Astronomy*, page 97. Kluwer Academic Publishers: Dordrecht, 1990.
72. Lucas, R. and Guélin, M. In Le Bertre, T., Lèbre, A., and Waelkens, C., editors, *IAU Symp. 191: Asymptotic Giant Branch Stars*, page 305. ASP: San Francisco, 1999.
73. Lucas, R., Guélin, M., Kahane, C., Audinos, P., and Cernicharo, J. *Ap&SS*, 224, 293, 1995.
74. Mackay, D. D. S. and Charnley, S. B. *MNRAS*, 302, 793, 1999.
75. Mamon, G. A., Glassgold, A. E., and Huggins, P. J. *ApJ*, 328, 797, 1988.
76. Mamon, G. A., Glassgold, A. E., and Omont, A. *ApJ*, 323, 306, 1987.
77. Markwick, A. J. *Chemistry in Dynamically Evolving Astrophysical Regions*. PhD thesis, UMIST, 2000.
78. Markwick, A. J. and Millar, T. J. *A&A*, 359, 1162, 2000.
79. Mauron, N. and Huggins, P. J. *A&A*, 349, 203, 1999.
80. Mauron, N. and Huggins, P. J. *A&A*, 359, 707, 2000.
81. Millar, T. J., Flores, J. R., and Markwick, A. J. *MNRAS*, 327, 1173, 2001.
82. Millar, T. J. and Herbst, E. *A&A*, 288, 561, 1994.
83. Millar, T. J., Herbst, E., and Bettens, R. P. A. *MNRAS*, 316, 195, 2000.
84. Monnier, J. D., Danchi, W. C., Hale, D. S., Tuthill, P. G., and Townes, C. H. *ApJ*, 543, 868, 2000.
85. Ohishi, M., Kaifu, N., Kawaguchi, K., et al. *ApJ*, 345, L83, 1989.
86. Olofsson, H. In van Dishoeck, E., editor, *IAU Symp. 178: Molecules in Astrophysics: Probes & Processes*, page 457. Kluwer Academic Publishers: Dordrecht, 1997.
87. Olofsson, H., Bergman, P., Lucas, R., et al. *A&A*, 353, 583, 2000.
88. Osterbart, R., Balega, Y. Y., Blöcker, T., Men'shchikov, A. B., and Weigelt, G. *A&A*, 357, 169, 2000.
89. Ryde, N., Eriksson, K., and Gustafsson, B. *A&A*, 341, 579, 1999.
90. Sahai, R. and Bieging, J. H. *AJ*, 105, 595, 1993.
91. Schröder, K.-P., Winters, J. M., and Sedlmayr, E. *A&A*, 349, 898, 1999.
92. Sharp, C. M. and Huebner, W. F. *ApJS*, 72, 417, 1990.
93. Turner, B. E. *ApJ*, 388, L35, 1992.
94. Weigelt, G., Balega, Y., Bloecker, T., et al. *A&A*, 333, L51, 1998.
95. Willacy, K. and Cherchneff, I. *A&A*, 330, 676, 1998.
96. Willacy, K. and Millar, T. J. *A&A*, 324, 237, 1997.
97. Willson, L. A. and Bowen, G. H. *Nature*, 312, 429, 1984.
98. Woods, P. M., Millar, T. J., Herbst, E., and Zijlstra, A. A. *A&A*, 402, 189, 2003.
99. Yamamoto, S., Saito, S., Kawaguchi, K., Kaifu, N., and Suzuki, H. *ApJ*, 317, L119, 1987.
100. Yamamura, I., de Jong, T., Onaka, T., Cami, J., and Waters, L. B. F. M. *A&A*, 341, L9, 1999.

6 Dynamics and Instabilities in Dusty Winds

Yvonne Simis[1] and Peter Woitke[2]

[1] Astrophysikalisches Institut Potsdam
[2] Zentrum für Astronomie und Astrophysik, Technische Universität Berlin

6.1 Introduction

Recent high-angular resolution observations of various astronomical objects, see Figure 6.1 for examples, reveal that astrophysical gases in general are not smooth and symmetric, but often exhibit a large variety of internal structures. Such inhomogeneities, i.e., a clumpy, patchy, or cloudy distribution of matter in time and space on different length scales, are particularly observed in those objects where dust formation takes place or has taken place in the past: in the circumstellar envelopes (CSEs) of AGB stars and their progeny [post-AGB objects, planetary nebulae (PNe)], in the explosive cooling flows of novae and supernovae, but also in the hostile environments of R Coronae Borealis and Wolf Rayet stars.

The cause of the spatio-temporal structure formation is usually ascribed to certain instabilities inherent in the considered physical system that amplify small initial perturbations arising, e.g., from fluctuations. Especially well known in astrophysics is the large family of *dynamical instabilities*, where disturbances of the velocity field play a crucial role (e.g., turbulence, gravitational and convective instabilities, Rayleigh–Taylor, Richtmyer–Meshkov, Kelvin–Helmholtz, see, e.g., [33]) . *Thermal instabilities* may arise from unstable or multiple solutions of the gas energy equation, which occur in particular when molecule formation or other phase transitions take place (e.g., [5]). For example, thermal bifurcations are discussed to occur in the outer atmospheres of cool stars due to molecule formation which reinforces radiative cooling [2, 9, 49]. Furthermore, *radiative instabilities* may be induced, e.g., by a temperature-dependent opacity and its nonlocal feedback on the radiation field and hence on the temperature structure of the medium [45, 64, 68]. These instabilities are often closely related to the instabilities of the thermal type.

> The natural response of nonlinear systems to instabilities is a self-organization of matter by the formation of spatio-temporal structures on various length scales

Considering the case of the AGB star winds, several physical mechanisms have been proposed to drive the slow and dense outflows of these objects: stellar pulsation combined with inefficient radiative cooling of the gas [7],

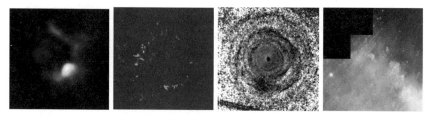

Fig. 6.1. Some examples of structure formation in dusty objects: the cloudy appearance of the innermost region of IRC+10216 [69] (scale $\approx 10^{15}$ cm), the patchy maser spots around TX Cam [11] (scale $\approx 10^{14}$ cm), the large concentric shells around IRC+10216 [48] (scale $\approx 10^{17}$ cm), and the "cometary knots" in the Helix nebula [HST picture STScI-PRC96-13a] (scale $\approx 10^{17}$ cm)

small-scale magneto-hydrodynamic or acoustic waves [54, 55], and, in particular, radiation pressure on dust grains that form in the wind. For cool AGB stars, on which we focus in this chapter, a combination of pulsation and radiation pressure on dust grains seems to be the most efficient mechanism [18, 29].

The outer atmospheres of AGB stars, where these winds are generated, are highly nonlinear physical systems. More and more complex molecules form in the outflow with increasing distance from the star. Created by the stellar pulsation, shock waves propagate through the atmospheres, which occasionally heat and compress the gas, thereby possibly destroying all molecules formed so far. Initiated by the chemical key process of seed particle formation (nucleation), dust grains may form in this time-dependent environment, a process that is capable of reinstalling optical thickness, thereby taking control of the radiative transfer and introducing an important global coupling. The chemical processes generally possess a strong temperature-dependence, often with threshold character, making a reliable temperature determination indispensable.

A schematic overview of the various couplings and interrelations between the relevant dynamical, chemical, thermal, and radiative processes is illustrated in Figure 6.2. Through the formation and presence of dust grains, new and important feedback mechanisms are introduced into this system that open up various possibilities for additional control and feedback loops not known in dust-free gases. From observations only, it is practically impossible to determine why and how certain structures form and evolve. In order to correctly interpret the observations and to make possible any predictions about the qualitative behavior of such systems, *self-consistent, time-dependent* model calculations are required, which must include the hydrodynamical equations based on a physical description of the driving forces, integrated with radiative transfer calculations, and a treatment of the chemical processes, in particular, a time-dependent theory of dust formation.

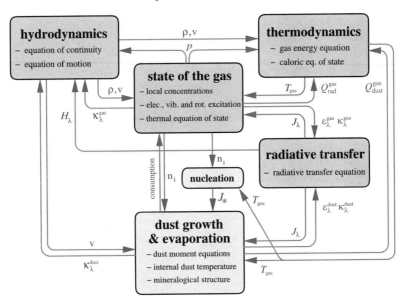

Fig. 6.2. Schematic overview of the interactions between the physical and chemical processes in the dust-forming AGB CSE. From [76]

In this chapter, emphasis will be put on instabilities known so far that are directly related to the formation and presence of dust grains. We focus on the actual physical mechanisms responsible for the self-organization, rather than on their observational appearance. We do not intend to be complete; the instabilities discussed here are examples for possible structure-formation processes that illustrate the enormous impact of the presence of dust on the AGB shells.

In Sect. 6.2.1 we discuss the basic stellar wind equation. Although this stationary approach will not suffice to discuss instabilities, it does already provide some basic insight into the wind-formation process. Next, in Sections 6.2.2 to 6.2.7, we introduce the specific descriptions of the physical processes involved and briefly discuss their numerical implementation. In Sect. 6.3 we discuss a number of instabilities typical for AGB winds that arise from the nonlinear interaction of the various physical processes. Finally, in Sect. 6.4 we make some conclusions and proposals for future work in this field.

6.2 Modeling the Dust-Driven AGB Wind

6.2.1 Basic Stellar Wind Equations

The rate at which a star loses mass and the time-dependent behavior of the wind are both connected to the position of the object in the HR diagram. In general, the mass-loss rate is low, $\dot{M} < 10^{-12}\,M_\odot\,\mathrm{yr}^{-1}$, for objects at the lower main sequence and increases with luminosity (up to 10^{-6}–$10^{-4}\,M_\odot\,\mathrm{yr}^{-1}$) on the AGB.

All types of stellar winds, hot or cool, fast or slow, dense or dilute, radiation-driven or pressure-driven, can be described by a single basic physical formulation. The mass-continuity equation dictates that in absence of sources, the matter that flows out of a certain shell has to flow into one of the adjacent shells; see also Chapter 4. Accordingly, the stationary mass-loss rate $\dot{M}$ is given by

$$\dot{M} = 4\pi r^2 \rho(r) v(r). \tag{6.1}$$

Hence, if there are no sources or sinks of matter, the mass-loss rate is constant throughout the CSE for a stationary outflow. In this section, we discuss only stationary solutions.

The equation of motion (the "Euler" equation), which was mentioned earlier, in Chapter 4, is

$$v\frac{dv}{dr} = -\frac{1}{\rho}\frac{dp}{dr} - \frac{GM}{r^2} + f(r), \tag{6.2}$$

where the first term on the right-hand side is the outward force due to the gas pressure, the second term is gravity, and the third term is an extra force, whose effect is studied below. Assuming that the atmospheric gas can be described by the ideal gas law, with μ=constant, the pressure derivative can be written as

$$\frac{1}{\rho}\frac{dp}{dr} = \frac{R}{\mu}\frac{dT}{dr} + \frac{RT}{\mu\rho}\frac{d\rho}{dr}. \tag{6.3}$$

where R is the gas constant. Differentiation of (6.1) and substitution of this derivative and (6.3) in (6.2) yields the "wind equation" [52]

$$\frac{dv}{dr} = v\left(\frac{2c_s^2}{r} - \frac{dc_s^2}{dr} - \frac{GM}{r^2} + f(r)\right) / \left(v^2 - c_s^2\right). \tag{6.4}$$

Here c_s is the isothermal sound speed:

$$c_s^2 = \frac{RT}{\mu}. \tag{6.5}$$

If $v^2 = c_s^2$, equation (6.4) has a singularity. Six different types of solutions for the wind equation (6.4) exist, but only one represents a stellar wind (see, e.g., [40]). This is the solution that starts subsonic and has a finite, supersonic velocity at infinity. It passes through the critical point r_c, the location of which is, in an isothermal CSE (c_T=const), given by

$$r_c = \frac{GM}{2c_s^2} - \frac{f(r_c)r_c^2}{2c_s^2} \quad \text{if} \quad v(r_c) = c_s. \tag{6.6}$$

Note that in the isothermal case the critical point coincides with the sonic point $v = c_s$, which need not be true in general [39].

The CSEs of AGB stars are definitely not isothermal but have a temperature gradient that is dependent on the distribution of gas and dust. These, in turn, depend on the temperature stratification in the CSE. Because of these, and other, interdependencies an analytical solution of the Euler equation (6.2) is usually not available, and the hydrodynamics equations should be solved numerically, either in stationary form, or fully time-dependent.

The isothermal case can be used to provide insight into the effect of the force $f(r)$, because it has an analytical solution. The expression for the location of the critical point (eq. (6.6)) shows that an outward force f pulls the critical point inwards. At the same time, the velocity gradient below the critical point decreases if a force f is acting (eq. (6.4)). Thus, everywhere below the critical point, r_c, addition of an outward force leads to an increase of the velocity. For stationary winds, in the absence of mass source terms, the mass-loss rate is constant throughout the atmosphere (see (6.1)). This implies that *an outward force results in an increase of the mass loss rate, when acting below the critical point.* On the other hand, the critical point does not move inward (compare eq. (6.6)) if the force becomes operational only further out, i.e., if

$$f(r) = 0 \quad \text{for } r < r_d,$$
$$f(r) > 0 \quad \text{for } r \geq r_d, \tag{6.7}$$

and $r_d > r_c$ (where r_d is the radius at which $f(r)$ starts to operate).

Hence, the force does *not* increase the mass-loss rate in that case. It does, however, increase the velocity of the wind, since for $r > r_c$, a positive force will increase dv/dr. Figure 6.3 illustrates the effect of a force $f(r) \propto \Gamma(r)/r^2$ on the location of the critical point and on the mass-loss rate. The latter increases if the force is nonzero in the subsonic region, as can be seen from the increased velocities (with respect to the case in which the force is not active in the subsonic region). The force due to radiation pressure on dust grains is also of the form (6.7). So, without having done any modeling one can already state that only the dust that forms below the critical point can lead to an increase in the mass-loss rate.[3]

For this reason it is extremely important that models for the wind formation cover the lower part of the CSE and not just the outer regions: The important physical processes occur between the photosphere and the sonic point.

---

[3] Hence, if the mass loss is driven by radiation pressure on dust, r_d represents the lower boundary of the dust-forming region, and the location of the critical point r_c almost coincides with r_d.

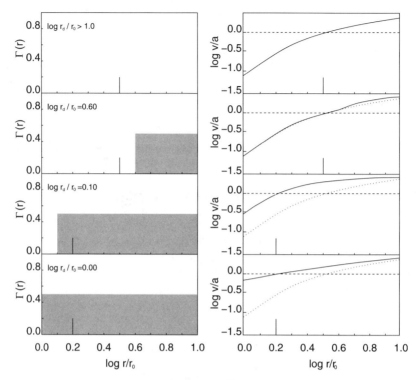

Fig. 6.3. The effect of a force $f(r) \propto \Gamma(r)/r^2$ with $\Gamma(r) = 0$ for $r < r_{\rm d}$ for various values of $r_{\rm d}$. The left-hand side shows $\Gamma(r)$. The right-hand side shows the actual flow velocity (solid line) and the flow velocity for the case $\Gamma(r) = 0 \ \forall \, r$ (dotted line) in terms of the velocity at the critical point (a). The location of this is annotated with a tick mark. This figure is a result of a numerical calculation and is adopted from [40] in a slightly modified form

6.2.2 Hydrodynamics of AGB Winds

The hydrodynamics equations are the continuity equations for the gas and the dust and the momentum equations for both. An energy equation for the gas could be added as well (see Chapter 7). Alternatively, because of the proximity of the luminous central star, one might solve the radiative transfer to calculate the temperature stratification, and omit the energy equation.

These equations form a system of *two-fluid hydrodynamics*; i.e., gas and dust are each described by their own set of hydrodynamics equations. The equations couple via the source terms.

The source term in each of the continuity equations for the gas and dust, $s_{\rm cond,g,d}$, covers the nucleation, growth, and evaporation of dust grains. In these processes matter is transferred from one state to another. Since the total mass is conserved, we have

$$\frac{\partial \rho_{g,d}}{\partial t} + \frac{1}{r^2}\frac{\partial}{\partial r}(r^2 \rho_{g,d} v_{g,d}) = s_{\mathrm{cond},g,d}, \qquad s_{\mathrm{cond},g} = -s_{\mathrm{cond},d}. \quad (6.8)$$

The condensation of gas into dust is discussed in Sect. 6.2.5, see also Chapters 4 and 5.

The momentum equations are

$$\frac{\partial}{\partial t}(\rho_g v_g) + \frac{1}{r^2}\frac{\partial}{\partial r}(r^2 \rho_g v_g^2) = -\frac{\partial P}{\partial r} + f_{\mathrm{drag},g} - f_{\mathrm{grav},g} + v_g s_{\mathrm{cond},g}, \quad (6.9)$$

$$\frac{\partial}{\partial t}(\rho_d v_d) + \frac{1}{r^2}\frac{\partial}{\partial r}(r^2 \rho_d v_d^2) = f_{\mathrm{rad}} + f_{\mathrm{drag},d} - f_{\mathrm{grav},d} - v_g s_{\mathrm{cond},g}. \quad (6.10)$$

The source terms of the momentum equation are the forces relevant for the CSE. Dust grains have no appreciable thermal motion, so that thermal dust pressure is not considered. We also neglect radiation pressure on the gas. The simplest force acting on both fluids is gravity:

$$f_{\mathrm{grav},d,g} = \frac{GM_* \rho_{d,g}}{r^2}. \quad (6.11)$$

The implementation of the gravitational force is straightforward, and will not be discussed further.

The radiative force acting on the dust grains, f_{rad}, is discussed below. The momentum equations couple via the viscous drag force of radiatively accelerated dust grains on the gas, f_{drag}. We will show below, when we discuss the drag force in more detail, that this force depends on the dust grain size, so that also the dust grain velocity depends on the dust grain size. Since at each point in the CSE dust grains of various sizes exist, we should use an n-fluid approach, rather than a two-fluid description, to model the flow very accurately, but this is numerically still too expensive at the moment. Moreover, as is shown in Sect. 6.2.5, we implement the formation of dust grains using a method that provides us with average dust grain sizes, rather than a full dust grain size distribution. Within this method expressions for the mean dust grain velocity are also available (cf. (6.36)), which can represent the dust grain velocity in (6.10).

The model is completed with the equation of state for ideal gases.

6.2.2.1 Radiation Pressure on Dust Grains

The radiative force exerted on the dust grains by the radiation field can be calculated as follows.[4] If L_λ is the luminosity of the source at wavelength λ,

---

[4] There are, of course, more methods to calculate the radiative force, each with its own underlying assumptions. We assume, for example, that the CSE is optically thin and that the dust grains are small with respect to the wavelength of the stellar photons: i.e., we adopt the small-particle limit of Mie theory [70].

this source emits $\lambda L_\lambda/hc$ photons of wavelength λ per second. The momentum associated with each of these photons is h/λ. The fraction of photons of wavelength λ that is intercepted by the dust grains at radial position r per unit volume is (cf. Chapter 4)

$$\frac{\kappa_\lambda \rho_d}{4\pi r^2} = \frac{1}{4\pi r^2} \int_0^\infty \pi a_{gr}^2 Q_{pr}(a_{gr}, \lambda) n(a_{gr}) da_{gr}, \qquad (6.12)$$

and the radiative force is

$$f_{rad} = \frac{\int L_\lambda \kappa_\lambda \rho_d d\lambda}{4\pi r^2 c}, \qquad (6.13)$$

where κ_λ is the mass extinction coefficient (in $g\,cm^{-2}$) of the dust, and $n(a)$ is the number density of dust grains with radii between a_{gr} and $a_{gr} + da_{gr}$. $Q_{pr}(a_{gr}, \lambda)$ is the efficiency factor for radiation pressure, which accounts for the amount of momentum that is removed from the incident beam and consists of a scattering and an absorption part. The absorbed radiation is removed from the beam completely, whereas the radiation that is scattered at angle θ with respect to the incident beam still contributes partially:

$$Q_{pr}(a_{gr}, \lambda) = Q_{ext}(a_{gr}, \lambda) - g_\lambda Q_{sca}(a_{gr}, \lambda)$$
$$= Q_{abs}(a_{gr}, \lambda) + Q_{sca}(a_{gr}, \lambda) - g_\lambda Q_{sca}(a_{gr}, \lambda), \qquad (6.14)$$

where g_λ is the mean cosine of the scattering angle θ. It turns out, see, e.g., [40], that for wavelengths beyond approximately 2 μm scattering is almost isotropic, so that $g_\lambda = 0$. Hence, in the following, we use the extinction efficiency factor Q_{ext} instead of the radiation pressure efficiency Q_{pr}.

We assume that the dust grain radius is small compared with the mean wavelength of the radiation field, so that the mass extinction coefficient can be calculated in the small-particle limit of Mie theory [18, 44] (and references therein). The integral in (6.12) can then be replaced by

$$\rho_d \kappa_d = \frac{3}{4} V_0 K_3 Q'(T), \qquad (6.15)$$

see [18]. Here V_0 is the volume per monomer for the dust grains, and K_3 represents the total amount of matter in the form of dust grains (see Sect. 6.2.5). For the extinction efficiency $Q'(T)$, e.g., the Planck mean or the Rosseland mean can be used as in [18].

With the grey dust opacity given in (6.15), the radiative force on the dust grains becomes

$$f_{rad} = \frac{L_* \kappa_d \rho_d}{4\pi r^2 c}. \qquad (6.16)$$

This force appears in the source term of the momentum equation for the dust. It is often assumed that the dust transfers all the momentum due to radiation pressure directly to the dust grains and that gas and dust grains have the same velocity. In that case, the dust momentum equation is omitted, and the radiative force is incorporated as a source term in the gas momentum equation, replacing the drag force in (6.9). We refer to this approach as *single-fluid* description. It is preferable to treat gas and dust as separate fluids (*two-fluid approach*) however, since if gas and dust are assumed to be completely coupled, various physical effects are suppressed.

6.2.2.2 Two-Fluid Flow and the Drag Force

Dust grains that are radiatively accelerated frequently collide with gas particles. By these collisions momentum is transferred to the gas, so that this is dragged outward to form the stellar wind. Since no momentum is lost, we have

$$f_{\text{drag,g}} = -f_{\text{drag,d}}. \tag{6.17}$$

The drag force is proportional to the rate of gas–grain collisions and the momentum exchange per collision. The exact formulation of the drag force depends on the nature of the collisions taken into account; a derivation can be found in, e.g., [58]. In general, the drag force is of the form

$$f_{\text{drag}} = \sigma_d n_g n_d m_g |v_{\text{coll}}| v_{\text{dr}}, \tag{6.18}$$

where $\sigma_d = \pi a_{\text{gr}}^2$ is the collisional cross section of a dust grain with radius a_{gr}, and $\sigma_d n_g v_{\text{coll}}$ is the average time between two subsequent collisions. When only inelastic gas–grain collisions are taken into account, one can use

$$v_{\text{coll}} = v_{\text{th}} \sqrt{\frac{64}{9\pi} + \left(\frac{v_{\text{dr}}}{v_{\text{th}}}\right)^2} \tag{6.19}$$

(see, e.g., [12]; see [58] for the general expression), where v_{th} is the thermal velocity of the gas and v_{dr} is the drift velocity of the dust grains with respect to the gas. Hence, v_{dr} determines the amount of momentum transfer per collision, and the rate of collisions depends on a weighted sum of the thermal velocity and the drift velocity. The expression in (6.19) provides a simple approximation to the full drag force (see [58]); for an overview of approximations in use see [57].

The drag force increases with increasing drift velocity. Therefore, an equilibrium value $\bar{v}_{\text{dr}}$ exists [12]. Two common ways to incorporate radiation pressure on dust and the subsequent transfer of momentum to the gas are based on assumptions about the drift velocity. Firstly, in *single-fluid* calculations $v_{\text{dr}} = 0$ is assumed; this assumption is called *position coupling* [26]. The momentum equation for the dust does not need to be calculated, and the drag force in the gas momentum equation is simply replaced by the radiative force, (6.16). The second method, called *momentum coupling* [26], assumes that dust grains always drift at their equilibrium drift velocity. This velocity can be calculated by equating the drag force and the radiative force. Since the relative velocity of the dust grains with respect to the gas is known, there is in this case no need to solve the momentum equation of the dust grains. In order to calculate the velocity of the gas, its momentum equation is solved by replacing the drag force by the radiative force.

The assumptions of position and momentum coupling seem justified if equilibrium drift is reached rapidly and/or if the drift velocity is small compared to the gas velocity. One should realize, however, that by replacing the drag force by the radiative force in the gas momentum equation, one in fact stipulates that the dust grains carry no mass. Also, in the single-fluid approximation, the system misses a degree of freedom, namely, the relative motion of the dust grains with respect to the gas. In the momentum-coupled case, the dust grains can move independently of the gas, though not freely.

In two-fluid hydrodynamics, the full system with both the gas and the dust momentum equations is solved, so that not necessarily all the momentum absorbed by the dust grains is transferred to the gas. Two-fluid models have been studied by a number of research groups.

It has been pointed out by [3, 4, 46] that for stationary and isothermal atmospheres, the assumption of complete momentum coupling breaks down at large distances above the photosphere and for small dust grains. Self-consistent, but stationary, two-fluid models, considering the dust grain-size distribution, dust formation, and the radiation field were presented in [36, 38]. Two-fluid time-dependent wind models were computed very early, in [79], and later in [47]. Neither model is self-consistent, since a simplified description of dust grain formation is adopted and a heuristic description of the inner 5–10 R_*, i.e., the subsonic region, is used. Time-dependent two-fluid radiation hydrodynamics, adopting a fixed dust grain size, is also studied in [65, 66, 67].

In the model setup presented in this chapter we combine the aspects mentioned above, and we show that a two-fluid time-dependent hydrodynamics code that self-consistently calculates the outflow under the assumption of spherical symmetry, for a given set of initial parameters $(M_*, L_*, T_*, \epsilon_{C/O})$, automatically generates a number of instabilities, which can be related to observed "structure" in the shell ejected by the AGB star.

6.2.3 Numerical Implementation of Hydrodynamics of AGB Winds

We have used the hydrodynamics equations in advection form:

$$\frac{\partial w}{\partial t} + \frac{\partial (vw)}{\partial r} = S(r,t), \qquad (6.20)$$

where $w = w(r,t)$ is the flow variable (i.e., the mass, momentum, or energy density), v is the transport velocity, and $S(r,t)$ is the source term. Note that here the coordinate r denotes a Cartesian coordinate and that we have limited ourselves to one-dimensional flow in this example.

The basic equations of hydrodynamics (6.20) form a nonlinear system of first-order hyperbolic equations. There are various ways to translate such equations into a finite difference scheme (FDS), i.e., to discretize them, which is needed to use them in a numerical calculation. One can, e.g., use (see [32])

$$w_n^{l+1} = w_n^l - v\frac{\Delta t}{2\Delta r}(w_{n+1}^l - w_{n-1}^l) + S\Delta t, \qquad (6.21)$$

where the subscripts and superscripts represent the position index and the time index, respectively. This FDS is an Euler scheme that uses *centered differencing*, and gives second-order accuracy in space. Furthermore, it is an explicit scheme; i.e., variables at time step $l+1$ do not depend on other variables at $l+1$:

$$w_n^{l+1} = f(\{w^l\}). \qquad (6.22)$$

Implicit schemes, on the other hand, allow w_n^{l+1} to depend on both values of w in nearby cells at time l and $l+1$:

$$w_n^{l+1} = g(\{w^l\}, \{w^{l+1}\}). \qquad (6.23)$$

In this case, a nonlinear algebraic system has to be solved at each grid point for each time step. Hence advancing one time step with an implicit scheme in general is more time-consuming than in an explicit scheme. On the other hand the time step to be taken in an explicit scheme is limited (see, e.g., [32]) whereas in an implicit scheme it can be chosen more freely.

A very complete overview of finite difference methods, as well as a thorough introduction to the principles of computational gas dynamics, can be found in [41]. A more concise review, in which a wealth of other useful references can be found, is [42]. Papers that provide an almost step-by-step manual to get started with the so-called FCT/LCD code are [6] and [32].

The implementation of sources and force terms in the dynamics equations can be straightforward, e.g., in the case of gravity, or more complicated, as is the case for the momentum transfer (drag force) from dust grains on the gas. In general, we can say that taking into account a certain physical or chemical process as a source term in the hydrodynamics equations becomes numerically more difficult if the characteristic time scale associated with it becomes small compared to the dynamical time scale.

6.2.3.1 The Drag Force

By using an explicit method, (6.18) for the drag force enforces very small numerical time steps. As a consequence of the quadratic dependence on the drift velocity, the characteristic time scale of the drag force can be many orders of magnitude smaller than the time scales of the other forces (gravity, radiation pressure) and the dynamical time scale. This means that in order to avoid errors, the time step should be adjusted to that of the drag force, which may be many orders of magnitude smaller than the dynamical time step. This is unwanted since taking small time steps leads to inaccuracy (because then more steps are needed to cover the same simulation time, and in each step there is the risk of systematic errors). In order to be able to perform two-fluid hydrodynamics in an explicit calculation one might study the behavior of the gas-grain system on small length and time scales, i.e., on a subgrid level. Doing so, we essentially derive an alternative expression to (6.18) that is based on the same physics but is suitable for implementation in an explicit calculation. The derivation of such an expression in the limit of supersonic drift is discussed in [62].

Implicit calculations allow, at the costs of a more expensive solving method and hence a shorter time coverage of the calculation, an implementation of the full drag force [58]; a clear overview of this can be found in [57].

6.2.3.2 Stellar Pulsation

The fact that AGB stars are long-period variables (LPVs) with a pulsation period of a few hundred up to a thousand days influences the wind. Through stellar pulsation, the atmosphere can be levitated, and relatively cool and dense regions that facilitate dust nucleation and growth may arise, see Chapter 4. Stellar pulsation is not included in our calculations: The star is located outside our computational grid (the innermost grid point is taken to be at or close to the stellar photosphere). Hence, if we want to study the influence of the stellar pulsation on the wind, we must incorporate it as an inner boundary condition. We prescribe a sinusoidal variation of the innermost gridpoint, simulating a piston. This indeed results in an important increase of the density scale height above the photosphere, which facilitates the formation of dust grains, since the gas density at larger radii and hence at lower temperatures increases.

It should be noted that this approach is a severe simplification. Light curves observed with, e.g., MACHO [1] show multiperiodicity, indicating that AGB stars pulsate in the first or higher overtone, see Chapter 2. However, the light curves of AGB stars are strongly affected by the time-dependent dust-formation process and are thus not a direct measure of the stellar pulsation. In fact, the occasional formation of dust layers can produce multiperiodic or chaotic light curves, even if the stellar pulsation is strictly sinusoidal [74].

Detailed theoretical models of pulsating atmospheres exist, but no connection between them and dynamical CSE calculations has yet been made.

6.2.4 Gas Phase Chemistry

In order to model the nucleation and growth of solids consistently, accurate information about the availability of the constituents is needed. Another reason to keep track of the molecular composition of the gas is its relevance for diagnostics; see [60]. Here, we describe a gas equilibrium chemistry calculation that is suitable for inclusion in a hydrodynamical wind model. Keep in mind, in this section and in the next, where dust grain processes are discussed, that the implementation in a model always brings about certain simplifications and assumptions. For an extensive discussion of the chemical and dust processes in the AGB CSE consult Chapter 5.

Among late-type stars, two types of chemically very different objects can be distinguished. In M-stars, oxygen is more abundant than carbon, and in C-stars the ratio is the opposite. The distinction between both classes is very strict, because the CO molecule is the first molecule to emerge in the flow, due to its high bond energy ($\approx 11.1\,\mathrm{eV}$). In the absence of dissociating UV photons, the least abundant of C and O will be blocked out, leaving the other for the formation of complex molecules and dust grains. Here, we focus on the case in which the C/O ratio is larger than unity. The motivation for this choice is the fact that the nucleation and growth of hydrocarbon-based dust grains is well studied and can be calculated relatively easily, which is an advantage when implemented in a time-dependent hydrodynamics code. Apart from CO, we include H, H_2, C, C_2, C_2H, and C_2H_2.

Dust-induced structure formation surely does not exclusively occur in CSEs with a carbon-rich chemistry. Also, in oxygen-rich CSEs the strongly nonlinear couplings and feedback mechanism caused by the presence of dust lead to instabilities in the outflow. The specific conditions (temperature, density) required for a certain instability may be very different for both kinds of chemical compositions, though.

A network of competing chemical reactions determines the formation and dissociation of molecules. Rate coefficients of these reactions depend on the abundances of the participating species and on the temperature. The abundances of each molecule can be found by simultaneously solving all rate equations. This, however, is computationally very expensive, since it implies solving a set of stiff nonlinear differential equations. An alternative method, more suitable for implementation in time-dependent hydrodynamics, is chemical equilibrium. Under the assumption of chemical equilibrium each reaction is balanced by its reverse. Now a much simpler system of coupled equations (for the partial pressures of each element) has to be solved. The resulting concentrations are no longer time-dependent. This is a problem if the characteristic time scale of the gas chemistry is larger than the dynamical time scale. The actual implementation and solution of the equilibrium gas chemistry are presented below.

Here, we present dissociation equilibrium chemistry [22] in a reduced form [12]. In order to assume chemical equilibrium in the AGB CSE we have to assume that large non-equilibrium effects due to shocks and UV radiation are not expected. For example, we assume that interstellar UV photons are absorbed by the dust and hence cannot penetrate into the dust-formation regions[5] and that late-type stars do not have active chromospheres [12].

Consider an equilibrium chemistry consisting of J elements $\mathcal{A}^j, j = 1, \ldots, J$, out of which molecules ($\mathcal{M}^k = \mathcal{A}^1_{l_{1,k}} \mathcal{A}^2_{l_{2,k}} \ldots \mathcal{A}^J_{l_{J,k}}$) can be formed via hypothetical chemical reactions of the form

$$\mathcal{M}^k \rightleftharpoons l_{1,k}\mathcal{A}^1 + l_{2,k}\mathcal{A}^2 + \cdots + l_{J,k}\mathcal{A}^J. \tag{6.24}$$

The partial pressures of the molecules and the atoms are related by the dissociation constants $\kappa_f^{\mathcal{M}^k}$, which depend on the temperature only:

$$\frac{P_{\mathcal{M}^k}}{\prod_{j=1}^{J}(P_{\mathcal{A}^j})^{l_{j,k}}} = \kappa_f^{\mathcal{M}^k}(T). \tag{6.25}$$

Here $l_{j,k}$ is the multiplicity of the atom species $\mathcal{A}^j$ in the molecule $\mathcal{M}^k$. For each element participating in the chemistry, the total number of atoms is given by

$$n_{\langle \mathcal{A}^j \rangle} = \epsilon_j n_{\langle H \rangle} = n_{\mathcal{A}^j} + \sum_k l_{j,k} n_{\mathcal{M}^k}, \quad j = 1, \ldots, J, \tag{6.26}$$

where ϵ_j denotes the abundance of the element $\mathcal{A}^j$ relative to hydrogen. Multiplication of (6.26) by kT then gives a set of J coupled equations for the partial pressures $P_{\mathcal{A}^j}$. With known dissociation constants $\kappa_{\mathcal{M}^k}$, temperature T, hydrogen density $n_{\langle H \rangle}$, and abundances ϵ_j, the system can be solved, and all densities $n_{\langle \mathcal{A}^j \rangle}$ and $n_{\mathcal{M}^k}$ can be calculated.

6.2.5 Nucleation and Growth of Dust Grains

Dust particles have an important influence on the dynamics and the radiative transfer in the outer atmospheres of AGB stars (see Figure 6.2). Consequently, a theoretical description of the formation process of these particles from the gas phase is one of the most challenging and crucial problems in understanding the dynamical behavior, the mass-loss mechanism, and the spectral appearance of these stars.

Assuming the dust particles to be spherical and to be composed of a uniform material[6] $\mathcal{M}$, a detailed description of the dust component is given by the size distribution function $f(N, \boldsymbol{r}, t)$[7], where N is the number of

---

[5] Interstellar UV radiation is important in the outer CSE regions [27], but we ignore this because we are mostly interested in the inner regions of the wind, where the composition of the gas, via the formation of dust grains, can influence the mass-loss rate. In calculating the gas abundances for diagnostic reasons it should be taken into account, though.

[6] A generalization of the following method to *dirty dust grains.* has been presented by [13]

[7] Not to be confused with the force $f(r)$ in Sect. 6.2.

monomers[8] contained by a dust grain. The temporal evolution of $f(N)$ due to surface chemical reactions[9] that populate and depopulate the size N (Figure 6.4) is given by the master equation

$$\frac{\partial f(N)}{\partial t} + \nabla\left(\boldsymbol{v}(N)\, f(N)\right) = R_\uparrow - R^\uparrow + R^\downarrow - R_\downarrow, \qquad (6.27)$$

where $\boldsymbol{v}(N) = \boldsymbol{v}_\mathrm{g} + \boldsymbol{v}_\mathrm{drift}(N)$ is the size-dependent velocity of the particles, which, because of the radiation pressure, will generally differ from the gas velocity $\boldsymbol{v}_\mathrm{g}$. For the most simple case of i-mer addition/evaporation reactions[10] of type $\mathcal{M}(N) + \mathcal{M}_i \rightleftharpoons \mathcal{M}(N+i)$ the rates can be expressed by

$$R^\uparrow = \sum_i f(N)\, A_N\, n_i\, v_{\mathrm{th},i}\, \alpha_i, \qquad (6.28)$$

$$R_\uparrow = \sum_i f(N-i)\, A_{N-i}\, n_i\, v_{\mathrm{th},i}\, \alpha_i, \qquad (6.29)$$

$$R^\downarrow = \sum_i f(N+i)\, A_N\, n_i\, v_{\mathrm{th},i}\, \alpha_i\, \frac{1}{(S)^i\, b_i^\mathrm{th}\, b_i^\mathrm{chem}}, \qquad (6.30)$$

$$R_\downarrow = \sum_i f(N)\, A_{N-i}\, n_i\, v_{\mathrm{th},i}\, \alpha_i\, \frac{1}{(S)^i\, b_i^\mathrm{th}\, b_i^\mathrm{chem}}, \qquad (6.31)$$

where A_N is the reactive surface area, and n_i, $v_{\mathrm{th},i}$, and α_i are the particle density, the (reduced) thermal velocity, and the sticking probability of the i-mer.[11] By means of detailed balance considerations [25, 53] the reverse (here spontaneous evaporation) rates in (6.30) and (6.31) have already been expressed in terms of their respective forward rates, introducing the supersaturation ratio[12] $S = n_1 kT_\mathrm{g}/p_\mathrm{vap}^\mathcal{M}(T_\mathrm{d})$, where $p_\mathrm{vap}^\mathcal{M}$ is the saturation vapor pressure at dust temperature T_d, that expresses the departure from phase equilibrium ($S \neq 1$), and further b-coefficients which express the departure from chemical equilibrium in the gas phase (e.g., $n_1 \neq \overset{\circ}{n}_1$) and from thermal equilibrium between gas and dust ($T_\mathrm{g} \neq T_\mathrm{d}$).

---

[8] The smallest unit building up the solid material, e.g., a carbon atom for graphite or an Al_2O_3-unit for corundum.

[9] Following [35], coagulation affects $f(N)$ on time scales that are considerably longer than the formation time scales and can be neglected in dust-forming winds.

[10] *Heterogeneous* growth reactions, e.g., $C(N) + C_2H_2 \rightleftharpoons C(N+2) + H_2$, which are essential for the growth of carbonaceous dust grains, can be included in a similar way [23, 53].

[11] Effects of the drift motion on the chemical surface reactions are ignored here. For a first-order treatment of such effects see [37].

[12] A solid phase is thermodynamically stable if the surrounding gas is saturated ($S = 1$); it will evaporate if the gas is undersaturated ($S < 1$) and will potentially grow if the gas is supersaturated ($S > 1$); disregarding chemical and thermal nonequilibrium effects here. Note, however, that the nucleation requires considerable supersaturation to initiate the condensation process.

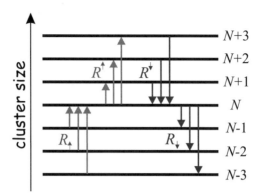

Fig. 6.4. Chemical surface reactions affecting the dust size distribution $f(N)$. Upward and downward arrows indicate growth and evaporation, respectively

Of course, a direct solution of the system of coupled equations (6.27) is not feasible, since μm-sized particles can easily contain 10^{12} atoms. However, the exact knowledge of $f(N)$ is actually not required in the models. It is usually sufficient to calculate only certain overall properties of the dust component, like the total dust surface per volume $4\pi a_0^2 K_2$ (necessary, e.g., to quantify chemical surface reactions and the overall drag force (see (6.18)) or the dust mass density $\rho_\mathrm{d} = 4\pi a_0^3 \rho_M K_3/3$. The moments of the size distribution function K_j are hereby defined as

$$K_j = \sum_{N_\ell}^{\infty} f(N)\, N^{j/3}\, dN \ . \tag{6.32}$$

Here a_0 and ρ_M are the hypothetical radius of one monomer in the solid compound and the mass density of the considered dust grain material, respectively. In particular, the total dust opacity can be expressed by K_3 or K_2 in the small- or large-particle limit of Mie theory (see, e.g., (6.15)). For the purpose of modeling, it seems therefore appropriate to calculate only the evolution of the dust moments K_j as function of time and space. A corresponding system of partial differential equations can be obtained by multiplying (6.27) by $N^{j/3}$ and summing up all equations from N_ℓ to ∞. After some algebraic manipulations[13] the result is [23]

$$\frac{\partial K_0}{\partial t} + \nabla\big(\langle v\rangle^0 K_0\big) = J(N_\ell), \tag{6.33}$$

$$\frac{\partial K_j}{\partial t} + \nabla\big(\langle v\rangle^j K_j\big) = N_\ell{}^{j/3} J(N_\ell) + \frac{j}{3\,\tau_\mathrm{gr}} K_{j-1} \quad (j=1,2,3,...), \tag{6.34}$$

where the growth time scale is defined as

---

[13] Some simplifications used during this derivation are $A_N = 4\pi a_0^2 N^{2/3}$ and $i \ll N$, which are reasonable approximations for macroscopic particles $N \geq N_\ell$.

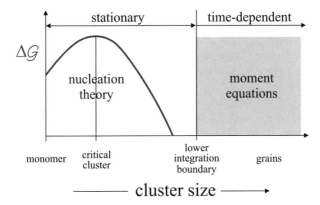

Fig. 6.5. Basic concept of the dust moment method. The chemical processes below the lower integration boundary N_ℓ, leading to the formation of seed particles, are treated by a stationary nucleation theory, whereas the temporal evolution of larger clusters (the *dust grains*) are treated by the time-dependent dust moment equations

$$\tau_{\rm gr}^{-1} = \sum_i i\, 4\pi a_0^2\, n_i\, v_{{\rm th}\,i}\, \alpha_i \left(1 - \frac{1}{(S)^i\, b_i^{\rm th}\, b_i^{\rm chem}}\right), \qquad (6.35)$$

which measures the time required to add/remove one layer of monomers to/from all existing surfaces. Appropriate means of the dust velocities

$$\langle v \rangle^j = \frac{1}{K_j} \sum_{N_\ell}^{\infty} v(N)\, f(N)\, N^{j/3}\, dN \qquad (6.36)$$

must be considered in order to include size-dependent drift effects.[14]

Only *macroscopic* particles larger than a certain minimum size N_ℓ are included in the dust moments. This lower integration boundary is chosen such that the chemical and radiative properties of the particles beyond this size can be calculated from macroscopic properties of the solid material. Processes affecting smaller particles (the *clusters*) are treated by a net flux in size-space $J(N_\ell)$ through the lower integration boundary (Figure 6.5). Following time-scale arguments [23], a stationary flow of clusters in size space up to N_ℓ can be assumed, making it possible to identify $J(N_\ell)$ with the nucleation rate $J_\star$ in case of net growth, i.e., the formation rate of seed particles in a supersaturated gas. Here, the full details of reaction kinetics among the clusters can be considered. Alternatively, the problem is simplified when the concept of classical nucleation theory is adopted,

$$J_\star = \frac{\overset{\circ}{f}(N_\star)}{\tau_{\rm gr}(N_\star)}\, \frac{1}{Z}, \qquad (6.37)$$

---

[14] In case of position coupling $\langle v \rangle^j \equiv v_{\rm d} \equiv v_{\rm g}$. In a two-fluid approach $\langle v \rangle^j \equiv v_{\rm d} \neq v_{\rm g}$ can be used as a first-order approximation.

where $N_\star$ is the size of the critical cluster, $\overset{\circ}{f}$ the equilibrium cluster size distribution function in the supersaturated gas, and Z the Zeldovic factor [22] . In classical nucleation theory [15, 16, 56], these quantities are related to the surface area and the *surface tension* of the bulk material. However, both quantities are ill-defined for small clusters typical of astrophysical environments [51]. More reliable results can be obtained when these quantities are adjusted to individual cluster data, in particular, the Gibbs free energies [15] of the clusters $\Delta_f^\ominus G(N)$ [8, 34]. This procedure is called *modified classical nucleation theory* [22].

In summary, the dust formation can be characterized by a two-step process of nucleation and growth. It can conveniently be described by a few partial differential equations in conservation form for the dust moments $K_j(\boldsymbol{r},t)$ (usually $j=0,1,2,3$). The dust-formation process shows very strong nonlinear dependencies on the gas and dust temperature with *threshold character*. Memory effects originate from the *nucleation barrier* and the strongly *density-dependent time scale for dust growth*. A typical overall dust-formation time scale amounts to months or years at densities $\approx 10^{11}\text{cm}^{-3}$ (see Chapter 4 and, e.g., Figure 3 in [75]). These long time scales immediately suggest that it is in fact essential to treat the dust-formation process in a time-dependent way in modeling the winds of AGB stars.

6.2.6 Radiative Transfer

Instead of solving the energy equation and having to calculate cooling and heating rates, we simply assume that the gas temperature is given by the local radiative equilibrium temperature T_eq. This temperature is found through semianalytically solving the radiative transfer, in a way similar to the one described in [18, 44]. Simultaneously solving the full radiative transfer problem and the hydrodynamics would be superior to this approach; cf. Chapter 4.

The transfer equation, taking into account coherent, isotropic scattering in a spherical, static CSE around a spherical radiation source, is [43]

$$\mu \frac{\partial I_\nu}{\partial r} + \frac{1-\mu^2}{r}\frac{\partial I_\nu}{\partial \mu} = -\kappa_\nu \rho (I_\nu - J_\nu). \qquad (6.38)$$

A modified version of the Eddington approximation that takes into account the curvature of the atmosphere is given by

$$J - 3K + \frac{1}{2}\mu_\star F = 0, \qquad (6.39)$$

where $\mu_\star = \sqrt{1-(R_\star/r)^2}$, and J, F, and K are the zeroth, first, and second moments of the intensity I, respectively. With this modified Eddington approximation the first two moments of the transfer equation, assuming a grey

---

[15] A thermodynamic potential defined as $G = E + pV - TS$. The total Gibbs free energy of any thermodynamical system minimizes in thermodynamical equilibrium under the constraints that temperature T and pressure p are kept constant.

atmosphere, can be integrated to obtain [44]

$$F(r) = (R_*/r)^2 \mathcal{F} \tag{6.40}$$

$$J(r) = \mathcal{F}\left[W(r) + \frac{3}{4}\int_0^\infty \kappa_{\text{tot}}\rho\left(\frac{R_*}{r}\right)^2 dr\right], \tag{6.41}$$

where $\mathcal{F}$ is the integrated flux at R_*. Here, κ_{tot} is the total mass extinction coefficient of gas and dust, $\kappa_{\text{tot}} = \kappa_{\text{g}} + \kappa_{\text{d}}$, and $W(r)$ is the geometrical dilution factor,

$$W(r) = \frac{1}{2}\left[1 - \left(1 - \left(\frac{R_*}{r}\right)^2\right)^{1/2}\right]. \tag{6.42}$$

If the atmosphere is in radiative equilibrium, we have $J = B$, where $B = (\sigma_R/\pi)T^4$ is the integrated Planck function. The integrated flux at R_* is $\mathcal{F} = \sigma_R T_*^4$. Combining this information, we obtain an expression for the equilibrium temperature in the CSE [18, 44],

$$T_{\text{eq}}^4(r) = \frac{1}{2}T_*^4\left[2W(r) + \frac{3}{2}\tau_{\text{tot}}\right], \tag{6.43}$$

with the modified total optical depth

$$\tau_{\text{tot}}(r) = \int_r^\infty \kappa_{\text{tot}}(r')\rho(r')\left(\frac{R_*}{r'}\right)^2 dr'. \tag{6.44}$$

The location of the photosphere, R_*, where by definition $T_{\text{eq}} = T_*$, is now given by the condition $\tau(R_*) = \frac{2}{3}$. The total mass extinction coefficient, κ_{tot}, is the sum of the extinction coefficients of the gas and the dust. The latter was calculated in the previous section, the gas extinction coefficient is assumed to be constant throughout the CSE and is adopted from [18], $\kappa_{\text{g}} = 2.0 \times 10^{-4}$ cm^2 g^{-1}.

6.2.7 Magnetic Fields, Rotation, Companions

We have discussed, in the current section several physical processes relevant for dust-driven winds. In Sect. 6.1 we mentioned that there are other mechanisms that can (contribute to) drive the outflow.

For completeness, we now discuss briefly the influence of magnetic fields, stellar rotation, and the presence of a (sub) stellar companion on the outflow. Note that neither of these is taken into account in the model setup we have sketched in this section.

Since we have, so far, limited ourselves to discussing dust-driven winds, we discuss only the influence of magnetic fields, rotation and companions on dust processes, although there is direct influence on the outflow as well.

Magnetic activity may lead to the formation of magnetic cool spots. These star-spots are characterized by a lower temperature than that of the surrounding stellar surface. Hence, also the the column of gas above the spot is relatively cool. This facilitates dust formation locally, and it is therefore a possible cause for deviations from spherical outflow, and thus the formation of structure. The analogue of the solar magnetic cycle, the periodic change of direction of the stellar magnetic field, enables periodic structure formation. This has been suggested as a mechanism in forming the concentric shells around some (post)-AGB stars and PNe [63].[16] The presence of magnetic fields in AGB stars has long been a point of discussion, but recent observations of masers show that they probably exist [71].

Most AGB stars are slow rotators, because of their expansion after leaving the main sequence. The centrifugal forces are by far not large enough to overcome gravity and to give rise to mass loss directly. However, rotation does somewhat reduce gravity, which leads to an increase of the density scale height in the CSE, around the equatorial plane. This results in an increased density at large radii and hence at low temperatures, so that dust formation can proceed more efficiently there. This way, a configuration of the circumstellar matter typical for stars at the final stage of the AGB arises, with more dust in the equatorial plane than near the poles [14].

If a companion, either a star or a massive planet, transfers angular momentum to the AGB star, the influence of rotation on the dust distribution becomes huge. These effects are possibly related to the axisymmetric appearance of many PNe. Chapters 9 and 10 of this book are devoted to binarity and evolution after the AGB.

6.3 Instabilities and Structure in the Outflow

6.3.1 Dust-Induced Shock Waves

The first dynamical instability presented here (see Figure 6.6) was already discovered by [17] and has meanwhile been confirmed by other independent calculations [29]. An initial compression of the gas enhances the chemical surface reaction rates responsible for dust growth and hence leads to an acceleration of the dust-formation process. The increase of the opacity causes a hydrodynamic acceleration of the dust-forming zones by radiation pressure on dust grains and frictional coupling between dust and gas. The larger velocities of the dust-forming zones lead to a further compression of the medium, which is situated ahead of these zones. Thus, an overall positive feedback is established (Figure 6.6), i.e., a self-amplifying control loop that amplifies

---

[16] The inversion of the magnetic field direction, although not in interaction with dust formation but through the magnetic pressure, may also provide a way to create shell structure [24].

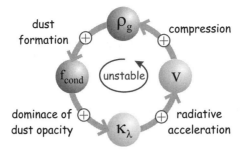

Fig. 6.6. Dust-induced shock waves as dynamical instability. ρ_g = gas density, f_{cond} = degree of condensation, κ_λ = spectral extinction coefficient, v = hydrodynamical velocity

initial perturbations in the system. The latter, nonlocal effect by compression causes an outward-directed wave like behavior of these disturbances.

The formation of dust-induced shock waves has been studied in the framework of time-dependent, spherically symmetric model calculations, which include hydrodynamics, chemical equilibrium, modified classical nucleation theory, a time-dependent treatment of dust growth and evaporation, and grey radiative transfer [18, 28, 29, 74]. The results of these model calculations indicate that the cool outflows of AGB stars are generally *not* expected to contain a homogeneous dust distribution (Figure 6.7). The pulsation of the central star creates acoustic waves, which steepen into shock waves during the passage through the atmosphere along its large density gradient. These shock waves accelerate the gas outward, and cause a time-dependent levitation of the outer atmosphere, i.e., a considerable increase of the densities as compared to static or stationary models. Thereby, the dust formation is substantially favored during certain limited time intervals related to the passage of pulsational shock waves. Consequently, the dust forms in a discontinuous way, resulting in a shell-like layered dust structure.

Amplified by the instability described above, the coupled phenomenon of dust formation and shock wave propagates outward. A characteristic feature of the dust-induced shock waves is the formation of local density inversions in the post shock regions, indicating the strong radiative acceleration of the forming dust behind the shock waves. Leaving aside nonspherical effects, these dust-induced shock waves provide the most prominent spatio-temporal structure-formation phenomenon to be expected on length scales of a few stellar radii in the CSEs of AGB stars.

The time-dependent formation, radiative acceleration, and radial dilution of the dust shells introduces a new characteristic internal time scale τ_{dust} to dust forming systems, which may interfere with the pulsational period P of the star. Consequently, the dynamical behavior of dust-forming CSEs is similar to a periodically driven oscillator characterized by its eigenfrequency. In the numerical models, the formation of dust shells close to the star occurs in

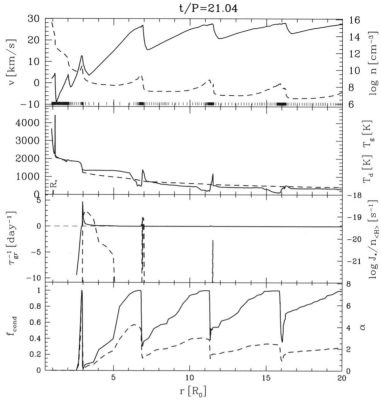

Fig. 6.7. Dust-induced shock waves in the CSE of a C-star according to a spherically symmetric model calculation [59]. In this model, the gas energy equation is solved time-dependently on the basis of various atomic and molecular non-LTE heating/cooling rates [77], which result in considerable deviations between T_g and $T_d = T_{eq}$. Note the common occurrence of dust, enhanced densities, and enhanced velocities. v = hydrodynamical velocity, n = gas particle density (dashed), T_g = gas temperature, T_d = dust temperature (dashed), τ_{gr}^{-1} = inverse dust growth time scale in units of atomic layers/day, J_* = nucleation rate per hydrogen nucleus (dashed), f_{cond} = degree of condensation of carbon, α = ratio of radiative acceleration to gravitational deceleration (dashed)

fact in monoperiodic, multiperiodic, or even chaotic time intervals, depending on the relation between τ_{dust} and P [19].

6.3.2 Drift Instability

When the two-fluid approach is used to model the dynamics of the CSE, the amount of momentum acquired by the dust grains due to radiation pressure is not necessarily entirely and immediately transfered to the gas. The dust grains can accelerate with respect to the gas and thereby "store" momentum

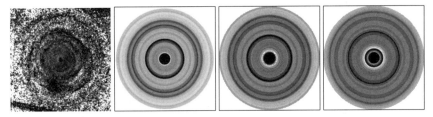

Fig. 6.8. Upper left frame: Composite $B+V$ image of IRC+10216, with an average radial profile subtracted to enhance the contrast (adapted from Mauron & Huggins (1999)). Note that a few patches in the image are residuals of the removal of the brightest background objects, and these should be ignored. Other frames: series of snapshots of our 1.5-fluid calculation. Plotted is the dust column density, also with an average radial profile subtracted. The average radial profiles are calculated for each snapshot separately, hence the slight difference in color from plot to plot. The theoretical profiles are shown for ages 44, 118, and 211 years. The size of our computational grid corresponds to the field of view of the observational image ($131'' \times 131''$) and a distance of 120 pc

that can be transfered to the gas when the drift velocity decreases again. This behavior introduces an extra degree of freedom that can lead to variations in the outflow velocity and the mass loss rate, as was suggested in [10]. Figure 6.8 shows the results of a two-fluid hydrodynamical calculation for the stellar parameters of the extreme carbon star IRC+10216, adapted from [73]. Without applying a time-dependent inner boundary, or any other time-dependent force, the calculations produce concentric shells around this object. Both the time scale of the modulation and the shell/intershell density contrast found in our models match the observations.

From these results, we infer that the shells can be formed by a hydrodynamical oscillation in the CSE, while the star is on the AGB.

Responsible for this effect is a subtle mechanism, involving an intricate nonlinear interplay between gas-grain drift, dust grain nucleation, radiation pressure, and CSE hydrodynamics. In previous calculations the shells were not found, presumably because either dust grain drift or a self-consistent description of the dust grain chemistry was not part of the model. A schematic representation of the mechanism driving the mass-loss variability is given in Figure 6.9. The physical processes mentioned in it relate to the subsonic parts of the CSE. Only there, changes in the driving forces have effect on the mass-loss rate (see Sect. 6.2.1). One observes two modes of outflow: low and high mass loss. The mass loss rate is low when the momentum transfer from dust grains to gas is inefficient. The velocity of the gas, in this phase, is low, and we refer to this phases as the *slow phase*. The drift velocity of the dust grains is high, so the passage of the dust grains through the zone in which dust grain growth is efficient is fast. This results in small dust grains. Hence, see (6.18), the drag force exerted on the gas is relatively weak. This results in a low gas density and hence an again higher drift velocity, because the frequency

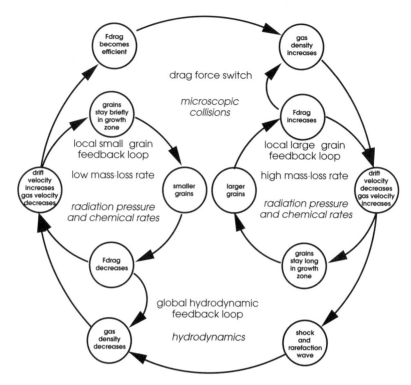

Fig. 6.9. Schematic representation of the mechanism driving the mass loss variability. For details, see Sect. 6.3.2

of gas-grain collisions decreases. So, during the phase of low mass loss, the average dust grain gradually becomes smaller and drifts faster.

If the mass-loss rate is high, exactly the opposite feedback loop determines the fate of the dust grains. The momentum transfer rate is high, so that the gas reaches high velocities (hence the *fast phase*), and the drift velocity of the dust grains is low. Hence, the dust grains move relatively slowly through the region of efficient dust grain growth. This results in rather large dust grains, with a large collision cross section and therefore a short mean free path. This, in turn, leads to a high momentum transfer rate and a further decrease of the drift velocity.

The two turn-around points, which mark the end of the slow phase and the onset of the fast phase and vice versa, turn out to be triggered by one single process: the sudden increase of the drag force at the end of the low-mass-loss (slow) phase. This increase is a result of the quadratic dependence of the drag force on the drift velocity: The increasing drift leads to an increase of the momentum transfer per gas-grain collision, *and* it prevents a further decrease of the rate of collisions due to the declining dust grain size. The new efficiency of the drag force heralds the onset of the fast phase with its high mass-loss rate. The suddenness of the transition involves the development of an outward shock and hence a rarefaction wave moving toward the stellar surface. This wave, which is a direct consequence of the onset of the drag force, brings a decrease of the gas density. As a result thereof, dust grains can no longer grow as efficiently as before, and their mean free paths increase again. Hence, the onset of the slow phase is triggered by the same process that ended it during the previous cycle. The time scale of the mass loss modulation is the time required by the rarefaction wave to cross the dust forming part of the subsonic region.

For the mechanism described above to work, incorporation of time-dependent dust formation and full two-fluid flow in the numerical model calculation are essential. The result in Figure 6.8 was generated with a version of the hydrodynamics code in which the piston that usually mimics the stellar pulsation was switched off, in order to guarantee that no period processes were present a priori. Also, the temperature stratification was kept constant. When taking into account these processes, the instability discussed in Sect. 6.3.1 starts to act as well [61]. In some cases this instability leads to oscillations of the mass loss-rate at a time scale much larger than the period of the stellar pulsation P, even if dust grain drift is ignored [74]. This, and the other causes for the formation of concentric shells on a 100-year time scale that were mentioned in Sect. 6.2.7, implies that hydrodynamical model calculations should contain as many physical processes as possible, for a full understanding of the mechanism that creates the periodic structure.

6.3.3 Radiative Instability of Dust Formation

One of the principal results of the self-consistent spherical models for the dust-formation process in AGB star winds (see Sect. 6.3.1) is the strong feedback of the dust formation on the temperature structure as a result of radiative transfer effects. In particular, the so-called *backwarming effect* has important consequences, i.e., the increase of the radiation field due to scattering and thermal reemission from the already condensed zones. In spherical symmetry, this leads to an *increase* of the temperatures inside a dust shell, which temporarily inhibits further nucleation and dust production in the innermost regions and may even cause a forming dust shell to partly re-evaporate *from the inside* (see negative τ_{gr} in Figure 6.7). If we abandon the assumption of spherical symmetry, the radiative transfer effects are more pronounced and even more complex.

> What are the important radiative transfer effects and how do they influence the dust formation in case of a nonspherical distribution?

This question leads us to another type of instability as sketched in Figure 6.10. In radiative equilibrium, a decrease of the mean spectral intensities J_λ is always accompanied by a decrease of the gas and dust temperatures $T_{\text{g}}, T_{\text{d}}$ ($\delta T_{\text{g,d}}/\delta J_\lambda > 0$, positive feedback, marked by $\oplus$). Thereby, the formation of

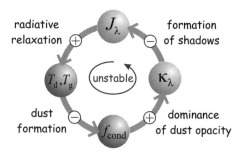

Fig. 6.10. Radiative instability of dust formation. Note the nonlocal character of the interaction between extinction and mean intensity. Positive feedbacks are marked by $\oplus$, negative feedbacks by $\ominus$. J_λ = mean spectral intensity, T_g = gas temperature, T_d = dust temperature, $f_{\rm cond}$ = degree of condensation, κ_λ = spectral extinction coefficient

dust is generally favored, resulting in an acceleration of the dust formation process ($\delta f_{\rm cond}/\delta T < 0$, negative feedback, marked by $\ominus$). A larger degree of condensation $f_{\rm cond}$, however, causes a larger local opacity κ_λ of the dust–gas mixture ($\oplus$), which influences the radiation field in a nonlocal way. In spherical symmetry and radiative equilibrium, an increase of the opacity strictly leads to an increase of the radiation field ($\oplus$, as, e.g., evident from the monotonically increasing $J(\tau)$-relation in grey stellar atmospheres). Consequently, the sign of the overall feedback in the control loop depicted in Figure 6.10 is usually negative in spherical symmetry ($\oplus \cdot \ominus \cdot \oplus \cdot \oplus = \ominus$); i.e., the control loop is stable: It damps and thus tends to level off perturbations.

However, if we abandon the assumption of spherical symmetry, the latter interaction may have the opposite effect [78], because a *local* enhancement of the opacity can result in the formation of a *shadow*. Inside this shadow, the mean intensities J_λ decrease ($\ominus$), leading to an overall positive feedback in the control loop ($\oplus \cdot \ominus \cdot \oplus \cdot \ominus = \oplus$). Thus, the dust-forming system is unstable; i.e., it will self-amplify perturbations of various kinds (e.g., $\delta f_{\rm cond}$) due to multidimensional radiative transfer effects.

In order to simulate this radiative instability and to study related structure-formation processes in C-star CSEs, we have developed an axisymmetric numerical model that combines the treatment of equilibrium chemistry, nucleation, and time-dependent dust formation as described in Sect. 6.2.4 and 6.2.5 with frequency-dependent radiative transfer. For the transfer part we use an accurate Monte Carlo method [50] that solves the axisymmetric radiative transfer problem in the dust CSE, assuming the dust grains to have a size-independent temperature T_d. The inner boundary is prescribed by a black body of effective temperature $T_{\rm eff}$. The gas is assumed to be optically thin and grey, resulting in $T_g \equiv T_{\rm bb}$, where $T_{\rm bb} = (\pi J/\sigma_R)^{1/4}$ is the black-body temperature. Both temperature structures $T_g(r, \phi, t)$ and $T_d(r, \phi, t)$ are hence direct results of the transfer part, where r is the radius and ϕ the azimuth angle measured from the equatorial plane. The calculations are performed in an iterative way, calculating the radiative transfer prior to each timestep of the dust evolution.

There are no velocity fields considered in this model. The density distribution is fixed (constant in time) and assumed to be given by $\rho_g(r,\phi) \propto \exp(-r/H+z)$, where $H = 0.25\,R_\star$ is a scale height and z are random numbers with Gaussian distribution of given variance. The initial state of the CSE at $t=0$ is chosen to be dust-free. Thereby, the quasi-static formation of dust in an inhomogeneous C-star environment is considered.

The results of this model calculation[17] are depicted in Figure 6.11. The denser regions close to the star begin to condense first, because of the strong density-dependence of the process of dust formation. Consequently, the spatial dust distribution at first resembles the inhomogeneous density distribution in the CSE (left-hand side of Figure 6.11) with a cutoff at the inner edge as a consequence of the temperature being too high for nucleation close to the star. The process of dust formation continues for a while in this way, until the first zones become optically thick. Each already optically thick zone casts a shadow into the CSE wherein the temperatures decrease by several 100 K, which improves the conditions for further dust formation therein. At the same time, scattering and reemission from the already condensed zones intensifies the radiation field in the regions that are situated between the optically thick zones. The stellar flux now escapes preferentially through these segments, which are still optically thin, thereby heating them up and worsening the conditions for further dust formation there. These two contrary effects amplify the initial spatial contrast of the degree of condensation introduced by an inhomogeneous density distribution.

As time passes, the optical depths in the circumstellar environment generally increase and may exceed unity. Consequently, the inner part of the CSE heats up considerably, and the forming dust "shell" starts to evaporate from the inside, as in the one-dimensional hydrodynamical models shown in Sect. 6.3.1. Thereby, a stable self-regulating mechanism takes control. The evaporation of the dust shell continues until the backwarming effect is reduced accordingly. Thus, the temperatures at the inner edge of the dust shell finally reach about the sublimation temperature (where the saturation vapor pressure $p_{\rm vap}(T_{\rm d})$ equals the partial pressure of the monomer in the gas, i.e., $S=1$)[18] of the dust grain material.

However, the sublimation temperature depends on the density. Denser regions are more resistant against evaporation, because the partial pressures in the gas are higher, and hence the supersaturation ratios are larger. Consequently, the denser parts of the inner edge of the dust shell remain, whereas the slightly thinner parts evaporate completely (right-hand side of Figure 6.11). In the shadowed regions behind the remaining "clouds," the

---

[17] See also the mpeg-movie at
http://astro.physik.tu-berlin.de/~woitke/sfb555.html

[18] Again, disregarding chemical and thermal nonequilibrium effects here.

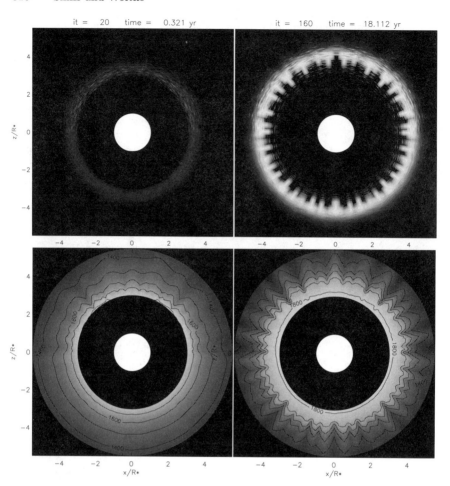

Fig. 6.11. Self-organization and spatio-temporal structure formation, triggered by radiative instabilities, during the formation of a dust shell around a carbon-rich AGB star. The figures show the dust temperature structure (lower row) and the logarithmic degree of condensation $\log(f_{\rm cond})=-3.5,\ldots,0$ (upper row) as contour plots in the x/z-plane at an early (l.h.s.) and a late (r.h.s.) phase as indicated on top. In the upper row, black means negligible condensation, white complete condensation. The white circles in the center of each figure mark the star. Black regions are not included in the model. A slightly inhomogeneous density distribution is considered. In the upper half of the model ($z>0$), the scatter in ρ is 0.15 dex, whereas the scatter in ρ is 0.05 dex in the lower half of the model. Parameters: $T_{\rm eff}=3600\,{\rm K}$, $R_\star=10^{13}\,{\rm cm}$ and C/O$=2$

conditions for the subsequent formation and the survival (i.e., the thermal stability) of the dust are improved due to a shielding effect. Thus, *radially aligned, cool, linear dust structures* finally result, oriented according to the preferred direction of the radiation field and surrounded by warmer almost dust-free regions.[19]

6.4 Conclusions and Outlook

The winds of cool AGB stars are generated in their outer atmospheric regions, which extend from the photosphere outward to a distance of a few stellar radii. In these layers, various physical and chemical processes interact, resulting in a highly nonlinear coupled system. Among these processes, such as hydrodynamics, chemistry, and radiative transfer, the formation of dust particles plays an outstanding role. Once formed from the gas phase, the dust particles take control over the dynamics of the system by radiative acceleration and momentum transfer to the gas via drag forces. Additionally, the dust component usually dominates the radiative transfer due to its large opacity. Thereby, the temperature structure is affected, which, in return, triggers the chemical processes leading to its formation.

The inclusion of dust formation in astrophysical flows thus introduces additional important feedback and control loops that may stabilize or destabilize the system. In this chapter, we have discussed a number of instabilities that demonstrate that large-scale structure-formation processes in AGB winds can be the result of instabilities due to the presence or the formation of dust grains.

- *Dust-induced shock waves* can result from a dynamical instability in the dust-formation zone, leading to radial structures (dust shells) on a length scale of a few stellar radii.

- *Drift instabilities* may enable two different dynamical states of the dust-gasmixture: Firstly, a high mass-loss state, characterized by a close dynamical coupling between gas and dust, and hence low drift velocities, and secondly, a low mass loss state, where the dust and gas components decouple dynamically. The system may switch between these states, leading to shell structures on timescales of a few hundred years that are observed at distances on the order of a thousand stellar radii [48].

- *Radiative instabilities in the dust-formation zone* can lead to the formation of dust clouds rather than dust shells due to multidimensional radiative

---

[19] Compare this result to the convective (Schwarzschild) instability where an adiabatic ascent of matter is unstable (self-amplifying). As a nonlinear result of this instability, the well-known large-scale convection rolls form. Here, the formation of dust is radiatively/thermally unstable, which leads to a fragmentation of dust shells into dust clouds.

transfer effects. In the shadow of previously condensed regions, the conditions for the subsequent formation of dust are improved, resulting in radially oriented, linear, cool dust structures surrounded by warmer, almost dust-free gases, through which the radiative flux preferentially escapes. These tangential structures are similar to certain shapes observed in PNe, suggesting a hypothetical link between the formation of dust clouds close to the stellar surface in the final AGB phase (e.g., IRC+10216, see Figure 6.1.a) and the "cometary knots" in PNe (see Figure 6.1.d).

This list is definitely not complete. The instabilities presented here have been discovered by *self-consistent* modeling where an *utmost complete* physical description is aspired, by taking into account all possible interrelations. No such things as a "condensation temperature," the "position of the sonic point," or the "mass-loss rate" are prescribed: All these quantities must result from a truly self-consistent model, where eventually only the fundamental stellar parameters $M_\star$, $L_\star$, T_{eff} and the element abundances are left to be prescribed.

It is to be expected that more instabilities will be identified when more physical processes are incorporated into the model calculations. Important effects can, e.g., be expected by the inclusion of magnetic fields or stellar rotation, as suggested by the mostly axisymmetric appearance of PNe. However, the physical processes included so far may already reveal much more self-organization if certain simplifying assumptions in the models are relaxed: Frequency-dependent radiative transfer will certainly improve our understanding of the energy balance of the gas in the dust-formation zone. The inclusion of time-dependent chemistry can be expected to introduce new timescales. A more detailed treatment of the radiative cooling can alter the thermodynamical structure and dynamical behavior of the propagating shock waves caused by the stellar pulsation.

Future model calculations should, in particular, also be capable of following (at least qualitatively) the formation of clumpy structures on various length scales, e.g., in order to model maser spots [30, 31], and to study the evolution of the objects beyond the AGB. For this, deviations from spherical symmetry are essential and should be accounted for. This requires a two- or three-dimensional radiation hydrodynamics code. Such codes exist (e.g., [21] or, for PNe, [20]), but not yet with all the relevant physical processes on the AGB incorporated.

Simulating the stellar pulsation by means of a "piston" as inner boundary condition is of course a rough simplification. With this approach, the desired increase of scale heights is achieved, but by neglecting the higher-frequency modes of the pulsation (more generally, the wave generation) probably too little momentum and energy are pumped into the CSE. Including the stellar interior, with its convective and driving zone for the stellar pulsation, should lead to more realistic results here.

Finally, in order to obtain the correct evolution of AGB stars, which is strongly affected by the mass-loss phenomenon, a better connection between stellar evolution calculations and the models for the wind generation in the outer atmospheres is necessary. The inclusion of mass-loss rates predicted from such model calculations into evolutionary calculations already show promising results [72], but generally speaking, these models should replace the much too simple outer boundary condition currently used in the stellar evolution codes.

Acknowledgments: We thank Christiane Helling for carefully reading the manuscript and providing many useful comments. YS thanks Vincent Icke and Carsten Dominik for their cooperation in the work described and Detlef Schönberner and Matthias Steffen for enabling the writing of this chapter under DFG grant SCHO 394/23-1. PW acknowledges support of the DFG, Sonderforschungsbereich 555, Komplexe Nichtlineare Prozesse, Teilprojekt B8.

References

1. Alard, C., Blommaert, J. A. D. L., Cesarsky, C., et al. *ApJ*, 552, 289, 2001.
2. Ayres, T. R. *ApJ*, 244, 1064, 1981.
3. Berruyer, N. *A&A*, 249, 181, 1991.
4. Berruyer, N. and Frisch, N. *A&A*, 126, 269, 1983.
5. Biermann, P., Kippenhahn, R., Tscharnuter, W., and Yorke, H. *A&A*, 19, 113, 1972.
6. Boris, J. P. NRL Mem. Rep., 3237, 1976.
7. Bowen, G. H. *ApJ*, 329, 299, 1988.
8. Chang, C., John, M., Patzer, A. B. C., and Sedlmayr, E. In Buttet, J., Châtelain, A., Monot, R., et al., editors, *Ninth International Symposium on Small Particles and Inorganic Clusters (ISSPIC 9)*, page 4.2, 1998.
9. Cuntz, M. and Muchmore, D. O. *ApJ*, 433, 303, 1994.
10. Deguchi, S. In Habing, H. J. and Lamers, H. J. G. L. M., editors, *IAU Symp. 180: Planetary Nebulae*, volume 180, page 151. Kluwer Academic Publishers: Dordrecht, 1997.
11. Diamond, P. J. and Kemball, A. J. In Le Bertre, T., Lèbre, A., and Waelkens, C., editors, *IAU Symp. 191: AGB stars*, page 195. ASP: San Francisco, 1999.
12. Dominik, C. *Formation of dust grains in the winds of cool giants*. PhD thesis, Technischen Universität Berlin, 1992.
13. Dominik, C., Sedlmayr, E., and Gail, H.-P. *A&A*, 277, 578, 1993.
14. Dorfi, E. A. and Höfner, S. *A&A*, 313, 605, 1996.
15. Draine, B. T. and Salpeter, E. E. *J. Chem. Phys.*, 67, 2230, 1977.
16. Feder, J., Russell, K. C., Lothe, J., and Pound, G. M. Adv. Phys., 15, 111, 1966.
17. Fleischer, A. J., Gauger, A., and Sedlmayr, E. *A&A*, 242, L1, 1991.
18. Fleischer, A. J., Gauger, A., and Sedlmayr, E. *A&A*, 266, 321, 1992.
19. Fleischer, A. J., Gauger, A., and Sedlmayr, E. *A&A*, 297, 543, 1995.

20. Frank, A., Balick, B., Icke, V., and Mellema, G. *ApJ*, 404, L25, 1993.
21. Freytag, B. In *Astronomische Gesellschaft Meeting Abstracts*, volume 18, page 144, 2001.
22. Gail, H.-P., Keller, R., and Sedlmayr, E. *A&A*, 133, 320, 1984.
23. Gail, H.-P. and Sedlmayr, E. *A&A*, 206, 153, 1988.
24. García-Segura, G., López, J. A., and Franco, J. *ApJ*, 560, 928, 2001.
25. Gauger, A., Gail, H.-P., and Sedlmayr, E. *A&A*, 235, 345, 1990.
26. Gilman, R. C. *ApJ*, 178, 423, 1972.
27. Glassgold, A. E. *ARA&A*, 34, 241, 1996.
28. Helling, C., Winters, J.-M., and Sedlmayr, E. *A&A*, 358, 651, 2000.
29. Höfner, S., Feuchtinger, M. U., and Dorfi, E. A. *A&A*, 297, 815, 1995.
30. Humphreys, E. M. L., Gray, M. D., Yates, J. A., et al. *MNRAS*, 282, 1359, 1996.
31. Humphreys, E. M. L., Gray, M. D., Yates, J. A., et al. *A&A*, 386, 256, 2002.
32. Icke, V. *A&A*, 251, 369, 1991.
33. Inogamov, N. A. *The Role of Rayleigh–Taylor and Richtmyer–Meshkov Instabilities in Astrophysics: An Introduction*, volume 10 of Astrophysics and Space Physics Reviews. Russian Academy of Sciences: Moscow, 1999.
34. Jeong, K. S., Chang, C., Sedlmayr, E., and Sülzle, D. *J. Phys. B*, 33, 3417, 2000.
35. Krüger, D. *A computational multi-component method for modeling the evolution of size distribution functions and its application to cosmic dust grains*. PhD thesis, Technische Universtät Berlin, 1997.
36. Krüger, D., Gauger, A., and Sedlmayr, E. *A&A*, 290, 573, 1994.
37. Krüger, D., Patzer, A. B. C., and Sedlmayr, E. *A&A*, 313, 891, 1996.
38. Krüger, D. and Sedlmayr, E. *A&A*, 321, 557, 1997.
39. Lamers, H. J. G. L. M. In De Greve, J. P., Blomme, R., and Hensberge, H., editors, *Stellar Atmospheres: Theory and Observations*, page 69. Springer-Verlag: Heidelberg, 1997.
40. Lamers, H. J. G. L. M. and Cassinelli, J. P. *Introduction to Stellar Winds*. Cambridge University Press: Cambridge, 1999.
41. Laney, C. B. *Computational Gasdynamics*. Cambridge University Press: Cambridge, 1998.
42. LeVeque, R. In Steiner, O. and Gautschy, A., editors, *Computational Methods for Astrophysical Fluid Flow*, page 1. Springer-Verlag: Berlin, Heidelberg, 1998.
43. Lucy, L. B. *ApJ*, 163, 95, 1971.
44. Lucy, L. B. *ApJ*, 205, 482, 1976.
45. MacDonald, J. and Mullan, D. *ApJ*, 481, 963, 1997.
46. MacGregor, K. B. and Stencel, R. E. *ApJ*, 397, 644, 1992.
47. Mastrodemos, N., Morris, M., and Castor, J. *ApJ*, 468, 851, 1996.
48. Mauron, N. and Huggins, P. J. *A&A*, 349, 203, 1999.
49. Muchmore, D. *A&A*, 155, 172, 1986.
50. Niccolini, G. and Woitke, P. and Lopez, B. *A&A*, 399, 703, 2003.
51. Nuth, J. A. and Donn, B. *J. Chem. Phys.*, 77(5), 2639, 1982.
52. Parker, E. N. *ApJ*, 128, 664, 1958.
53. Patzer, A. B. C., Gauger, A., and Sedlmayr, E. *A&A*, 337, 847, 1998.
54. Pijpers, F. P. and Habing, H. J. *A&A*, 215, 334, 1989.
55. Pijpers, F. P. and Hearn, A. G. *A&A*, 209, 198, 1989.
56. Salpeter, E. E. *ApJ*, 193, 579, 1974.

57. Sandin, C. and Höfner, S. *A&A*, 398, 253, 2003.
58. Schaaf, S. A. In *Handbuch der Physik*, volume VIII/2, page 591. Springer-Verlag: Berlin, Göttingen, Heidelberg, 1963.
59. Schirrmacher, V. *Nichtgleichgewichts-Strahlungskühlung in pulsierenden Staubhüllen*. Master's thesis, TU Berlin, Germany, 2000.
60. Sedlmayr, E. and Winters, J.-M. In De Greve, J. P., Blomme, R., and Hensberge, H., editors, *Stellar Atmospheres: Theory and Observations*, page 89. Springer-Verlag: Heidelberg, 1997.
61. Simis, Y. J. W., Dominik, C., and Icke, V. *A&A*, submitted, 2002.
62. Simis, Y. J. W., Icke, V., and Dominik, C. *A&A*, 371, 205, 2001.
63. Soker, N. *ApJ*, 540, 436, 2000.
64. Spiegel, E. A. *ApJ*, 126, 202, 1957.
65. Steffen, M. and Schönberner, D. *A&A*, 357, 180, 2000.
66. Steffen, M., Szczerba, R., Men'shchikov, A., and Schönberner, D. *A&AS*, 126, 39, 1997.
67. Steffen, M., Szczerba, R., and Schönberner, D. *A&A*, 337, 149, 1998.
68. Trujillo Bueno, J. and Kneer, F. *A&A*, 232, 135, 1990.
69. Tuthill, P. G., Monnier, J. D., Danchi, W. C., and Lopez, B. *ApJ*, 543, 284, 2000.
70. van de Hulst, H. C. *Light Scattering by Small Particles*. John Wiley & Sons: New York, 1957.
71. Vlemmings, W., Diamond, P. J., and van Langevelde, H. J. *A&A*, 375, L1, 2001.
72. Wachter, A. C., Schröder, K.-P., Winters, J.-M., Arndt, T. U., and Sedlmayr, E. *A&A*, 384, 452, 2002.
73. Winters, J.-M., Dominik, C., and Sedlmayr, E. *A&A*, 288, 255, 1994.
74. Winters, J.-M., Le Bertre, T., Jeong, K. S., Helling, C., and Sedlmayr, E. *A&A*, 361, 641, 2000.
75. Woitke, P. In Diehl, R. and Hartmann, D., editors, *Astronomy with Radioactivities*, page 163. MPE Report 274, 1999.
76. Woitke, P. *Rev. in Mod. Astr.*, 14, 185, 2001.
77. Woitke, P., Krüger, D., and Sedlmayr, E. *A&A*, 311, 927, 1996.
78. Woitke, P., Sedlmayr, E., and Lopez, B. *A&A*, 358, 665, 2000.
79. Woodrow, J. E. J. and Auman, J. R. *ApJ*, 257, 247, 1982.

7 Circumstellar Envelopes

Hans Olofsson

Stockholm Observatory

7.1 Mass Loss and Circumstellar Envelopes

The gas and bulk (in the form of microscopic grains) particles that escape the stellar gravitational attraction, i.e., the stellar wind, form an expanding envelope around the AGB star, a *circumstellar envelope* (CSE). Since the mass-loss and acceleration efficiencies substantially vary depending on the properties of the mass-losing star, the characteristics of the CSEs also vary considerably, e.g., in terms of opaqueness, geometry, kinematics, and chemistry. It is observationally established that mass-loss rates up to $10^{-4}\,\mathrm{M}_\odot\,\mathrm{yr}^{-1}$ exist, i.e., the star can lose a solar mass of material in about 10^4 years; a time scale that is very short even for an AGB star. Such a high mass-loss rate also means that the central star will be highly obscured; i.e., essentially all of the stellar radiation emitted at visual and near-infrared wavelengths will eventually be absorbed in the dusty CSE and reemitted at longer wavelengths. In general, the mass-loss rates are lower than this, and the stellar and circumstellar characteristics can be studied simultaneously, although they are not always easy to disentangle. The AGB stellar winds are slow by astronomical standards; the large majority are found in the range 3–$30\,\mathrm{km\,s}^{-1}$, and typical velocities lie in the range 5–$15\,\mathrm{km\,s}^{-1}$. The dust constitutes about one percent or less of the total mass, but probably plays a major dynamic role, and the evidence is good that it is responsible for at least the most intense mass loss. The chemical compositions of the gas and the dust reflect the chemical composition of the stellar atmosphere at the time when the material was ejected, as outlined in Chapter 5. In particular, the relative abundances of C and O are important, and throughout this chapter we often describe separately the characteristics of CSEs with C/O<1 (hereafter referred to as O-CSEs) and >1 (C-CSEs). In addition to these global characteristics the mass loss may have a directional dependence, resulting in CSEs that deviate more or less from overall spherical symmetry.

Even though the existence, and the profound importance, of intense mass loss on the AGB is well established, many uncertainties remain. In particular, dependences on time and such important stellar characteristics as mass and metallicity have turned out to be difficult to establish observationally. In addition, it has proven difficult to derive the mass-loss characteristics of an AGB star from first principles, but increasingly sophisticated numerical models are being developed; see Chapter 4.

A CSE has the interesting property that the radial dependences of various components carry temporal information on the mass loss, as well as on chemical composition changes in the stellar atmosphere; i.e., part of the stellar evolution history is imprinted in the CSE, although not necessarily in a straightforward way. The properties of the CSE also provide the starting conditions of the circumstellar post-AGB evolution, and in particular the formation of planetary nebulae (PNe); see Chapter 10.

In this chapter we describe the observed properties of CSEs, the methods used, and the theoretical background required to derive quantitative results from the observational data. A number of simple analytical results are derived, which serve the purpose of giving some physical insight, as well as providing order-of-magnitude estimates. The main probes of the CSE characteristics are atomic and molecular lines in emission and absorption, mainly at radio and infrared wavelengths, and mm to infrared continuum and spectral feature emission of the dust.

Different probes highlight different regions and different physical/chemical processes in a CSE. An essential part of the art lies in the synthesis of the wealth of information over a broad wavelength range. The existence and the properties of CSEs are the results of a complex interplay between physical and chemical processes having different characteristic time scales. Hence, many branches of physics and chemistry are required for the interpretation: hydrodynamics, thermodynamics, molecular and solid state physics, radiation theory, etc.; Chapters 4, 5, and 6.

We summarize our present knowledge of the global characteristics, such as geometry and kinematics, the small-scale structure of the circumstellar medium, and the gas/dust composition of AGB CSEs. Finally, we discuss the mass-loss characteristics of AGB stars in different evolutionary stages, as well as the mass-loss rate-dependence on stellar characteristics, such as mass, metallicity, and pulsational behavior.

7.2 A "Standard" Gaseous AGB CSE

We introduce here the physical and chemical structure of a "standard" gaseous CSE around an AGB star (basically along the lines of Goldreich and Scoville [77]); i.e., a spherical cloud of smoothly distributed gas particles, which is formed by a constant stellar mass-loss rate, and which expands with a constant velocity outside the limited acceleration region. The radial density distributions of different species (the chemistry is outlined in Chapter 5) and the radiation fields of interest are introduced, and the radial kinetic temperature distribution is derived. The properties of a standard dusty CSE are introduced in Section 7.4. The most important basic characteristic of a CSE is its expansion, and this is where we will start.

7.2.1 A Dust-Driven Wind

The different possible driving mechanisms of the wind of an AGB star have been outlined in Chapter 4 (see also [159]). It is quite possible that the wind mechanism may be different for an early and a late AGB star, a low- and a high-mass AGB star, during and in between thermal pulses, etc. However, in this chapter we need only a reasonable qualitative and quantitative description of an AGB wind. We have therefore chosen the dust-driven wind scenario, since this lends itself to a relatively simple description, and there are reasons to believe that this is the mechanism that determines the final characteristics of winds of AGB stars with intermediate to high mass-loss rates and luminosities. In this scenario the momentum of the stellar photon gas is transferred to the grains and eventually, via friction, to the gas particles. Stationary wind models based on radiation pressure on dust were extensively described in [63, 230], and dynamical models, in which the effects of the pulsating star on the atmospheric structure are taken into account, are presented in Chapter 4.

A simplified equation of motion for a dust-driven wind (e.g., ignoring pressure terms, assuming that the drag force has only an r^{-2} dependence, and one-size, nonevolving grains) has the solution [101]

$$v_{\rm g}(r) = v_\infty \sqrt{1 - (1 - (v_0/v_\infty)^2)(r/r_0)}, \qquad (7.1)$$

where

$$v_\infty = \sqrt{2GM(\Gamma - 1)/r_0} \qquad (7.2)$$

is the asymptotic value at large r (the terminal velocity), r_0 the radius at which the acceleration starts (roughly the dust condensation radius), v_0 the gas velocity at r_0 (roughly the sound speed), G the gravitational constant, M the stellar mass, and Γ the ratio of the drag force and the gravitational force on the gas, given by

$$\Gamma = \frac{3(Q_{\rm p,F}/a_{\rm gr})L\psi}{16\pi GcM\rho_{\rm gr}} \frac{v_{\rm g}}{v_{\rm d}}, \qquad (7.3)$$

where $Q_{\rm p,F}$ is the radiation pressure efficiency of a grain averaged over the stellar spectrum, $a_{\rm gr}$ the grain radius, $\rho_{\rm gr}$ the density of a grain, $v_{\rm d}$ the dust velocity, L the stellar luminosity, ψ the dust-to-gas mass ratio (defined as the ratio of the dust and gas mass-loss rates, $\dot{M}_{\rm d}/\dot{M}_{\rm g}$), and c the speed of light. These formulae give an indication of how the kinematics depend on luminosity (note that r_0 is roughly proportional to $L^{0.5}$), metallicity (through ψ), condensation radius, grain radiative properties, etc. With the reasonable input values $M = 1\,{\rm M}_\odot$, $L = 10^4\,{\rm L}_\odot$, $r_0 = 2\times10^{14}$ cm, $\rho_{\rm gr} = 2\,{\rm g\,cm}^{-3}$, $a_{\rm gr} = 10^{-5}$ cm, $Q_{\rm p,F} = 0.02$, $\psi = 0.005$, and $v_{\rm g}/v_{\rm d} = 1$, we obtain $v_\infty \approx 16\,{\rm km\,s}^{-1}$. Expression (7.1) describes a fairly rapid rise to the final velocity; e.g., $0.9v_\infty$ is reached at $r \approx 5r_0$, since $v_0/v_\infty \ll 1$.

The dust particles move faster outwards than the gas particles do; i.e., the former drift through the latter with a drift velocity $v_{\rm dr} = v_{\rm d} - v_{\rm g}$. An approximate expression for this drift velocity can be obtained by equating the radiation force on the grains and the drag force due to gas–grain collisions in the limit of supersonic motion [155],

$$v_{\rm dr} = \sqrt{v_\infty Q_{\rm p,F} L/c\dot{M}}. \tag{7.4}$$

This drift velocity can be substantial, e.g., $\dot{M} = 10^{-7}\,{\rm M}_\odot\,{\rm yr}^{-1}$, $v_\infty = 10\,{\rm km\,s}^{-1}$, and the same values as above result in $v_{\rm dr} \approx 20\,{\rm km\,s}^{-1}$, but at high mass-loss rates it becomes small, e.g., $\dot{M} = 10^{-5}\,{\rm M}_\odot\,{\rm yr}^{-1}$ and $v_\infty = 20\,{\rm km\,s}^{-1}$ give $v_{\rm dr} \approx 3\,{\rm km\,s}^{-1}$. We expect grain destruction if $v_{\rm dr}$ becomes too high, say about $30\,{\rm km\,s}^{-1}$.

7.2.2 The Gas Density Distribution

The gas kinematics, v_g, and the gas density distribution, ρ_g, in the CSE, and the gas mass-loss rate of the central star, $\dot{M}_g$, are related to each other, to a first approximation, via the continuity equation

$$\dot{M}_{\rm g}\left(t - \int_{r_{\rm i}}^{r}\frac{{\rm d}r}{v_{\rm g}(r)}\right) = 4\pi r^2 v_{\rm g}(r)\rho_{\rm g}(r). \tag{7.5}$$

Thus, we expect the mass-loss history of the star to be imprinted in the radial density structure. Relation (7.5) is, in an astrophysical context, a well-defined constraint on the large-scale density distribution, but it is of course only an approximation to the real physical situation, albeit in many cases a good one.

There are a number of reasons why (7.5) is just an approximate relation. First of all, the expansion velocity may vary in time such that matter ejected at different times run into each other; i.e., apart from the obvious problem of two winds with different velocities at the same point, we get interactions and the development of shocks. In particular, CSEs around post-AGB objects will be affected by this (Chapter 10). Besides, the joint expansion of two fluids, gas and dust, may lead to instabilities and hence alterations in the kinematics and the density structure; see Chapter 6. There is every reason to expect that these instabilities do not occur only in the radial direction, and hence inhomogeneities in the density distribution, "a clumpy medium," will exist. This may occur already in the region where the wind is initiated due to the formation of molecules and dust [53]. In the case of a highly clumped medium one may expect that (7.5) gives an adequate description of the density distribution of clumps, while the density inside a clump is not easily related to the expansion of the CSE. Rotation will invalidate (7.5) if it is differential and substantial. Presently, no evidence exists for rotation of AGB CSEs, nor is it theoretically expected.

In the case of a standard gaseous CSE, which mainly consists of molecular hydrogen, the number density distribution of H$_2$, $n_{\text{H}_2}(r)$, is given by

$$n_{\text{H}_2}(r) = \frac{X\dot{M}}{8\pi m_\text{H} v_\infty r^2} \approx 10^6 \left[\frac{\dot{M}}{10^{-6}}\right]\left[\frac{15}{v_\infty}\right]\left[\frac{10^{15}}{r}\right]^2 \text{ cm}^{-3}, \quad (7.6)$$

where X is the mass fraction of H, m_H the weight of a hydrogen atom, and $\dot{M}$ is measured in units of M$_\odot$ yr^{-1} (from now on we drop the subscript g), v_∞ in km s^{-1}, and r in cm (these are the units that will be used throughout this chapter for these quantities). This means a considerable range in the number density within a single CSE; e.g., $\dot{M} = 10^{-5}$ M$_\odot$ yr^{-1} results in $n_{\text{H}_2}(10^{14} \text{ cm}) \approx 10^9$ cm^{-3} and $n_{\text{H}_2}(10^{17} \text{ cm}) \approx 10^3$ cm^{-3}.

The extinction from a radius r and outwards can be crudely estimated using the relation between H$_2$ column density, i.e., in this case $n_{\text{H}_2}(r)r$, which is obtained by integrating (7.6) from r and outwards, and visual extinction, that seems appropriate for interstellar clouds, $N_{\text{H}_2} = 10^{21} A_\text{V}$ cm^{-2} mag^{-1} [22]. The result is

$$A_\text{V} \approx \left[\frac{\dot{M}}{10^{-6}}\right]\left[\frac{15}{v_\infty}\right]\left[\frac{10^{15}}{r}\right] \text{ mag.} \quad (7.7)$$

7.2.3 Molecular Abundance Distributions

Individual molecular density distributions may differ substantially from the total gas density distribution due to chemical processes in the CSE. The by far most abundant element, hydrogen, appears in AGB CSEs mainly in molecular form, H$_2$, and its destruction radius is larger than that of any other species, but its chemistry is quite complicated; see Chapter 5. This light and nonpolar molecule is not easily observed under the conditions prevalent in AGB CSEs. The atomic form of hydrogen has been searched for in the 21 cm line with mainly negative results, due to the difficulty of discriminating the weak circumstellar emission against the strong, and ubiquitous, interstellar emission. Thus, other atomic species or, even more importantly, other molecular species have to be used as probes. The latter often have radiative transitions that are easily excited in CSEs.

We define the *fractional abundance* of species A as

$$f_\text{A}(r) = \frac{n_\text{A}(r)}{n_{\text{H}_2}(r)}. \quad (7.8)$$

The radial dependence of the molecular fractional abundance depends on the origin of the species in question and the radius at which photodissociation or chemical destruction processes set in. The photodissociation depends on the shielding against the interstellar UV radiation (or the radiation from a nearby bright star or stellar association), i.e., on the outward column density of dust (or molecules in the case of photodissociation in lines), and hence on the mass-loss rate. As a quantitative example we here provide an approximate expression for the calculated photodissociation radius (defined as the radius where the abundance has decreased by a factor of two from its initial value) of CO, a photospheric species, which is dissociated in lines, [57, 178]:

$$R_{\rm ph,CO} = 10^{16} \left(0.8 \left[\frac{v_\infty}{15} \right] + 6 \left[\frac{\dot{M}}{10^{-6}} \right]^{0.7} \left[\frac{15}{v_\infty} \right]^{0.6} \left[\frac{f_{\rm CO}}{10^{-3}} \right]^{0.6} \right) \text{ cm}, \quad (7.9)$$

where the first term gives the photodissociation radius when there is no shielding. All other species, except H_2, have significantly smaller photodissociation radii. The amount of circumstellar mass inside the photodissociation radius of CO is roughly given by

$$M_{\rm ph,CO} = 2 \times 10^{-3} \left[\frac{\dot{M}}{10^{-6}} \right]^{1.7} \left[\frac{15}{v_\infty} \right]^{1.6} \left[\frac{f_{\rm CO}}{10^{-3}} \right]^{0.6} \text{ M}_\odot; \quad (7.10)$$

i.e., CSEs are in general low-mass objects. It requires a mass-loss rate of about 5×10^{-5} $M_\odot$ yr^{-1} to get $M_{\rm ph,CO} \approx 1$ $M_\odot$.

The effectiveness of the chemical processes depends on the density. Thus, we expect a kind of onion structure in the CSE. Some molecular abundance distributions peak close to the star and stretch outward to a radius that differs from species to species, while others form shells whose radii and widths also differ from species to species. In addition, the different sizes depend on the mass-loss rate of the star. In the outskirts of CSEs most of the gas appears in atomic or ionized form, but here the conditions are usually such that atoms and ions do not produce observable lines. As usual, reality is more complicated than assumed, and the presence of, e.g., inhomogeneities and the development of shocks will produce different, occasionally very different, results.

7.2.4 The Radiation Fields

The radiation fields to be considered in the context of the physical and chemical properties of a CSE emanate from three sources. First, the radiation from the central star will have characteristics that, depending on the stellar evolutionary state, fall in the range between those of a luminous ($\approx 10^4$ $L_\odot$) and cool (≈ 2500 K) red giant evolving up the AGB and a much less luminous ($\lesssim 10^2$ $L_\odot$) but hot ($\lesssim 10^5$ K) white dwarf. It may be that chromospheric radiation from the red giant also plays a role here, i.e., a substantially higher UV flux than expected from a cool AGB-star. Second, the dust in the CSE contributes scattered starlight, and in particular, its own heat radiation. This component strongly depends on the mass-loss rate. Third, the interstellar radiation field is of particular importance since it determines, through photodissociation, the radial distributions of the various circumstellar species.

This component may vary drastically depending on the location of the star with respect to, e.g., the Galactic plane and OB associations. It will introduce an uncertainty in the mass-loss rate estimates based on molecular lines, and in the abundance estimates of various species.

7.2.5 The Energy Balance of the Gas

This section describes the energy balance of the gas in a standard CSE. We give quantitative expressions for the most important heating and cooling processes, and derive results for the kinetic temperature distribution [84].

7.2.5.1 Heating and Cooling Processes

The local kinetic energy (i.e., excluding the kinetic energy of the expansion) of the gas is determined by a balance between heating (H) and cooling (C) processes according to (in a steady-state situation)

$$\frac{dT_k}{dr} = \frac{\gamma - 1}{n_{H_2} k v_\infty} \left(\sum_i H_i - \sum_i C_i \right), \qquad (7.11)$$

where γ is the adiabatic index (in the following we set $\gamma = \frac{5}{3}$, i.e., we ignore internal excitation of H_2), and k is Boltzmann's constant. The dominant heating is due to collisions with the grains streaming through the gas. An approximate expression is obtained by assuming that each grain transfers the energy $\frac{1}{2} m_{H_2} v_{dr}^2$ at a collision. This results in

$$H_{dg} = m_H n_{H_2} n_d \pi a_{gr}^2 v_{dr}^3, \qquad (7.12)$$

where n_d is the number density of grains. In addition, there is heating due to collisions with electrons photoejected from the grains (the photoelectric effect), heat transfer from grains to the gas particles, photodissociation of abundant molecules, photoionization of C, cosmic ray-ionization of H_2, etc. Some of these processes may dominate in certain parts of a CSE, e.g., the photoelectric effect in the external parts.

One of the dominant cooling terms is due to the adiabatic expansion of the gas,

$$C_{ad} = 2 n_{H_2} k v_\infty \left(T_k / r \right). \qquad (7.13)$$

There is also a substantial contribution from line cooling, mainly CO and H_2O in O-CSEs, and CO and HCN in C-CSEs.

7.2.5.2 The Kinetic Temperature Radial Distribution

We can solve (7.11) for the dust–gas heating term and the adiabatic cooling term separately. The former dominates in the inner CSE, and the solution is given by

$$T_{\mathrm{k}}(r) = T_0 + C \frac{a_{\mathrm{gr}}^2 \psi Q_{\mathrm{p,F}}^{3/2} L^{3/2}}{\rho_{\mathrm{gr}} v_\infty^{1/2} \dot{M}^{1/2}} \left(\frac{1}{R_0} - \frac{1}{r} \right), \quad (7.14)$$

where C is a constant, while the latter dominates in the external regions, and the solution is given by

$$T_{\mathrm{k}}(r) = T_0 \left(R_0/r \right)^{4/3}. \quad (7.15)$$

This shows that the kinetic temperature in the inner region of the CSE scales roughly as $\dot{M}^{-0.5}$ (and also the dependence on other parameters); i.e., we expect thin CSEs to be warmer than thick CSEs. The reason is that the heating is sensitively dependent on the drift velocity, which decreases considerably with an increasing density, i.e., with increasing mass-loss rate; see (7.4). More detailed treatments including, e.g., molecular line cooling, now exist [84, 227, 299]. These show, when compared with observational data, that the dependence on $\dot{M}$ is much less pronounced, probably as a consequence of the grain properties changing with the mass-loss rate. Figure 7.1 shows the result of a numerical model for a C-CSE formed by a mass-loss rate of $10^{-5}\,\mathrm{M_\odot\,yr^{-1}}$. The following radial distribution for the kinetic temperature roughly applies in this case:

$$T_{\mathrm{k}}(r) = 400 \left[\frac{10^{15}}{r} \right]^{0.9} \mathrm{K}. \quad (7.16)$$

The external parts of high-$\dot{M}$ CSEs may become very cool if the expansion velocity of the envelope is so high that no heating processes can counteract the adiabatic cooling. An interesting example of this is provided by the post-AGB object the Boomerang Nebula, where the molecular gas has been accelerated up to $\approx 160\,\mathrm{km\,s^{-1}}$, and where the gas kinetic temperature is estimated to be below that of the cosmic microwave background [224].

7.2.5.3 Excitation of Atoms and Molecules

We will only briefly discuss the excitation of circumstellar atoms/molecules, i.e., the processes that determine their energy-level populations. For more details see Chapter 4. In CSEs a non-LTE behavior is very common, and a detailed treatment of the excitation is therefore called for. Here, we limit ourselves to a discussion of radiative excitation due to the stellar radiation, the dust radiation, and the line radiation itself, and collisional excitation. These processes are introduced via radiative and collisional cross sections. The heavier and more complex a molecule is, the more energy levels are involved and the more difficult the excitation calculations become. Note that this may occur also for simple molecules, e.g., H_2O, which has millions of lines that are important in atmosphere models of M-type AGB stars.

The excitation calculation is nonlocal in the sense that the radiation field falling on an atom/molecule depends on the environment, i.e., the excitation

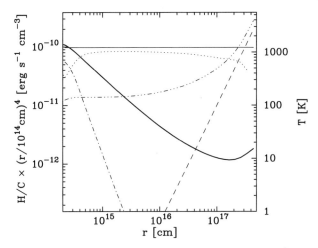

Fig. 7.1. Results of a numerical circumstellar model ($\dot{M}=10^{-5}\,M_\odot\,\mathrm{yr}^{-1}$, $v_\infty=15\,\mathrm{km\,s^{-1}}$) showing the kinetic temperature distribution (thick solid line) due to gas–grain collision (thin solid) and photoelectric (dashed) heating and adiabatic (dot-dot-dot-dashed), CO-line (dotted), and H_2-line (dot-dashed) cooling. Note the scaling with r^4

of other atoms/molecules in the CSE, and not only on the radiation source. Also, the excitation depends on the ability of photons emitted in the lines to escape, i.e., if this probability is low, the excitation efficiency of the line is effectively increased (often called trapping). If the atoms/molecules have lines that are separated in velocity by less than the gas expansion velocity, there will also be an interaction between atoms/molecules in different parts of the CSE, quite a common phenomenon among molecular radio lines. The nonlocal nature is partly, but not altogether, limited by the kinematics, i.e., the expansion at a velocity much larger than the local velocity spread effectively disconnects different parts of the CSE.

The collisional cross sections are normally poorly known, even with the most common species H_2. The extrapolation to higher temperatures, which may be of importance in CSEs, is usually very uncertain. This may lead to substantial uncertainties also in more sophisticated models.

A number of approximative methods have been used to deal with the nonlocal problem. Foremost among these is the escape probability or large velocity gradient method [34], which requires steep velocity gradients. However, this method works particularly poorly in the radial direction of a CSE, i.e., in the direction of the photons from the central source, a fact that may or may not be a problem depending on the emitting species and the thickness of the CSE. Other methods are now frequently used [263].

7.2.6 Line Emission, Scattering, and Absorption

The radiative transfer in lines in AGB CSEs is usually quite complicated, since the conditions are such that the excitation of energy levels is normally in non-LTE. However, a simple analysis can provide considerable insight into, e.g., the structure of the emitting region and the optical depth in the line, and it can be used to get a reasonable abundance estimate. We will here assume a standard gaseous CSE.

7.2.6.1 Column Densities

We start by introducing the column density, i.e., the number of atoms or molecules in a column of a given surface area. The *radial column density* from an inner radius R_i to an outer radius R_e of species A is given by, see (7.5),

$$N_{A,r} = \frac{X\dot{M}}{8\pi m_H} \int_{R_i}^{R_e} \frac{f_A(r)}{v_g(r)r^2} dr. \qquad (7.17)$$

In the case of a constant expansion velocity and abundance this reduces to

$$N_{A,r} = \frac{f_A X \dot{M}}{8\pi m_H v_\infty} \left(\frac{1}{R_i} - \frac{1}{R_e}\right) \approx n_A(R_i) R_i ; \qquad (7.18)$$

i.e., in this simple case the column density from a given radius and outward is the product of the radius and the molecular number density at this radius. This radial column density is obtainable, e.g., when one is observing absorption lines toward the central object.

Most line observations of CSEs, however, are done in emission lines by radio telescopes. In this case the radial column density can be obtained only if an (uncertain) assumption of R_i is done. Therefore, column densities obtained from radio data are often given in terms of the *source-averaged column density*. This means an average of all the column densities (along the line of sight) over the source area. Such a definition leads to the result

$$N_{A,s} = \frac{f_A X \dot{M}}{2\pi m_H v_\infty} \frac{R_e - R_i}{R_e^2} \approx 4 n_A(R_i) \frac{R_i^2}{R_e} ; \qquad (7.19)$$

i.e., $N_{A,s} = 4(R_i/R_e) N_{A,r}$, and for $R_i \ll R_e$ these two column densities are very different. This should always be kept in mind in comparing column densities obtained using different methods.

A *beam-averaged column density* is sometimes introduced, where the column densities through the CSE are weighted with the telescope response pattern,

$$N_{A,b} = \frac{2\ln 2\, f_A X \dot{M}}{\pi m_H v_\infty \theta_b D}, \qquad (7.20)$$

where θ_b is the full width at half power of a Gaussian response pattern (in radians), and D is the distance to the source. This is defined only for a single value of θ_b, and hence of limited use.

In the following analysis we also need the column density along the line of sight of the atoms/molecules that are able to interact with the same radiation, i.e., the ones that lie within a Doppler shift of each other that is smaller than the local line width δv (due to thermal and microturbulent motion). When large-scale motions are present it is *the incremental column density*, at a given line-of-sight velocity, v_z (throughout this chapter given with respect to the systemic velocity, v_*), and at an offset p from the center of the CSE, which is of interest, i.e.,

$$dN_A(p, v_z) \approx n_A(p, v_z)\delta z(p, v_z), \qquad (7.21)$$

where z is the direction along the line of sight, and δz the distance over which the atoms/molecules are resonant, i.e., the interaction length; Figure 7.2. We have here assumed that the velocity gradient is large enough that the density is essentially constant over the interaction length. We assume further that the local line shape is rectangular, i.e., $\phi(v) \propto 1/\delta v$, and therefore $\delta z(p, v_z) = \delta v \, [dv(p, v_z)/dz]^{-1}$. In a spherical CSE, which expands with a constant velocity, the particles that move with a line-of-sight velocity v_z lie on a cone with the opening angle $\theta = \arccos v$ (with respect to the line of sight through the center), where we have introduced $v = v_z/v_\infty$. In this case the relation between r, p, and v is particularly simple, $p = r(1 - v^2)^{1/2}$, and the velocity gradient along the line of sight on this cone and at an offset p from the center is given by $dv_z(p,v)/dz = (1 - v^2)^{3/2} v_\infty/p$; i.e., the incremental column density at an offset p from the center of the CSE, and at the line-of-sight velocity v_z, is given by

$$dN_A(p, v) = (\delta v/v_\infty) \, n_A(p) \, p \left(1 - v^2\right)^{-1/2}, \qquad (7.22)$$

where we have used $n_A(p, v) = n_A(p)(1 - v^2)$. In this simple analysis (7.22) becomes divergent for $v=1$, i.e., along the line of sight through the CSE center.

7.2.6.2 Emission Lines

In this section we derive results for the expected line profile and brightness distribution of an emission line from a standard CSE at a distance D. The atoms/molecules are assumed to have a population distribution described by a Boltzmann distribution at a single temperature $T_{ex}(r)$ inside a CSE of inner radius R_i and outer radius R_e. It is further assumed that the interaction length is short, in a relative sense, i.e., density and temperature may be approximated by their values at the middle of the interaction range and (7.22) applies.

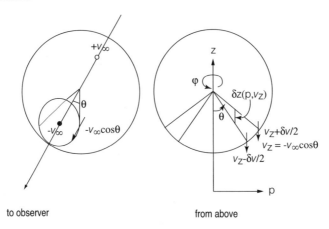

Fig. 7.2. Geometry of expanding spherical CSE

Line Profiles and Brightness Distributions

The brightness distribution for the line $u \to l$ is given by

$$b_{ul}(p,v) = \frac{2h\nu_{ul}^3/c^2}{\exp(h\nu_{ul}/kT_{\text{ex}}(p,v))-1}\left[1-e^{-\tau_{ul}(p,v)}\right], \quad (7.23)$$

where ν_{ul} is the transition frequency. The optical depth is obtained from the incremental column density (7.22) as

$$\tau_{ul}(p,v) = \frac{c^3}{8\pi\nu_{ul}^3}g_u A_{ul}\frac{f_A(p,v)X\dot{M}}{8\pi m_H v_\infty^2}\frac{e^{-E_l/kT_{\text{ex}}(p,v)}}{\mathcal{Z}(T_{\text{ex}}(p,v))}$$
$$\times \left(1-e^{-h\nu_{ul}/kT_{\text{ex}}(p,v)}\right)\frac{1}{p\sqrt{1-v^2}}, \quad (7.24)$$

where g_u is the degeneracy of the upper state, A_{ul} the Einstein A-coefficient, $\mathcal{Z}$ the partition function, and E_l the energy of the lower state.

This brightness distribution is often observed using a telescope having a sensitivity $P(x,y)$ that varies considerably over the emitting region. Here we assume that $P(x,y)$ can be approximated with a circular two-dimensional Gaussian with a full width at half maximum equal to θ_b. The observed flux density at an offset p is given by

$$S_l(p,v) = \int b_{ul}(p',\varphi,v)\,P(p,p',\varphi)p'\mathrm{d}p'\mathrm{d}\varphi, \quad (7.25)$$

where p' and φ are polar coordinates with an origo at the center of the CSE. This expression can be simplified if we normalize to the values at $v=0$ and $p=\theta_b D$, i.e., the values at the line center and at a full beam width offset, we assume that the criterion for the Rayleigh–Jeans regime, i.e., $T_{\mathrm{ex}} \ll h\nu_{ul}/k$, is fulfilled, the fractional abundance is constant, and we integrate over radius rather than p'. The result is, after some straightforward algebra,

$$S_1(x_0, v) = 2\pi\theta_{\mathrm{b}}^2(1-v^2)e^{-4\ln 2\, x_0^2(1-v^2)}$$
$$\times \int_{x_i}^{x_e} b_{ul}(x, v)e^{-4\ln 2\, x^2(1-v^2)} I_0\left[8\ln 2\, xx_0\right] x\, dx\,, \quad (7.26)$$

where $x = r/\theta_{\mathrm{b}} D$, $x_{i,e} = R_{i,e}/\theta_{\mathrm{b}} D$, $x_0 = p/\theta_{\mathrm{b}} D$, I_0 is the modified Bessel function of order zero, and

$$b_{ul}(x, v) = \frac{2kT_{\mathrm{ex}}(x)\nu_{ul}^2}{c^2} \left[1 - \exp\left(-\frac{\tau_1 t(x)}{x(1-v^2)}\right)\right], \quad (7.27)$$

where $\tau_1 = \tau_{ul}(1, 0)$, and

$$t(x) = \frac{T_{\mathrm{ex},1}}{T_{\mathrm{ex}}(x)} \frac{\exp-[E_l/kT_{\mathrm{ex}}(x)]}{\exp-[E_l/kT_{\mathrm{ex},1}]} \frac{\mathcal{Z}(T_{\mathrm{ex},1})}{\mathcal{Z}(T_{\mathrm{ex}}(x))}, \quad (7.28)$$

where $T_{\mathrm{ex},1} = T_{\mathrm{ex}}(1)$. The latter is still a complicated expression, but for sufficiently high temperatures it is possible to simplify further. For instance, in the case of a linear rotor molecule where the partition function is particularly simple, we may write $t(x) = [T_{\mathrm{ex},1}/T_{\mathrm{ex}}(x)]^2$. We will use this approximation in what follows, but note that in general, $t(x)$ is roughly constant if T_{ex} varies only slightly within the radial range R_i to R_e. We expect that the excitation temperature decreases with radius. It is therefore reasonable to parameterize the excitation temperature distribution as

$$T_{\mathrm{ex}}(x) = T_{\mathrm{ex},1} x^{-\beta}. \quad (7.29)$$

The final expression for the flux density with the telescope pointed toward the center of the CSE is given by

$$S_1(0, v) = 2\pi\theta_{\mathrm{b}}^2(1-v^2)B(T_{\mathrm{ex},1}) \int_{x_i}^{x_e} x^{1-\beta}$$
$$\times \left[1 - \exp\left(-\frac{\tau_1 x^{2\beta-1}}{(1-v^2)}\right)\right] e^{-4\ln 2\, x^2(1-v^2)}\, dx\,, \quad (7.30)$$

where we have introduced $B(T_{\mathrm{ex},1}) = 2k(\nu_{ul}/c)^2 T_{\mathrm{ex},1}$, which reduces to

$$S_1(0, v) = \begin{cases} 2\pi\theta_{\mathrm{b}}^2 B(T_{\mathrm{ex},1})\tau_1\, I_{\mathrm{bf}}(\beta, v), & \tau \ll 1, \\ 2\pi\theta_{\mathrm{b}}^2(1-v^2)B(T_{\mathrm{ex},1})\, I_{\mathrm{bf}}(1-\beta, v), & \tau \gg 1, \end{cases} \quad (7.31)$$

where

$$I_{\mathrm{bf}}(\zeta, v) = \int_{x_i}^{x_e} x^\zeta\, e^{-4\ln 2\, x^2(1-v^2)}\, dx\,. \quad (7.32)$$

Expressions (7.31) can be solved for arbitrary β in the case of an unresolved source, i.e., $x_e \ll 1$. The results when $x_i \ll x_e$ are

$$S_l(0,v) = \begin{cases} \frac{2\pi\theta_b^2}{1+\beta} B(T_{\text{ex},1})\tau_1 x_e^{1+\beta}, & \tau \ll 1, \\ \frac{2\pi\theta_b^2}{2-\beta} B(T_{\text{ex},1})(1-v^2) x_e^{2-\beta}, & \tau \gg 1 \end{cases} \quad (7.33)$$

(in the limit of small optical depth $\beta > -1$, otherwise the inner radius becomes important, and $\beta < 2$ when the optical depth is high). That is, depending on the optical depth of the transition, and independent of the value of β, we expect a line profile with a shape that lies in between that of a rectangle (optically thin emission) and a parabola (optically thick emission). Indeed, this is also what is normally observed; see Figure 7.3.

However, in some cases the size of the emitting region is comparable to or larger than the beam, and it is possible to solve expressions (7.31) also in this case, but only for the particular values $\beta=0$ (applies when the emission comes from a narrow radial range) and 1 [close to the exponent of the kinetic temperature law as given in (7.16)]. For optically thin emission the results for an envelope with $x_i \approx 0$ and $x_e \gg \theta_b$ are

$$S_l(0,v) = \begin{cases} \frac{\pi\theta_b}{(1-v^2)^{1/2}} B(T_{\text{ex},1})\tau_1, & \tau \ll 1, \beta = 0, \\ \frac{1}{1-v^2} B(T_{\text{ex},1})\tau_1, & \tau \ll 1, \beta = 1. \end{cases} \quad (7.34)$$

That is, we obtain a line profile with peaks at $v = \pm 1$, and they are substantially more pronounced when the excitation temperature drops rapidly with radius. For optically thick emission the results are

$$S_l(0,v) = \begin{cases} B(T_{\text{ex},1}), & \tau \gg 1, \beta = 0, \\ \pi\theta_b(1-v^2)^{1/2} B(T_{\text{ex},1})\tau_1, & \tau \gg 1, \beta = 1. \end{cases} \quad (7.35)$$

That is, the line profile becomes flat-topped, but the limiting shape is a "smoothed parabola" in the $\beta=1$ case.

It is also interesting to give the results for a geometrically thin shell, i.e., $(x_e - x_i) \ll (x_e + x_i)/2$, where we can expect $\beta=0$, i.e., a constant excitation temperature with radius. The results are

$$S_l(0,v) = \begin{cases} 2\pi\theta_b^2 B(T_{\text{ex},1})\tau_1 \Delta x e^{-4\ln 2\, x_p^2(1-v^2)}, & \tau \ll 1, \\ 2\pi\theta_b^2 (1-v^2) B(T_{\text{ex},1}) \Delta x x_p e^{-4\ln 2\, x_p^2(1-v^2)}, & \tau \gg 1, \end{cases} \quad (7.36)$$

where $\Delta x = (x_e - x_i)$ and $x_p = (x_e + x_i)/2$. The optically thin line profile is double-peaked, and this becomes more pronounced than for the corresponding line profile of an extended CSE when $x_p \approx 1$. Even the optically thick line profile becomes double-peaked (when $x_p \approx 1$), but the peaks are not so sharp and lie inside the $v = \pm 1$ points unless x_p becomes sufficiently large.

Finally, we give the result for a geometry that may be of importance in the case of post-AGB objects, a thin (i.e., the width $\Delta h \ll \theta_b D$), edge-on, cylindrical disk with an inner radius R_i and an outer radius R_e, which expands with a constant velocity v_∞. The mass-loss rate into and the abundance within the disk are assumed to be constant. The results for an unresolved source ($x_i \approx 0$, and $x_e \ll 1$) are

$$S_l(0,v) = \begin{cases} \frac{\theta_b^2}{1+2\beta} \frac{1}{(1-v^2)^{1/2}} B(T_{\text{ex}},1)\tau_1 \frac{\Delta h}{\theta_b D} x_e^{1+2\beta}, & \tau \ll 1, \\ \frac{\theta_b^2}{1-\beta} (1-v^2)^{1/2} B(T_{\text{ex}},1) \frac{\Delta h}{\theta_b D} x_e^{1-\beta}, & \tau \gg 1 \end{cases} \quad (7.37)$$

(in the limit of small optical depth $\beta > -\frac{1}{2}$, and $\beta < 1$ when the optical depth is high). That is, in the case of a disk the optically thin line profile becomes double-peaked even though the emission is unresolved. The optically thick line profile is a smoothed parabola, i.e., the same result as that obtained for a highly resolved, spherical CSE with $\beta = 1$. This also shows that the line profile depends on the geometry.

Observed Emission-Line Profiles

In Figure 7.3 examples are given of observed line profiles, and it is striking how well the results obtained for a standard CSE describe the first four of them: (a) SiO is not abundant in the C-CSE of IRC+10216, and it is excited only close to the star; i.e, optically thin, spatially unresolved emission results in a rectangular line profile; (b) SiO is abundant in the O-CSE of RX Boo, but it is excited only close to the star; i.e, optically thick, spatially unresolved emission results in a parabolic line profile; (c) the rarer isotopomer ^{13}CO is not abundant in the C-CSE of IRC+10216, but it is easily excited to large radii; i.e, optically thin, spatially resolved emission results in a double-peaked line profile; (d) ^{12}CO is abundant in the C-CSE of IRC+10216, and it is easily excited to large radii; i.e, optically thick, spatially resolved emission results in a smoothed parabolic line profile. The highly double-peaked line profile in Figure 7.3e, can be explained in terms of emission from a geometrically thin shell; see (7.36). The double-component line profile in Figure 7.3f comes from a CSE of unknown properties, and the line profiles in Figure 7.3g and h are due to strong maser emission. Figure 7.4 shows a comparison of the results of a detailed numerical modeling and the observational data of CO rotational lines from the CSE of the C-star IRC+10216.

Most line profiles may be well fitted by the simple expression

$$S(v_z) = S(v_*) \left[1 - (v_z - v_*)^2/v_e^2\right]^\alpha, \quad (7.38)$$

where $\alpha = 1$ means a parabolic, $0 < \alpha < 1$ a more flat-topped, and $\alpha < 0$ a double-peaked line profile. For $\alpha > 1$ a "Gaussian-like" line profile is obtained. From the above analysis we do not expect to see Gaussian line profiles. However, should the line mainly originate from regions close to the star, where the gas is still accelerating, more Gaussian-like, or even triangular, line profiles can be formed. The v_e (in this chapter used for observational estimates) obtained in this way is a good estimate of the gas terminal velocity v_∞.

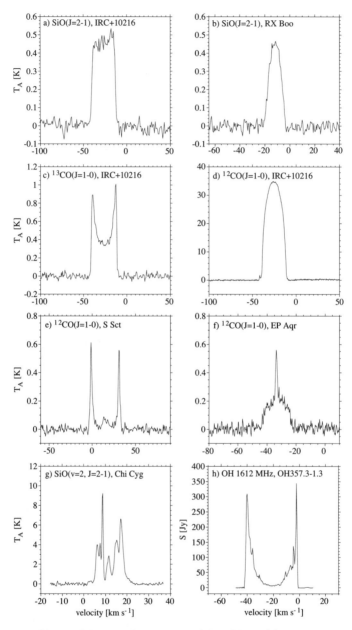

Fig. 7.3. Observed line profiles toward AGB CSEs: (a) optically thin, spatially unresolved emission, (b) optically thick, spatially unresolved emission, (c) optically thin, spatially resolved emission, (d) optically thick, spatially resolved emission, (e) emission from geometrically thin, spatially resolved shell, (f) double-component line profile, (g) maser emission-line profile, and (h) maser emission-line profile

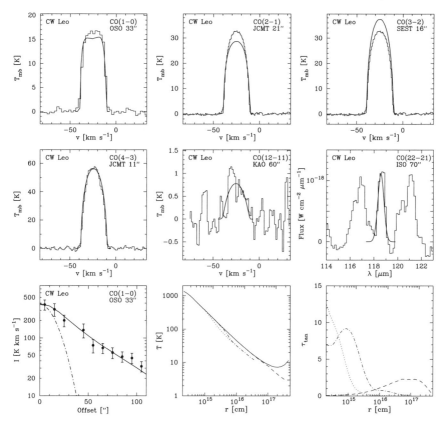

Fig. 7.4. Comparison between the results of a circumstellar CO model (solid lines) and observed CO rotational line profiles (histograms) of the C-star IRC+10216 (variable name CW Leo; assumed distance 130 pc). The observational data are from radio telescopes (OSO 20m, JCMT, SEST), the KAO, and the ISO (note that the ISO line is not spectrally resolved). The lowest panels show the model (solid line) and observed radial CO ($J=1\to 0$) intensity distribution (the beam is dot-dashed) (left), the radial distributions of the model kinetic temperature (solid) and the model excitation temperatures (middle), and the tangential optical depths (right) of the $J=1\to 0$ (dashed), $J=12\to 11$ (dot-dashed), and $J=22\to 21$ (dotted) lines

An effect not covered by this analysis is selfabsorption, which becomes stronger the higher the optical depth. On the front side of the envelope the emission from the cooler outer region is dominating (since the emission from the inner region is absorbed here), while on the rear side the emission from the inner warmer region is dominating. This leads to less emission in the blueshifted wing, and hence a line asymmetry that shifts the line with respect to the stellar velocity. This effect is seen in the optically thick ^{12}CO spectrum in Figure 7.3d (compare with the width of the ^{13}CO spectrum).

The Emitting Region

Before leaving this section we will show that it is not only photodissociation and chemistry that determine the region where the emission comes from. Also the excitation requirements play an important role. We illustrate this with the example of a linear rotor molecule for which the optical depth takes on a simple expression. Its dependence on the rotational quantum number of the upper energy level, J, is given by (in the Rayleigh–Jeans approximation, and a Boltzmann distribution for the level populations)

$$\tau \propto J^2 e^{-BJ(J+1)/T_{ex}}, \tag{7.39}$$

where B is the rotational constant (in K). This expression has a maximum for $J_m \approx 0.5\sqrt{T_{ex}/B}$; i.e., for the temperature T_{ex} the maximum optical depth occurs in the $J_m \to J_m-1$ transition. This can be converted into the radius $r_{l,J}$ at which this transition has its maximum excitation by using expression (7.16) and assuming LTE-excitation,

$$r_{l,J} \approx 10^{17} B^{-1.1} J^{-2.2} \text{ cm}. \tag{7.40}$$

As an example, the $J=1 \to 0$ and $J=4 \to 3$ transitions of CO (for which $B \approx 2.8\,\text{K}$) have a maximum excitation at 4×10^{16} cm and 2×10^{15} cm, respectively. This indicates that different transitions probe different regions of a CSE, even though the brightness is strongly affected also by the column density ($\propto r^{-1}$) and the excitation temperature. A proper modeling of CO rotational line emission shows this effect; Figure 7.4. The two lowest, rightmost panels show the radial distributions of the excitation temperatures and tangential optical depths of three CO lines coming from levels of varying excitation energy. Clearly, the higher the energy of the transition, the closer to the star it becomes subthermally excited, and the closer to the star peaks the optical depth of the corresponding line.

7.2.6.3 Maser Emission

We will only briefly discuss the particular characteristics of maser emission (see [60] for details). The physical conditions in a CSE are such that deviations from LTE are the rule rather than the exception, and population inversion (i.e., the number density per sublevel in the upper state is higher than that of the lower state) is common. Indeed, a fair fraction of the "normal" line profiles is actually due to optically thin maser emission. However, sometimes maser action leads to clearly observable characteristics such as narrow, time-variable components that form part of the total line profile, or in extreme cases, such as for a number of transitions in OH, H_2O, and SiO, the emission is totally dominated by maser action, and the line profiles may become quite complex; Figure 7.3g. An exception to this are the line profiles of the strong OH 1612 MHz masers, which show an exceptional regularity in the form of a double-peaked profile, at $\pm v_\infty$, due to strong radial amplification in the rear and the front parts of the CSE along the line of sight to the star. This produces strong emission at $\pm v_\infty$, and a very characteristic double-peaked profile, Figure 7.3h. Another characteristic of strong maser emission is the high brightness temperatures, much higher than the local temperature. Measurements of brightness temperatures in excess of 10^{12} K have been done [213].

The difference between maser emission and normal emission lies in the fact that the stimulated emission produces a cascade of photons when a transition has a population inversion. In mathematical terms, the population inversion means a negative optical depth in the radiative transfer equation, and the solution becomes (as long as τ does not depend on I)

$$I_\nu = I_{\mathrm{bg},\nu}\, e^{|\tau_\nu|} + S_\nu \left(e^{|\tau_\nu|} - 1\right), \tag{7.41}$$

where I_{bg} is the background radiation and S is the source function. Thus, the masing species can either amplify the background radiation or the internally generated radiation. It is clear that for sizable optical depths the maser intensity is very sensitive to any changes in optical depth, e.g., if $|\tau| \approx 20$, a 10% change in optical depth will lead to a change in intensity by a factor of ten. This also means a strong enhancement of emission from regions of large optical depth, and hence a highly distorted image of the source. It is impossible to produce more maser photons per unit time than the number of inversions per unit time provided by the excitation. When the maser intensity reaches this level, the maser action is saturated. Analytical studies of maser emission from spheres, shells, disks, and cylinders have been done, and they give useful insight [60, 61].

7.2.6.4 Scattered Light

An important, but not yet fully explored, method for the study of CSEs is based on the scattering of stellar (or interstellar) light by circumstellar atoms/molecules and dust. In this way the high angular resolutions achievable at optical/infrared wavelengths can be used in the study of the cool AGB CSEs. Scattered light can also be highly polarized and this provides additional information. A simple model is introduced here.

The most intuitive derivation is obtained if the scattering is expressed in terms of a scattering cross section [99]. The scattered flux density, at a frequency corresponding to the line-of-sight velocity v_z, from an area ΔA, and at an offset p from the star is given by, in the optically thin limit,

$$S_{\mathrm{sc},\nu}(p) = \frac{\Delta A}{4\pi D^2}\, \sigma_\mathrm{s} \phi(\nu) F_{*,\nu} \int_{-\infty}^{\infty} \left(\frac{R_*}{r}\right)^2 n_{\mathrm{sc}}(p, v_z)\, \mathrm{d}z, \tag{7.42}$$

where σ_s is the scattering cross section, $\phi(\nu)$ the local line factor (assumed to be rectangular for line scattering, $\phi(\nu) = c/\nu\delta v$, and $\phi(\nu) = 1$ for continuum scattering), $F_{*,\nu}$ the flux density at the stellar surface, and R_* the stellar radius. The integral results in

$$\int_{-\infty}^{\infty} \left(\frac{R_*}{r}\right)^2 n_{\rm sc}(p, v_z)\,{\rm d}z = \begin{cases} R_*^2 n_{\rm sc}(p) \frac{\delta v}{v_\infty} \frac{1}{p} \left[1-v^2\right]^{1/2}, & \text{line,} \\ R_*^2 n_{\rm sc}(p) \frac{\pi}{2} \frac{1}{p}, & \text{continuum.} \end{cases} \tag{7.43}$$

We can rewrite this using the stellar flux density, $S_{*,\nu} = F_{*,\nu}(R_*/D)^2$, as

$$\frac{S_{{\rm sc},\nu}(p)}{S_{*,\nu}} = \frac{\Delta A}{8} \sigma_{\rm s} \frac{n_{\rm sc}(p)}{p} g(v), \tag{7.44}$$

where

$$\sigma_{\rm s} = \begin{cases} \frac{c^3}{8\pi\nu^3 \delta v} \frac{g_u A_{ul}}{g_l}, & \text{line,} \\ Q_{\rm s} \pi a_{\rm gr}^2, & \text{dust,} \end{cases} \tag{7.45}$$

$$n_{\rm sc}(p) = \begin{cases} \frac{f_A X \dot{M}}{8 m_{\rm H} v_\infty p^2} \frac{g_l e^{-E_l/kT_{\rm ex}}}{\mathcal{Z}(T_{\rm ex})}, & \text{line,} \\ \frac{3 M_{\rm d}}{16\pi^2 v_{\rm d} \rho_{\rm gr} a_{\rm gr}^3 p^2}, & \text{dust,} \end{cases} \tag{7.46}$$

$$g(v) = \begin{cases} \frac{2c}{\pi \nu v_\infty} \left[1-v^2\right]^{1/2}, & \text{line,} \\ 1, & \text{continuum,} \end{cases} \tag{7.47}$$

where we have assumed grains of a single radius, $a_{\rm gr}$, with a scattering efficiency $Q_{\rm s}$.

In the case of constant mass-loss rate and expansion velocity, i.e., $n_{\rm sc}(p) \propto p^{-2}$, we find that the scattered radiation declines as p^{-3}, and that for line scattering a "softened" parabolic line shape is obtained. An example of circumstellar KI scattering in a C-CSE is shown in Figure 7.5.

7.2.6.5 Absorption Lines

A circumstellar absorption line is formed when the relatively cool gas in the CSE absorbs the strong background radiation from the central star, or from the developing HII-region in a young PN. The lineformation process is relatively complicated for several reasons. In many cases the absorbing species are located close to the star, and we may expect that the gas has still not reached the state of constant-velocity outflow; i.e., the details of the acceleration are important. The excitation of the emitting particles is likely to have a strong dependence on their distance to the star. Furthermore, there is dust in the envelope that scatters, absorbs, and reemits the stellar light. This introduces some additional complexity since it contributes to the line-of-sight absorption, to the excitation of the emitting particles, and weakens the absorption lines through emission.

The detailed line profile is determined by the various radial dependences, which will determine the sampling of the velocity curve. The decrease in

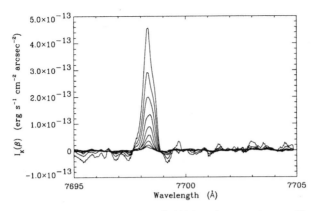

Fig. 7.5. Circumstellar KI resonance line (at 770 nm) scattering profiles at different angular offsets from the C-star R Scl. The upper spectrum represents the emission at $4''$ offset, and the following spectra are displaced consecutively further out by $0''\!.8$. The line intensity decreases roughly as the third power of the angular offset. From [99]

excitation temperature with radius will shift the line towards the blue edge, but this is counteracted by the rapidly falling density with the distance from the star. For a species that is not of photospheric origin, the line will also become blueshifted. Finally, the velocity gradient determines the interaction length, and this will move the line toward the stellar velocity if the slope of the velocity law is small, and toward the blue edge if the acceleration to the final velocity is very rapid. Of course, once all these complications are mastered, the absorption lines, e.g., high-resolution molecular ro-vibration line profiles, may be a very powerful diagnostic tool for probing the structure and dynamical behavior of the inner CSE [142, 284]; Figure 7.6.

7.2.7 Abundance Estimates

In principle, a fairly detailed modeling is required to obtain the abundance of a circumstellar atom or molecule from emission lines. However, for most species this is a too cumbersome approach; e.g., radiative transition rates and collisional cross sections are missing. A simpler method, which appears to lead to reasonable results for optically thin radio lines, is usually adopted. We outline here the *rotation-temperature diagram* method.

It is clear from the previous discussion that the formation of an absorption line is a complicated phenomenon. Hence, to estimate abundances from absorption lines is not an easy task, and there exists no simple formula that can be used to obtain even rough estimates.

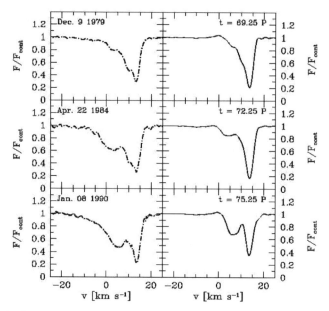

Fig. 7.6. High-resolution spectra of a CO ro-vibration line [$v=0 \rightarrow 2$ P(7)] in absorption toward the C-star IRC+10216 at different epochs in the variability cycle (observed profiles to the left, and model profiles to the right). From [284]

7.2.7.1 Abundance Formula

An estimate of the abundance of an atom/molecule can be obtained by using the optically thin limit of (7.31). We here choose $\beta=0$, since this leads to a weaker dependence on the usually unknown size of the emitting region [i.e., x_i and x_e; see (7.33)]. Since the following expressions are usually encountered in connection with radioastronomical observations, we will convert from the flux density scale to the antenna temperature scale using the relation $S(0,0) = 2kT_A\Omega_A/\lambda^2$, where $\Omega_A = \pi/4\ln 2\,\theta_b^2$ for a Gaussian beam. The final result is

$$f_A = I_{ul}\frac{4\pi^2 m_H k}{\ln 2\, hc^3}\frac{v_\infty \theta_b D}{X\dot{M}}\frac{\mathcal{Z}(T_{ex})\nu_{ul}^2}{g_u A_{ul}}\frac{e^{E_l/kT_{ex}}}{\int_{x_i}^{x_e} e^{-4\ln 2\, x^2}\,dx}, \quad (7.48)$$

where we have introduced the intensity integrated over the line profile, $I = 2v_\infty T_A$. The disadvantage with this expression is that it relies on a single line, and that the excitation temperature has to be assumed or estimated in some way. The latter is usually done by observing several lines of the same species, and in this case we may rewrite (7.48) in a more useful form.

7.2.7.2 The Rotation-Temperature Diagram

The excitation temperature T_{ex} and the abundance of a species can be estimated simultaneously by observing two or more optically thin lines from the

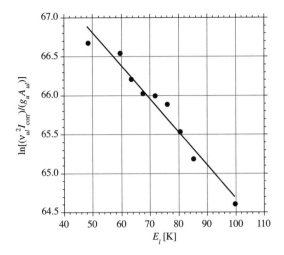

Fig. 7.7. A rotation-temperature diagram of HC_5N, where the data are obtained toward the C-star IRC+10216 (I_{corr} is the observed intensity corrected for the beam filling). The best-fit line corresponds to a rotation temperature of 24 K and a source-averaged column density of 5.5×10^{14} cm^{-2}

same species. Expression (7.48) may be rewritten in the form

$$\ln\left(\frac{\nu_{ul}^2 I_{ul}/\eta}{g_u A_{ul}}\right) + \ln\left(\frac{8\pi k}{hc^3}\frac{\mathcal{Z}(T_{\mathrm{ex}})}{N_{\mathrm{A,b}}}\right) = -\frac{E_l}{kT_{\mathrm{ex}}}, \qquad (7.49)$$

where $\eta = \int_{x_i}^{x_e} e^{-4\ln 2\, x^2}\, dx$ is the beam-filling factor, and $N_{\mathrm{A,b}}$ the beam-averaged column density (7.20). However, most results on CSEs are based on the more traditional form of (7.49) used in the study of the ISM, where

$$\eta = \begin{cases} \frac{\theta_s^2}{\theta_s^2 + \theta_b^2}, & \text{if Gaussian source,} \\ 1 - e^{-4\ln 2\,(R_e/\theta_b)^2}, & \text{if disk source,} \end{cases} \qquad (7.50)$$

and θ_s is the source size (full width at half maximum). This is not the correct beam-filling factor, and hence in this case, the column density rather takes the form of the source-averaged column density, $N_{\mathrm{A,s}}$ (7.19), but the analysis is no longer fully stringent.

In the rotation-temperature diagram method one plots the data points in a $(\ln[(\nu_{ul}^2 I_{ul}/\eta)/(g_u A_{ul})], F_l/k)$-diagram. If the assumptions are fulfilled, i.e., the energy-level populations can be described by a single excitation temperature, T_{rot}, the lines are optically thin, etc., we expect the points to lie close to a straight line according to (7.49). That is, a fit of a straight line to the data points results in an estimate of $N_{\mathrm{A,s}}$ and T_{rot}. An example of a rotation-temperature diagram is shown in Figure 7.7.

7.2.7.3 Detectability

It is generally true that the larger and the more complex a molecule becomes, the larger the partition of populations over many energy levels is, and the more difficult it becomes to detect. We will present a crude estimate of the detectability of molecules using the linear rotor as an example. By inverting (7.48) we find that the antenna temperature has the frequency dependence $T_A \propto \nu^4 \exp[-h\nu^2/4BkT_{ex}]$ when $\nu \gg 2B$, and the emission is unresolved [B is given in frequency units; for a linear rotor $\nu_{J,J-1} = 2BJ$, and $E_{J-1} = hBJ(J-1)$]. This expression has a maximum for

$$\nu_{opt} = \sqrt{8BkT_{ex}/h}\,; \tag{7.51}$$

i.e., a species with the rotational constant B should preferentially be observed at this frequency. The decline in intensity for observing frequencies higher than this is rapid, $T_A(2\nu_{opt})/T_A(\nu_{opt}) \approx 0.04$, while the dependence is less critical for lower frequencies, $T_A(\nu_{opt}/2)/T_A(\nu_{opt}) \approx 0.2$. As an example we will take the long carbon-chain molecules HC_nN, where the $n=1$, 3, 5, 7, and 9 species have been detected in C-CSEs. For an excitation temperature of 20 K we find the optimum frequencies (and transitions) to be 123 GHz ($J=14 \rightarrow 13$) for $n=3$, and 43 GHz ($J=38 \rightarrow 37$) for $n=7$.

Using (7.48) we can also calculate the abundance required to produce an 0.1 K line at the optimum frequency in a 30 m telescope from a CSE at a distance of 500 pc with $\dot{M} = 10^{-5}$ $M_\odot$ yr^{-1} and $v_\infty = 15$ km s^{-1} (assuming an appropriate source size, $R_e = 3 \times 10^{16}$ cm). We find that $f = 2 \times 10^{-7}$ and 7×10^{-6} for $n=3$ and 7, respectively; i.e., when compared with the estimated abundances in Table 7.1 it is clear that already at the molecular size of HC_7N there is a detection problem even for a high-mass-loss-rate source at a distance of 500 pc. The required abundance scales as $T_{A,n} v_\infty^2 D^2 / \dot{M} R_e d$, where d is the telescope diameter and $T_{A,n}$ the noise level. Considering this, it is likely that detections of more complicated species will come through, e.g., sub-mm and far-IR observations of ro-vibrational lines of various bending and flopping modes.

7.3 Circumstellar Line Observations

This section gives an overview of the observational properties of AGB CSEs as inferred from atomic/molecular line data. It summarizes the atomic and molecular species detected at radio, infrared, and optical wavelengths, and provides estimates of their abundances.

7.3.1 Atomic Species

The observational information on the atomic composition of AGB CSEs is meager. The reason for this is that for a given sensitivity, a minimum amount of emitting atoms is required to produce a detectable line. This means a high mass-loss rate, which is normally associated with cool stars whose CSEs are mainly in molecular form. In such CSEs atoms appear only in the external parts where the molecules become photodissociated, and here the excitation is often not enough to produce any emission. A better alternative would be a warmer star, still with a relatively high mass-loss rate, but this combination is found only for supergiants.

The 21 cm hyperfine line of HI is the obvious choice in a search for the most abundant element. However, such observations are severely hampered by the presence of strong interstellar HI emission in essentially all directions. Hitherto, the M-star o Ceti, the S-star RS Cnc, and the C-star IRC+10216 have been the only AGB stars detected in this line [26, 75, 162].

Fine-structure lines are inherently much stronger than hyperfine lines, and this has made it possible to detect CI, but only in one or possibly two AGB CSEs, through its $^3P_1 \rightarrow {}^3P_0$ line at 609 μm (492 GHz) [143, 151]. These data provide an important test of the physical/chemical models, since ultimately all the C locked in molecular form should be gradually released as the gas recedes from the star. The high-resolution ISO spectrometers have allowed a few detections of fine-structure emission lines from FeI, FeII, SI, and SiII [1].

A somewhat different way of studying the atomic component of CSEs has proved to be relatively successful. It uses light scattered in strong resonance lines of circumstellar atoms. Both NaI and KI have been detected in AGB CSEs in this way [99, 181].

7.3.2 Molecular Species

There are essentially two distinct ways in which circumstellar molecules are detected, either through absorption lines in the near-infrared where the star is bright, or through emission lines in the mid-IR to radio wavelength range where the transitions are sufficiently excited despite the low temperatures in the CSEs and the background radiation is weak. Observations of line-scattered stellar light provide another method.

7.3.2.1 Radio Emission Lines

The first circumstellar molecule to be detected at radio wavelengths, apparently in maser emission at 18 cm, was OH in 1968 toward the supergiant NML Cyg [282]. Initially, the progress was slow, but a few years later CO was detected, apparently in "normal" emission at a wavelength of 2.6 mm, toward the C-star IRC+10216 [238]. It was quickly realized that a CSE around an evolved star, in particular a C-star, can be extremely rich in different molecular species. The CSE of IRC+10216 has turned out to be a gold mine, since this object probably is the most nearby C-star ($\approx$130 pc), and it happens to have a very high mass-loss rate ($\approx 2 \times 10^{-5}$ M$_\odot$ yr^{-1}); Figure 7.8. At

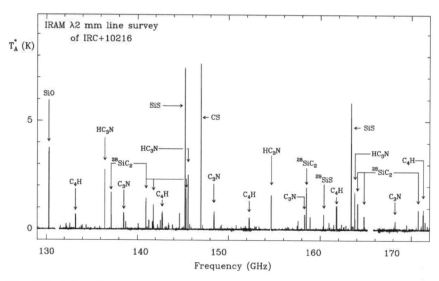

Fig. 7.8. A broadband spectrum, centered at 150 GHz, of the C-star IRC+10216 shows a wealth of circumstellar molecular emission lines. From [40]

this epoch 52 molecules have been detected at radio wavelengths in AGB CSEs, 43 of these in the CSE of IRC+10216; Table 7.1. This appears impressive, but some of them ($\approx$20) were detected only in IRC+10216, and most of them (>30) in fewer than five sources. There are only a dozen objects with more than 10 molecular species detected. It is notable that as opposed to the situation for the interstellar medium, only one ion, HCO^+, has been detected. Only toward IRC+10216 have unbiased searches for molecules been performed [40, 93, 120].

It is clear that in the C-CSEs a large number of relatively complicated polyatomics exists. No doubt, most of them are a result of an efficient C-based circumstellar chemistry; see Chapter 5. Among the most interesting are the long carbon chains, the cyanopolyynes HC_nN ($n = 1, 3, 5, 7, 9$), the related hydrocarbons C_nH ($n = 2, 3, 4, 5, 6, 7, 8$) [39], and the carbenes H_2C_n ($n = 3, 4, 6$) [37]. Also nitrogen–carbon chains C_nN ($n = 1, 3, 5$) [96], silicon–carbon chains C_nSi ($n = 1, 4$) [198], and sulphur–carbon chains C_nS ($n = 1, 2, 3$) are present. Simple ring molecules, such as the triangular SiC_2 and cyclic C_3H_2 and the rhomboidal SiC_3 [3], have been detected, but branched molecules are notably rare. On the spectacular side, we have the detections of the refractory species NaCl, AlCl, AlF, KCl, MgNC, MgCN, NaCN, and AlNC [38, 141, 252, 298], which have so far been detected only in the circumstellar medium. Most likely, even more complicated species exist in AGB CSEs and their detections may come through sub-mm and far-IR observations of ro-vibrational lines of various bending and flopping modes.

The O-CSEs are apparently less rich in molecules, but the situation may have been different if there had been an M-star equivalent in mass-loss rate and proximity to IRC+10216. Here OH, H_2O, and SiO dominate, but also S-bearing species, in particular SO, SO_2, and H_2S are frequently found [210]. Quite unexpected detections of C-bearing species, HCN, CS, and CN, have been made [208].

7.3.2.2 Radio Maser Emission Lines

The presence of circumstellar maser emission is fascinating, but leads to interpretational problems (Section 7.2.6.3). In some cases the entire line profile is dominated by maser emission (here called strong masers), but more common is the presence of maser features in a line profile that is otherwise dominated by "thermal" emission (here called weak masers). We will not discuss the large number of line profiles that are produced by optically thin maser emission (particularly common in low-lying rotational transitions), e.g., in low-mass-loss-rate CSEs the CO $(J = 1 \to 0)$ transition has a population inversion over a large radial range, but the emission is optically thin and produces a "normal" line profile [227].

Strong maser emission from the $^2\Pi_{3/2}$, $J = \frac{3}{2}$, Λ-doubling transitions of OH at 18 cm is quite ubiquitous among O-CSEs. Of particular importance has been the satellite line at 1612 MHz, in which a number of surveys have been performed; Section 7.3.3 [100]. In this connection we also mention the possibility to measure magnetic field strength and orientation using the OH emission [246]. Likewise, strong maser emission from the 6_{16}–5_{23} rotational transition at 22 GHz of H_2O has commonly been found in O-CSEs. In addition, several ground-state rotational lines and two vibrationally-excited rotational H_2O lines have been detected in strong maser emission [80, 185]. SiO provides a remarkable example where rotational transitions up to $J = 8 \to 7$ in vibrationally-excited states up to $v=4$ (corresponding to about 7000 K) have been seen in maser emission in the main isotopomer [83], and where also the rare isotopomers ^{29}SiO and ^{30}SiO exhibit strong maser emission in ground as well as excited vibrational states [78]. In this category we also place HCN, for which strong maser emission, in C-CSEs, has been detected in five vibrationally-excited rotational transitions [16]. Finally, strong maser emission has been found in vibrationally-excited rotational lines of CS and SiS towards IRC+10216.

Weak maser emission in the form of narrow features appears to be common in low lying rotational transitions of HCN, SiO, CS, and SiS.

7.3.2.3 Infrared Lines

Initially, we note that it is not always straightforward to determine whether an infrared line has a circumstellar origin or whether it emerges from the

Table 7.1. Molecules detected in AGB CSEs at radio wavelengths. The (rough) number of sources detected in each species is given (Σ), as well as abundances with respect to H_2 (O: C/O<1; C: C/O>1; $k(l) = k \times 10^l$)

Molecule	Σ	Chemistry O	Chemistry C	Molecule	Σ	Chemistry O	Chemistry C
2-atomic:							
AlCl	1		2(−7)	OH	2000	2(−4)	
AlF	1		4(−8)	PN	1		?
CO	600	5(−4)	1(−3)	SiC	2		4(−8)
CN	40	2(−7)	5(−6)	SiN	1		2(−8)
CP	1		2(−8)	SiO	500	5(−6)	1(−7)
CS	35	1(−7)	1(−6)	SiS	15	7(−7)	2(−6)
KCl	1		2(−9)	SO	20	2(−6)	
NaCl	1		1(−9)				
3-atomic:							
AlNC	1		1(−9)	MgCN	1		1(−9)
C_2H	20		4(−6)	MgNC	1		2(−8)
C_2S	5		1(−6)	NaCN	1		2(−8)
HCN	120	4(−6)	2(−5)	SiC_2	5		3(−7)
H_2O	300	3(−4)	1(−6)	SiCN	1		4(−9)
H_2S	300	1(−5)		SO_2	15	2(−6)	
HNC	10	1(−7)	1(−7)				
4-atomic:							
ℓ-C_3H	2		4(−8)	HC_2N	1		8(−9)
C_3N	5		3(−7)	NH_3	5	4(−6)	1(−7)
C_3S	1		3(−8)	SiC_3	1		3(−9)
5-atomic:							
C_4H	5		3(−6)	HC_3N	10		1(−6)
C_4Si	1		3(−9)	HC_2NC	1		2(−9)
c-C_3H_2	5		3(−8)	H_2C_3	1		2(−9)
6-atomic:							
C_5H	1		6(−8)	CH_3CN	5		3(−9)
C_5N(?)	1		9(−9)	H_2C_4	1		5(−9)
7-atomic:							
C_6H	1		8(−8)	HC_5N	5		2(−7)
8-atomic:							
C_7H	1		3(−9)	H_2C_6	1		?
9-atomic:							
C_8H	1		1(−8)	HC_7N	2		4(−8)
11-atomic:							
HC_9N	1		1(−8)				
Ions:							
HCO^+	2		1(−9)				

(extended) atmosphere. We have tried to select absorption and emission lines that originate in the expanding CSE, but since many of the observations were made by ISO, no kinematical information exists that can be used as a guide. In addition, the proposed warm molecular layer should be regarded as an intermediate region [251].

CO was first detected in circumstellar absorption in the gas outflow of IRC+10216 [76], and a few other such stars were found in the coming years [69]. Of special importance here is the possibility of observing symmetric species that do not emit radiation in the radio regime, and such a molecule, C_2H_2 (abundant in the photosphere), was indeed found in circumstellar absorption toward IRC+10216 [214]. Subsequently, CH_4, SiH_4, C_4H_2, and the carbon chains C_3 and C_5 were detected, while the 2 μm ro–vibration lines of H_2 were not detected, toward IRC+10216 [142]. C_2H_2 and HCN absorption lines below about 15 μm, observed by ISO, probably originate from the very inner parts of a CSE; Figure 7.9 [2, 43]. At this epoch 15 different molecules have been observed in circumstellar absorption, the majority of them detected only in IRC+10216.

The circumstellar infrared lines turn into emission at ≈ 10 μm, where the stellar flux drops markedly and the size of the emitting region increases. ISO opened up the possibility for this type of observation; Figures 7.9 and 7.10. Of particular interest are the detections of CO_2 in a handful of O-CSEs [132, 217], HCN toward a few C-CSEs [2, 43], SO_2 in a few O-CSEs [290], and C_3 toward IRC+10216 [36]. At least the three former may probe an intermediate layer between the photosphere and the CSE [33]. In terms of detailed modeling the most important observations are those of a large number of H_2O lines in the 30–200 μm range in two O-CSEs [250], and a large number of lines in the same wavelength range due to high rotational transitions of CO and HCN in two C-CSEs [35, 219].

The cool parts of a CSE can also be observed at infrared wavelengths in lines strong enough to scatter the stellar light. So far, this method has been successfully applied to spectrally resolved CO ro-vibration lines in a few cases [218].

All circumstellar molecules detected at infrared wavelengths are summarized in Table 7.2.

7.3.2.4 Optical Lines

The optical spectra of AGB stars very likely contain narrow features due to circumstellar molecular absorption. These features are not easily identified among the wealth of photospheric features, and in addition, very high spectral resolution is required to get the kinematical and temperature information needed to establish a circumstellar origin. Nevertheless, a detection of C_2 toward IRC+10216 has been reported [6].

Table 7.2. Molecules detected in AGB CSEs at infrared and optical wavelengths. The (rough) number of sources detected in each species is given (Σ), as well as abundances with respect to H_2 (O: C/O<1; C: C/O>1; $k(l) = k \times 10^l$)

Molecule	Σ	Chemistry O	Chemistry C	Molecule	Σ	Chemistry O	Chemistry C
2-atomic:							
C_2	1		2(−6)	CS	1		3(−7)
CN	1		1(−5)	SiO	1		8(−7)
CO	10	?	4(−4)	SiS	1		4(−6)
3-atomic:							
C_3	1		1(−6)	HCN	3		2(−5)
C_2H	1		3(−6)	H_2O	2	1(−5)	
CO_2	15	3(−7)		SO_2	15	?	
4-atomic:							
C_2H_2	7		5(−5)	NH_3	2		2(−7)
5-atomic:							
C_5	1		1(−7)	SiH_4	1		2(−7)
CH_4	1		4(−6)				
6-atomic:							
C_2H_4	1		1(−8)				

7.3.2.5 Molecules in Post-AGB CSEs

In young post-AGB CSEs one finds the same species as those detected in AGB CSEs; see also Chapter 10. Good examples are the C-rich objects AFGL618 and AFGL2688, where more than 20 species have been detected [4]. In both cases, the CSEs are very thick, and they must have been formed by a high mass-loss rate on the AGB. Another (probable) post-AGB object is OH231.8+4.2, which has an O-CSE with an impressive setup of molecular species, including the only circumstellar detection of OCS [220].

The more evolved objects are expected to show a different molecular setup, since the increasing UV-flux and the presence of shocks should have an effect on the chemistry [288]. This is most dramatically shown by the detections of polyacetylenes in AFGL618 and AFGL2688, as well as methylpolyynes and the benzene ring in AFGL618 [41, 42]. A number of ionic species have been detected only in the CSE around a young PN, NGC7027 [160]. In addition, the abundant H_2 molecule becomes readily detectable in PNe of the bipolar type [52]. The molecular species detected only in post-AGB CSEs are CH, CH^+, CO^+, H_2, N_2H^+, OCS, H_2CO, HC_4H, HC_6H, CH_3C_2H, CH_3C_4H, and C_6H_6.

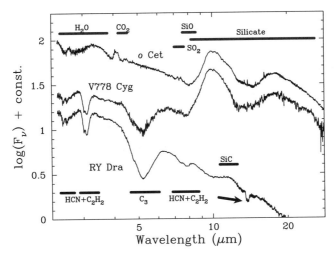

Fig. 7.9. ISO short-wavelength spectrometer (SWS) spectra of an M-star (*o* Ceti; upper), where the molecular absorption bands due to SiO, H_2O, and CO_2, and the dust emission features at ≈ 10 and $18\,\mu$m, are marked, and of a C-star (RY Dra; lower), where the molecular absorption bands due to HCN, C_2H_2, and C_3, and a dust emission feature at $\approx 11\,\mu$m, are marked. The star V778 Cyg (middle) represents an interesting case where features indicating both a C- and an O-rich chemistry are present. From [291]

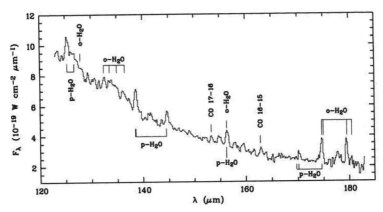

Fig. 7.10. ISO long-wavelength spectrometer (LWS) spectrum of the M-star W Hya, where rotational lines due to H_2O and CO are present (note that in this low-mass-loss-rate object the H_2O lines are stronger than those of CO). From [8]

7.3.3 Surveys

As in all other parts of astronomy, surveys, in particular full-sky surveys, are very important, not only for obtaining numbers and characteristics of already known populations, but also for the detection of new unexpected objects, which may lead to a giant leap forward in our understanding.

No unbiased surveys of the sky exist for circumstellar atomic/molecular line emission. Essentially all line surveys are based on optical and/or infrared input data, e.g., IRAS-color selected samples, and they have been carried out at radio wavelengths. In addition, most surveys are based on incomplete samples, which makes it difficult to draw definite conclusions on general population characteristics, such as mass-loss rate and expansion velocity distributions. Nevertheless, important results can be obtained.

Surveys of SiO, H_2O, and OH maser emission toward objects in different evolutionary stages have been made [64, 102, 165, 168, 196, 247]. There are also some impressive unbiased surveys of limited regions of the sky: the VLA OH 1612 MHz surveys of the Galactic Center [173] and the Galactic plane [20], and the ATCA/VLA OH 1612 MHz survey of the region $|\ell|\leq 45°$ and $|b|\leq 3°$ [231, 232] and the northern galactic plane [233]. Although not unbiased, we also mention in this connection the SiO 43 GHz maser surveys of the Galactic Bulge [115], the Galactic Center [55], and the Galactic Bar [116], and the 86 GHz maser survey of the inner Galaxy [186].

Only maser lines have been detected from extragalactic CSEs. In the LMC there are seven known OH/IR-stars, of which two may be supergiants [287]. The two supergiants are detected also in H_2O 22 GHz emission, and one of them has a detectable SiO maser [261].

Surveys have been made of CO radio line emission toward objects in different evolutionary stages [86, 88, 123, 148, 195, 204, 292, 300] (a compilation of older data can be found in [176]). Objects as distant as the Galactic Center have been detected [283].

7.3.4 Brightness Distributions

Spatial information on line brightness $I_{A,ul}$ is very important for identifying the physical, chemical, kinematical, and geometrical structures of a CSE. Unfortunately, information on line-brightness distributions in AGB CSEs exist only at radio wavelengths and is quite meager, even though some individual results are quite impressive. The main reason is that observations with single radio telescopes do not provide enough angular resolution. Even in the favorable case of CO emission the angular size of a CSE is quite small. According to (7.9) the CO photodissociation (angular) radius is given by $\theta_{\rm ph,CO} \approx 10'' [\dot{M}/10^{-6}]^{0.6}[0.5\,{\rm kpc}/D]$, and all other molecular emission regions are much smaller than this. This can be compared with the beam widths $\theta_{\rm b} = 50'' [100\,{\rm GHz}/\nu][15\,{\rm m}/d]$, where d is the telescope diameter. Hence, detailed observations have to be performed in interferometric mode, a method that is technically rather complex, and it has therefore taken some time to develop.

Only maser emission produced high enough brightnesses to be detected by the first interferometer instruments. The "images" produced were fairly crude, consisting of the location, line-of-sight velocity, and brightness of a number of point sources. In addition, a significant fraction of the emission was resolved out. This, combined with the difficulty in relating maser emission to physical conditions and even geometry, made them of limited use. The first detailed images came with the observations of OH 1612 MHz maser emission in the CSEs of two high-mass-loss-rate AGB stars [9, 23]. The progress in technical ability since then has been substantial, and impressive brightness distribution images now exist for both maser and "normal" emission. These will be further discussed in connection with the geometry of AGB CSEs; Section 7.6.1.

7.3.5 Molecular Abundances

One of the primary goals of observing a circumstellar molecular species is to determine its fractional abundance. The uncertainty of such estimates are often substantial, and usually abundance ratios are preferred in the comparison with theoretical predictions. From a stellar point of view elemental abundances are important measures of evolution, in particular isotope ratios, an area in which circumstellar data gradually are being used with some success. Most of the results presented here are based on radio line data.

7.3.5.1 Fractional Abundances

Fractional abundance estimates of the detected species are presented in Tables 7.1 and 7.2. These are in some cases averages over many objects, and there is reason to expect that the abundance may vary from object to object. The estimates are obtained using a number of methods, and hence form a heterogeneous set. In a few cases the estimates are based on only one or two lines using formulae like (7.48). In most cases a number of lines were observed, and a more reliable abundance estimate can be obtained from a rotation-temperature diagram, Section 7.2.7.2. Finally, in a few cases a more sophisticated radiative transfer analysis, based on a more detailed circumstellar model, has been carried out. This applies, for instance, to H_2O in O-CSEs [79], HCN and CN in C-CSEs [172], and a number of the species showing infrared absorption lines [142]. The majority of the estimates apply to IRC+10216, and are based on spectral scans [40, 93] or more directed searches. Surveys of various samples in different molecular line emissions provide data in some cases [5, 74, 197, 205]. Note that the fractional abundances all rely on an assumed CO fractional abundance through the mass-loss rate, unless another mass-loss-rate estimator was used.

By comparing different studies, we estimate that in general, the abundances are accurate to within a factor of ten for individual sources. The uncertainty decreases to within a factor of five for those molecules that are well observed and/or for which a more sophisticated analysis was performed.

However, even a sophisticated analysis is uncertain if the size of the emitting region is unknown, the radiative transfer through the inner CSE is important, the optical depths are very high, the molecular data are uncertain, etc. Also, comparisons with theory must be regarded as only tentative as long as no detailed maps of line brightness distributions exist. An interesting example of this is the overlap of the C_2H, C_3H, and C_4H regions in IRC+10216, Figure 7.18, a surprising result if the two latter species are formed, through reactions with different time scales, from C_2H [94].

Of particular interest are two studies, where the authors observed the same molecular species in seven high-mass-loss-rate C-stars with similar characteristics [197, 289]. Using (7.48) and theoretical estimates of the photodissociation radii, they concluded that, in general, the abundances vary between the sources by no more than a factor of five. This gives credence to the estimates, at least as long as the comparison is made within a homogeneous data set and the CSEs are relatively similar.

Another interesting example is provided by the presence of various C-species in O-CSEs (e.g., HCN), where they are not expected to be produced by the stellar atmosphere chemistry [207]. The question is the origin of the C, and whether the chemistry is atmospheric or circumstellar; Chapter 5. On the other side of the coin we have the detection of H_2O in the extreme C-star IRC+10216, a detection so surprising that a scenario with an evaporating Kuiper belt of icy bodies (comets) was invoked to explain it [184].

7.3.5.2 Abundance Ratios

Abundance ratios are expected to be more reliable than individual abundances. As an example we provide the relative abundances of long-carbon-chain molecules in the CSE of IRC+10216, Figure 7.11. For the cyanopolyynes, HC_nN, the abundance decreases by almost four orders of magnitude in the sequence $n=1$ to 9. The same applies to the related species C_nH when n goes from 2 to 8, but the odd-number species are about an order of magnitude less abundant than the even-numbered. These results are in reasonable agreement with theoretical predictions; Chapter 5. Ratios involving isomers, e.g., $HC_2NC/HC_3N \approx 0.003$, and branched species, e.g., $CH_3CN/HCN \approx 0.0003$ and $CH_3CN/HC_2N \approx 0.3$, provide further constraints on the models.

7.3.5.3 C/O Ratio

The elemental composition of AGB stars with very thin CSEs is preferably estimated using visual and near-IR high-resolution spectra combined with detailed stellar atmosphere models; Chapter 4. This procedure runs into difficulties as the opacity of the CSE increases. The elemental composition of the most extreme objects can be inferred only from circumstellar data. Elemental abundances are obtained from the molecular abundances only if the

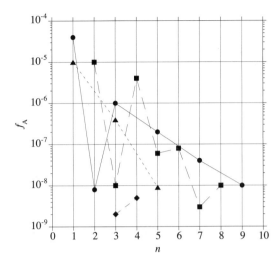

Fig. 7.11. Fractional abundances of carbon-chain species, HC_nN (circles), C_nH (squares), C_nH_2 (diamonds), and C_nN (triangles) in the CSE of the C-star IRC+10216

circumstellar chemistry is known in detail. In addition, the circumstellar gas-phase elemental composition may differ from that of the atmosphere due to selective depletion of atoms on grains.

The simplest problem is to distinguish between O- and C-rich objects. An example is provided by SiO and HCN radio line intensity ratios. In O-CSEs we expect SiO to be abundant, while only small amounts of HCN should be present, since most of the C is tied to CO. In C-CSEs we expect the opposite behavior. Both species are expected to respond in the same manner to different excitation conditions. It has been shown that $I_{\rm HCN}/I_{\rm SiO}$ ratios discriminate clearly and unambiguously between the two types of CSEs [31, 207].

Next in line stands a more detailed determination of the C/O ratio, which is a considerably more difficult problem. The effects of trends in the C/O ratio are likely to be much smaller than those due to whether C/O is <1 or >1. Unfortunately, one has to conclude that there is still a long way to go before circumstellar data can be used to obtain detailed information about the elemental composition.

7.3.5.4 Isotope Ratios

Another, and very important, side of the elemental composition concerns the isotopic ratios of various elements (see Chapter 2, and Chapter 5 and the discussion on meteoritic grains). Here the observational data are usually more easily converted into astrophysically interesting quantities, since one expects

Table 7.3. Isotope ratios in the CSE of IRC+10216

Isotope ratio	Ratio	IRC/Solar	Species used
$^{12}C/^{13}C$	45	0.5	CS, SiC$_2$
$^{12}C/^{14}C$	>63000		CO
$^{14}N/^{15}N$	5300	20	HCN
$^{16}O/^{17}O$	840	0.3	CO
$^{16}O/^{18}O$	1300	3	CO
$^{28}Si/^{29}Si$	>15	>0.8	SiS, SiC$_2$
$^{28}Si/^{30}Si$	>20	>0.7	SiS, SiC$_2$
$^{32}S/^{33}S$	>60	>0.5	SiS
$^{32}S/^{34}S$	22	1	CS, SiS
$^{35}Cl/^{37}Cl$	2.3	0.7	AlCl, NaCl, KCl
$^{24}Mg/^{25}Mg$	8	1	MgNC
$^{24}Mg/^{26}Mg$	7	1	MgNC

the corresponding lines from the isotopomers to respond in the same way to the physical conditions. In most cases one simply lets line intensity ratios also reflect the isotope ratios, i.e., $^nA/^mA = I(^nAB)/I(^mAB)$. Minor corrections for the small differences in beam size and transition probability are usually introduced. This method is fairly reliable, provided that (*i*) the observed lines from the two isotopomers correspond to the same transition, (*ii*) great care is taken in selecting optically thin lines, and (*iii*) there is no difference in the spatial distribution and in the excitation of the two isotopomers. To this we add that since the lines are weak, there is an increased possibility of blending with other lines, or even misidentifications.

Only in IRC+10216 does there exist a reasonable setup of isotope ratios; Table 7.3 [135, 136, 273]. The uncertainties are of order ±25% or less. Major deviations from the terrestrial values are found for N (^{15}N is highly underabundant) and O (^{18}O is underabundant and ^{17}O is overabundant). These data show the potential, but also the problem that the weakness of the lines severely limits the number of sources that can be studied.

In principle, one would expect the results for the $^{12}C/^{13}C$ ratio to be relatively good, since there is a fair number of detections of ^{12}CO and ^{13}CO in the same objects. However, there are two main problems. Isotope-selective photodissociation and fractionation are problematic in the case of CO, and it is expected that ^{12}CO and ^{13}CO are differently excited due to the differences in optical depths. Nevertheless, it has been shown in a detailed study, including a comparison with photospheric $^{12}C/^{13}C$ ratios, that reliable circumstellar $^{12}CO/^{13}CO$ ratios are possible to obtain, even though the corrections for optical depth effects may be substantial [226].

7.4 A "Standard" Dusty AGB CSE

In this section we introduce the properties of a "standard" dusty AGB CSE. The basic assumptions are the same as for the "standard" gaseous CSE; Section 7.2. In one sense, the radiative transfer of radiation through a CSE of dust particles is more complicated than the line radiative transfer. It is highly nonlocal, since there is no resonance condition that limits the interaction region. On the other hand, there is no complicated excitation analysis. We limit ourselves to thermal dust emission, scattering by dust was covered in Section 7.2.6.4.

7.4.1 The Energy Balance of the Dust

In an AGB CSE the dust temperature is locally determined by a balance between the absorption of the infalling radiation (at shorter wavelengths) and the subsequent reemission (at longer wavelengths). We assume that the infalling energy is distributed within the particle, and ignore other processes like gas impact and chemical reaction heating. The heating term depends on the total radiation that falls on a dust grain at a radius r; i.e., it depends on the radiative transfer through the entire CSE for all frequencies. Since this is a complicated problem, we present a simplified analysis in the optically thin regime for illustrative purposes.

7.4.1.1 Heating and Cooling Processes

The heating term is given by (assuming one-size, spherical grains)

$$H_{\text{abs}} = \int F_\nu(r) n_{\text{d}}(r) \pi a_{\text{gr}}^2 Q_{\text{a},\nu} \text{d}\nu, \tag{7.52}$$

where F_ν is the energy flux density due to the central source, and the cooling term is given by

$$C_{\text{em}} = \int n_{\text{d}}(r) 4\pi a_{\text{gr}}^2 Q_{\text{a},\nu} \pi B_\nu(T_{\text{d}}(r)) \text{d}\nu, \tag{7.53}$$

where B_ν is the blackbody brightness, T_{d} the dust temperature, and $Q_{\text{a},\nu}$ the dust absorption efficiency. Henceforth we will assume that the latter has a frequency dependence according to $Q_{\text{a},\nu} \propto \nu^s$. Note that when the central star is very hot, the absorption occurs at short wavelengths where $s \approx 0$ (the grains are larger than the wavelength), while s may take on a very different value for emission, which occurs at longer wavelengths.

7.4.1.2 The Grain Temperature Radial Distribution

In the optically thin case the incident radiation can be related directly to the flux of the central star. Equating the heating and cooling terms, and assuming that the dust absorption efficiency does not depend on temperature, gives

$$T_\mathrm{d}(r) = \frac{h}{k}\left(\frac{c^2}{8\pi h}\right)^{\frac{1}{4+s}} \left(\frac{\int_0^\infty \nu^s F_{*\nu} d\nu}{\int_0^\infty \frac{x^{3+s}}{e^x-1}dx}\right)^{\frac{1}{4+s}} \left(\frac{R_*}{r}\right)^{\frac{2}{4+s}}, \qquad (7.54)$$

where $F_{*\nu}$ is the energy flux density at the stellar surface [the integral in the denominator has the approximate value $(s+3)!$]. This expression can be further simplified if we assume that the star emits as a blackbody with a radius R_* and temperature T_*. Using this we get

$$T_\mathrm{d}(r) = T_* \, (R_*/2r)^{\frac{2}{4+s}}. \qquad (7.55)$$

The observational data suggest that $s \approx 1$, and hence for an optically thin CSE around an AGB star with $R_* = 3 \times 10^{13}$ cm and $T_* = 2700$ K we expect a radial distribution of the grain temperature according to

$$T_\mathrm{d}(r) = 500 \left[\frac{10^{15}}{r}\right]^{0.4} \mathrm{K}. \qquad (7.56)$$

The more optically thick the CSE becomes, the more of the stellar radiation is converted into longer-wavelength radiation close to the star, and the cooler we expect the dust envelope to be, but it appears that the radial functional dependence given by (7.54) applies also for thick CSEs [113].

We can use (7.56) and Wien's displacement law $\lambda_\mathrm{max} T \approx 2900\,\mathrm{\mu m\,K}$ to crudely estimate the radius at which the peak of the blackbody emission occurs at the wavelength λ. The result is

$$r_{\mathrm{d},\lambda} \approx 4 \times 10^{15} \left[\frac{\lambda}{10}\right]^{2.5} \mathrm{cm}, \qquad (7.57)$$

where λ is measured in µm. That is, for $\lambda = 10\,\mathrm{\mu m}$ and $1\,\mathrm{mm}$ the results are $r_{\mathrm{d},\lambda} \approx 4 \times 10^{15}$ cm and $\approx 4 \times 10^{20}$ cm, respectively. Although this is not the same as the radius from which most of the flux at the wavelength λ originates, it is clear that dust emissions at different wavelengths probe different parts of the CSE. This, combined with the fact that the grains are difficult to destroy, means that dust emission can be used as an important probe of the long-term mass-loss history.

7.4.2 Spectral Energy Distribution

We here introduce results that apply in the optically thin limit for a standard dust CSE. We start by deriving the frequency-dependence of the luminosity, i.e., the *spectral energy distribution* (SED; observed as $F_\nu = L_\nu/4\pi D^2$) from a dust CSE. This is given by

$$L_{\mathrm{d},\nu} = \int_{R_\mathrm{i}}^{R_\mathrm{e}} 4\pi r^2 n_\mathrm{d}(r) 4\pi a_{\mathrm{gr}}^2 Q_{\mathrm{a},\nu} \pi B_\nu(T_\mathrm{d}(r))\, \mathrm{d}r\,, \qquad (7.58)$$

where the number density distribution for the grains is obtained from $n_\mathrm{d}(r) = \dot{M}_\mathrm{d}/\rho_{\mathrm{gr}} 4\pi v_\mathrm{d} r^2$, and the grain temperature from (7.55), $T_\mathrm{d}(r) = T_0(R_0/r)^{2/(4+s)}$. This expression can be rewritten as

$$L_{\mathrm{d},\nu} = \frac{4\pi(4+s)h}{c^2 \nu_0^s} \left(\frac{kT_0}{h}\right)^{\frac{4+s}{2}} \frac{\chi_{\mathrm{gr},0}\dot{M}_\mathrm{d} R_0}{v_\mathrm{d}} \nu^{\frac{2+s}{2}} I_{\mathrm{bb}}(y_\mathrm{i}, y_\mathrm{e}, s)\,, \qquad (7.59)$$

where

$$I_{\mathrm{bb}}(y_\mathrm{i}, y_\mathrm{e}, s) = \int_{y_\mathrm{i}}^{y_\mathrm{e}} y^{\frac{2+s}{2}} (e^y - 1)^{-1}\, \mathrm{d}y\,, \qquad (7.60)$$

where we have introduced the grain cross section per unit mass $\chi_{\mathrm{gr},\nu} = 3Q_{\mathrm{a},\nu}/4a_{\mathrm{gr}}\rho_{\mathrm{gr}}$ [$\chi_{\mathrm{gr},\nu} = \chi_{\mathrm{gr},0}(\nu/\nu_0)^s$], and the limits of the integral are given by $y_{\mathrm{i,e}} = (R_{\mathrm{i,e}}/R_0)^{2/(4+s)}(h\nu/kT_0)$. $I_{\mathrm{bb}}(0,\infty,s)$ is ≈1.8 for $s=1$, and depends only weakly on s, but in general, it has a frequency-dependence. For intermediate frequencies, where $y_\mathrm{i} \to 0$ and $y_\mathrm{e} \to \infty$, we obtain $L_{\mathrm{d},\nu} \propto \nu^{(2+s)/2}$, while for low frequencies, where y_i and y_e are small and only the external part of the CSE contributes, we get $L_{\mathrm{d},\nu} \propto \nu^{2+s}$, and for high frequencies, where y_i and y_e are large and only the inner part contributes, we get $L_{\mathrm{d},\nu} \propto \nu^{2+s} e^{-h\nu/kT_\mathrm{d}(R_\mathrm{i})}$. Thus, with $s=1$ the SED increases steeply at low frequencies (the Rayleigh–Jeans regime), ν^3, has a more moderate increase at intermediate frequencies, $\nu^{1.5}$, and a steep decline at high frequencies (the Wien regime); Figure 7.12. The optically thin SED is broader than a single-temperature blackbody, and it peaks at the frequency given by Wien's displacement law for the temperature at the inner radius. An interesting aspect of the steep decline at low frequencies is that eventually the stellar photosphere will start to dominate again, since its flux density decreases as ν^2. This effect is evident for low-mass-loss-rate objects [272].

Optical depth effects will become increasingly important the higher the frequency, since χ_{gr} increases with frequency; i.e., we can expect the sub-mm emission to be optically thin even at high mass-loss rates. The details of the SED depend on $\chi_{\mathrm{gr},\nu}$, which in turn depends on the dust composition (Section 7.5.1) and the grain-size distribution (Section 7.5.4.2). The latter is important because the absorption and scattering (which we have ignored here) strongly depend on the grain size. To analyze this in full detail requires sophisticated numerical models, and pioneering work on circumstellar SED modeling was done by [216]. A dust radiative transfer code is now publicly available, DUSTY [114]. Figure 7.13 shows observed spectral energy distributions for optically thin and thick AGB CSEs.

We also need expressions for the flux density from extended emission observed with an instrument having an angular-dependent sensitivity. With

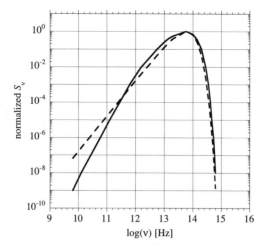

Fig. 7.12. Expected spectral energy distribution for optically thin circumstellar dust emission with a dust temperature at the inner radius of 1000 K (solid line). The dashed line shows a blackbody curve for the same temperature

the same assumptions as above, the observed flux density through a circular point spread function $P(\rho)$ at an offset p is given by

$$S_{d,\nu}(p) = \frac{\chi_{gr,\nu} \dot{M}_d}{4\pi v_d D^2} \int B_\nu(T_d(r)) P(p, p', \theta) \frac{p'}{r^2} \, dp' \, d\theta \, dz . \quad (7.61)$$

To get an analytical solution we have to make some simplifying assumptions: (i) the Rayleigh–Jeans approximation is valid, (ii) the beam is boxcar-shaped with a full width θ_b, (iii) the dust temperature distribution is given by (7.55), (iv) only emission from inside the radius $\theta_b D/2$ is considered, and (v) the inner radius is negligible. In this case the result toward the center of the CSE is

$$S_{d,\nu}(0) = \frac{(4+s) 2k\nu^2}{(2+s) c^2} \frac{\chi_{gr,\nu} \dot{M}_d}{v_d D^2} T_0 R_0 \left(\frac{\theta_b D}{2 R_0} \right)^{\frac{2+s}{4+s}} . \quad (7.62)$$

It is important to realize that the observed flux densities are dependent on the angular resolution of the observation, and hence corrections for this may have to be done in producing, e.g., SEDs. At sufficiently large offsets we may ignore the variation of the dust temperature along the line of sight as well as over the beam, and the result is (for $s = 1$)

$$S_{d,\nu}(p) \approx 0.06 \, S_\nu(0) \, (\theta_b D/p)^{1.4} ; \quad (7.63)$$

i.e., the emission is sharply peaked toward the center of the CSE, and the decline with offset goes as $p^{-1.4}$.

In case the Rayleigh–Jeans approximation is not applicable, the calculations become more complicated, but a useful expression is obtained using (7.59) in the case that the source is much smaller than the beam,

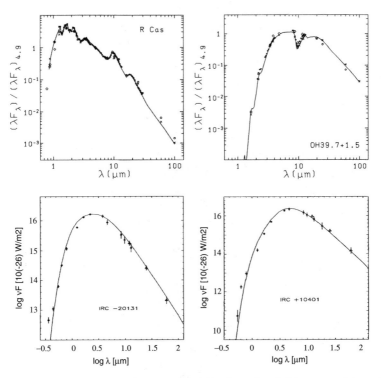

Fig. 7.13. Observed spectral energy distributions for AGB stars with optically thin and thick O-CSEs (upper panels: R Cas with $\tau_{10\,\mu m}=0.2$, and OH39.7+1.5 with $\tau_{10\,\mu m}=10$; adapted from [12]) and C-CSEs (lower panels: IRC–20131 with $\tau_{1\,\mu m}=0.8$, and IRC+10401 with $\tau_{1\,\mu m}=5$; adapted from [161])

$$S_{d,\nu}(0) = \frac{(4+s)h}{c^2} \frac{\chi_{gr,\nu} \dot{M}_d R_0}{v_d D^2} \left(\frac{kT_0}{h}\right)^{\frac{4+s}{2}} \nu^{\frac{2-s}{2}} I_{bb}(y_i, y_e, s). \quad (7.64)$$

We will use this expression to derive a simple formula for estimating the dust mass-loss rate in Section 7.7.2.1.

7.4.3 Spectral Signatures

The dust particles do not only provide broad-band emission/absorption due to internal collective motions of the atoms. The possibility of groups of atoms in the lattice to perform various vibrational motions leads to spectral signatures, particularly in the wavelength range 2–100 μm. As examples we mention the Si–O stretch and O–Si–O bending modes, that give rise to features at ≈ 10 and 18 μm, respectively, in O-CSEs; Figure 7.14. These features are much broader than atom/molecule spectral lines, typically $\lambda/\Delta\lambda \approx 10$, since the atomic motions are strongly distorted by the environment, which in turn

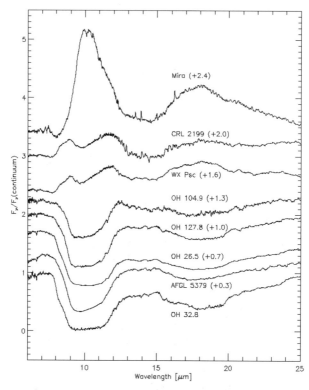

Fig. 7.14. ISO SWS continuum-divided spectra showing the 10 and 18 μm features, attributed to amorphous silicates, in O-CSEs of different opaqueness. From [243]

depends on the degree of order in the lattice structure (from amorphous with broad features to crystalline with "narrow" features). These spectral signatures are vital for identifying the composition of the grains. In Table 7.4 we list some of the best known dust features and their suggested origins. However, one should note that different particles may nevertheless show very similar features, and unique identifications are therefore difficult.

There is also a grey zone between large molecules and solid particles, but no clear demarcation of what is what exists. Based on the radiative properties, i.e., well-defined spectral features or bands, a reasonable dividing line lies at ≈ 100 atoms and a size of ≈ 10 Å. A number of relatively narrow features, e.g., at 3.3, 3.4, 6.2, 7.7, 8.6, and 11.3 μm, in post-AGB CSEs suggests that large molecules are present, e.g., polycyclic aromatic hydrocarbons (PAHs) based on the benzene ring [14].

The different dust signatures require different grain temperatures to become excited and hence visible; e.g., a feature at 10 μm requires about 300 K. In particular, the rich spectrum of crystalline dust appears to be a good temperature probe [28, 27]. In addition, individual features, such as the well-

Table 7.4. Dust spectral signatures in AGB CSEs

Feature [μm]	Identification
3.1	stretching of O–H bond in amorphous H_2O ice, O-CSEs
9.7	stretching of Si–O bond in amorphous silicate, O-CSEs
11	amorphous Al_2O_3(?), O-CSEs
11.3	phonon mode in SiC lattice, C-CSEs
13	spinel(?), O-CSEs
15–50	>40 features in crystalline silicates, such as olivines (e.g., forsterite) and pyroxenes (e.g., enstatite), O-CSEs
17	spinel(?), O-CSEs
18	bending of O–Si–O bond in amorphous silicate, O-CSEs
19.5	magnesiowustite (Mg,Fe)O(?), O-CSEs
20	TiC(?), "21 μm feature," post-AGB C-CSEs
30	MgS(?), peaks in the range 26–33 μm, C-CSEs
31.8	spinel(?), O-CSEs
43	crystalline H_2O ice, O-CSEs
62	crystalline H_2O ice, O-CSEs
62	dolomite, PN
92	calcite, PN

known 10 μm feature, may show a variety of different shapes [229]. It is not clear whether this is due to compositional variations, different condensation histories, a temperature effect, or combinations thereof. In summary, the dust emission contains a wealth of very interesting information, which is not easily extracted.

7.5 Circumstellar Dust Observations

In much the same way as is done for line emission we summarize in this section the observations of dust continuum and feature emission from AGB CSEs.

7.5.1 Dust Compositions

The composition of the circumstellar dust is reflected in the SED, although in such a way that it is not always easy to disentangle. Nevertheless, this is an important problem to solve, since the dust composition not only carries information on the chemistry of the central star, but also on the physical/chemical conditions in the stellar atmosphere, and, in particular, in the region where the mass loss is initiated. Of particular importance to the identification are the dust spectral signatures; Table 7.4 and Figure 7.14. SEDs that stretch over more than two orders of magnitude are now available [243].

The gross characteristics of the SEDs can be explained using "dirty" (clean silicates are too transparent, but a small contamination with, e.g., Fe or Al may be enough) amorphous silicates for O-CSEs (including the 10 and 18 μm features) and amorphous carbon for C-CSEs (featureless) [113]. There appears to be no significant difference between the dust emission from CSEs of S-stars and O-CSEs [122]. In addition, the 11.3 μm emission feature observed in C-CSEs can be well modeled if a small amount of SiC dust is added, of order 10% [175].

As instruments improved, new features of dust origin were identified, and we are slowly entering a new era in our study of the dust composition, a mineralogy, at least in O-CSEs. A 13 μm emission feature, present in many O-CSEs [237], has not yet been conclusively identified, but a probable carrier is spinel, $MgAl_2O_4$ (i.e., a nonsilicate), which would also explain tentative features at 17 and 32 μm [68]. Quite a number of relatively narrow features beyond 20 μm in O-CSEs suggests the presence of crystalline material, e.g., silicates like pyroxenes and olivines [187, 276]. The relation between circumstellar amorphous and crystalline grains is certainly not understood, but it appears that the fraction of the latter increases with $\dot{M}$, although this may be only an apparent effect due to the lower temperature of the crystalline dust [145]. There are also about 20 C-rich objects that show crystalline silicate features, but these are attributed to an O-rich dust disk [156, 188, 291]. We note here that crystalline features have not yet been detected in the ISM, but around young stellar objects. In cold O-CSEs features appear at ≈43 and 62 μm, which are attributed to crystalline ice.

The carriers of the 21 μm feature in post-AGB C-CSEs, and the 30 μm feature in AGB and post-AGB C-CSEs have not yet been identified, although TiC and MgS may play a role for the former [270] and the latter [119], respectively. The 21 μm feature appears to have an intrinsic profile and a peak wavelength of 20.1 μm [267], while the peak of the 30 μm feature varies [268], possibly as a function of the evolutionary stage of the object. The properties of dust in post-AGB objects are discussed also in Chapter 10.

A disturbing problem, but also an interesting challenge, is the not unreasonable expectation that the dust properties will change as a star evolves, and perhaps even differ from star to star with reasonably similar characteristics due to the sensitivity of the grain formation to the physical conditions. However, this is a difficult problem, and detailed models are required to sort out effects of composition, grain size, optical depths, etc.

7.5.2 Surveys

As opposed to the case for the line surveys, there exist a number of full-sky, or at least part of it, surveys for continuum emission. The first step came with the Two Micron Sky Survey (listing about 5000 objects [194]); there is also a southern sky extension [67]), which among its important results provided the first detection of the extreme C-star IRC+10216 [11]. This revealed that

AGB-stars can be heavily dust-enshrouded, and essentially invisible at optical wavelengths. Eventually, this led to the important IRAS mission, a mid- to far-IR all-sky survey [193], resulting in the Point Source Catalogue with measurements at 12, 25, 60, and 100 µm (about 10^5 objects are estimated to be AGB or post-AGB objects). Follow-up studies of IRAS-sources in the near-IR were made [72, 97, 146, 147]. ISO was not a survey instrument, but its archive contains a large amount of data, where also the 100–200 µm band is covered. Also, the ISOGAL survey, which sampled the Bulge and the inner Galactic disk, was performed with ISO [209].

In the case of CSEs, two-band, and even better three-band, surveys are particularly important. There are two major surveys in the near-IR in progress: the Two Micron All Sky Survey (2MASS; [150]) and the DEep Near-Infrared Survey (DENIS; [65]). They are mapping the sky in three wavelength bands; J (1.25 µm), H (1.65 µm), and K_s (2.15 µm) for 2MASS, and I (0.8 µm), J, and K_s for DENIS. These surveys should jointly detect essentially all AGB stars in the Galaxy ($\approx 2 \times 10^5$ sources) and in the Magellanic Clouds (DENIS data contain about 33,000 and 7,700 AGB stars in the LMC and SMC, respectively) [47]. In the pipeline are also a number of space missions, e.g., SIRTF, SOFIA, Herschel, and NGST.

Low-resolution spectroscopic information in the range about 8–23 µm is available in the IRAS Low Resolution Spectrograph data base [200], and ISO, though not in survey mode, covered the range 2–200 µm at different spectral resolutions.

7.5.3 Brightness Distributions

The angular resolutions required to study the dust brightness distributions in the inner parts of AGB CSEs can in general be obtained only with interferometric techniques. The first such observations were done in the range 5–13 µm using two subapertures on a single telescope [183], and by a heterodyne system working at 10 µm and two small telescopes [241]. Also speckle interferometry was used quite early to obtain one-dimensional brightness distributions [58, 73].

The instruments and the observational methods have improved significantly over the years, and a number of interesting results have been obtained. The methods of heterodyne-interferometry, speckle and aperture-masking imaging at near- to mid-IR wavelengths have produced detailed images of dust in the near vicinity of a few stars. Further results from observations of dust brightness distributions are discussed in Section 7.6.1.

7.5.4 Dust Properties

We have in Section 7.5.1 already discussed the dust compositions that have been more or less well identified. In this section we discuss other important dust characteristics such as the dust-to-gas mass ratio, the grain-size distribution, and the ability to use the dust emission to determine the circumstellar C/O ratio.

7.5.4.1 Dust-to-Gas Mass Ratio

Despite its importance for the study of CSEs, one should realize that the dust constitutes only a minor fraction of the total mass. There is a physical upper limit to the dust-to-gas mass ratio, ψ, which is given by the number of atoms that can condense into grains. In the case of 100% carbon grains the limiting factor is given by the amount of available C after CO formation. This results in

$$\psi \leq 12\, X \left(\frac{C}{O} - 1\right) \frac{O}{H} \frac{v_d}{v_\infty}. \qquad (7.65)$$

With C/O < 4, O/H = 5×10^{-4}, $X = 0.7$, and $v_d/v_\infty = 1.5$, the result is $\psi \lesssim 0.02$, but note that C/O ratios as low as 1.1 are quite common [158], and hence we expect $\psi \lesssim 0.001$. For O-CSEs, where the silicates dominate, the situation is somewhat more complicated, but, e.g., forsterite, Mg_2SiO_4, can result in $\psi \lesssim 0.005$ if all the Mg is condensed into this mineral.

Estimates of reliable dust-to-gas mass ratios are rare, but have been obtained by modeling simultaneously the dust emission and the dynamics of a dust-driven wind; see Section 7.7.2.3. For a sample of C-CSEs a trend in ψ was found in the sense that the lower luminosity objects ($L \lesssim 8000\, L_\odot$) have $\psi \approx 0.0025$, while for the higher luminosities it increases to ≈ 0.015 [92]. In a study of three thick O-CSEs; ψ-values in the range 0.003 to 0.006 were obtained [133].

7.5.4.2 Grain Sizes

Also the grain-size distribution gives important information on the physical/chemical conditions in the upper stellar atmosphere and inner CSE (e.g., [59, 155]). In addition, it is of great interest to compare with interstellar grain-size distributions and the properties of presolar grains (mainly in the form of carbon grains), since presumably the AGB grains play an important role as seeds for further processing. However, to determine the grain-size distribution is even more difficult than identifying the composition of the grains. The main difficulty is that this requires a synthesis of observations at different wavelengths and of different processes (e.g., absorption and scattering efficiencies depend strongly, and differently, on the grain size).

Assuming, to a first approximation, that the grains are spherical, the following expression is often used to give the number of grains with the radius a_{gr},

$$n(a_{gr})\mathrm{d}a_{gr} = n_0\, a_{gr}^{-\alpha}\, \exp(-a_{gr}/a_{gr,0})\, \mathrm{d}a_{gr}, \qquad (7.66)$$

where $a_{gr,0}$ gives an estimate of the largest grain size. The best current result for interstellar silicate grains is obtained with $\alpha = 3.0$ and $a_{gr,0} = 0.14\,\mu\text{m}$ and for interstellar carbon grains with $\alpha = 3.5$ and $a_{gr,0} = 0.28\,\mu\text{m}$ [149]. Similar studies of O- and C-CSEs resulted in $\alpha = 3.0$ and $a_{gr,0} \leq 0.14\,\mu\text{m}$, and $\alpha = 3.5$ and $a_{gr,0} \approx 0.1\,\mu\text{m}$ apply, respectively [125, 126].

It is hard to put a lower limit on the grain sizes, since the distinction between a large molecule and a solid particle is not clear. The presence of relatively narrow features in the mid-IR in post-AGB CSEs suggests that large molecules, e.g., PAHs, are present, although the excitation is not enough to make them visible in AGB CSEs [14]. Larger grains than $\approx 0.2\,\mu\text{m}$ certainly exist, but probably not as a consequence of grain formation in stellar atmospheres. Rather, they seem to have their origin in circumstellar disks, where there is time enough to allow considerable grain growth [131, 274].

In reality, things are even more complicated than this. Not only the grain size is important, but also the three-dimensional form of a grain affects its radiative properties [190, 215].

7.5.4.3 C/O Ratio

The dust emission is not suitable for estimating the elemental composition of a CSE, but with considerable efficiency it can be used for discriminating whether C/O is <1 or >1. This can be achieved by looking for characteristic dust signatures, but the main tool in this context is color–color diagrams.

The color–color diagram based on the IRAS [12–25] and [25–60] colors is a very important working tool and certainly produces a discrimination between O- and C-CSEs, but there is considerable confusion in some areas of the diagram [255]. A combination of mid-IR color (e.g., [12–25] or [K–12]) and a near-IR color (e.g., [H–K] or [K–L]) has turned out to be more effective, since the O-CSEs show a larger excess in the 10–20 μm range, relative to the 1–5 μm range, than the C-CSEs; Figure 7.15 [97]. Also in a near-IR color–color diagram, such as will be obtained by the ongoing near-IR surveys, there is a separation between O- and C-CSEs, although less distinct than in the near- and mid-IR counterparts.

7.6 Geometry and Kinematics of AGB CSEs

The geometry, i.e., the three-dimensional density distribution, and the kinematics of AGB CSEs have implications for our understanding of the mechanism that produces the mass loss, and for the geometry of the successors, the PNe. These properties are also, to a first approximation, related to each other via the continuity equation (7.5).

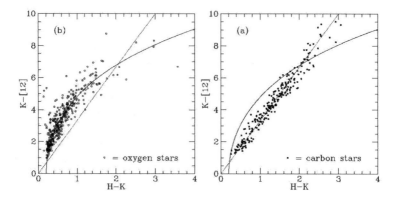

Fig. 7.15. Color–color diagrams based on the [H–K] and [K–12] colors for a sample of O-CSEs (left), and a sample of C-CSEs (right). From [262]

7.6.1 Geometry

The determination of the geometry is not a simple observational task, since the conversion from a 2D brightness distribution to a 3D density distribution can be quite complicated. Here, radiative (e.g., saturation, maser action), excitation, chemistry, and observational (e.g., the lack of interferometer sensitivity to extended emission) effects play an important role. Indeed, these complications in combination with a lack of observational data with sufficiently high angular resolution have the effect that our knowledge of the geometry of AGB CSEs is limited. We begin by discussing the evidence in favor of overall spherical symmetry. This is followed by a presentation of data that suggest a more complicated geometry. It is fair to say that no consensus has been reached.

7.6.1.1 Evidence of Overall Spherical Symmetry

In studies of the region close to the star we are restricted to interferometric observations of molecular maser line emission and near- to mid-IR dust emission. It is for various reasons difficult to draw any definite conclusions about the geometry and kinematics of CSEs from these data. The observations are often performed with instruments consisting of relatively few individual elements, which limits the quality of the images. In addition, the interferometers are mainly sensitive to gradients in the brightness distribution, and extended emission may be missed. For maser emission we have the additional problem of substantially enhanced emission from regions of large optical depth. Nevertheless, rather sophisticated models are now being developed in order to interpret the maser data ([110, 111]).

The evidence at small scales, i.e., $\leq 10^{15}$ cm, is in favor of an essentially isotropic mass loss, although not in the form of a smooth wind. Spectacular

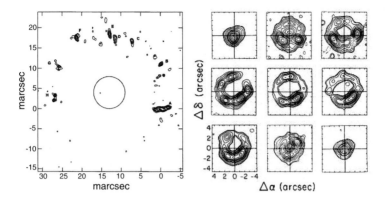

Fig. 7.16. Left: An SiO ($v=1$, $J=1 \to 0$) image of the M-star TX Cam (VLBA; from [56]). The circle is an estimate of the stellar size, and it is placed at the center of the maser emission. **Right:** OH 1612 MHz brightness distributions, in selected velocity intervals, toward the OH/IR-star OH26.5+0.6 (VLA; adapted from [25]). This produces slices in planes, orthogonal to the line of sight, through the OH shell along the line of sight. The upper leftmost panel shows emission from molecules moving directly toward us, while the lower rightmost panel shows emission from molecules moving directly away from us

results have been obtained on SiO masers, which are located within a few stellar radii of the star. The images reveal roughly circular ring structures of diameters $\approx 10^{14}$ cm in a region that probably lies inside the dust CSE; Figure 7.16. Proper motions related to motions in the upper atmosphere have been found [21]. The H_2O and OH maser data lead to less conclusive results, but do not obviously suggest major departures from sphericity in the inner regions [45, 51, 244, 245]. Important to note is that at these small scales ($\lesssim 0''.1$) the relative position between the star and the maser emission is often not accurate enough, and the location of the star, as well as its size, has to be assumed. Toward the M-star W Hya the position and size of the star were measured simultaneously with the H_2O maser emission (at 22 GHz), and the result is an approximately spherical shell, ≈ 3 stellar radii in radius, centered on the star [211]. In this context we mention the possiblity of using maser emission to estimate the magnetic field as a distance from the star [144, 244, 265], and to do stellar parallax measurements of also highly obscured stars [259, 266]. So far, no useful line probes for C-CSEs at these small scales are available. Thus, it is difficult to say whether there are any differences in the structure and kinematics of the innermost regions of O- and C-CSEs.

At larger scales, $\gtrsim 10^{16}$ cm, the OH (1612 MHz) maser line outlines, more or less, overall spherical shells in about 15 O-CSEs, but with considerable patchiness in the brightness distributions. The highest-quality data are those

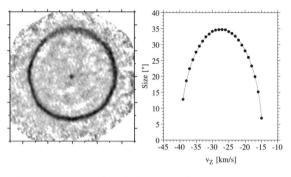 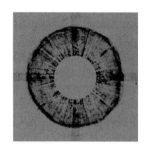

Fig. 7.17. Left: A CO ($J=1\rightarrow 0$) map of the carbon star TT Cyg in a narrow velocity interval centered at the systemic velocity, $\approx -27\,\mathrm{km\,s^{-1}}$ (left; IRAM PdB interferometer). The brightness distribution is very well fitted by a circular ring, of diameter 68″, with a Gaussian distribution in the radial direction. **Middle:** The radii of such rings fitted to the brightness distributions in 25 velocity intervals of $1\,\mathrm{km\,s^{-1}}$ width, and a fit of relation (7.67) to these data (see text for details; adapted from [202]). **Right:** Stellar light scattered in a circumstellar shell (86″ in diameter) around the carbon star U Ant (the star lies at the center behind the artificial obscuration; from [82])

of OH 26.5+0.6; Figure 7.16 (the shell may be somewhat displaced with respect to the star, an effect likely due to an anisotropic, external radiation field). The evidence for sphericity is further strengthened by good fits to the data of the angular size versus line-of-sight velocity relation (applicable to a spherical shell expanding at the constant velocity v_∞)

$$\theta(v_z) = \theta(v_*) \left[1 - (v_z - v_*)^2/v_\infty^2\right]^{1/2}, \qquad (7.67)$$

where $\theta(v_z)$ is the measured size in a narrow velocity range around v_z, and $\theta(v_*)$ is the size at the systemic velocity [105].

At roughly the same size scale crude interferometer maps of CO radio line emission toward ≈ 45 objects (O- and C-CSEs) show geometries that are in general consistent with overall spherical symmetry [191]. At slightly larger scales, $\gtrsim 5\times 10^{16}$ cm, about 25 CSEs have now been mapped in various CO radio lines using single telescopes. At these resolutions, where the objects are just barely resolved and the dynamic ranges of the maps are limited, the results are reasonably consistent with overall spherical symmetry [30, 137].

Observations of four C-stars in CO radio line emission clearly reveal spatially resolved CSEs in the form of geometrically thin shells at the scale 10^{17}–10^{18} cm [171, 202]. These have a remarkable overall spherical symmetry, e.g., in the case presented in Figure 7.17 the deviations from a spherical form are less than a few percent. The results are strongly corroborated by images of such shells in scattered stellar light [81, 82], and in deconvolved IRAS images of dust emission [117, 275].

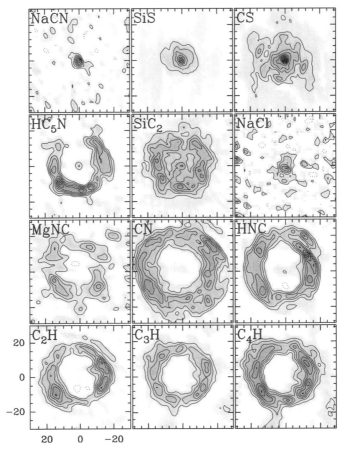

Fig. 7.18. Molecular line brightnesses, in a narrow velocity interval centered on the systemic velocity, observed toward IRC+10216 (IRAM PdB interferometer; arcsec scale). The observed lines all lie around 100 GHz. From [95]

Only the C-CSE of IRC+10216 has been observed at high-angular resolution in emission from a fair number of molecular species, ≈15. The high-quality data presented in Figure 7.18 sample the geometry at different size scales (and show that species like NaCN, NaCl, CS, and SiS, as expected, are of photospheric origin in a C-rich object, while the others are produced in the CSE, e.g., CN by the photodissociation of HCN). Figure 7.19 shows images of the IRC+10216 CSE at different size scales. Quite remarkable are the images of multiple-shell structures, in the form of essentially circular arcs superposed on the, apparently circularly symmetric, reflection nebulosity, in dust scattered light [182]. The combined picture is a CSE with an overall spherical symmetry on large scales, with possible evidence of an opening of the CSE along an axis, but asymmetries in the CSE have been found at small

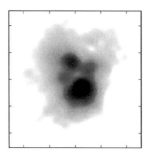 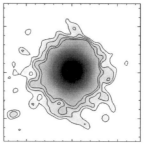 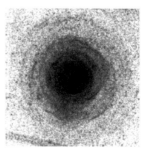

Fig. 7.19. Images of the CSE of the C-star IRC+10216 at 2.15 μm (left; 1″ square; from [278]), at 1.3 mm (middle; 2″.5 square; from [91]), and in the B+V filters (right; 2″.2 square; from [182])

scales; Section 7.6.1.2. This may indicate that IRC+10216 is approaching the tip of the AGB.

As outlined in Section 7.4.1.2, long-wavelength dust emission is potentially a very good probe of the mass-loss history and the CSE geometry. This normally requires the detection of weak, extended emission. The IRAS 60 μm images can provide relevant data, but the angular resolution is low. Nevertheless, ≈15% of the sources in a sample were, at least partly, resolved [294]. Some individual objects have been clearly resolved, e.g., the M-star W Hya, where the brightness distribution appears essentially circularly symmetric [104].

Of relevance here is also the fact that the remnant AGB CSEs of young post-AGB objects possesses overall spherical symmetry in the CO line emission [18, 249], and circular multiple-shell structures in scattered light are found also around this type of objects [157, 223]. Data from ISO show the existence of very extended, apparently spherical, dust CSEs around post-AGB objects [239], and numerous PNe have spherical halos [7].

7.6.1.2 Evidence of Departures from Overall Spherical Symmetry

There certainly exists evidence, of varying quality, for departures from overall spherical symmetry in AGB CSEs. Imaging of dust emission has led to impressive results on IRC+10216; Figure 7.19a. Within 10 stellar radii the dust distribution is highly inhomogeneous, although an axial symmetry is probably present, and proper motions of individual dust clumps have been detected [253, 279]. These structures are traced on larger scales through coronographic imaging [189]. The axial symmetry is seen also in polarimetric images, at a scale of ≲10″ [140]. Similar studies of a few O-CSEs certainly suggest asymmetries, clumpiness, and even time-variable mass loss, but with the present data it is difficult to obtain unique solutions for the structures [107, 174]. The presence of infrared polarization in a number of cases suggests departures from sphericity [121].

In the only sample of CO radio line interferometer data there is a significant fraction (≈30%) of CSEs with indications of nonspherical geometries [191]. They appear to have an inner asymmetric component surrounded by a larger symmetric envelope. There are CO data suggesting bipolar outflows in a handful of objects and elongations in a few cases [138, 152, 153]. In particular, the o Ceti data are convincing, and they outline a spherical CSE disrupted by a bipolar outflow [124]. In this case, the companion probably plays an important role in the shaping of the CSE [180]. Also, about ten objects with definitely peculiar line profiles exist. They are best described as a narrow central feature centered on a broader component [148, 154]; see Figure 7.3f. Only in one case do high-angular-resolution data exist, and they suggest a disk-like gas distribution [15]. In addition, a few cases are characterized by very narrow CO radio line profiles, and emission from a circumbinary disc is suggested [127]; see also Chapter 9.

Finally, the message from studies of post-AGB objects and PNe is clear: At some point a marked axisymmetry must develop; see Chapter 10.

7.6.1.3 Distance from OH 1612 MHz Maser Emission

We end this section with a discussion of an interesting property of the OH 1612 MHz emission. Most of this emission comes from regions in the front and the rear parts of the CSE along the line of sight through the star, i.e., where the amplification path along the line of sight has maxima. The maser intensity varies along with the pump line, and hence with the light variations of the central star [256]. This has the interesting effect that due to the light travel time, the emission from the back of the CSE lags that from the front by a time that depends on the diameter of the OH shell, usually about a few weeks. Hence, a combination of the linear diameter measured in this way and the angular diameter gives an accurate method for estimating the distances to these stars (often better than 10%; [258]).

7.6.2 Kinematics

The information on the kinematics is much better than that on the geometry, since the gas expansion velocity can be estimated directly from the line profile. However, a problem is the lack of complete samples as a basis for the determination of terminal velocity distributions. Studies of the acceleration region and the 3D kinematical structure suffer from the same lack of enough angular resolution as described for the study of the geometry.

7.6.2.1 Terminal Velocity Distributions

A number of quite extensive studies exist today that provide terminal velocities for AGB CSEs with different characteristics. In Figure 7.20 we have

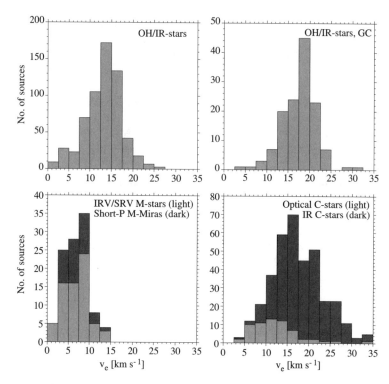

Fig. 7.20. Gas expansion velocity distributions for selected samples of O- and C-CSEs (see text for data references)

collected some of the results: (*i*) a compilation of stellar OH masers [46], (*ii*) a Galactic Center OH maser survey [173], (*iii*) CO surveys of M-type irregular, semiregular, and short-period Mira variables [148, 292], and (*iv*) CO surveys of bright carbon stars [204] and infrared carbon stars [89].

In all cases the distributions are rather sharply peaked. The result from the stellar OH maser sample is a v_e-distribution for O-CSEs with a median velocity of $13.6\,\mathrm{km\,s^{-1}}$ (935 objects), which stretches about an order of magnitude ≈ 3 to $30\,\mathrm{km\,s^{-1}}$. The CO surveys of M-type variables pick up the O-CSEs with lower expansion velocities; the medians are $7.0\,\mathrm{km\,s^{-1}}$ for both the irregular/semiregular (69 objects) and Mira (36 objects) samples. The Galactic Center sample is clearly dominated by high-v_e sources, a median of $18.1\,\mathrm{km\,s^{-1}}$ (133 objects). The same trend is seen for the C-CSEs; the sample dominated by irregular/semiregular variables has CSEs with lower expansion velocities (median $11.4\,\mathrm{km\,s^{-1}}$, 61 objects), while the infrared objects sample the high-v_e end (median $17.8\,\mathrm{km\,s^{-1}}$, 309 objects; the classification as C-object is more uncertain for this sample). The impression is that given the same mass-loss rate and pulsational behaviour, a C-CSE will have a higher expansion velocity than an O-CSE. In particular, about 30% of the M-type ir-

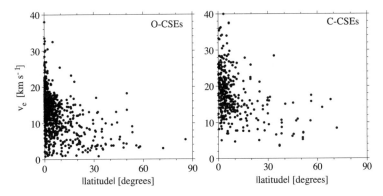

Fig. 7.21. Gas expansion velocity versus the absolute value of the galactic latitude for O and C-CSEs (see text for data reference)

regular/semiregulars have velocities lower than $5.0\,\mathrm{km\,s^{-1}}$; the corresponding value among the bright C-star irregulars/semiregulars is only about 10%.

7.6.2.2 Terminal Velocity Dependence on Mass and Metallicity

There is rather good evidence that part of the spread in v_e is due to the stars having different masses. In the above sample of Galactic Center OH/IR-stars (*ii*) the latitude distribution of the high-v_e sources is more concentrated to the Galactic plane than that of the low-v_e sources [170]. In general, O- and C-CSEs with high v_e tend to lie closer to the Galactic Plane than the low-v_e sources, as shown in Figure 7.21, where the results for the samples (*i*) and (*iv*) are shown. Since we expect the more massive stars to be concentrated to the plane, these findings suggest that there is a tendency for v_∞ to increase with main-sequence mass, probably as an effect of increasing luminosity.

Another indication of a mass dependence is provided by the correlation between v_e and pulsation period for both O- and C-CSEs in the range 300^d–600^d; Figure 7.22 (the period is very likely positively correlated with the stellar mass; Section 7.8.3.2). There is no obvious difference between the O- and C-CSEs, but two other interesting results on the O-CSEs are apparent. The semiregulars have the same velocity distribution as the shorter-period Miras, despite having periods shorter by a factor of about two. Also, it appears that for periods longer than 600^d the expansion velocity levels out at $\approx 20\,\mathrm{km\,s^{-1}}$.

There is some evidence of a v_e dependence on metallicity, as is expected from the dependence of v_∞ on ψ for a dust-driven wind [relations (7.2) and (7.3)]. The four AGB OH/IR-stars detected in the lower-metallicity LMC have expansion velocities close to $10\,\mathrm{km\,s^{-1}}$, i.e., low, although not excessively so, when compared with similar Galactic stars [297]. An estimate in the other direction can be obtained from Galactic Center stars, which presumably have

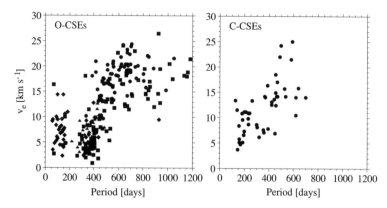

Fig. 7.22. Gas expansion velocity versus pulsational period for different samples of O-CSEs [Semiregulars (diamonds), short-period Miras (triangles), OH/IR-stars (squares), Galactic Center OH/IR-stars (circles)], and a sample of C-CSEs (Semiregulars and Miras) (see text for data references)

higher metallicities (by a factor $\gtrsim 2$) than stars in the solar neighborhood. Indeed, Galactic Center stars with periods longer than 300^d have higher v_e than otherwise similar stars in the Bulge and the Solar neighborhood [286]. This result is evident in the v_e–P diagram for the O-CSEs. Likewise, the gas expansion velocities of C-CSEs show a trend of a decrease with galacto-centric distance in the solar neighborhood [89].

7.6.2.3 The Acceleration Region

The region in which acceleration occurs is presumably quite small, and hence it presents severe observational difficulties. A very suitable method is high spectral resolution observations of infrared absorption lines of different excitation (Section 7.2.6.5). Its potential has been shown in studies of CO vibrational lines [284]. Another possibility, although more crude, is to compare line widths of probes originating from different regions. The comparisons made, using data of high enough quality, point to only marginal differences [17], except in a few cases [44, 299]. Also, the shape of the line is affected if it is formed in the acceleration region. In particular, the ground-state SiO radio lines show different profiles from those of the CO radio lines [32]. However, interferometer studies have led to inconclusive results on the importance of acceleration in the SiO line-emitting region [177, 222].

That is, at least for AGB CSEs the data support a scenario where the terminal velocity is reached within a few tens of stellar radii.

7.6.3 Small-Scale Structure

Whether the circumstellar medium is smooth or clumpy is important not only for the understanding of the mass-loss mechanism, but also for the excitation of the atoms/molecules, the radiative transfer of line as well as continuum emission, and the ability of the molecules to survive the dissociating radiation. Thus, it may have a profound effect on our modeling of the CSE emission. It has traditionally been assumed that the mass loss occurs in the form of a steady and smooth stellar wind, mainly because this simplifies the models and our ability to derive quantitative results.

There are other interesting aspects that distinguishes a clumpy medium from a smooth one. The former tends to produce brightness distributions that are closer to circularly symmetric (if v_∞ is isotropic) because the photodissociation, which normally limits the brightness distribution, is determined more by the clump properties than by the geometry of the mass loss. Furthermore, we expect the sizes of the molecular CSEs to be less dependent on the mass-loss rate, since the photodissociation is determined by the clump properties. Also, the outer cut-offs in the molecular distributions will be smoother than otherwise expected (provided there is a distribution of clump properties). Regions in which the chemistry has developed differently may become mixed, and unexpected spatial overlaps of some molecular species may occur.

7.6.3.1 Estimates of Clumpiness

The SiO, H_2O, and OH maser maps of the inner regions consist of many, often unresolved, brightness spots. VLBI observations of SiO masers suggest spot sizes as small as a few 10^{12} cm [50]. Assuming a clump size of 10^{13} cm and $n_{H_2} \leq 10^{11}$ cm^{-3} (the SiO masers are quenched at higher densities) leads to a clump mass of $\approx 10^{-6}$ M$_\odot$. In fact, for reasonable mass-loss rates the gas within 10^{14} cm of the star must be substantially clumped to provide the densities required for the SiO masers; see (7.6). Also H_2O masers show spot sizes as small as a few 10^{12} cm [112]. The characteristic OH maser spot size appears to be a few 10^{14} cm at a radius of a few 10^{15} cm [45]. The high-angular-resolution data obtained at infrared wavelengths are consistent with a highly inhomogeneous medium in this region; Section 7.6.1.2. The OH (1612 MHz) masers outline shells that are not uniformly filled with emission; Figure 7.16. Thus, the maser data support a (already from the start) nonuniform CSE, but it is difficult to give any quantitative statements about the fraction of the matter that is clumped and at what size scale this clumping occurs. Theoretical attempts have been made to infer clumpiness from the OH 1612 MHz line profile, but the results are inconclusive [49, 257]. Additional guidance may be obtained from the high-quality H_2O maser data on the supergiant S Per, where true maser clump sizes of a few 10^{14} cm have been measured [213]. In addition, the water maser clouds are about 30 times denser than the average wind density, and about five clumps of mass $\approx 10^{-5}\,M_\odot$ are ejected per pulsational cycle.

Most of the brightness distributions toward IRC+10216 show a very patchy structure; Figure 7.18. In many cases, species with quite different excitation characteristics have emission peaks in the same region, suggesting real variations in column density rather than changes in excitation mechanism.

The CO emission is probably the best probe of the true density distribution. Its main drawback is the large size of the CO envelopes, which causes significant averaging along the line of sight and saturation. Therefore, the large, geometrically thin CO shells around a few C-stars are excellent objects in a study of the small-scale structure; Section 7.6.1. In the case presented in Figure 7.17 a number of arguments favor a clumpy medium, and the final estimate is a total of ≈ 7000 clumps of mass $\approx 10^{-6}\,M_\odot$, size 4×10^{15} cm, H_2 density $\approx 10^4$ cm^{-3}, and kinetic temperature ≈ 20 K [202].

There is also evidence of clumpiness in the observations of post-AGB objects. In general, the patchiness of the CO brightness maps increases with the evolutionary stage of the central object [18, 293]. The most extreme example is provided by the direct observations of CO emission from cometary globules in the Helix nebula [109]. The estimated clump properties are $M = 5\times 10^{-6}\,M_\odot$, $n_{H_2} \lesssim 10^5$ cm^{-3}, and $T_k = 25$ K.

7.6.3.2 Morphological Evolution of a Clumpy AGB CSE

We will give a possible, although at this point speculative, explanation to the observational appearance of CSEs, and also to the apparent drastic change in morphology from an essentially spherical AGB CSE to a neutral envelope around a post-AGB object with, as it seems, a substantial equatorial density enhancement and a marked axi-symmetric structure.

We adopt the view that the circumstellar medium is highly clumped already at the ejection. The true density distribution will probably have structures at many different scales, but a crude estimate of the clump-ejection rate is given by $\dot{N} = 1\,(10^{-6}/M_{cl})\,(\dot{M}/10^{-6})$, where the clump mass is measured in units of $M_\odot$; i.e., only a few clumps per cycle are ejected. An ejected clump will expand due to internal motions while receding from the star. We expect that the SiO masers are located inside the expanding CSE, and there are at most about ten clumps in this area. In the region where the H_2O masers are located there are still too few clumps—up to about one hundred, but only a fraction provide detectable maser emission—to give any reliable information on the morphology. This is also the area probed by the different IR high-angular-resolution observations. The same applies partly to the OH maser region for low $\dot{M}$, while for higher $\dot{M}$ accidental overlaps of clumps, combined with unsaturated amplification, may give a highly perturbed view of the geometry. In the region where the OH 1612 MHz masers (for high $\dot{M}$) are located the clump overlap becomes significant due to the large number of (inflated) clumps. The CO radio lines originate in a region in which the overlap is substantial and the brightness contrast is consequently limited. Therefore, an overall spherical symmetry may be inferred, even though a moderate equatorial density enhancement could be present. Any asymmetry, if existing, is most easily seen in the region close to the star or in emission from geometrically thin shell regions.

The AGB CSE may have a weak equatorial density enhancement, at least toward the end of the AGB. Mechanisms for producing such nonsphericity exist; see Chapter 10. During the early post-AGB evolution a (perhaps isotropic) high-velocity wind commences. Even if the equatorial density enhancement is only moderate, the clump acceleration in the polar direction is more effective than in the equatorial direction where the density of clumps is higher. This quickly accentuates the departure from sphericity, and the appearance of an axi-symmetric structure. A cavity develops, and clumps at successively lower latitudes are accelerated to intermediate velocities, while the material in the polar regions gradually disappears. In this way an elliptical shell or an hour-glass morphology develops depending on the strength of the original equatorial density enhancement. At this point the UV-flux from the central star may drastically enhance the contrast in the brightness distribution as a result of increased photodissociation in directions of low optical depth. Finally, only the equatorial density enhancement remains observable in molecular line emission, and the medium is very clumpy and becomes affected by the pressure of the developing nebulosity.

7.7 Mass-Loss-Rate Estimators

It is crucial to our understanding of AGB stars to obtain reliable estimates of the stellar mass-loss rate. This is defined as $\dot{M} = -dM_*/dt$, but it is difficult to estimate the mass-loss rate through this relation. Instead, estimates based on circumstellar emission in some form appear to be the most reliable, since the physics/chemistry of a CSE is reasonably well established. They also have the widest applicability, and cover essentially the entire AGB and beyond. As outlined in Sections 7.2 and 7.4 a CSE provides an extensive variety of physical conditions, from the inner dense and warm region close to the stellar photosphere to the tenuous and cool outer parts where it gradually merges with the interstellar medium. Hence, different methods, utilizing data obtained at different wavelengths, will necessarily probe the mass-loss rate at different epochs. In the following two sections we discuss methods for estimating mass-loss rates based on circumstellar molecular radio line and dust continuum emission.

7.7.1 Mass-Loss Rates from Molecular Spectral Lines

In principle, all molecular species listed in Table 7.1 can be used as mass-loss-rate estimators, but in practice, this is far from the case. There is a number of criteria that must be fulfilled by a good mass-loss-rate estimator: (i) It should be applicable to O- and C-CSEs, as well as (ii) to a wide range of mass-loss rates and objects in different evolutionary stages; (iii) the lines should be strong and provide a large observational space; (iv) the radiative transfer should be manageable [which is simplified if (iva) the energy-level diagram is simple and the radiative transition rates and collisional cross sections are known, and (ivb) there are observable transitions of quite varying excitation energies]; and (v) the molecular abundance should be known or easily estimated. The usefulness of a molecule as a mass-loss-rate estimator depends on the extent to which these criteria are fulfilled. It is clear that CO is outstanding. OH has been frequently used despite the fact that it certainly does not fulfil criteria (i), (ii), and (iv). Its main advantage lies in the large observational space, its relatively simple chemistry, and the regular behavior of its 1612 MHz maser emission. We discuss here the validity and applicability of simple formulae for estimating the mass-loss rate based on CO and OH radio line emission.

7.7.1.1 The Principle

The principle for estimating mass-loss rates from optically thin circumstellar line emission is rather simple. The mass-loss rate is given by the amount of mass inside a certain radius divided by the time it has taken the star to eject the material, i.e., the time taken to reach this radius, $\dot{M} = 2 m_\mathrm{H} M_\mathrm{A} v_\infty / X m_\mathrm{A} f_\mathrm{A} R_\mathrm{e,A}$, where M_A is the total mass of species A inside the radius $R_\mathrm{e,A}$, and m_A is its molecular weight. In the optically thin limit, assuming spatially unresolved emission in the Rayleigh–Jeans regime from a standard CSE, and an excitation temperature that is constant with radius, we can use (7.33), with $\tau \ll 1$ and $\beta = 0$, to obtain the relation between mass-loss rate and observed intensity

$$\dot{M} = \frac{8\pi^2 m_\mathrm{H} k \nu_{ul}^2 \mathcal{Z}(T_\mathrm{ex}) e^{E_l/kT_\mathrm{ex}}}{\ln 2 \, X h c^3 g_u A_{ul}} \frac{T_\mathrm{A} v_\infty^2 \theta_\mathrm{b}^2 D^2}{f_\mathrm{A} R_\mathrm{e,A}}, \qquad (7.68)$$

where we have converted from flux density scale to antenna temperature scale as outlined in Section 7.2.7.1. This shows the expected dependency on intensity, expansion velocity, distance, source size, and abundance. A detailed calculation is required to obtain $R_\mathrm{e,A}$, to handle optical depth effects, and to determine the excitation temperature and its variation with radius.

One should bear in mind here that $R_\mathrm{e,A}$ depends on the mass-loss rate in the case of photodissociation-limited emission. This relation between the size of the emitting region and the mass-loss rate can also be used to estimate the latter, but it requires a detailed photodissociation model, which also includes the shielding properties of the dust.

7.7.1.2 CO Radio Line Emission

It is generally agreed that mass-loss-rate estimates obtained from CO rotational line emission at radio wavelengths are among the most reliable. Therefore, a simple formula, similar to (7.68), would be of great value. However, the CO line intensities dependence on mass-loss rate, gas expansion velocity, CO abundance, etc., is rather complicated. We give here a formula based on extensive modeling (see [227] for details). The result is (for an unresolved CSE)

$$\dot{M} = s_J\, I_{CO,J}^{a_J}\, v_\infty^{b_J}\, f_{CO}^{-c_J}\, \theta_b^2\, D^2 \ \ M_\odot\, \text{yr}^{-1}, \qquad (7.69)$$

where $I_{CO,J}$ is the integrated intensity of the $J \to J{-}1$ line given in K km s^{-1}, v_e in km s^{-1}, D in kpc, and θ_b in arcsec. For the $J=1\to 0$ line the constants are $s_1 = 5.6\times 10^{-13}$, $a_1 = 0.70$, $b_1 = 0.71$, and $c_1 = 0.83$. The corresponding values for the $J=2\to 1$ and $J=3\to 2$ lines are 9.7×10^{-12}, 0.86, 0.63, 0.66, and 1.3×10^{-11}, 0.97, 0.56, 0.52, respectively. This formula gives values within a factor of two of the modeling in the ranges $10^{-7} \leq \dot{M} \leq 10^{-5}\, M_\odot\,\text{yr}^{-1}$, $5 \leq v_e \leq 20\,\text{km s}^{-1}$, and $10^{-4} \leq f_{CO} \leq 10^{-3}$. The line intensities become more dependent on the mass-loss rate for lower values than this, and at higher values the line intensities saturate (due to a combination of high optical depths and the fact that an increase in the mass-loss rate leads to a cooling of the CSE, which compensates for the larger number of CO molecules). There are (substantial) uncertainties due to, e.g., the treatment of CO photodissociation, the strong dependence on luminosity at low mass-loss rates, and the strong dependence on the gas–grain collision heating at high mass-loss rates [84, 139, 221]. It is clear from (7.69) that for low mass-loss rates the higher-frequency lines are much stronger; e.g., at $10^{-7}\, M_\odot\,\text{yr}^{-1}$ the result is $I_{3-2}/I_{1-0} \approx 20$ when both transitions are observed with the same telescope.

Uncertainties not directly dependent on the modeling are introduced through D and f_{CO}. The former creates a problem that this method shares with most other methods for estimating mass-loss rates. Assumptions about the latter usually center on a few $\times 10^{-4}$ and 10^{-3} for O- and C-CSEs, respectively, but there have been no direct measurements of this quantity. This is a relatively well-defined problem: If CO is fully associated, the CO abundance in an O-CSE is determined by the abundance of C, while in a C-CSE it is determined by the abundance of O. However, it remains to be shown what are the true circumstellar CO abundances, and to what extent they vary from star to star.

The effect of clumpiness in the CSE may be an additional problem in the mass-loss-rate estimate. The accumulation of gas into clumps increases the local density, but not necessarily the collisional excitation, since presumably the heating of the gas (through the dust streaming) becomes less efficient. It will also lead to a higher shielding against photodissociation for a given mass-loss rate.

Each mass-loss-rate estimator provides an estimate of the average mass-loss rate during a certain epoch. This is determined either by excitation or the abundance distribution. In the case of the CO ($J=1\to 0$) line, the molecules are approximately excited out to the photodissociation limit (7.9). In general, it is the CO molecules in the outer regions that contribute most of the observed intensity, and we conclude therefore that the CO ($J=1\to 0$) line measures the mass-loss rate at the retarded time

$$t_{\rm CO} \approx 10^3 \left[\frac{\dot{M}}{10^{-6}}\right]^{0.7} \left[\frac{15}{v_\infty}\right]^{1.6} \left[\frac{f_{\rm CO}}{10^{-3}}\right]^{0.6} {\rm yr}. \quad (7.70)$$

Higher transitions come from increasingly smaller regions, and hence probe the mass-loss rate at increasingly smaller retarded times; Section 7.2.6.2.

We use (7.69) to estimate the observational space for CO radio line observations of AGB CSEs. For the $J=2\to 1$ line the result is

$$D_{\rm CO} \approx 1 \left[\frac{\dot{M}}{10^{-6}}\right]^{0.5} \left[\frac{1}{I_{\rm CO,2}}\right]^{0.4} \left[\frac{f_{\rm CO}}{10^{-3}}\right]^{0.3} \left[\frac{15}{v_\infty}\right]^{0.3} \left[\frac{20}{\theta_{\rm b}}\right] {\rm kpc}, \quad (7.71)$$

with the same units as in (7.69). This shows that for reasonable mass-loss rates and sensitivities the observational space of CO radio lines is quite limited. In the case $I_{\rm CO,2} = 1\,{\rm K\,km\,s^{-1}}$, $\dot{M} = 10^{-5}\,{\rm M_\odot\,yr^{-1}}$, $f_{\rm CO} = 10^{-3}$, and $\theta_{\rm b} = 10''$, we obtain $D \approx 6\,{\rm kpc}$ for the $J=2\to 1$ line, i.e., about a tenth of the distance to the LMC. Note that for $\dot{M} = 10^{-8}\,{\rm M_\odot\,yr^{-1}}$ we are limited to stars within 0.2 kpc.

7.7.1.3 OH 1612 MHz Maser Line Emission

The OH 1612 MHz line, which is seen only in maser emission toward O-CSEs, has been extensively used to estimate mass-loss rates with some success. This is at first sight somewhat surprising, since the molecule, in principle, fulfills only criteria (*ii*) and (*iii*) above, but can be explained as follows. The maser mechanism works only below a critical density, i.e., for a given mass-loss rate only outside a certain radius, due to collisional quenching of the inversion. Furthermore, OH is formed, through photodissociation of H_2O, outside a radius roughly determined by the UV optical depth from the outside approaching one. Thus, when taking into account the steep density gradient in an expanding CSE, the OH number density forms a relatively thin shell, which, when located outside the radius determined by the critical density, will give rise to maser emission. The radius of this shell, essentially where $\tau_{\rm UV} \approx 1$, is determined by the mass-loss rate. Hence a measure of the shell size or the OH luminosity (which is determined by the size if the maser is saturated, i.e., the surface brightness varies little from star to star) enables an estimate of the mass-loss rate. Another way of looking at it is that the luminosity of a saturated maser is determined by the number of pump photons, i.e., not the number of masing molecules. In the case of the OH 1612 MHz maser this means 35 μm photons, and these are emitted by the circumstellar dust [62].

Hence, the OH 1612 MHz maser emission is actually a measure of the dust mass-loss rate. Of importance here is that ISO observations detected for the first time the proposed pump line at 35 μm in absorption and the ensuing rotational cascade lines in emission [242], and theoretical modeling managed to fit reasonably well the observed intensities [248].

The radial dependence of the OH abundance has been calculated using a photodissociation model for H_2O and OH, which includes also dust- and self-shielding [192]. As expected, the OH number density reaches a peak at a radius R_{OH}, and the mass-loss rate is related to this radius via

$$\dot{M} = 4 \times 10^{-6} \left[\frac{R_{OH}}{10^{16}} \right]^{1.4} \left[\frac{v_\infty}{15} \right]^{0.6} M_\odot \, \text{yr}^{-1}. \quad (7.72)$$

A mass-loss rate estimate based on this has the main advantage of being distance-independent if R_{OH} can be obtained from phase-lag measurements; Section 7.6.1.3. Of course, the main disadvantage is that it requires a measurement of the emitting region, a quantity available for only a few AGB CSEs.

A relation that involves the observed intensity would be more useful. As outlined above, we can expect a relation between the OH luminosity and the size of the emitting region for a saturated maser. Using this, and the fact that a minimum value of the OH column density is required for saturated maser emission (and the somewhat shaky assumption that all OH 1612 MHz masers have this minimum column density), the relation

$$\dot{M} = 7 \times 10^{-11} \frac{S_{OH}^{0.5} v_\infty D}{f_{OH}} \, M_\odot \, \text{yr}^{-1} \quad (7.73)$$

applies, where S_{OH}, the geometric mean of the peak flux densities of the two emission features, is given in Jy, v_∞ in km s^{-1}, and D in kpc [10]. A value of 1.6×10^{-4} is frequently used for f_{OH}. The mass-loss rates estimated from the OH 1612 MHz maser size and luminosity rest on much looser ground than the CO estimates. Accuracies better than an order of magnitude can hardly be expected.

An estimate of the mass-loss-rate epoch probed by the OH 1612 MHz emission is obtained from (7.72)

$$t_{OH} = 10^2 \left[\frac{\dot{M}}{10^{-6}} \right]^{0.7} \left[\frac{15}{v_\infty} \right]^{1.4} \text{yr.} \quad (7.74)$$

That is, roughly speaking, the OH emission gives information on a time scale ten times shorter than does the CO ($J = 1 \to 0$) emission. In fact, it is probably more complicated since the OH emission also probes the time scale defined by the 35 μm emission.

Likewise, we can use (7.73) to estimate the observational space for the OH 1612 MHz emission. The result is

$$D_{\rm OH} = 0.5 \left[\frac{\dot{M}}{10^{-6}}\right] \left[\frac{0.1}{S_{\rm OH}}\right]^{0.5} \left[\frac{f_{\rm OH}}{10^{-4}}\right] \left[\frac{15}{v_\infty}\right] \text{ kpc.} \qquad (7.75)$$

In the case $S_{\rm OH}=0.1$ Jy and $\dot{M}=10^{-5}\,{\rm M}_\odot\,{\rm yr}^{-1}$ we obtain $D_{\rm OH}=5\,{\rm kpc}$, i.e., a result similar to that for CO, but note that $D_{\rm OH}$ scales as $\dot{M}$, while $D_{\rm CO}$ scales only as $\dot{M}^{0.5}$.

7.7.2 Mass-Loss Rates from Dust Emission

An alternative to using line emission for estimating the mass-loss rate is provided by the circumstellar dust emission. This has several advantages: The observational space and basis are usually much larger, and there is no dependence on complicated line-transfer effects such as maser emission and photodissociation. On the other hand, the emitting properties of the dust grains are poorly known at far-IR and longer wavelengths, and so is the dust-to-gas mass ratio required to convert to the total mass-loss rate. Furthermore, our knowledge of the dust compositions of different types of objects is quite limited. A clear disadvantage is that the dust emission provides no kinematical information.

We illustrate the principle by deriving some simple, but useful, results in the optically thin limit. The more elaborate method of fitting SEDs is also discussed, as well as relations between mass-loss rate and distance-independent colors.

7.7.2.1 Monochromatic Fluxes

At sufficiently long wavelengths the Rayleigh–Jeans approximation is valid, and the flux density is obtained from (7.62) in the optically thin limit. This expression gives, for $s=1$, a dust mass-loss-rate estimate

$$\dot{M}_{\rm d} = 1.0 \times 10^{-3} \frac{\lambda^2}{\chi_{{\rm gr},\nu}} \frac{S_{{\rm d},\nu} v_{\rm d} D^{1.4}}{\theta_{\rm b}^{0.6} T_{0,15}} \; {\rm M}_\odot\,{\rm yr}^{-1}, \qquad (7.76)$$

where $S_{{\rm d},\nu}$ is given in Jy, $v_{\rm d}$ in $\text{km\,s}^{-1}$, D in kpc, λ in mm, $\chi_{\rm gr}$ in $\text{cm}^2\,\text{g}^{-1}$, $\theta_{\rm b}$ in arcsec, and $T_{0,15}$ is the dust temperature at 10^{15} cm. At shorter wavelengths, where the source can be assumed to be unresolved, the more general expression (7.64) can be used, and the result is

$$\dot{M}_{\rm d} = 1.9 \times 10^{-3} \frac{1}{\chi_{{\rm gr},\nu}\lambda^{0.5}} \frac{S_{{\rm d},\nu} v_{\rm d} D^2}{T_{0,15}^{2.5}} \; {\rm M}_\odot\,{\rm yr}^{-1}, \qquad (7.77)$$

in the same units as above, except for λ, which is given in µm (we have used $I_{\rm bb}=1.8$). Note the substantial dependence on the uncertain quantity $T_{0,15}$. An estimate of $T_{0,15}$ can be obtained using (7.54), $T_{0,15}=130\,(L/\lambda_0)^{0.2}$ K, where λ_0 is the wavelength (in µm) of the peak of the stellar SED, and the stellar luminosity, L, is given in $L_\odot$. Both these mass-loss-rate estimates contain the dust expansion velocity, v_d, which is not a directly measurable quantity. A common assumption is that v_d lies in the range of $(1-2)v_e$. In addition, an estimate of the total mass-loss rate requires an assumption of the dust-to-gas mass ratio, ψ. The mass-loss rates obtained in this way are accurate only to within an order of magnitude.

A rough estimate of the mass-loss-rate epoch probed by dust emission at the wavelength λ is obtained using (7.57),

$$t_{\text{dust}} \approx 10^2 \left[\frac{\lambda}{10}\right]^{2.5} \left[\frac{15}{v_d}\right] \text{ yr}, \qquad (7.78)$$

where λ is given in μm; i.e., the dependence on λ is substantial. As described in Section 7.4.1.2, this will depend on the mass-loss rate once optical depth effects become important.

7.7.2.2 Colors

Another way of estimating the mass-loss rate is to use colors as measures of the thickness of a CSE. These are distance-independent, a substantial advantage, and the reason why this works is the fact that the CSEs become redder the thicker they are; i.e., to a first approximation the shape of the SED depends only on the optical depth.

Useful relations can be obtained through detailed modeling, or by comparing colors with mass-loss rates obtained using a reliable method, i.e., the colors are secondary estimators. Particularly important are relations between mass-loss rate and near-IR colors, since a number of surveys are presently in progress. A tight relation between mass-loss rate (estimated from SED fits and assumed dust-to-gas mass ratios) and [K–L] have been found [161, 164],

$$\log\left[\frac{\dot{M}}{10^{-6}}\right] = \begin{cases} -\frac{2.75}{[K-L']} + 2.25, & 0.7 \leq [K-L'] \leq 3.0, \text{ O-CSE} \\ -\frac{9.0}{[K-L]+1.4} + 2.75, & 1.0 \leq [K-L] \leq 8.0, \text{ C-CSE} \end{cases} \qquad (7.79)$$

(note that the detailed relation depends on the filters used). The applicability of such relations is limited by the facts that their dependence on color is very steep at the low-mass-loss-rate end, they saturate when the mass-loss rate approaches $10^{-5} M_\odot \text{ yr}^{-1}$, and the colors may change during the pulsational cycle, and may depend on metallicity. The spreads around the fits suggest uncertainties on the order of a few in the regions where the methods are applicable, excluding the uncertainties in the comparison mass-loss rates.

In principle, one would expect that better mass-loss-rate estimates are obtained from far-IR fluxes, since these measure the more optically thin circumstellar dust emission and there is little contamination from the central star. However, comparisons between far-IR colors and CO, OH, and dust mass-loss-rate estimates show, somewhat surprisingly, considerable scatter, and the reason for this is not clear. Possibly, colors that combine near- and mid-IR fluxes are better, e.g., [K–12].

7.7.2.3 Spectral Energy Distributions

The most sophisticated way to obtain the dust mass-loss rate is a detailed modeling of the entire SED, which normally requires, at least, measured fluxes between 1 and 100 μm. Of course, the necessary physical input to such a model is substantial (e.g., dust condensation radius or temperature, dust velocity law, dust chemical composition, grain sizes, and radiative properties), and a fair fraction of the crucial data are uncertain.

The principle is simple in the sense that once the model has been defined, the SED is calculated and the measured fluxes are compared with computed fluxes calculated for the relevant observing equipment. The outcome of the model is a dust radial optical depth at an arbitrary frequency, which is related to the dust mass-loss rate. If we assume that the dust expansion velocity follows the same functional form as the gas expansion velocity, (7.1), the following expression results:

$$\dot{M}_d = \frac{2\pi}{\chi_{gr,\nu}} v_{d,\infty} r_0 \tau_{d,\nu} . \tag{7.80}$$

Model fits to SEDs of O-CSEs [12, 113, 229] and C-CSEs [92, 113, 161] have been done. Examples are given in Figure 7.13.

The derived dust mass-loss rates are normally converted to total mass-loss rates using an assumed dust-to-gas mass ratio. Alternatively, one can rely upon a kinematical model for a dust-driven wind, which relates the gas expansion velocity to the drag force between the gas and the dust; Section 7.2.1. It is possible to show that in this case we have [92]

$$\dot{M} = \tau_F \frac{L}{cv_\infty} \left(1 - \frac{1}{\Gamma}\right), \tag{7.81}$$

where τ_F is the flux-averaged dust optical depth obtained from the fit to the SED, and Γ is given in (7.3). This formula essentially says that the momentum rate of the gas, $\dot{M}v_\infty$, should equal that of the photon flow, L/c, times the effectiveness of the scattering, i.e., as measured by the dust optical depth. In this way a dust-to-gas mass-ratio estimate is also obtained. Assuming $\tau_F = 5$, $L = 20{,}000$ L$_\odot$, $v_\infty = 20$ km s^{-1}, and $\Gamma \gg 1$ we obtain $\dot{M} \approx 10^{-4}$ M$_\odot$ yr^{-1}; i.e., a dust-driven wind can result in substantial mass loss.

7.7.2.4 Dust Features

In principle, one can use the spectral features present in the dust emission to estimate the mass-loss rate in much the same way as one uses spectral lines. The main difference is that atomic/molecular lines are uniquely identified, while a dust feature may be the composite result of many components, whose identification and relative abundances are uncertain. This disadvantage is partly avoided if one uses the dust features as secondary estimators; i.e., their strengths are calibrated using mass-loss rates obtained by other methods. Of course, such a procedure is meaningful only if the secondary estimator can provide mass-loss rates for large samples. One problem, though, remains. The dust feature is expected to be seen in emission for low mass-loss rates, and also to increase in strength with increasing mass-loss rate. However, as the dust optical depth increases it is possible to reach a point where the strength of the dust feature actually decreases with increasing mass-loss rate, and eventually it is seen in absorption. The latter certainly applies to the 10 μm feature in O-CSEs (as a rule of thumb it reaches its maximum relative strength at $\approx 10^{-6}\,M_\odot\,\mathrm{yr}^{-1}$, and is seen in absorption above $10^{-5}\,M_\odot\,\mathrm{yr}^{-1}$; Figure 7.14), but not to the 11.3 μm feature in C-CSEs. These are the only two features that are strong and ubiquitous enough to be used as mass-loss-rate estimators. A relation between the optical depth of the 10 μm feature and the dust mass-loss rate of an O-CSE [229], and relations between estimated mass-loss rates and measures of the strengths of the 10 and 11.3 μm features in C-CSEs exist [235, 236]. We can expect, at best, to get order-of-magnitude estimates using these methods.

7.8 Mass-Loss Rate

One of the most important tasks in the study of stellar evolution is to estimate the mass-loss rate of stars on and beyond the AGB [281]. This is because the mass loss from the surface of the star eventually proceeds at a rate much faster than that at which matter is added to the inert core ($\approx 10^{-7}\,M_\odot\,\mathrm{yr}^{-1}$), and hence it appears to be the sole factor determining the AGB lifetime once heavy mass loss has started. The post-AGB mass loss is crucial to the time scale of evolution of the stellar remnant, and hence the formation of PNe. In addition, the matter lost during the AGB represents a significant fraction of the total metal-enriched gas-return to the interstellar medium. This material also includes the seeds of the interstellar grains, and these particles probably facilitate the star formation directly through their cooling capabilities and/or indirectly by protecting molecular coolants like CO and H_2O.

It is not enough to determine the mass-loss-rate distributions of various samples. We need also to obtain the mass-loss-rate dependence on time, main-sequence mass, and metallicity, i.e., $\dot{M}(t, M_{MS}, Z)$, for the evolution of populations of stars with different characteristics to be modeled; see Chapters 2, 3, and 8. This is not a simple task since for individual stars it is impossible to know the time since the evolution on the AGB started, and very difficult to determine main-sequence mass and metallicity. By necessity, this means that most of the conclusions in this section will be qualitative. A healthy skepticism on the side of the reader is called for.

Of course, dependences on mass, time, and metallicity have their roots in the mass-loss rate's dependence on, e.g., luminosity, pulsational regularity and strength, and dust-formation efficiency.

7.8.1 Mass-Loss-Rate Distributions

The lack of complete volume-limited samples of AGB-stars with determined mass-loss rates means that we will have to settle for mass-loss-rate distributions of samples selected and observed in different ways. In particular, these samples contain stars of different masses and at different stages in the AGB evolution. It is possible, though, to select samples where the metallicity is relatively uniform.

In Figure 7.23 we present mass-loss-rate distributions for two samples of M-stars and two samples of C-stars. The M-type irregular and semiregular variables probe the low-mass-loss-rate end. The mass-loss rates were determined through detailed modeling of the circumstellar CO radio line emission [206]. The median mass-loss rate is about $3\times10^{-7}\,M_\odot\,\mathrm{yr}^{-1}$. For the Galactic Center OH/IR-stars the mass-loss rates were estimated using the OH 1612 MHz emission. The median mass-loss rate of this sample is substantially higher, $3\times10^{-5}\,M_\odot\,\mathrm{yr}^{-1}$. One of the C-star samples consists of a volume-limited sample (≤ 1 kpc) of mainly optically bright stars. The mass-loss rates were determined by modeling the CO radio line emission [227]. The results for a subsample, consisting of essentially all the C-stars within 500 pc, are indicated separately. The median mass-loss rate is $4\times10^{-7}\,M_\odot\,\mathrm{yr}^{-1}$ ($2\times10^{-7}\,M_\odot\,\mathrm{yr}^{-1}$ for the stars within 500 pc), but there is a tail toward much higher mass-loss rates: $\approx 25\%$ and $\approx 5\%$ of the stars have mass-loss rates in excess of $10^{-6}\,M_\odot\,\mathrm{yr}^{-1}$ and $10^{-5}\,M_\odot\,\mathrm{yr}^{-1}$, respectively. The second sample consists of mainly highly obscured C-stars within about 6 kpc, and the mass-loss rates are substantially higher, with a median of $1\times10^{-5}\,M_\odot\,\mathrm{yr}^{-1}$ [89]. The mass-loss rates were determined using a simplified version of formula (7.69) and less accurate distances, and hence are rather uncertain.

7.8.2 Mass-Loss Rate and Circumstellar Kinematics

There are two main characteristics of the mass-loss process: the stellar mass-loss rate and the circumstellar gas expansion velocity. The mass-loss rate is to a large extent determined by the conditions at the transonic point, while the gas expansion velocity is determined by the acceleration beyond this point. Hence, these two properties do not necessarily correlate with each other. However, in the dust-driven wind scenario we expect that the mass-loss rate and the terminal gas velocity are correlated, $\dot{M} \propto v_e^{0-3}$ depending on the strength of the radiation pressure and gravity [63]. Therefore, a comparison between the two provides important results, which any mass-loss mechanism model must be able to explain. In Figure 7.24 we present the mass-loss rates and gas expansion velocities for the samples of irregular and semiregular M- and bright C-stars discussed in the previous section, with the results for some Miras added. The uncertainties in the expansion velocities are rather low, typically less than $\pm 10\%$. The mass-loss rates were determined through

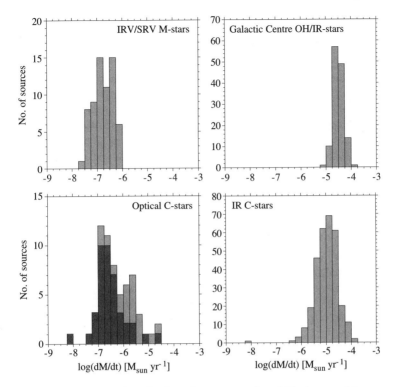

Fig. 7.23. Mass-loss-rate distributions for two samples of M-stars, and two samples of C-stars. For the optical C-stars, the subsample with stars within about 500 pc is darker (see text for data references)

modeling of CO radio line emission, and they are estimated to have an uncertainty of about ±50% (within the adopted circumstellar model). Hence, we conclude that for at least these samples, the two mass-loss characteristics correlate well ($\dot{M} \propto v_e^{2.5-3.5}$; similar results have been obtained in other studies [154, 292]), but also that there is a substantial spread, about an order of magnitude, in the mass-loss rates for a given expansion velocity.

7.8.3 Mass-Loss Rate and Stellar Properties

Based on observations of RGB stars, Reimers introduced a relation describing the mass-loss-rate dependence on stellar characteristics such as radius and luminosity [212]. Extensive observations of AGB-stars have shown that this relation gives a far from adequate description of AGB mass loss. Subsequent work has produced a number of empirical relations between mass-loss rate and stellar properties [29, 264]. Attempts have also been made to derive mass-loss rate relations from physical arguments [13], or from numerical simulations [108, 271, 285]. They indicate that in the dust-driven regime we

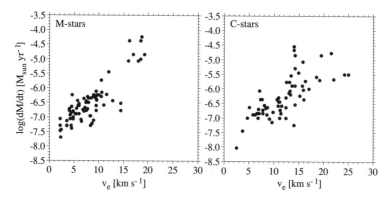

Fig. 7.24. Mass-loss rate versus circumstellar gas expansion velocity for M-stars and C-stars (see text for data references)

can expect a mass-loss rate that increases strongly with luminosity, decreases with increasing mass (which is more than compensated for by the luminosity dependence), decreases strongly with increasing temperature, is essentially independent of C/O (only C/O > 1 models are considered), and is almost unaffected by the period and amplitude of pulsation [271], but, considering the complexity of the problem, these results must be regarded as highly tentative. Below some critical values a dust-driven wind is not possible, and only low mass-loss rates and gas expansion velocities are produced [285]. Even though these results are interesting, we are still a long way from deriving the mass-loss rate from first principles.

7.8.3.1 Dependence on Stellar Temperature and C/O

The effective temperature and elemental setup (i.e., in this context the C/O ratio) are stellar properties that are reasonably well determined for low and intermediate mass-loss-rate stars, and the latter can be determined from circumstellar characteristics for high-mass-loss-rate objects. Hence, there is a possibility to make a meaningful study of the mass-loss dependence on these properties, but the effects of other properties confuse the interpretation.

Only a few studies of the temperature dependence have been performed. The conclusion is that for low and intermediate mass-loss-rate stars there is no apparent mass-loss-rate dependence on stellar temperature [206, 227]. A weak increase in mass-loss rate with decreasing temperature is seen for the C-stars, but only when the Miras are included [227].

It appears that all AGB C-stars have mass loss, at a rate in excess of $10^{-8}\,M_\odot\,\mathrm{yr}^{-1}$ [204]. This certainly does not apply to the AGB M-stars. At the high-mass-loss-rate end most mass-loss-rate estimates for C-stars are based on CO radio line emission, which saturates strongly at mass-loss rates in excess of $10^{-5}\,M_\odot\,\mathrm{yr}^{-1}$. The extreme M-stars, the OH/IR-stars, reportedly have mass loss rates approaching 10^{-4} $M_\odot$ yr^{-1}, but these estimates are certainly less accurate than the CO estimates. Therefore, presently, there is nothing that contradicts the conclusion that C- and M-stars reach about the same peak mass-loss rates. It should be noted, though, that the hot bottom burning process prevents the formation of C-stars more massive than about $4\,M_\odot$ (Chapter 2), and this can make a difference if the peak mass-loss rate increases with increasing mass.

7.8.3.2 Dependence on Stellar Mass

It is beyond doubt established that the total amount of mass lost during the AGB evolution, M_{MS}–M_{WD}, depends on the main-sequence mass, since stars up to $\approx 8\,M_\odot$ leave stellar remnants with masses M_{WD} less than about $1\,M_\odot$ (Chapter 3), and only a few tenths of a solar mass are lost before the AGB evolution. If the time scale for mass loss on the AGB depends only weakly on the mass, we may conclude that the average mass-loss rate on the AGB increases with main-sequence mass. However, we need to be more detailed than this.

The fact that samples contain stars of different masses and in different evolutionary stages means that any possible dependences on mass and time are difficult to disentangle. It is therefore important to identify main-sequence mass estimators, since to estimate the mass of an individual AGB star is very difficult. It appears that the best choice here is the period of pulsation, which is related to the (present) mass via the fundamental pulsation equation, $P = QR^{3/2}M^{-1/2}$, where Q is the pulsational constant, which depends on the mode of pulsation; see Chapter 2. There is good evidence of the period increasing with the main-sequence mass; see Chapter 8. However, as a cautionary remark, considerable mass loss will decrease the mass to such an extent that the period becomes significantly longer. There is observational evidence of this "lengthened period" effect, and also of the period depending weakly on metallicity [286].

In Figure 7.25 we show mass-loss rates versus periods for three samples of M-stars (semiregular variables, Miras, and Galactic Center OH/IR-stars), and a sample of bright C-stars of semiregular and mira variability. The positive trend with period is clear in the period range 300^{d} to 600^{d}, and it suggests that the mass-loss rate increases with the stellar mass. The semiregulars with periods less than 200^{d} cover essentially the same mass-loss-rate range as the Miras in the period range 200^{d} to 400^{d}. It also appears that stars with periods longer than 600^{d} have reached their maximum mass-loss rate, except for possibly brief periods of enhanced mass loss. A mass-loss-rate dependence on mass is probably, at least partly, a luminosity effect, but any evidence for a luminosity dependence of the mass-loss rate is only weak.

The mass-loss-rate trend with period may have a more complicated origin. For instance, in the case of the bright C-stars there appears to be a distinct difference in mass-loss rate between the semiregularly variable, shorter-period

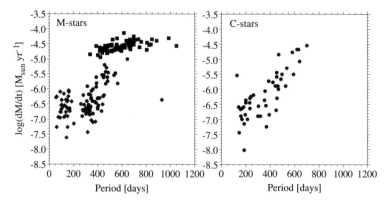

Fig. 7.25. Mass-loss rate versus period for three samples of M-stars [semiregulars (diamonds), Miras (circles), and Galactic Center OH/IR-stars (squares)], and a sample of optically bright C-stars (see text for data references)

stars and the regularly variable, longer-period stars, an effect that seems natural to attribute to the difference in pulsational pattern. Clearly, the regularity and strength of the pulsations, which may depend on mass, must play an important role in determining the mass-loss rate, and not only the luminosity. In addition, the effect of "lengthened period" could mean that part of the trend in Figure 7.25 may be an effect of the mass-loss rate increasing with evolutionary age, Section 7.8.4, and hence decreasing mass due to intense mass loss.

7.8.3.3 Dependence on Metallicity

A mass-loss-rate dependence on metallicity is not necessarily expected. If anything, we expect an effect on the dust-to-gas mass ratio in O-CSEs where Si is important, an element not produced in AGB stars. In C-CSEs, where carbon grains dominate, the effect of a lower metallicity is more uncertain, since C is definitely produced, and brought to the surface, in carbon stars. Clearly, the dust may play a dynamical role, and hence lead to a metallicity effect in v_∞ (Section 7.6.2), but not necessarily on the mass-loss rate, whose magnitude is determined by the conditions close to the transonic point. At least, from the similarity of the $M_{\rm WD}$-distributions in the solar neighborhood and the LMC [277], one may conclude that the total amount of mass lost does not depend (strongly) on the metallicity.

The metallicity is difficult to estimate for an individual AGB star. However, here we are fortunate, since samples with clearly different metallicities can be identified, e.g., the Magellanic Clouds ($Z_{\rm LMC} \approx 0.5\,Z_\odot$, $Z_{\rm SMC} \approx 0.2\,Z_\odot$), and the Galactic Center ($Z \gtrsim 1.5\,Z_\odot$).

There are indications that the dust content is lower in O-CSEs of lower metallicity. For instance, LMC M-stars have bluer near-IR colors than Galactic M-stars with similar periods and pulsation properties; i.e., the dust opacity is lower [100]. In a detailed study of three M-type stars, with similar characteristics except for the metallicity, it was found that the dust optical depths of the O-CSEs are in the ratios 1:0.1:0.06 (Galaxy:LMC:SMC), in all likelihood an effect of ψ decreasing with metallicity [90]. In Figure 7.25 it is clear that in the period range 300^d–600^d the mass-loss rates of the Galactic Centre OH/IR-stars are much higher than those of the M-Miras. The mass-loss rates of the two samples were estimated differently, but the difference is large enough to suggest a higher mass-loss rate for a given mass in the metal-richer Galactic Center sample. Similar studies have not been performed for C-CSEs.

There are no studies that conclusively show that the mass-loss rate depends on the metallicity, e.g., when comparing results for the SMC and the LMC [260, 297].

7.8.4 Dependence on Time

We can expect variations in the mass-loss rate on time scales from a year up to the total AGB lifetime, and different probes are required for the different time scales. To identify any mass-loss-rate variation, irrespective of the time scale, is a difficult problem. On the long term the age on the AGB is not an easily measurable quantity for an individual star, and statistical studies, where dependences on mass and time usually are difficult to separate, are used. On time scales shorter than about 10^4 years it is possible to get information from individual stars, but difficult to draw general conclusions.

7.8.4.1 Long-Term Changes

We start by looking at the long-term changes during evolution along the AGB. Statistical studies have been performed, e.g., based on an OH maser luminosity function [10] or on the distribution of AGB-stars (and their CSEs) in a far-IR color–color diagram [254]:

$$\dot{M}(t) = \dot{M}_{\min} \left(1 - t/t_{\mathrm{ML}}\right)^{-\beta} \qquad (7.82)$$

was obtained, with $\beta \approx 0.5$–0.75, and t_{ML} is the time scale of AGB mass loss. This points toward a drastic increase in mass-loss rate at the end of the AGB evolution, a so-called superwind phase. Such a behavior has support from theoretical studies [13, 24, 108, 271], as well as from other observational studies. For instance, a number of highly evolved OH/IR stars show CO $J = 2 \rightarrow 1$ and $J = 1 \rightarrow 0$ line intensity ratios much higher than expected from a constant mass-loss rate [106]. A reasonable explanation is a recent increase in the mass-loss rate, which has affected only the $J = 2 \rightarrow 1$ emission, which comes from a region smaller than that of the $J = 1 \rightarrow 0$ emission [85]. A similar conclusion has been obtained by comparing OH, CO, and dust mass-loss-rate estimates [70, 134]. Also for C-stars evidence exists of increasing mass-loss rate at the end of the AGB phase [269].

It must be said that a different interpretation of the far-IR color–color diagram is possible: The distribution of stars in this diagram is instead determined by the stellar mass distribution, and the time scale to shed all the stellar envelope mass, if there is a strong mass-loss-rate dependence on mass, and essentially no mass-loss-rate evolution with time for individual stars [66].

There are a number of other indications that the mass-loss rate increases in the long term on the AGB [100]. Stars with "lengthened" periods, very likely an effect of mass loss, must be in a later evolutionary stage than stars with the same luminosity but shorter periods. There is evidence of these "lengthened-period" stars having larger mass-loss rates, and hence their mass-loss rates having increased with time. Also, the spreads in $(\dot{M},P)$ diagrams are large for a given period, suggesting that stars with the same mass have very different mass-loss rates, and hence the mass-loss rate changes with time, although not necessarily in an increasing way. A recent analysis of a far-IR-color distribution suggests an exponential increase with time (although the time scale remains uncertain), i.e., not too different from (7.82), in the range 10^{-6}–10^{-5} $M_\odot$ yr^{-1} [118].

At the end of the AGB evolution the star must decrease its mass-loss rate substantially in a short time. The termination of the AGB is reached when the stellar envelope mass becomes very low, say $\approx 0.001\,M_\odot$. If the mass-loss rate is as high as $10^{-4}\,M_\odot$ yr^{-1}, it must decrease with orders of magnitude within decades. Some evidence exists for stars in, or just after, this evolutionary phase [169, 295].

7.8.4.2 Episodic Mass Loss

It is very likely that the long-term evolution of the mass loss is modulated on shorter time scales. Fortunately, mass-loss-rate variations on time scales in the range 10^2–10^4 years can be studied for individual objects. The first indications of such variations were indirect, e.g., anomalous line-intensity or broad-band-flux ratios. In the latter case, quite a number of AGB stars were shown to have high, in a relative sense, 60 μm IRAS fluxes, and these were interpreted as due to the presence of cold dust at large distances from the stars and a lack of warm dust close to the stars [280, 296]. However, there is a considerable uncertainty in the IRAS 60 μm fluxes, due to the presence of interstellar cirrus emission, and excess emission at 60 μm is an unreliable indicator of variable mass loss [113]. Other evidence in favor of variations comes from the analysis of SEDs of M-type Miras and non-Miras, in this case on a time scale of 10^2 years [179]. On even shorter time scales, interferometry mapping at 10 μm found two categories of sources in terms of dust condensation radius, those with small (a few stellar radii) and those with large (essentially detached CSEs) dust condensation radii [54].

More direct evidence of episodic mass loss is provided by the detections of CO radio line emission from geometrically thin shells around five carbon stars [171, 201, 202], and subsequent images obtained in scattered stellar light [81]. These shells indicate substantial mass-loss-rate changes, by more than two orders of magnitude, over time scales of a few hundred years; see Figure 7.17. Detached dust shells have been found toward three C-stars [117, 275], and possibly one M-star [103]. The physical relation between the gas and the dust shells is not clear, but it has been suggested that at least the gas shells are connected to He-shell flashes [203]. The same explanation has been used to explain data on OH/IR-stars [166, 167]. Theoretical models, which follow the mass-loss-rate evolution through this stage of AGB evolution, have been presented [19, 228, 264], and also hydrodynamical calculations of the evolution of the CSE structure [240]. It is presently difficult to put these scarce observations into the context of the general AGB evolution. However, CO observations of a relatively complete sample of C-stars suggest that at least for C-stars, highly episodic mass loss may be a common phenomenon during part of their final evolution [204]. The same may apply to highly evolved M-stars [166].

Recently, images in dust-scattered stellar or interstellar light have been obtained of a few C-rich AGB and post-AGB objects. These show arcs (in one case also visible in CO radio line data [71]), i.e., very likely parts of concentric shells centered on the star, of enhanced dust density superposed on a general r^{-2} density decline due to constant mass loss [157, 182, 225]; Figure 7.19. These results suggest a semiregular modulation of the mass-loss rate on time scales in the range 10^2–10^3 years. Alternatively, these density variations can be entirely attributed to effects of the circumstellar dynamics [234] (see Chapter 6). Likewise, many PNe show multiple halos, which suggests variations on time scales of a few hundred years [7].

7.8.5 Gas/Dust Return

Estimates of gas/dust mass returns in the solar neighborhood and in the whole Galaxy of various stellar populations are mainly uncertain due to the lack of complete samples and the difficulty in obtaining reliable space distributions. What we are aiming for is to calculate the expression

$$\dot{M}_{\text{tot}} = 2\pi \int_0^\infty \Sigma(R) \langle \dot{M}(R) \rangle R \, dR \qquad (7.83)$$

for the total mass-return rate, where $\Sigma(R)$ is the surface density of the stellar population, and $\langle \dot{M}(R) \rangle$ the average mass-loss rate of this population, at the galacto-centric radius R, and consequently $\Sigma(R_0)\langle \dot{M}(R_0) \rangle$ is the local column mass-return rate. Both the surface density distribution and the average mass-loss rate are quantities that are subject to substantial uncertainties.

We start with the well-studied sample of optically bright C-stars discussed above. This population has a surface density of $\approx 45 \, \text{kpc}^{-2}$ [48] ([87]

advocate a value about two times higher). If we use the median mass-loss rate (which is a measure of the mass-loss rate for the majority of the stars) of the bright C-star subsample within 500 pc, $2\times10^{-7}\,M_\odot\,yr^{-1}$, we find a local column mass-return rate of $\approx 10^{-5}\,M_\odot\,yr^{-1}\,kpc^{-2}$. That is, a single star within 1 kpc with a mass-loss rate of $\approx 10^{-5}\,M_\odot\,yr^{-1}$ will provide an equal amount, e.g., IRC+10216 at a distance of about 130 pc and with $\dot{M} \approx 2\times10^{-5}\,M_\odot\,yr^{-1}$. Clearly, the large number of low-mass-loss-rate objects contributes very little to the local mass-return rate from C-stars. The surface density of C-stars losing more than $10^{-5}\,M_\odot\,yr^{-1}$ is $\approx 10\,kpc^{-2}$, but this is particularly uncertain due to incomplete samples and distance uncertainties [87, 98, 130]. The result is a local column mass-return rate of at least $10^{-4}\,M_\odot\,yr^{-1}\,kpc^{-2}$ for this population of C-stars. Crude estimates suggest that the local mass return from M-type stars is comparable to that of the C-stars [128, 199], and also for these stars the mass return is dominated by the rare high-mass-loss-rate objects. Hence, the local mass-return rate from AGB-stars is $\gtrsim 2\times10^{-4}\,M_\odot\,yr^{-1}\,kpc^{-2}$, and this can certainly be too low by a factor of a few. As a comparison we provide the (very uncertain) local mass-return rates for M supergiants and W-R stars, $\approx 2\times10^{-5}\,M_\odot\,yr^{-1}\,kpc^{-2}$ and $\approx 6\times10^{-5}\,M_\odot\,yr^{-1}\,kpc^{-2}$, respectively [129].

The total mass-return rate in the Galaxy due to C-stars can be crudely estimated assuming that their surface density is constant out to 10 kpc, beyond which it drops rapidly [98]. The result is $\gtrsim 0.03\,M_\odot\,yr^{-1}$. For the M-stars we assume an exponential distribution with a scale length of 3 kpc, and the result for the total mass return becomes $\gtrsim 0.2\,M_\odot\,yr^{-1}$, i.e., much higher than that of the C-stars due to the very different spatial distrbution [163]. This leads to a crude estimate of the total mass-return rate of AGB-stars in the Galaxy of $\gtrsim 0.3\,M_\odot\,yr^{-1}$, and the true value can certainly be higher than this by a factor of a few.

7.9 Conclusions and Outlook

We have a rather good overall picture of stellar mass loss on the AGB. All AGB C-stars appear to lose matter at a rate in excess of about $10^{-8}\,M_\odot\,yr^{-1}$. The same is not necessarily true for the M-type AGB stars, but once they have established themselves on the TP-AGB they have mass-loss rates comparable to those of C-stars. The large majority of the wind velocities fall in the range 3–$30\,km\,s^{-1}$, and it appears that, given the same characteristics, a C-star wind has a higher expansion velocity than the wind of an M-type star. The mass loss is relatively isotropic, except perhaps during the final AGB evolution, and, very likely, occurs in the form of clumps. The time-averaged (say over a time scale of 10^4 yr) mass-loss rate probably increases gradually as the star evolves, and its maximum probably increases with the stellar main-sequence mass. At the end it may reach a value of $10^{-4}\,M_\odot\,yr^{-1}$ or even higher. Its dependence on metallicity appears weak, but this has been tested only in a narrow range of metallicities.

We also have a rather good basic understanding of the CSEs produced by the mass loss. They consist of a mixture of gas and dust in about the same proportions as in the interstellar medium. The composition of the gas is well established qualitatively in the case of the simpler and most abundant species. The dust composition is more uncertain, but also the most abundant types of grains seem well identified. The most important processes that govern the physical structure of a CSEs are known, and the same is true for the chemical structure, although the detailed formation routes remain uncertain, in particular for the more complex species and the formation of the original solid nuclei out of which the dust grains emerge.

Despite these achievements, a number of important issues still remain. An interplay between observations and theory is required to make progress, even though the following list is divided into observational and theoretical tasks:

- Observationally establish the structure of the inner CSE, i.e., the region where the final characteristics of the mass loss and the initial density structure and chemical composition are established. This applies to both the gaseous and the dusty medium.
- Observationally establish the geometry of the mass loss (in different phases). A dedicated study of a reasonably large number of O- and C-rich CSEs using different observing methods is required.
- Observationally establish how the density structure, as well as the microstructure, e.g., gas/dust temperature, microturbulence, etc., of a CSE evolve as the gas and the dust flow outward.
- Observationally establish the mass-loss-rate dependences on mass, pulsational activity, C/O, metallicity, and time. Carefully selected samples observed with a variety of methods are called for. Detection of extragalactic CSEs in molecular line emission is on the wish list.
- Observationally establish the evolution during and after the brief "superwind" phase.
- Observationally establish the circumstellar molecular setup and its spatial structure. Information based on only one object (IRC+10216) is far from enough. The same applies to the dust composition and its spatial structure.
- Observationally establish the composition of the grey zone between molecules and bulk particles.
- Observationally establish the gas/dust return, and its composition, of AGB stars.
- Theoretically derive the mass-loss characteristics from first principles.
- Theoretically describe in detail the evolution of the expanding two-fluid medium.

- Theoretically describe the grain nucleation processes, the grain growth processes, and the evolution into amorphous and crystalline grains.
- Theoretically describe the build-up of the full molecular setup for different elemental compositions and stellar characteristics.
- Theoretically calculate the emerging radiation from a CSE of gas and dust with all its complexities. To estimate detailed elemental setups must be a goal.

Acknowledgments: I am grateful to T. Posch and M. Lindqvist for comments on the manuscript, to F. Schöier for producing some of the figures, and to M. West for supplying the OH 1612 Mhz spectrum reproduced in Figure 7.3.

References

1. Aoki, W., Tsuji, T., and Ohnaka, K. *A&A*, 333, L19, 1998.
2. Aoki, W., Tsuji, T., and Ohnaka, K. *A&A*, 350, 945, 1999.
3. Apponi, A. J., McCarthy, M. C., Gottlieb, C. A., and Thaddeus, P. *ApJ*, 516, L103, 1999.
4. Bachiller, R., Forveille, T., Huggins, P. J., and Cox, P. *A&A*, 324, 1123, 1997.
5. Bachiller, R., Fuente, A., Bujarrabal, V., et al. *A&A*, 319, 235, 1997.
6. Bakker, E. J., van Dishoeck, E. F., Waters, L. B. F. M., and Schoenmaker, T. *A&A*, 323, 469, 1997.
7. Balick, B., Wilson, J., and Hajian, A. R. *AJ*, 121, 354, 2001.
8. Barlow, M. J., Nguyen-Q-Rieu, Truong-Bach, et al. *A&A*, 315, L241, 1996.
9. Baud, B. *ApJ*, 250, L79, 1981.
10. Baud, B. and Habing, H. J. *A&A*, 127, 73, 1983.
11. Becklin, E. E., Frogel, J. A., Hyland, A. R., Kristian, J., and Neugebauer, G. *ApJ*, 158, L133, 1969.
12. Bedijn, P. J. *A&A*, 186, 136, 1987.
13. Bedijn, P. J. *A&A*, 205, 105, 1988.
14. Beintema, D. A., van den Ancker, M. E., Molster, F. J., et al. *A&A*, 315, L369, 1996.
15. Bergman, P., Kerschbaum, F., and Olofsson, H. *A&A*, 353, 257, 2000.
16. Bieging, J. H. *ApJ*, 549, L125, 2001.
17. Bieging, J. H., Shaked, S., and Gensheimer, P. D. *ApJ*, 543, 897, 2000.
18. Bieging, J. H., Wilner, D., and Thronson, H. A. *ApJ*, 379, 271, 1991.
19. Bloecker, T. *A&A*, 297, 727, 1995.
20. Blommaert, J. A. D. L., van Langevelde, H. J., and Michiels, W. F. P. *A&A*, 287, 479, 1994.
21. Boboltz, D. A., Diamond, P. J., and Kemball, A. J. *ApJ*, 487, L147, 1997.
22. Bohlin, R. C., Savage, B. D., and Drake, J. F. *ApJ*, 224, 132, 1978.
23. Booth, R. S., Norris, R. P., Porter, N. D., and Kus, A. J. *Nature*, 290, 382, 1981.

24. Bowen, G. H. and Willson, L. A. *ApJ*, 375, L53, 1991.
25. Bowers, P. F. and Johnston, K. J. *ApJ*, 354, 676, 1990.
26. Bowers, P. F. and Knapp, G. R. *ApJ*, 332, 299, 1988.
27. Bowey, J. E., Barlow, M. J., Molster, F. J., et al. *MNRAS*, 331, L1, 2002.
28. Bowey, J. E., Lee, C., Tucker, C., et al. *MNRAS*, 325, 886, 2001.
29. Bryan, G. L., Volk, K., and Kwok, S. *ApJ*, 365, 301, 1990.
30. Bujarrabal, V. and Alcolea, J. *A&A*, 251, 536, 1991.
31. Bujarrabal, V., Fuente, A., and Omont, A. *ApJ*, 421, L47, 1994.
32. Bujarrabal, V., Gomez-Gonzales, J., and Planesas, P. *A&A*, 219, 256, 1989.
33. Cami, J. *Molecular gas and dust around evolved stars*. PhD thesis, University of Amsterdam, 2002.
34. Castor, J. I. *MNRAS*, 149, 111, 1970.
35. Cernicharo, J., Barlow, M. J., Gonzalez-Alfonso, E., et al. *A&A*, 315, L201, 1996.
36. Cernicharo, J., Goicoechea, J. R., and Caux, E. *ApJ*, 534, L199, 2000.
37. Cernicharo, J., Gottlieb, C. A., Guélin, M., et al. *ApJ*, 368, L43, 1991.
38. Cernicharo, J. and Guélin, M. *A&A*, 183, L10, 1987.
39. Cernicharo, J. and Guélin, M. *A&A*, 309, L27, 1996.
40. Cernicharo, J., Guélin, M., and Kahane, C. *A&AS*, 142, 181, 2000.
41. Cernicharo, J., Heras, A. M., Pardo, J. R., et al. *ApJ*, 546, L127, 2001.
42. Cernicharo, J., Heras, A. M., Tielens, A. G. G. M., et al. *ApJ*, 546, L123, 2001.
43. Cernicharo, J., Yamamura, I., González-Alfonso, E., et al. *ApJ*, 526, L41, 1999.
44. Chapman, J. M. and Cohen, R. J. *MNRAS*, 220, 513, 1986.
45. Chapman, J. M., Sivagnanam, P., Cohen, R. J., and Le Squeren, A. M. *MNRAS*, 268, 475, 1994.
46. Chen, P. S., Szczerba, R., Kwok, S., and Volk, K. *A&A*, 368, 1006, 2001.
47. Cioni, M.-R., Loup, C., Habing, H. J., et al. *A&AS*, 144, 235, 2000.
48. Claussen, M. J., Kleinmann, S. G., Joyce, R. R., and Jura, M. *ApJS*, 65, 385, 1987.
49. Collison, A. J. and Nedoluha, G. E. *ApJ*, 442, 311, 1995.
50. Colomer, F., Graham, D. A., Krichbaum, T. P., et al. *A&A*, 254, L17, 1992.
51. Colomer, F., Reid, M. J., Menten, K. M., and Bujarrabal, V. *A&A*, 355, 979, 2000.
52. Cox, P., Huggins, P. J., Maillard, J.-P., et al. *A&A*, 384, 603, 2002.
53. Cuntz, M. and Muchmore, D. O. *ApJ*, 433, 303, 1994.
54. Danchi, W. C., Bester, M., Degiacomi, C. G., Greenhill, L. J., and Townes, C. H. *AJ*, 107, 1469, 1994.
55. Deguchi, S., Fujii, T., Izumiura, H., et al. *ApJS*, 128, 571, 2000.
56. Diamond, P. J., Kemball, A. J., Junor, W., et al. *ApJ*, 430, L61, 1994.
57. Doty, S. D. and Leung, C. M. *ApJ*, 502, 898, 1998.
58. Dyck, H. M., Zuckerman, B., Leinert, C., and Beckwith, S. *ApJ*, 287, 801, 1984.
59. Egan, M. P. and Leung, C. M. *ApJ*, 444, 251, 1995.
60. Elitzur, M. *Astronomical Masers*. Kluwer Academic Publishers (Astroph. and Sp. Sci. Lib. Vol. 170): Dordrecht, 1992.
61. Elitzur, M. and Bujarrabal, V. *A&A*, 308, 330, 1996.
62. Elitzur, M., Goldreich, P., and Scoville, N. *ApJ*, 205, 384, 1976.

63. Elitzur, M. and Ivezić, Ž. *MNRAS*, 327, 403, 2001.
64. Engels, D. and Lewis, B. M. *A&AS*, 116, 117, 1996.
65. Epchtein, N., de Batz, B., Copet, E., et al. *Ap&SS*, 217, 3, 1994.
66. Epchtein, N., Le Bertre, T., and Lepine, J. R. D. *A&A*, 227, 82, 1990.
67. Epchtein, N., Le Bertre, T., Lepine, J. R. D., et al. *A&AS*, 71, 39, 1987.
68. Fabian, D., Posch, T., Mutschke, H., Kerschbaum, F., and Dorschner, J. *A&A*, 373, 1125, 2001.
69. Fix, J. D. and Mulhern, M. G. *ApJ*, 430, 824, 1994.
70. Fong, D., Justtanont, K., Meixner, M., and Campbell, M. T. *A&A*, 396, 581, 2002.
71. Fong, D., Meixner, M., and Shah, R. Y. *ApJ*, 582, L39, 2003.
72. Fouque, P., Le Bertre, T., Epchtein, N., Guglielmo, F., and Kerschbaum, F. *A&AS*, 93, 151, 1992.
73. Foy, R., Chelli, A., Lena, P., and Sibille, F. *A&A*, 79, L5, 1979.
74. Fuente, A., Cernicharo, J., and Omont, A. *A&A*, 330, 232, 1998.
75. Gérard, E. and Le Bertre, T. *A&A*, 397, L17, 2003.
76. Geballe, T. R., Wollman, E. R., and Rank, D. M. *ApJ*, 183, 499, 1973.
77. Goldreich, P. and Scoville, N. *ApJ*, 205, 144, 1976.
78. González-Alfonso, E., Alcolea, J., and Cernicharo, J. *A&A*, 313, L13, 1996.
79. González-Alfonso, E. and Cernicharo, J. *ApJ*, 525, 845, 1999.
80. González-Alfonso, E., Cernicharo, J., Alcolea, J., and Orlandi, M. A. *A&A*, 334, 1016, 1998.
81. González Delgado, D., Olofsson, H., Schwarz, H. E., Eriksson, K., and Gustafsson, B. *A&A*, 372, 885, 2001.
82. González Delgado, D., Olofsson, H., Schwarz, H. E., et al. *A&A*, 399, 1021, 2003.
83. Gray, M. D., Humphreys, E. M. L., and Yates, J. A. *MNRAS*, 304, 906, 1999.
84. Groenewegen, M. A. T. *A&A*, 290, 531, 1994.
85. Groenewegen, M. A. T. *A&A*, 290, 544, 1994.
86. Groenewegen, M. A. T. and de Jong, T. *A&A*, 337, 797, 1998.
87. Groenewegen, M. A. T., de Jong, T., van der Bliek, N. S., Slijkhuis, S., and Willems, F. J. *A&A*, 253, 150, 1992.
88. Groenewegen, M. A. T., Sevenster, M., Spoon, H. W. W., and Pérez, I. *A&A*, 390, 501, 2002.
89. Groenewegen, M. A. T., Sevenster, M., Spoon, H. W. W., and Pérez, I. *A&A*, 390, 511, 2002.
90. Groenewegen, M. A. T., Smith, C. H., Wood, P. R., Omont, A., and Fujiyoshi, T. *ApJ*, 449, L119, 1995.
91. Groenewegen, M. A. T., van der Veen, W. E. C. J., Lefloch, B., and Omont, A. *A&A*, 322, L21, 1997.
92. Groenewegen, M. A. T., Whitelock, P. A., Smith, C. H., and Kerschbaum, F. *MNRAS*, 293, 18, 1998.
93. Groesbeck, T. D., Phillips, T. G., and Blake, G. A. *ApJS*, 94, 147, 1994.
94. Guélin, M., Lucas, R., and Cernicharo, J. *A&A*, 280, L19, 1993.
95. Guélin, M., Lucas, R., and Neri, R. In P. A. Shaver, editor, *Science with Large Millimetre Arrays*, page 276. Springer-Verlag: Berlin, 1996.
96. Guélin, M., Neininger, N., and Cernicharo, J. *A&A*, 335, L1, 1998.
97. Guglielmo, F., Epchtein, N., Le Bertre, T., et al. *A&AS*, 99, 31, 1993.
98. Guglielmo, F., Le Bertre, T., and Epchtein, N. *A&A*, 334, 609, 1998.

99. Gustafsson, B., Eriksson, K., Kiselman, D., Olander, N., and Olofsson, H. A&A, 318, 535, 1997.
100. Habing, H. J. A&A Rev., 7, 97, 1996.
101. Habing, H. J., Tignon, J., and Tielens, A. G. G. M. A&A, 286, 523, 1994.
102. Haikala, L. K., Nyman, L.-Å., and Forsström, V. A&AS, 103, 107, 1994.
103. Hashimoto, O., Izumiura, H., Kester, D. J. M., and Bontekoe, T. R. A&A, 329, 213, 1998.
104. Hawkins, G. W. A&A, 229, L5, 1990.
105. Herman, J., Baud, B., Habing, H. J., and Winnberg, A. A&A, 143, 122, 1985.
106. Heske, A., Habing, H. J., Forveille, T., Omont, A., and van der Veen, W. E. C. J. A&A, 239, 173, 1990.
107. Hofmann, K.-H., Balega, Y., Blöcker, T., and Weigelt, G. A&A, 379, 529, 2001.
108. Höfner, S. and Dorfi, E. A. A&A, 319, 648, 1997.
109. Huggins, P. J., Bachiller, R., Cox, P., and Forveille, T. ApJ, 401, L43, 1992.
110. Humphreys, E. M. L., Gray, M. D., Yates, J. A., et al. A&A, 386, 256, 2002.
111. Humphreys, E. M. L., Yates, J. A., Gray, M. D., Field, D., and Bowen, G. H. A&A, 379, 501, 2001.
112. Imai, H., Sasao, T., Kameya, O., et al. A&A, 317, L67, 1997.
113. Ivezić, Ž. and Elitzur, M. ApJ, 445, 415, 1995.
114. Ivezić, Ž., Nenkova, M., and Elitzur, M. *User Manual for DUSTY*. Univ. Kentucky Internal Rep., 1999.
115. Izumiura, H., Catchpole, R., Deguchi, S., et al. ApJS, 98, 271, 1995.
116. Izumiura, H., Deguchi, S., Fujii, T., et al. ApJS, 125, 257, 1999.
117. Izumiura, H., Waters, L. B. F. M., de Jong, T., et al. A&A, 323, 449, 1997.
118. Jackson, T., Ivezić, Ž., and Knapp, G. R. MNRAS, 337, 749, 2002.
119. Jiang, B. W., Szczerba, R., and Deguchi, S. A&A, 344, 918, 1999.
120. Johansson, L. E. B., Andersson, C., Ellder, J., et al. A&A, 130, 227, 1984.
121. Johnson, J. J. and Jones, T. J. AJ, 101, 1735, 1991.
122. Jorissen, A. and Knapp, G. R. A&AS, 129, 363, 1998.
123. Josselin, E., Loup, C., Omont, A., et al. A&AS, 129, 45, 1998.
124. Josselin, E., Mauron, N., Planesas, P., and Bachiller, R. A&A, 362, 255, 2000.
125. Jura, M. ApJ, 472, 806, 1996.
126. Jura, M. In T. J. Bernatowicz, E. Zinner, editor, *Astrophysical Implications of the Laboratory Study of Presolar Materials*, page 379. AIP: Woodbury, N.Y., 1997.
127. Jura, M. and Kahane, C. ApJ, 521, 302, 1999.
128. Jura, M. and Kleinmann, S. G. ApJ, 341, 359, 1989.
129. Jura, M. and Kleinmann, S. G. ApJS, 73, 769, 1990.
130. Jura, M. and Kleinmann, S. G. ApJ, 364, 663, 1990.
131. Jura, M., Webb, R. A., and Kahane, C. ApJ, 550, L71, 2001.
132. Justtanont, K., Feuchtgruber, H., de Jong, T., et al. A&A, 330, L17, 1998.
133. Justtanont, K., Skinner, C. J., and Tielens, A. G. G. M. ApJ, 435, 852, 1994.
134. Justtanont, K., Skinner, C. J., Tielens, A. G. G. M., Meixner, M., and Baas, F. ApJ, 456, 337, 1996.
135. Kahane, C., Cernicharo, J., Gomez-Gonzalez, J., and Guelin, M. A&A, 256, 235, 1992.
136. Kahane, C., Dufour, E., Busso, M., et al. A&A, 357, 669, 2000.
137. Kahane, C. and Jura, M. A&A, 290, 183, 1994.

138. Kahane, C. and Jura, M. *A&A*, 310, 952, 1996.
139. Kastner, J. H. *ApJ*, 401, 337, 1992.
140. Kastner, J. H. and Weintraub, D. A. *ApJ*, 434, 719, 1994.
141. Kawaguchi, K., Kagi, E., Hirano, T., Takano, S., and Saito, S. *ApJ*, 406, L39, 1993.
142. Keady, J. J. and Ridgway, S. T. *ApJ*, 406, 199, 1993.
143. Keene, J., Young, K., Phillips, T. G., Buettgenbach, T. H., and Carlstrom, J. E. *ApJ*, 415, L131, 1993.
144. Kemball, A. J. and Diamond, P. J. *ApJ*, 481, L111, 1997.
145. Kemper, F., Waters, L. B. F. M., de Koter, A., and Tielens, A. G. G. M. *A&A*, 369, 132, 2001.
146. Kerschbaum, F. and Hron, J. *A&AS*, 106, 397, 1994.
147. Kerschbaum, F., Lazaro, C., and Habison, P. *A&AS*, 118, 397, 1996.
148. Kerschbaum, F. and Olofsson, H. *A&AS*, 138, 299, 1999.
149. Kim, S., Martin, P. G., and Hendry, P. D. *ApJ*, 422, 164, 1994.
150. Kleinmann, S. G., Lysaght, M. G., Pughe, W. L., et al. *Ap&SS*, 217, 11, 1994.
151. Knapp, G. R., Crosas, M., Young, K., and Ivezić, Ž. *ApJ*, 534, 324, 2000.
152. Knapp, G. R., Dobrovolsky, S. I., Ivezić , Z., et al. *A&A*, 351, 97, 1999.
153. Knapp, G. R., Jorissen, A., and Young, K. *A&A*, 326, 318, 1997.
154. Knapp, G. R., Young, K., Lee, E., and Jorissen, A. *ApJS*, 117, 209, 1998.
155. Krüger, D. and Sedlmayr, E. *A&A*, 321, 557, 1997.
156. Kwok, S., Volk, K., and Bidelman, W. P. *ApJS*, 112, 557, 1997.
157. Kwok, S., Volk, K., and Hrivnak, B. J. *A&A*, 350, L35, 1999.
158. Lambert, D. L., Gustafsson, B., Eriksson, K., and Hinkle, K. H. *ApJS*, 62, 373, 1986.
159. Lamers, H. J. G. L. M. and Cassinelli, J. P. *Introduction to Stellar Winds*. Cambridge University Press: Cambridge, 1999.
160. Latter, W. B., Walker, C. K., and Maloney, P. R. *ApJ*, 419, L97, 1993.
161. Le Bertre, T. *A&A*, 324, 1059, 1997.
162. Le Bertre, T. and Gérard, E. *A&A*, 378, L29, 2001.
163. Le Bertre, T., Tanaka, M., Yamamura, I., and Murakami, H. *A&A*, in press, 2003.
164. Le Bertre, T. and Winters, J. M. *A&A*, 334, 173, 1998.
165. Le Squeren, A. M., Sivagnanam, P., Dennefeld, M., and David, P. *A&A*, 254, 133, 1992.
166. Lewis, B. M. *ApJ*, 533, 959, 2000.
167. Lewis, B. M. *MNRAS*, 576, 445, 2002.
168. Lewis, B. M., Eder, J., and Terzian, Y. *ApJ*, 362, 634, 1990.
169. Lewis, B. M., Oppenheimer, B. D., and Daubar, I. J. *ApJ*, 548, L77, 2001.
170. Lindqvist, M., Habing, H. J., and Winnberg, A. *A&A*, 259, 118, 1992.
171. Lindqvist, M., Olofsson, H., Lucas, R., et al. *A&A*, 351, L1, 1999.
172. Lindqvist, M., Schöier, F. L., Lucas, R., and Olofsson, H. *A&A*, 361, 1036, 2000.
173. Lindqvist, M., Winnberg, A., Habing, H. J., and Matthews, H. E. *A&AS*, 92, 43, 1992.
174. Lopez, B., Danchi, W. C., Bester, M., et al. *ApJ*, 488, 807, 1997.
175. Lorenz-Martins, S., de Araújo, F. X., Codina Landaberry, S. J., de Almeida, W. G., and de Nader, R. V. *A&A*, 367, 189, 2001.
176. Loup, C., Forveille, T., Omont, A., and Paul, J. F. *A&AS*, 99, 291, 1993.

177. Lucas, R., Bujarrabal, V., Guilloteau, S., et al. *A&A*, 262, 491, 1992.
178. Mamon, G. A., Glassgold, A. E., and Huggins, P. J. *ApJ*, 328, 797, 1988.
179. Marengo, M., Ivezić, Ž., and Knapp, G. R. *MNRAS*, 324, 1117, 2001.
180. Marengo, M., Karovska, M., Fazio, G. G., et al. *ApJ*, 556, L47, 2001.
181. Mauron, N. and Caux, E. *A&A*, 265, 711, 1992.
182. Mauron, N. and Huggins, P. J. *A&A*, 359, 707, 2000.
183. McCarthy, D. W. and Low, F. J. *ApJ*, 202, L37, 1975.
184. Melnick, G. J., Neufeld, D. A., Ford, K. E. S., Hollenbach, D. J., and Ashby, M. L. N. *Nature*, 412, 160, 2001.
185. Menten, K. M. and Melnick, G. J. *ApJ*, 341, L91, 1989.
186. Messineo, M., Habing, H. J., Sjouwerman, L. O., Omont, A., and Menten, K. M. *A&A*, 393, 115, 2002.
187. Molster, F. J., Waters, L. B. F. M., and Tielens, A. G. G. M. *A&A*, 382, 222, 2002.
188. Molster, F. J., Yamamura, I., . M. Waters, L. B. F., et al. *A&A*, 366, 923, 2001.
189. Murakawa, K., Tamura, M., Suto, H., et al. *A&A*, 395, L9, 2002.
190. Mutschke, H., Andersen, A. C., Clément, D., Henning, T., and Peiter, G. *A&A*, 345, 187, 1999.
191. Neri, R., Kahane, C., Lucas, R., Bujarrabal, V., and Loup, C. *A&AS*, 130, 1, 1998.
192. Netzer, N. and Knapp, G. R. *ApJ*, 323, 734, 1987.
193. Neugebauer, G., Habing, H. J., van Duinen, R., et al. *ApJ*, 278, L1, 1984.
194. Neugebauer, G. and Leighton, R. B. *Two-Micron Sky Survey. A Preliminary Catalogue*. NASA SP-3047: Washington, 1969.
195. Nyman, L.-Å., Booth, R. S., Carlstrom, U., et al. *A&AS*, 93, 121, 1992.
196. Nyman, L.-Å., Hall, P. J., and Le Bertre, T. *A&A*, 280, 551, 1993.
197. Nyman, L.-Å., Olofsson, H., Johansson, L. E. B., et al. *A&A*, 269, 377, 1993.
198. Ohishi, M., Kaifu, N., Kawaguchi, K., et al. *ApJ*, 345, L83, 1989.
199. Olivier, E. A., Whitelock, P., and Marang, F. *MNRAS*, 326, 490, 2001.
200. Olnon, F. M., Raimond, E., Neugebauer, G., et al. *A&AS*, 65, 607, 1986.
201. Olofsson, H., Bergman, P., Eriksson, K., and Gustafsson, B. *A&A*, 311, 587, 1996.
202. Olofsson, H., Bergman, P., Lucas, R., et al. *A&A*, 353, 583, 2000.
203. Olofsson, H., Carlström, U., Eriksson, K., Gustafsson, B., and Willson, L. A. *A&A*, 230, L13, 1990.
204. Olofsson, H., Eriksson, K., Gustafsson, B., and Carlström, U. *ApJS*, 87, 267, 1993.
205. Olofsson, H., Eriksson, K., Gustafsson, B., and Carlström, U. *ApJS*, 87, 305, 1993.
206. Olofsson, H., González Delgado, D., Kerschbaum, F., and Schöier, F. *A&A*, 391, 1053, 2002.
207. Olofsson, H., Lindqvist, M., Nyman, L.-Å., and Winnberg, A. *A&A*, 329, 1059, 1998.
208. Olofsson, H., Lindqvist, M., Winnberg, A., Nyman, L.-Å., and Nguyen-Q-Rieu. *A&A*, 245, 611, 1991.
209. Omont, A. and et al. In M. D. Bicay, R. M. Cutri, and B. F. Madore, editor, *ASP Conf. Ser. 177: Astrophysics with Infrared Surveys: A Prelude to SIRTF*, page 261. ASP: San Francisco, 1999.

210. Omont, A., Lucas, R., Morris, M., and Guilloteau, S. A&A, 267, 490, 1993.
211. Reid, M. J. and Menten, K. M. ApJ, 360, L51, 1990.
212. Reimers, D. Memoires of the Societe Royale des Sciences de Liege, 8, 369, 1975.
213. Richards, A. M. S., Yates, J. A., and Cohen, R. J. MNRAS, 306, 954, 1999.
214. Ridgway, S. T., Hall, D. N. B., Wojslaw, R. S., Kleinmann, S. G., and Weinberger, D. A. Nature, 264, 345, 1976.
215. Rouleau, F. and Martin, P. G. ApJ, 377, 526, 1991.
216. Rowan-Robinson, M. and Harris, S. MNRAS, 202, 767, 1983.
217. Ryde, N., Eriksson, K., Gustafsson, B., Lindqvist, M., and Olofsson, H. Ap&SS, 255, 301, 1997.
218. Ryde, N., Gustafsson, B., Hinkle, K. H., et al. A&A, 347, L35, 1999.
219. Ryde, N., Schöier, F. L., and Olofsson, H. A&A, 345, 841, 1999.
220. Sánchez Contreras, C., Bujarrabal, V., and Alcolea, J. A&A, 327, 689, 1997.
221. Sahai, R. ApJ, 362, 652, 1990.
222. Sahai, R. and Bieging, J. H. AJ, 105, 595, 1993.
223. Sahai, R., Hines, D. C., Kastner, J. H., et al. ApJ, 492, L163, 1998.
224. Sahai, R. and Nyman, L.-Å. ApJ, 487, L155, 1997.
225. Sahai, R., Trauger, J. T., Watson, A. M., et al. ApJ, 493, 301, 1998.
226. Schöier, F. L. and Olofsson, H. A&A, 359, 586, 2000.
227. Schöier, F. L. and Olofsson, H. A&A, 368, 969, 2001.
228. Schröder, K.-P., Winters, J. M., and Sedlmayr, E. A&A, 349, 898, 1999.
229. Schutte, W. A. and Tielens, A. G. G. M. ApJ, 343, 369, 1989.
230. Sedlmayr, E. and Dominik, C. Space Sci. Rev., 73, 211, 1995.
231. Sevenster, M. N., Chapman, J. M., Habing, H. J., Killeen, N. E. B., and Lindqvist, M. A&AS, 122, 79, 1997.
232. Sevenster, M. N., Chapman, J. M., Habing, H. J., Killeen, N. E. B., and Lindqvist, M. A&AS, 124, 509, 1997.
233. Sevenster, M. N., van Langevelde, H. J., Moody, R. A., et al. A&A, 366, 481, 2001.
234. Simis, Y. J. W., Icke, V., and Dominik, C. A&A, 371, 205, 2001.
235. Skinner, C. J. and Whitmore, B. MNRAS, 231, 169, 1988.
236. Skinner, C. J. and Whitmore, B. MNRAS, 234, 79P, 1988.
237. Sloan, G. C., Levan, P. D., and Little-Marenin, I. R. ApJ, 463, 310, 1996.
238. Solomon, P., Jefferts, K. B., Penzias, A. A., and Wilson, R. W. ApJ, 163, L53, 1971.
239. Speck, A. K., Meixner, M., and Knapp, G. R. ApJ, 545, L145, 2000.
240. Steffen, M. and Schönberner, D. A&A, 357, 180, 2000.
241. Sutton, E. C., Storey, J. W. V., Townes, C. H., and Spears, D. L. ApJ, 224, L123, 1978.
242. Sylvester, R. J., Barlow, M. J., Liu, X. W., et al. MNRAS, 291, L42, 1997.
243. Sylvester, R. J., Kemper, F., Barlow, M. J., et al. A&A, 352, 587, 1999.
244. Szymczak, M., Cohen, R. J., and Richards, A. M. S. MNRAS, 297, 1151, 1998.
245. Szymczak, M., Cohen, R. J., and Richards, A. M. S. MNRAS, 304, 877, 1999.
246. Szymczak, M., Cohen, R. J., and Richards, A. M. S. A&A, 371, 1012, 2001.
247. Te Lintel Hekkert, P., Caswell, J. L., Habing, H. J., et al. A&AS, 90, 327, 1991.
248. Thai-Q-Tung, Dinh-v-Trung, Nguyen-Q-Rieu, et al. A&A, 331, 317, 1998.
249. Truong-Bach, Nguyen-Q-Rieu, Morris, D., and Deguchi, S. A&A, 230, 431, 1990.

250. Truong-Bach, Sylvester, R. J., Barlow, M. J., et al. *A&A*, 345, 925, 1999.
251. Tsuji, T., Ohnaka, K., Aoki, W., and Yamamura, I. *A&A*, 320, L1, 1997.
252. Turner, B. E., Steimle, T. C., and Meerts, L. *ApJ*, 426, L97, 1994.
253. Tuthill, P. G., Monnier, J. D., Danchi, W. C., and Lopez, B. *ApJ*, 543, 284, 2000.
254. van der Veen, W. E. C. J. *A&A*, 210, 127, 1989.
255. van der Veen, W. E. C. J. and Habing, H. J. *A&A*, 194, 125, 1988.
256. van Langevelde, H. J., Janssens, A. M., Goss, W. M., Habing, H. J., and Winnberg, A. *A&AS*, 101, 109, 1993.
257. van Langevelde, H. J. and Spaans, M. *MNRAS*, 264, 597, 1993.
258. van Langevelde, H. J., van der Heiden, R., and van Schooneveld, C. *A&A*, 239, 193, 1990.
259. van Langevelde, H. J., Vlemmings, W., Diamond, P. J., Baudry, A., and Beasley, A. J. *A&A*, 357, 945, 2000.
260. van Loon, J. T. *A&A*, 354, 125, 2000.
261. van Loon, J. T., Zijlstra, A. A., Bujarrabal, V., and Nyman, L.-Å. *A&A*, 368, 950, 2001.
262. van Loon, J. T., Zijlstra, A. A., Whitelock, P. A., et al. *A&A*, 329, 169, 1998.
263. van Zadelhoff, G.-J., Dullemond, C. P., van der Tak, F. F. S., et al. *A&A*, 395, 373, 2002.
264. Vassiliadis, E. and Wood, P. R. *ApJ*, 413, 641, 1993.
265. Vlemmings, W. H. T., Diamond, P. J., and van Langevelde, H. J. *A&A*, 394, 589, 2002.
266. Vlemmings, W. H. T., van Langevelde, H. J., and Diamond, P. J. *A&A*, 393, L33, 2002.
267. Volk, K., Kwok, S., and Hrivnak, B. J. *ApJ*, 516, L99, 1999.
268. Volk, K., Kwok, S., Hrivnak, B. J., and Szczerba, R. *ApJ*, 567, 412, 2002.
269. Volk, K., Xiong, G., and Kwok, S. *ApJ*, 530, 408, 2000.
270. von Helden, G., Tielens, A. G. G. M., van Heijnsbergen, D., et al. *Science*, 288, 313, 2000.
271. Wachter, A., Schröder, K.-P., Winters, J. M., Arndt, T. U., and Sedlmayr, E. *A&A*, 384, 452, 2002.
272. Walmsley, C. M., Chini, R., Kreysa, E., et al. *A&A*, 248, 555, 1991.
273. Wannier, P. G., Andersson, B.-G., Olofsson, H., Ukita, N., and Young, K. *ApJ*, 380, 593, 1991.
274. Waters, L. B. F. M., Cami, J., de Jong, T., et al. *Nature*, 391, 868, 1998.
275. Waters, L. B. F. M., Loup, C., Kester, D. J. M., Bontekoe, T. R., and de Jong, T. *A&A*, 281, L1, 1994.
276. Waters, L. B. F. M., Molster, F. J., de Jong, T., et al. *A&A*, 315, L361, 1996.
277. Weidemann, V. *A&A*, 188, 74, 1987.
278. Weigelt, G., Balega, Y., Blöcker, T., et al. *A&A*, 333, L51, 1998.
279. Weigelt, G., Balega, Y. Y., Blöcker, T., et al. *A&A*, 392, 131, 2002.
280. Willems, F. J. and de Jong, T. *A&A*, 196, 173, 1988.
281. Willson, L. A. *ARA&A*, 38, 573, 2000.
282. Wilson, W. and Barrett, A. *Science*, 161, 778, 1968.
283. Winnberg, A., Lindqvist, M., Olofsson, H., and Henkel, C. *A&A*, 245, 195, 1991.
284. Winters, J. M., Keady, J. J., Gauger, A., and Sada, P. V. *A&A*, 359, 651, 2000.

285. Winters, J. M., Le Bertre, T., Jeong, K. S., Helling, C., and Sedlmayr, E. A&A, 361, 641, 2000.
286. Wood, P. R., Habing, H. J., and McGregor, P. J. A&A, 336, 925, 1998.
287. Wood, P. R., Whiteoak, J. B., Hughes, S. M. G., et al. ApJ, 397, 552, 1992.
288. Woods, P. M., Millar, T. J., Herbst, E., and Zijlstra, A. A. A&A, 402, 189, 2003.
289. Woods, P. M., Schöier, F. L., Nyman, L.-Å., and Olofsson, H. A&A, 402, 617, 2003.
290. Yamamura, I., de Jong, T., Onaka, T., Cami, J., and Waters, L. B. F. M. A&A, 341, L9, 1999.
291. Yamamura, I., Dominik, C., de Jong, T., Waters, L. B. F. M., and Molster, F. J. A&A, 363, 629, 2000.
292. Young, K. ApJ, 445, 872, 1995.
293. Young, K., Cox, P., Huggins, P. J., Forveille, T., and Bachiller, R. ApJ, 522, 387, 1999.
294. Young, K., Phillips, T. G., and Knapp, G. R. ApJ, 409, 725, 1993.
295. Zijlstra, A. A., Gaylard, M. J., Te Lintel Hekkert, P., et al. A&A, 243, L9, 1991.
296. Zijlstra, A. A., Loup, C., Waters, L. B. F. M., and de Jong, T. A&A, 265, L5, 1992.
297. Zijlstra, A. A., Loup, C., Waters, L. B. F. M., et al. MNRAS, 279, 32, 1996.
298. Ziurys, L. M., Savage, C., Highberger, J. L., et al. ApJ, 564, L45, 2002.
299. Zubko, V. and Elitzur, M. ApJ, 544, L137, 2000.
300. Zuckerman, B. and Dyck, H. M. ApJ, 304, 394, 1986.

8 AGB Stars as Tracers of Stellar Populations

Harm J. Habing[1] and Patricia A. Whitelock[2]

[1] Leiden Observatory
[2] South African Astronomical Observatory

8.1 Introduction

The AGB stars that we see now were formed some time ago, and their presence in a particular galaxy informs us about the history of that galaxy. The brightest AGB stars (30 000 $L_\odot$) are perhaps 100 Myr old, whereas 3 000 $L_\odot$ AGB stars are found in globular clusters with an age exceeding 10 Gyr. Other types of evolved stars, e.g., RR Lyrae variables, may also be used as tracers of galactic history, but AGB stars have two important advantages: (i) In any galaxy they are the most luminous red stars (except for the even rarer and much younger red supergiants), and (ii) they are at their brightest in the near-infrared, where the effect of interstellar extinction is often unimportant. Unfortunately, AGB stars are short-lived and therefore rare, which is clearly a disadvantage if their progenitor population is small, as it is in dwarf galaxies. In this chapter we will discuss only the most luminous AGB stars, a still rarer species: If, for example, we restrict ourselves to AGB stars more luminous than stars at the tip of the red giant branch (and that is close to what we do), only one luminous AGB star is predicted per 12 000 main-sequence stars of 1 $M_\odot$ and 1 per 240 stars of 5 $M_\odot$ [190]; the metallicity Z has a minor influence on these values.

What is a *population*? Clearly, it is a sample of stars selected by certain criteria. Assume that a star is specified by three parameters: its main-sequence mass, M_{MS}, its age, a, and its atomic composition or "metallicity," Z. The first, frequently used, definition is that a population consists of stars with a range of masses, all of which were formed at roughly the same time, so that they have a common age, a, and a common metallicity, Z. Most stellar clusters are examples of such a population: The stars in a cluster differ only in their mass, M_{MS}. In the literature this type of population is also referred to as a "single stellar population" or SSP. The color-magnitude diagram (CMD) is a useful tool for identifying and characterizing particular populations.

In the second definition of *population* one considers the phase-space distribution function $f(x, v, t)$, i.e., the number of stars as a function of their location in the galaxy, x, and of their velocity, v. Suppose that we can decompose f into a set of "building blocks," functions that each describe meaningfully the distribution in space and in velocity of a group of stars then each building block may be called a *population*. Later in this chapter we use this definition when discussing the Mira variables in the solar neighborhood: the longer their periods, the more the stars are confined to the Galactic plane and the closer they follow Galactic rotation. In this, and in other Milky Way applications, dynamical equilibrium is usually assumed; i.e., f does not vary in time ($\partial f/\partial t = 0$). The assumption is not necessary, nor is it always ensured; e.g., the Magellanic Clouds are continuously deformed by each other and by passage close to the MWG, and are therefore not in dynamical equilibrium. Many of the dwarf spheroidals may also lack dynamical equilibrium. Obviously, we expect that these two definitions of population will be equivalent, and result in the selection of identical samples of stars from any larger group or galaxy.

Historically, the word *population* was used for the first time by Baade (1944, [14]) after he had succeeded in detecting individual stars in the inner part of M 31 on red-sensitive photographic plates. He noticed that these red stars were somewhat fainter than the O and B stars seen before in the outer parts of M 31, but that they were comparable to the known stars of "high velocity" close to the Sun and to the stars in globular clusters. Baade then proposed the existence of two "populations" of stars that differ in their distribution in the Hertzsprung-Russell diagram. Technically, Baade's analysis is the same as the present-day decomposition of color-magnitude diagrams; of course, today the size and the quality of the databases are much greater. Furthermore, our understanding of stellar evolution, although still incomplete, has advanced dramatically from Baade's time when the astrophysical connection between main-sequence stars and red giants was not yet understood.

Practicalities

We will consider only "luminous AGB stars," that is, red stars above the tip of the red giant branch (tRGB), for which we adopt the value of 2900 $L_\odot$ (or $M_{bol} = -3.9$) [152]. This value depends somewhat on metallicity, but not enough to affect our discussion. For I(tRGB) we will use -4.0, a value that is practically independent of metallicity. We also will use K(tRGB) $= -6.6$, although this quantity depends on the color of the star: When $(J - K)$ varies from 1.0 to 10.4, K(tRGB) varies from -6.3 to -7.1 [135].

We thus exclude AGB stars with a luminosity below 2900 $L_\odot$, and we do this for convenience . First, in a sample of red giants with $L < 2900 L_\odot$ the RGB stars are much more frequent than the early AGB stars and since the characteristics of the two groups are very similar, it will be very difficult to separate the two groups. Second, all *red* stellar objects with $L > 2900 L_\odot$ will be AGB stars (except for a few red supergiants). This selection by luminosity excludes stellar groups in which the AGB terminates below the tRGB, as probably happens in metal-poor globular clusters. Figure 8.17 shows that the tRGB marks a very strong discontinuity in the luminosity function of the red giants in the Magellanic Clouds and accentuates the rarity of luminous AGB stars.

AGB stars of all masses will eventually experience thermal pulsations, and this happens when they approach their maximum luminosity. This does not mean, however, that all our "luminous" AGB stars suffer from thermal pulses, although many, perhaps most, will do that. Theoretical work [190] suggests that the TP-AGB phase begins at $M_{bol} = -3.1$ for a star of $1\,M_\odot$, at -3.3 for $2\,M_\odot$, and at -5.7 for $5.0\,M_\odot$; these numbers depend a bit on metallicity and on the way mass-loss is treated. Because stars of a mass between 1 and $2\,M_\odot$ have thermal pulses at a luminosity below that of the tip of the RGB, it follows that intrinsic C-stars may be found with luminosities below the tRGB.

AGB stars close to their final luminosity often have one or more of the following characteristics: They are (i) (intrinsic) carbon stars, (ii) long-period variables (LPVs) with large amplitudes ($\Delta I > 0.5$); these include Miras and maser stars (see below) and (iii) stars with thick circumstellar envelopes (CSEs) (these include the maser stars). Note that a star may be a C star *and* an LPV *and* have a thick CSE, all at the same time: CW Leo (IRC +10216) is an example. Many of the carbon stars detected in, e.g., spectroscopic surveys will be *extrinsic*, and thus not be on the AGB; for the difference between intrinsic and extrinsic carbon stars see chapter 9. Variability is often, and perhaps always, present in luminous AGB stars, but the amplitude may be as small as a few tenths of a magnitude. Small-amplitude variables (say $0.1 < \Delta I < 0.3$) are usually irregular, while the larger-amplitude Miras have much better defined periods. AGB stars do not pulsate in the regular and predictable manner of, e.g., the Cepheids and RR Lyraes. In fact, the recent monitoring programs in search of microgravitational deflections, programs such as MACHO, EROS, and OGLE, have given huge databases with accurate photometric measurements, and this has led to much attention to the pulsation modes of long-period variables; for more details the reader is referred to Chapter 2.

The derivation of M_{bol} from photometry in a number of spectral bands is straightforward if the maximum in the flux distribution is covered by the bands. For stars with very thick CSEs the maximum is between 5 and 20 μm, which is beyond the range commonly studied.

A brief word about maser stars. There are three molecules known to form strong circumstellar masers: several lines of SiO and H_2O, and three lines of OH. Circumstellar masers have been found only when the mass-loss rate is higher than about $10^{-7}\,M_\odot\,\mathrm{yr}^{-1}$. Whenever a maser is found, we will call the star a maser star. A special subclass are the OH/IR stars, in which the 1612 MHz OH line dominates strongly (often it is the only OH line detected), whereas simultaneously the circumstellar matter obscures the star at visual wavelengths, and sometimes even in the K band. We will assume that all maser stars detected in 1612 MHz-line surveys are OH/IR stars. The 1612 MHz maser line has a regular profile with two peaks. The difference, ΔV, between the velocities of the two peaks is an indicator of the initial mass of the star (see below). $\Delta V/2$ provides a good estimate of the gas expansion velocity of the CSE (see Chapter 7); in this chapter denoted by V_{exp}.

We will have to estimate the ages of the AGB stars that we find. How? The more massive the star, the higher the luminosity it will reach. Model calculations [190] predict that a star of $1\,M_\odot$ will ultimately reach $M_{\text{bol}} = -4.0$, and a star of $5\,M_\odot$ will reach $M_{\text{bol}} = -5.8$; these values depend remarkably little on the metallicity, Z, but the models do not extend to very high metallicities. If we measure M_{bol} for a single star, we cannot assume that it has reached its maximum brightness; it may still be increasing in luminosity, and it probably *is*. Thus, its present M_{bol} tells us only an upper limit to its age.

In a few cases we may go further. First, if we observe different stars in a cluster (which are all of the same age) the highest luminosity is closer to that of an AGB star with the age of the cluster; for an application see Figure xx in Chapter 2. Second, for M-type LPVs the period, P, is a measure of the stellar age (longer period, younger age); empirical proof is shown in Figures 8.3 and 8.4. We will, when appropriate, follow [90]: (i) $P < 255^{\text{d}}$, age > 5 Gyr; (ii) $255 < P < 450^{\text{d}}$, age between 1 and 3 Gyr; and (iii) $P > 450^{\text{d}}$, age < 1 Gyr. A histogram of the periods found in a sample of stars tells us the distribution of the stellar ages. However, we note that such ages are established by a comparison with low-metallicity models and that at high metallicities the ages could be *considerably* higher; this remains a serious problem in interpreting the ages of bulge populations in particular. Third, for OH/IR stars the velocity difference, ΔV, is, as we will see later, an age indicator – albeit a rough one; empirical proof may be found in [166].

We will frequently use color-magnitude diagrams (CMDs) in which stars are specified by a photometric color, e.g., $I - K$, along the X-axis and by a photometric magnitude, e.g., K, along the Y-axis.

Some AGB stars will have developed from close binaries that went through a "common envelope" phase and will now exist of one AGB star with a mass close to the sum of the two individual masses. For a discussion of these "blue stragglers" see Chapter 9. As an example, consider a binary of two $1\,M_\odot$ stars. The system needs at least 10 Gyr before the first star turns off the main sequence, but after merging with its companion it may reach the luminosity of a $2\,M_\odot$ AGB star, to which one would assign an age of about 1.2 Gyr. The fraction of the luminous AGB stars that originate in binaries is unclear, but such stars could make an important contribution to the populations observed in dense regions, such as the centers and bulges of galaxies [146]. Nevertheless, we assume in the following discussion that AGB stars are, in general, born as single stars.

Surface brightness fluctuations offer a potential method of determining distances to elliptical galaxies; the method has been applied to galaxies with red shifts up to $10\,000$ km s^{-1} [99]. This technique relies on the fact that the smoothness or otherwise of identical galaxies is proportional to their relative distances. In intermediate-age populations the brightest AGB stars will influence these measurements, particularly at near-IR wavelengths. This provides a rather different reason for understanding the AGB and its dependence on age and metallicity in extragalactic systems.

When the galaxy is too far away to distinguish individual stars, the presence of AGB-stars may still show up in the integrated spectrum [106]. The so-called AGB phase transition provides a useful tool in identifying populations in the age range of about 400 Myr to 2 Gyr [116]. In this age bracket the luminous AGB evolutionary phase makes a major contribution to the integrated luminosity, and significantly alters the $(V-K)$-colors, of any population. Stars younger than this proceed rather rapidly through their AGB evolution due to hot bottom burning, while older stars spend little time on the AGB, which terminates just above the tRGB. At the very end of this chapter we will discuss briefly a recent study that claims to have detected AGB stars in the integrated spectrum of a galaxy 64 Mpc away.

The Timeliness of the Study of AGB Stars

In an external galaxy the brightest evolved stars will be the supergiants, followed by AGB stars. The advent of adaptive optics on large telescopes, together with the dramatic improvements of infrared arrays, now makes it possible to detect and study individual AGB stars in ever more distant galaxies. The availability of the NICMOS camera on HST is contributing greatly to such studies. This chapter summarizes the information to date, but given the intense observational activity in this area we can be certain that parts of it will be out of date by the time the book is published.

Outline of the Remainder of this Chapter

We will discuss mainly AGB stars in galaxies of the Local Group (LG) and we will use van den Bergh's book [183] as the source of quantitative information. In Section 2 we discuss the Milky Way Galaxy (MWG), in Section 3 the Magellanic Clouds. Section 4 is for M 31 and its immediate companions, M 32 and NGC 205, and for M 33. Section 5 will deal with AGB stars in the other galaxies of the LG, and Section 6 is for galaxies outside of the LG; this last section will be brief, in view of our ignorance.

A word about citations: The literature on the subject of this chapter is vast and growing rapidly. We limit the citations to those from recent times and refer the reader to van den Bergh's book [183] for further details. The reader who wants more recent information is advised to use the Simbad database (http://simbad.u-strasbg.fr).

8.2 The Milky Way Galaxy (MWG)

The Milky Way is a spiral galaxy that is conveniently, but simplistically, subdivided into four principal components: nucleus, bulge, thick and thin disk, and halo. Observations of other spiral galaxies show that the relative sizes and masses of these components vary considerably from one system to another. One hopes that a detailed study of these components and their interactions will help us to understand how, in general, these components originate and evolve with time. There is much ongoing research on the MWG stimulated by various factors including the probable black hole at its center, the growing evidence for a bar, and by new observational possibilities, notably the search for gravitational microlenses. The mass of the MWG is $(1.8 - 3.7) \times 10^{11} M_\odot$ [182].

Interest in the Milky Way has increased over the last 10 years or so. There are at least two different reasons for this development: On the observational side, large-scale surveys have been made at X-ray, IR, and radio wavelengths; on the theoretical side, the presence of a large black hole in the center, of a bar in the inner MWG, and the large scale warping of the Galactic disk in the outer parts, make the MWG a desired prototype for studies of other galaxies. For the study of AGB stars the infrared surveys (ISOGAL, MSX, DENIS, 2MASS) are of the greatest importance. Finding masers associated with these stars, one measures their radial velocities: a fundamental complement to the IR data. Two large surveys have been made at 1612 MHz hunting for OH/IR stars [162]: one survey in the north, using the VLA, and one in the south, using the ATCA array. One may also hunt at 43 or at 86 GHz for SiO maser lines. SiO maser lines have the advantage that they are found more frequently than 1612 MHz OH masers and the disadvantage that too large an amount of observing time is needed to make an unbiased blind survey: One has to start from a list of candidate objects. Using the IRAS database a large search for stellar 43 GHz masers has been made with the Nobeyama telescope [96, 48, 47]. The IRAS survey, however, is strongly affected by confusion near the Galactic center. Recently a search has been started using the 30m IRAM telescope at 86 GHz based on the DENIS/ISOGAL survey [126].

The red giants in the MWG dominate the emission at near-IR wavelengths (2 to 5 μm). The color picture on the cover of this book shows a full-sky, edge-on composite image obtained at various wavelengths by the 2MASS survey. Notice the small size of the bulge and the yellow streaks in the plane, which indicate interstellar clouds. Figure 8.1 shows an older, less spectacular version of the color picture: the distribution on the sky of the IRAS point sources with the colors of mass-losing AGB stars. A reanalysis [98] of the stars in Figure 8.1 shows that the IRAS survey sampled AGB stars out to several kpc beyond the Galactic center, except in crowded areas close to this center. The total number of AGB stars in our MWG is an interesting item. Considering only stars with a CSE that makes the star brighter at 25 μm than at 12 μm, the total number within the MWG is estimated to be about 200,000. The stellar disk has a rapid cut-off in the anticenter direction; the IRAS database does not contain AGB stars beyond 12 kpc from the center. The average luminosity of these stars is another useful piece of information.

8 AGB Stars as Tracers of Stellar Populations   417

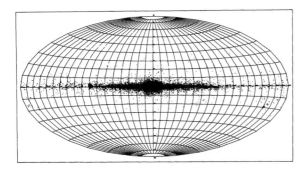

Fig. 8.1. The distribution in galactic coordinates of IRAS point sources with the colors of mass-losing AGB stars. This Figure may be compared with the picture on the cover that displays the distribution of RGB stars. From [80]

Assuming that the distribution of the stars along the line of sight peaks at the distance of the Galactic center, and that this distance is 8.0 kpc it follows that most of these AGB stars have a luminosity of 3500 $L_\odot$ (or $M_{bol}= -4.1$) which is about the maximum luminosity reached by an AGB star of about 1.0 $M_\odot$, a solar abundance and a main sequence life time of 11 Gyr [190]. The sometimes expressed notion that AGB stars have a mass significantly larger than that of the Sun and an age of a few hundred Myr is true for only the most luminous AGB stars.

Figure 8.1 shows the distribution of AGB stars over longitude and latitude (l, b). One is tempted to convert this map into one of the distribution in space. Axial symmetry is not compatible with three facts: (1) the existence of spiral arms, (2) the existence of a bar, and (3) that the most distant sources are only a few kpc beyond the center of the MWG. One more consideration: A complete description of the MWG implies knowledge not only of the spatial distribution of the stars but also of their motions: The measurement of the stellar velocity is an essential complement to that of the stellar positions. Turn to Figure 8.2, where the dots show the distribution of radial velocities of OH/IR stars. The stellar velocities are shown on a grey background that represents the radial velocities of the interstellar CO molecule. In the disk of the MWG the stellar and gas velocities overlap, but for absolute longitudes within 30° from the center a large fraction of the stellar velocities exceed those of the gas by more than 100 $km\,s^{-1}$. Of course, the radial velocities are only one component of the velocity vector v. The other two components can in principle be measured through the proper motions of the maser stars. Right now this is outside of our range of possibilities, but within a few years the VERA radio synthesis array will come on line in Japan and will measure proper motions even for stars at the distance of the Galactic center.

We will begin with a discussion of the solar neighborhood: a prototypical site where several different stellar populations coexist. Then the MWG will be discussed sequentially starting at the center and then moving outward.

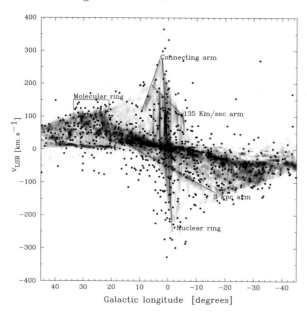

Fig. 8.2. The dots are radial velocities of OH/IR stars as a function of galactic longitude [161, 162, 166]. Underneath the dots is the distribution of velocities measured for the interstellar CO molecule [41]. Straight lines indicate structures in the distribution of the CO gas, taken from [66]. Diagram courtesy of M. Messineo

8.2.1 AGB Stars in the Solar Neighborhood

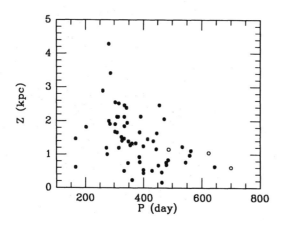

Fig. 8.3. The distance, Z, from the Galactic plane for Mira variables, shown as a function of the period. Closed and open circles are M- and C-stars. From [200]

At distances less then one kpc all Miras are very bright in the near-IR (if $P > 100^d$ then $K \lesssim 4^m$), and we will assume that practically all Miras are known. Even those in the solar neighborhood are faint at visual wavelengths: $V - K$ is typically between 5 and 12 at mean light, and significantly more if circumstellar material is present. Miras are useful, not only because they are AGB stars, but also because their absolute magnitude is narrowly correlated

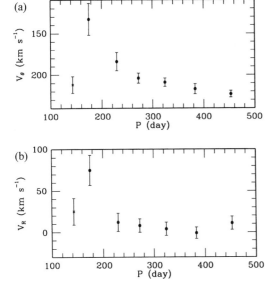

Fig. 8.4. The kinematics of long-period variables depend on the period. Taking the measured line-of-sight velocity as a starting point, the average value is determined for the component (**upper panel**) V_θ, in the direction of galactic rotation, $(l, b) = (90°, 0°)$ and for the component (V_R) in the anticenter direction (**lower panels**). From [59]

with their period and thus their distances can be accurately determined; see Figure 2.47.

The distribution of Mira variables perpendicular to the Galactic plane depends on the period of the pulsation (see Figure 8.3). Stars with short periods (around 200^d) reach greater heights than Miras with long periods (500 to 600^d). This shows that stars with longer periods have larger main-sequence masses than stars of shorter period. In OH/IR stars, with a double peak in their 1612 MHz line, the velocity difference between these peaks, ΔV, plays a similar role: stars with smaller ΔV have a shorter pulsation period; see Figure 13 in [207], which is quite comparable to Figure 8.3, except that P is replaced by ΔV.

The same conclusions are drawn when the 3D distribution of the space velocities of local Miras is analyzed (Figure 8.4). Distances derived from the $\log P/K$ relation are combined with proper motion and radial velocity measurements so that the full velocity vector is known. The statistical properties of this vector may then be derived as a function of period, P. Consider first the stars of $P > 200^d$. $\langle V_\theta \rangle$ decreases slowly but gradually for increasing $\log P$, whereas $\langle V_R \rangle$ is constant. This identifies the Miras with periods longer than 200^d as disk stars; stars with larger values of P have higher main-sequence masses. The remarkable points, however, are those for $P < 200^d$. They divide these short-period (=SP) variables into two groups with different photometric and kinematic properties: SP-red and SP-blue [59]; see also [101]. The velocities of the SP-blue stars have a strong component in the anticenter direction, $74 \pm 18 \, \text{km s}^{-1}$, and the component in the direction of galactic rotation is only

$133 \pm 19 \,\mathrm{km\,s^{-1}}$, roughly $100 \,\mathrm{km\,s^{-1}}$ below the circular speed. These SP-blue stars are probably Miras of low initial mass and the field star equivalents of the Miras detected in globular clusters of the MWG. They may have been formed deep inside the MWG, and now be on elongated orbits extending, as it were, the bar. The SP-red stars have kinematics similar to those of local Miras of the Galactic disk (i.e., with much longer periods) and are suggested to be in an earlier evolutionary phase; i.e., they are not at the tip of the AGB.

Another important local sample is made up of 58 AGB stars with thick CSEs, i.e., with high mass-loss rates ($\dot{M} > 10^{-6} \,\mathrm{M_\odot \,yr^{-1}}$) [140]. All those with sufficient observations are variable, and the periods, when known, are long. Of 47 stars with periods, 36 have $P > 500^d$. Half of the variables are O-rich and half C-rich. Eliminating the three O-rich stars with periods over 1000^d, which will have intermediate-mass progenitors, the O-rich and C-rich stars have similar periods between 300 and 800^d, an average progenitor mass of 1.3 $\mathrm{M_\odot}$, and will give birth to a white dwarf with a mass of about 0.6 $\mathrm{M_\odot}$. The surface density of these stars is about 15 per kpc^2, to be compared with $(2.2 \pm 0.6) \times 10^6$ main-sequence stars of similar mass per kpc^2. With an average mass-loss rate of around $1.5 \times 10^{-5} \,\mathrm{M_\odot \,yr^{-1}}$, and a lifetime, in this high-mass-loss phase, of about $4 \times 10^4 \,\mathrm{yr}$, each star will lose approximately 0.5 $\mathrm{M_\odot}$. The total rate of mass returned to the interstellar medium from these stars will be about $2 \times 10^{-4} \,\mathrm{M_\odot \,kpc^{-2} \,yr^{-1}}$, a factor of ten greater than the estimated return from supergiants; see also Chapter 7.

OH/IR stars with $P > 1000^d$ are very rare, and only one of those with a well-determined distance might fall within 1 kpc of the Sun; OH 357.3–1.3 (IRAS 17411–3154) has a phase-lag distance of 1.2 ± 0.4 kpc (West 1998 [194]) and a PL distance of 1.0 kpc [140].

8.2.2 The Galactic Nucleus: the Center and Its Immediate Surroundings

The Galactic center coincides with the small radio source SgrA*, which has coordinates $(l, b) = (-0.056°, -0.046°)$ [150]; this is about 10 pc from the origin of the galactic coordinate system adopted by the IAU in the 1970s when the position of SgrA* was still uncertain by a few arcminutes. The relative position of the SgrA* radio source and the surrounding cluster of infrared stars has been determined using SiO masers. the accuracy achieved is a few milliarcsec [125].

We assume a distance $R_0 = 8.0 \,\mathrm{kpc}$ ($m - M = 14.51$) [144] to the Galactic center, thus 1 arcminute corresponds to 2.3 pc. The extinction in the V band, A_V, reaches 40 magnitudes in some areas, and is patchy; see e.g., [71]. Because $A_K \approx 0.09 A_V$ one needs an infrared survey reaching $K = 8 + 0.1\, A_V$ to reveal all AGB stars. Even for $A_V = 30$ this is achieved in the DENIS and 2MASS surveys. A search in the K band will miss, of course, the stars with CSEs

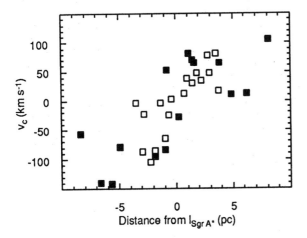

Fig. 8.5. (l, V) diagram of OH/IR stars near the Galactic center (open squares) and of M-giants (filled squares) as discussed in [122]. The longitude is measured from SgrA* and converted into parsecs assuming a distance of 8 kpc. From [115]

thick enough to absorb heavily in that waveband. The stars have also been detected at 7.8 μm in the ISOGAL survey.

Within a Few Parsecs from the Center

This area is different from all other areas in our MWG. A black hole with a mass of $3 \times 10^6 M_\odot$ coincides with the radio source SgrA* and is surrounded by (infra)red stars. Some of these stars are luminous and young, e.g., Wolf-Rayet stars, but there is also a concentration of AGB stars. For eighteen OH/IR stars within 5 pc projected distance from SgrA* the radial velocity increases linearly with longitude (Figure 8.5): $V_{rad} = A \times l$, where $A = 30 \,\mathrm{km\,s^{-1}\,pc^{-1}}$. A similar gradient in radial velocity is derived from 15 M-giants studied in the near-IR [122]. If we assume that V_{rad} is the circular speed around SgrA*, then ρ, the mass density, is uniform and equals 1.3×10^5 $M_\odot\,\mathrm{pc^{-3}}$. It is of some importance that SgrA* coincides with the peak of the concentration of the OH/IR stars to within 0.3 pc, providing independent confirmation that SgrA* is indeed the dynamical center and not just the position of the black hole. The brightest OH/IR star in the infrared, OH 359.762+0.120, cannot be a foreground star because its maser image is smeared out by interstellar scintillation, and that is known to occur only close to SgrA*. Located at a distance of 8 kpc this star has $M_{bol} = -6.5$, an initial mass $> 5\,M_\odot$, and an age below 80 Myr [21]. Nine SiO maser stars have been detected in an area of 16×30 pc^2 around SgrA*[133]; six of those can be identified with OH/IR stars confirming the impression that at the same distance from us SiO masers are found more frequently than OH masers.

The coexistence of Wolf-Rayet stars with blue and red supergiants and with luminous AGB stars suggests continuous, although possibly stochastic, star formation. A discussion of the complex dynamics immediately around

SgrA* is beyond the scope of this book, but has been (and will continue to be) the subject of workshops and symposia.

Between a Few and a Few Hundred Parsecs from the Center

We turn now to the distribution around SgrA* on a scale of 100 to 200 pc. A survey in the JHK bands down to $K = 12.1$ of an area of $140 \times 280 \, \mathrm{pc}^2$ around SgrA* resulted in the detection of some 15,000 stars brighter than $K_0 = 8$, where K_0 is corrected for interstellar extinction [28]. The tip of the RGB lies at $K_0 = 7.9$, and thus the majority of these stars are on the AGB. Their distribution is elliptical with its center coincident with the Galactic center, and with the long axis in the Galactic plane and an axial ratio of 2.2. The most luminous stars have a stronger concentration toward the Galactic center.

A survey for OH/IR stars in a smaller area around the Galactic center resulted in the detection of 134 stars (see Figure 8.6, [115]). In the central $20 \times 20 \, \mathrm{pc}^2$ around SgrA* 21 additional OH/IR stars were found in a deeper survey [168]. Because all 155 stars have been detected at $\lambda = 18$ cm, the search was not affected by interstellar extinction. Monitoring at K has resulted in (long-period) light curves for 75 stars [207]; 21 stars were too faint to detect in the K band but these have all a counterpart at 7 and at 15 µm [141]. They are faint at K because of a thick CSE. The monitoring program also led to the serendipitous detection of 29 new non-OH LPVs. The bolometric magnitudes (corrected for interstellar extinction) indicate that the LPVs are AGB stars because $m_{\mathrm{bol}} < 10.7$, the value that corresponds to the luminosity of the tip of the RGB.

The period distribution of the 75 stars and of those from a few other samples are shown in Figure 8.7. LPVs with periods above 600^{d} are rare in other parts of the Milky Way, including Baade's windows. Periods larger than 450^{d} are associated with stars younger than 1 Gyr at low metallicity; the ages of such stars in this metal-rich environment are less clear. Figure 8.7 thus shows that a significant fraction of the maser stars are younger than this age, in contrast to the LPVs in Baade's NGC 6522 window (see below). In another, slightly less deep, survey in the K band 409 LPVs were discovered in an area of $56 \times 56 \, \mathrm{pc}^2$ around SgrA* [71].

Using the expansion velocity, $(\Delta V/2)$, the OH/IR stars may be divided into an older group ($\Delta V < 36 \, \mathrm{km \, s^{-1}}$; older than about 4 Gyr) and a younger group ($\Delta V > 38 \, \mathrm{km \, s^{-1}}$; less than about 1 Gyr). The older group has a distribution in (l, b) similar to that of the AGB stars found in the JHK photometrical survey just mentioned; both groups represent the same population. The younger group, however, is distributed much more narrowly in l and in b. In the (l, V_{rad})-plane the two groups have the same line of regression, with a slope of $dV/dl = 2.5 \, \mathrm{km \, s^{-1} \, pc^{-1}}$ (again l is measured in parsec from SgrA*). Assuming that dV/dl measures a gradient in circular velocity, one concludes that the mass distribution is homogeneous and the density given

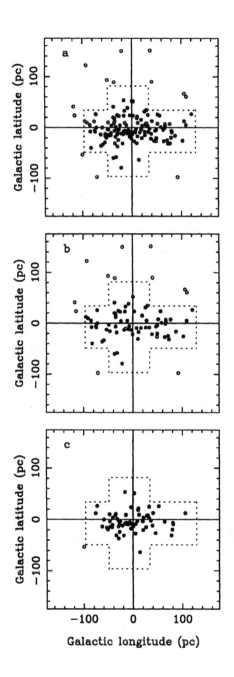

Fig. 8.6. The distribution of OH/IR stars near the center of the MWG. The filled circles are the stars discovered in a survey with the VLA; the dashed lines outline the limits of this survey. The open circles are stars detected in an earlier survey [81] that was less sensitive near the Galactic center. Diagram (a) shows the distribution of all stars, diagram (b) shows the stars with lower outflow velocities and diagram (c) shows the stars of higher outflow velocities. "Higher" and "lower" velocities are separated by a velocity of 18 km s^{-1}. From [115]

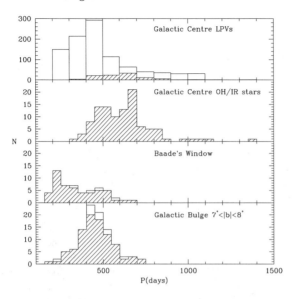

Fig. 8.7. Histograms of the pulsation periods of different samples of LPVs. The top diagram contains *all* variables found in [207]. The second diagram shows the histogram of the Galactic center OH/IR stars. The third shows the variables in the Baade NGC 6522 window, and the fourth diagram those in the bulge, between 7 and 8° from the Galactic plane. From [207]

by $\rho = 1.1 \times 10^4$ $M_\odot$ pc^{-3}, ten times less than the value of ρ closer to the center and discussed above. The two age groups differ, however, in the dispersion around this line of regression: The younger/more luminous group has a significantly smaller dispersion.

A more extensive analysis of the radial velocities of the OH/IR stars over the whole galaxy, using 3D models [165], shows that the majority of the OH/IR stars belong to one spatial/kinematic family: Between bulge and disk there is no clear separation. The younger group, mentioned above, is kinematically quite distinct, and its existence requires a different origin. Most likely, this younger group formed from a gas cloud of, say, 10^7 $M_\odot$ that fell into the center and had picked up the local galactic rotation velocity before it began to form stars. Within 0.4° from SgrA* there are three stars with exceptionally large radial velocities, -309, -342, and -355 km s^{-1}. They probably form a separate group on very elongated orbits and may be at their pericenter [186].

Figure 8.8 shows the distribution of $V_{\exp}(= \Delta V/2)$ for different samples. The distributions reflect the differences in the age (and perhaps metallicity) of the various samples and emphasizes that near the Galactic center, the OH/IR star population is significantly younger or significantly more metal-rich (or both).

As noted before, SiO masers appear to be more frequent than 1612 MHz OH masers at the same distance. Some ninety 43 GHz masers have been found in a recent survey [48]. The analysis of the radial velocity leads to similar results as the analysis of the 1612 MHz radial velocities, although the rapidly rotating nuclear disk has not been found. The disk appears in a more recent study based on ISOGAL sources [126]. Three SiO maser stars with

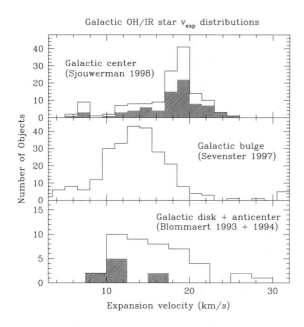

Fig. 8.8. Histograms of the velocity of the stellar wind of OH/IR stars (labeled as $V_{\rm exp}$) for three samples of OH/IR stars. The Galactic center sample top diagram is from [168]; the shaded histogram indicates the number of stars with a known bolometric magnitude. The middle diagram is for bulge OH/IR stars [163]. The lowest panel frame contains a sample of stars from the Galactic disk and from the anticenter direction [20, 22]; the shaded part shows the anticenter data

radial velocity $> +280$ km s^{-1} neatly complement the three OH/IR stars with large negative radial velocities.

8.2.3 The Inner MWG , $-30° < l < +30°$; the Bar and the Bulge

The (l, V) diagrams of interstellar CO and of the OH/IR stars (Figure 8.2) give good reasons to consider the longitude range $-30° < l < +30°$ as the "inner MWG." The MWG bar and its bulge are seen in this longitude range.

Because the Sun is practically in the Galactic plane, it is difficult to detect structure in the same plane. The first explicit suggestion of a bar in the MWG was made in 1964 by de Vaucouleurs [46], but it received little attention, presumably because of the limited information available. Since around 1990 the issue has been seriously debated; the discussion, however, is not yet over, but appears to have stalled again! All evidence taken together makes it very likely that a bar exists, but does not lead to an accurate determination of the basic properties of the bar, e.g., its total mass, its shape, the angle it makes with the line from the Sun to the Galactic center, its age, and whether the pattern rotates around the Galactic center.

Numerous AGB stars have been detected by the infrared surveys. Additional data are provided by programs that monitor the variability and by surveys for OH and SiO masers, e.g., [47, 96].

Evidence of the Bar from AGB Stars

Details of near-IR maps of the MWG made by the COBE satellite, such as that illustrated in Figure 8.9, show clearly that near the Galactic center the inner-most contours are skewed to positive longitudes; i.e., the red giants that emit this radiation seem to be closer to us at positive than at negative longitudes, providing evidence of a bar-like distribution for these stars.

An analysis of the distribution of AGB stars selected from the IRAS point source catalog shows an asymmetry consistent with the presence of a bar [193, 192], but does not constrain the parameters of the bar tightly. An analysis of the magnitudes of stars in two rectangular strips parallel to the Galactic plane ($-15° < l < +15°, 7° <| b |< 8°$) shows that the Miras at positive longitudes are significantly closer to the Sun than those at negative longitudes [197].

In the surveys for OH/IR stars at 1612 MHz more stars have been detected at positive longitudes than at the corresponding negative longitudes. This is explained by the collaboration of two factors: the limited sensitivity of the 1612 MHz survey and the presence of a bar: Maser stars have been detected only out to a distance of 10 to 12 kpc from the Sun. If there is a bar, the survey will have detected most OH/IR stars in the near end of the bar and will have missed most of those in the far end [162].

There is an interesting region of intense star formation between $l = 20°$ and $l = 30°$ where the bar "joins" a spiral arm [162, 105]. The radial velocities of 61 SiO maser stars with $15° < l < 25°$ show that the stars near the tip of the bar have velocities differing strongly from galactic rotation [49].

A few hundred 43 GHz SiO maser stars have been detected during studies of various parts of the bulge. Their radial velocities are consistent with streaming motions in a barred potential [47, 48, 96].

Figure 8.2 shows the distribution in (l, V) of some 800 OH/IR stars between $l = -45°$ and $+45°$ and between $b = -4°$ and $b = +4°$ [163, 164, 166]. The diagram contains many stars with non circular orbits. N-body simulations yield a mass estimate for the bar of 2×10^{10} M$_\odot$ and an orientation of 45° from the plane of the sky. Figure 8.10 shows a model for the interior of the MWG.

8.2.4 Our View through Baade's Windows

The inner Galaxy is shielded from us by heavy extinction, but there exist a few "windows" where the extinction happens to be much lower. The best known are two of the three windows selected by Baade (1963 [15]): the "Sgr I-window," $(l, b) = (1.37°, -2.63°)$ and the "NGC 6522-window," $(l, b) = (1.03°, -3.83°)$. The latter contains two globular clusters, NGC 6522 and NGC 6528, and is often referred to as "THE Baade Window," an unfortunate name because the Sgr I and Sgr II fields are also "Baade windows." The total extinction in the NGC 6522 window has been mapped in detail;

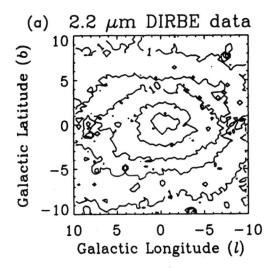

Fig. 8.9. The contours of the distribution of near-IR emission at 2.2 μm around the Galactic center. From [54]

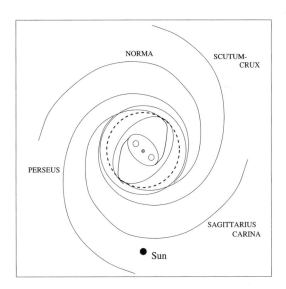

Fig. 8.10. Top-view sketch of the MWG. The larger ellipse outlines the bar; the smaller ellipse the inner Lindblad resonance. The dashed circle is the location of the inner-ultraharmonic resonance. Four spiral arms are indicated. From [162]

A_V varies between 1.5 and 2.8 [170]; the distribution of the dust along the line of sight is unknown. The distribution of the stars along the same line of sight is quite complex because the contributions by the disk, bulge, and halo are all comparable, [177] and to differentiate between the stars belonging to different components requires more information than is usually available.

ISOGAL sources in the Sgr I- and NGC 6522-windows have been cross correlated with M-giants found in photographic surveys [69]. Many ISOGAL

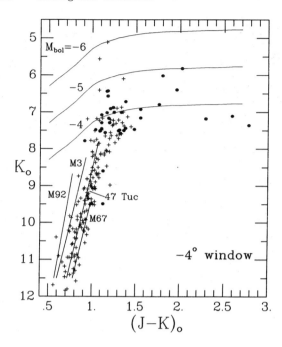

Fig. 8.11. An extinction-corrected IR color-magnitude diagram for M-giants in the NGC 6522 window. Most of the stars above the tip of the RGB ($M_{bol} = -3.8$) are on the AGB; a few may be foreground objects. Dots are Mira variables, crosses are M-giants, with, at most, small pulsation amplitudes. In drawing this diagram a distance modulus of $(m-M) = 14.2$ has been assumed; in this chapter we use $(m-M) = 14.5$. From [65]

sources are Mira variables, but there are also a small number of equally luminous stars with no, or low amplitude, variation and with large $(K-[15])$-colors indicating high mass-loss rates. Only a small fraction of all the M-giants in the inner bulge are AGB stars, i.e., $-3.8 > M_{bol}$, as can be seen from Figure 8.11. A serendipitous result of all surveys in Baade's Windows has been the *non*detection of intrinsic C-stars. This is generally understood to be a consequence of the old age, perhaps combined with the high metallicity. The low-luminosity C-stars found in the direction of the bulge are most probably products of binary evolution [196].

The bulge population has many similarities to those of globular clusters, but there are also important differences. In the clusters the stars more luminous than the tRGB are all Miras, and these are clearly AGB stars. Miras are numerous in all Baade windows, with periods up to 700^d. This is much longer than the maximum of around 300^d found for Miras in globular clusters, e.g., [73]. These longer periods imply that the population in the bulge contains more massive stars, and/or stars of much higher metallicity than are found in the MWG globular clusters. It seems from [56] that if the period distribution is interpreted in terms of metallicity spread, the result is entirely consistent with the metallicities determined from K giants [124]. A high mean metallicity also follows from the finding that the Miras in Sgr I are redder in $[H-K]$ and bluer in $[J-H]$ than those in the LMC.

We can get some indication about the ages of the stars in the bulge using the PL-relation. First, it should be clear that there is no evidence for

AGB stars undergoing hot bottom burning, of the kind found in the LMC and SMC, with $P > 450^d$ (see Section 2.2); because such stars would have initial masses in excess of 3 or 4 $M_\odot$, their absence in the bulge is hardly surprising. The longest-period stars found in the bulge have pulsation periods of around 700^d, but there are few of them, and they plausibly represent the progeny of binary mergers. We therefore take $P \approx 600^d$ as the upper limit, and $M_{bol} > -5.6$. These luminous stars are probably the descendants of the most metal-rich fraction of the bulge population, and we can estimate that their initial masses may have been around 2 $M_\odot$, but without AGB models for metal-rich stars it is difficult to link this to an age.

8.2.5 The Bulge at Higher Galactic Latitudes

Although the IRAS survey was confused in the Galactic plane, especially near the Galactic center, above a few degrees of latitude there is no problem. Near-IR photometry and some monitoring show that there are 113 Miras among the 141 IRAS point sources in two strips parallel to the Galactic plane at $7° < |b| < 8°$ and across the Galactic bulge, $-15° < l < 15°$ [199]. The sample also contains 22 M-giants, of which half are in the foreground. Periods in the range 104 to 722^d were determined for the 104 Miras. Such long periods could imply a significant population with ages less than 1 Gyr, but again the conclusion is made uncertain by the unknown effect of high metallicity.

Color-magnitude diagrams ($J - K$ versus K) of red giants in six fields with $l \approx 0°$ and at $b = -3, -6, -8, -10, -12°$ show that most stars are on the RGB, but a few stars are brighter than the tRGB ($K = 8$) [64]. These may be AGB stars, although some could be foreground stars. Among the brightest objects there are several LPVs. All fields have the same luminosity function that terminates around $M_{bol} = -4.2$, as do the metal-rich globular clusters [55].

8.2.6 The Outer Galactic Disk: $30° < l < 330°$

We assume that the Galactic disk is the dominant stellar component of the MWG at $30° < l < 330°$. From the COBE measurements a model has been derived for the disk [62] in which the emissivity $\rho(R, z)$ varies in the Galactic plane $\propto \exp(-R/h_R)$ ($h_R = 2.6$ kpc) and perpendicular to the plane $\rho \propto \mathrm{sech}^2(z/h_z)$ ($h_z = 330$ pc); h_z does not vary with R. There is no separate thick disk component, i.e., one with $h_z \approx 0.8$ kpc. The disk has an outer edge at $R > 12.1$ kpc, beyond which the radial scale length height, h_R, is only 0.5 kpc.

The analysis of the databases from the infrared surveys show a low surface density of AGB stars in the anticenter direction, so that the Sun must be close to a cut-off of the Galactic disk [79]. There is also a change in the character of the AGB stars in the outer MWG: A survey for OH/IR stars [175] and one for SiO maser stars [100] both show that the detection probability of maser stars is markedly lower in the outer MWG than in the inner MWG. This may have been caused by a lower metallicity in the outer MWG. Already in 1965 many more M-giants had been detected in the direction of the center of the MWG than in the anticenter, while this is not so for the C-stars [19]. Thus the ratio of C-type to M-type giants is higher in the outer Galaxy.

8.2.7 Globular Clusters

Globular clusters are among the oldest objects in the MWG, and the globular cluster system is a very important tracer of its structure and history. Low metallicity clusters, [Fe/H] $\lesssim -1$, occur at all latitudes and appear to be part of the halo [211]; clusters with [Fe/H] $\gtrsim -1$ are found at low latitudes only and are probably related to the bulge [129]. Some of the clusters may have originated in dwarf galaxies and subsequently merged with the Milky Way [37], and some might have been formed during the merger [12].

The stars we now see in globular clusters had main-sequence masses around 0.8 $M_\odot$. Intrinsic C-stars are not found in these clusters. There are a few M-type stars with luminosities above the tRGB, but only in clusters with a sufficiently high metallicity ([Fe/H] $\gtrsim -1$); these M-type stars are (almost) all Mira variables with $-3.6 \lesssim M_{bol} \lesssim -4.8$ (Figure 8.12). The maximum period of the Miras in a given cluster and the cluster metallicity are correlated (Figure 8.13). Furthermore, the Miras obey the same $\log P/K$ relation as do similar stars in the LMC [55]. In clusters of lower Fe-abundance the AGB evolution probably terminates at luminosities below the tRGB. We also remind the reader of the "SP-blue" Miras discussed earlier, since these may be the equivalents in the field of the Miras seen in globular clusters.

8.2.8 Halo

It has become increasingly clear over the last 20 years that mergers of fully formed dwarf galaxies with the MWG have had a significant effect on the observed structure of the thick disk, the bulge, and in particular the halo.

Seventy-five C-rich giants have been found at distances between about 16 and 60 kpc from the Galactic center [93, 94, 178]. More than half of these stars originally belonged to the Sagittarius dwarf galaxy, which experiences severe tidal stripping during its interaction with the MWG. Most of the remaining C-stars appear to be outlying members of the Magellanic Clouds (see also [137]). Repeated observations reveal many of them to be Mira variables. Five newly discovered dust-enshrouded C-stars, now in the halo, probably also originated in the Sagittarius dwarf galaxy [78, 113]. A new catalogue contains 403 faint high-latitude C-stars [29]; some of these may be AGB stars, although rather few have thick dust shells.

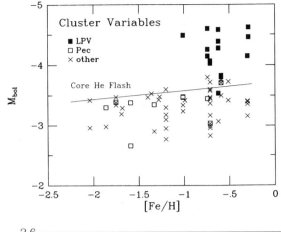

Fig. 8.12. Metallicity versus bolometric magnitude for variables in globular clusters. The straight line labeled "Core He Flash" shows the location of stars at the tip of the RGB. From [63]

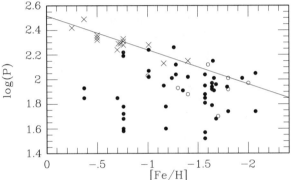

Fig. 8.13. The pulsation period of variables (Mira and semiregular) in globular clusters as a function of [Fe/H]. The crosses represent Mira variables; the filled circles are semiregular (SR) variables. From [56]

8.3 The Magellanic Clouds

The Magellanic Clouds (MCs) form a double system that is a satellite of the much more massive MWG. This does not imply that the MCs are small in an absolute sense: Both outshine the dwarf spheroidal companions of the MWG by several magnitudes (see Table 8.1). Because the MCs are nearby, they are large on the sky: The LMC covers an area of roughly 100 square degrees, the SMC about 30. Both clouds have an irregular appearance on maps of the interstellar gas (H I, Hα, dust) but this is no longer the case when one inspects maps that contain only post-main-sequence stars: The LMC and the SMC appear now to be remarkably smooth; see Figures 8.15 and 8.16. The inhomogeneities are therefore associated with only the youngest objects, and with the most recent history of the clouds. Another illustration of the same eefect is the prominence of the LMC bar on blue and yellow photographs, and its absence in the AGB star distribution (see below). In an H I-map of the SMC the contrast between young and old is particularly striking because the H I distribution is full of holes [171].

The three-body system MWG/LMC/SMC is so compact that the gravitational interactions produce noticeable effects: The LMC and SMC have a common envelope of H I gas, and they are connected by a bridge of gas and stars. From the SMC, in the direction opposite to the LMC, a long tail of gas extends over 100° on the sky: the Magellanic Stream. The dynamics are complex because the bodies are extended; little is actually known about the orbits of the clouds. Nevertheless, several dynamical models have been proposed, e.g., [114, 68], and of course, as always, the smallest, the SMC, is hit hardest: The gas in the stream is mainly drained from the SMC. The models agree that the Clouds are now at their closest approach of the MWG and will move to their farthest distance from the MWG, about 100 kpc in the next few Gyr. These dynamic models assume a mass for each of the three bodies; these are about 10^{10} $M_\odot$ for the LMC and 2×10^9 $M_\odot$ for the SMC. The star-formation rate (SFR) of the Magellanic Clouds has been a "hot topic" ever since Butcher [26] suggested a "star-formation burst" occurred only a few Gyr ago. While more recent studies broadly support this idea, the details remain controversial; see [30] and more recent discussions.

We may learn about the Magellanic Clouds by studying their AGB stars, and the reverse is true as well: Figure 3 in Chapter 2 shows $m_{\rm bol}$ for the brightest stars in clusters of both clouds arranged according to their SWB-type, a morphological classification that correlates well with the cluster age. From an upper envelope in the figure for $m_{\rm bol}$ we find respectively $M_{\rm bol}=$ −6.3, −5.9, −5.7, −5.3, −5.0, −4.0 for SWB-types II, III, IV, V, VI, VII or ages 0.04, 0.12, 0.37, 1.1, 3.3, 10 Gyr. An often-used model for the development of AGB stars, which includes mass-loss [190] and assumes $Z = 0.008$, predicts luminosities for these ages of $M_{\rm bol}=$ −6.4, −5.8, −5.3, −5.1, −4.8, −4.2, a fair, but not excellent, agreement.

8.3.1 AGB Stars Identified in Near-IR Surveys

The DENIS-survey in the IJK_s bands and the 2MASS-survey in the JHK_s bands contain tens of thousands of stars within the boundaries of the clouds; Figure 8.14 shows a color-magnitude diagram of all the stars seen by DENIS within the contours of the LMC. We assume that the sample of stars in area B contains only AGB stars, which, although not 100% correct, is a very good approximation.

Contour maps of the surface density of the stars in area B of Figure 8.14 are shown in Figures 8.15 and 8.16. At first sight the distribution of the stars in the LMC appears to correspond to a circular disk seen at an inclination of 30 to 40°; but see below. The center, $\alpha = 5^h 20^m, \delta = -69°$, is the same as that of H I [102] and of various other types of stars [195]; it is 0.6° north of the center of the bar. A careful analysis of the magnitudes of the AGB stars in the LMC, when one moves around the center in rings, shows that the average magnitude in each azimuthal segment varies systematically by about 0.25^m, leading to the conclusion that the stars on the western side are

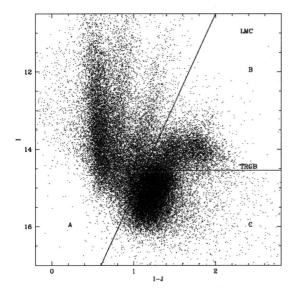

Fig. 8.14. A color-magnitude diagram of stars in the LMC detected in the DENIS survey. The stars in region A are foreground objects, those in area C are RGB stars belonging to the LMC and area B contains AGB stars in the LMC. From [31]

12% closer than those on the east [185]. Further analysis of the DENIS and 2MASS data [184] reveals that the LMC is elliptical ($\epsilon = 0.2$) with its long axis in the direction of the MWG center and perpendicular to the Magellanic Stream, suggesting that the gravitational interaction with the MWG is the cause of both ellipticity and Magellanic Stream. This study also confirms the limited thickness of the LMC disk.

Figure 8.16 shows that the distribution of the AGB stars in the SMC is as smooth as that of the LMC. There are two central peaks, and in contrast to the LMC one finds that the SMC has a considerable depth (several kpc) along the line of sight [181].

The distribution, LF, of the luminosities is shown in Figure 8.17 separately for the two clouds. In both galaxies the LF is discontinuous at the magnitude of the tRGB (LMC: $K = 12.0$; SMC: $K = 12.7$). Since the tRGB is probably a constant independent of metallicity, one derives the distance modulus of each cloud: 18.55 ± 0.08 for the LMC and 18.99 ± 0.08 for the SMC [33].

8.3.2 Long-Period Variables

The search for variable stars in the MCs began early in the twentieth century with photographic techniques and therefore in the blue part of the spectrum. Large-scale searches for red variables had to wait for instruments sensitive at (infra-)red wavelengths (photographic emulsions for the I band and solid-state detectors for the JHK bands). One of the first results was the discovery of a linear relation between the K magnitude and the log of the period, the

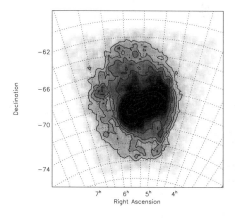

Fig. 8.15. The distribution of AGB stars in the LMC. From [33]

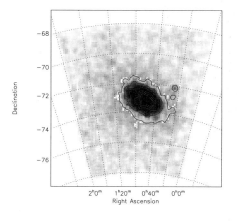

Fig. 8.16. The distribution of AGB stars in the SMC. From [33]

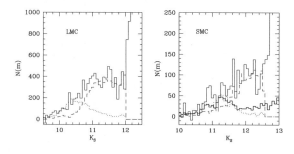

Fig. 8.17. The number of stars detected by DENIS versus K_s magnitude, in selected areas of the LMC and SMC. (Dashed line:) O-rich AGB stars; (dotted line:) C-rich AGB stars. Notice the sharp drop-off in the number of stars at $K_s = 12.2$ in the LMC and at 12.7 in the SMC, caused by the helium core flash. From [31]

"log P/K relation', [70]; see Figure 2.47. A similar dependence on the period is found for the bolometric luminosity, e.g., [57, 90, 187].

In the LMC hundreds of LPVs have been identified that have amplitudes $\Delta I \geq 0.5$ [72, 88]. The tRGB in the LMC is at $I = 14.5$, and since these surveys have sensitivity limits considerably fainter than this magnitude, most of the large-amplitude variables should have been found in the fields surveyed. Exceptions are stars missed due to crowding, high circumstellar extinction (see below), and an unfortunate distribution of the survey plates over the phase of the light curve. Reference [143] contains a list of 332 LPVs detected in a 16 square degree field; assuming that this list is 50% complete, as the authors suggest, we should expect on the order of 4000 large-amplitude LPVs in the 100 square degrees of the full LMC.

Large-scale surveys and monitoring programs in the SMC, however, are missing. Much of what we know about the red variables comes from early work at Harvard, e.g., [142]. Information about LPVs in the SMC has to be found from small samples that are included in studies of LMC stars. One such study, [206], presents JHK observations of 42 LPVs in the SMC and 48 in the LMC; a fair fraction of the SMC stars are Harvard variables. Seventeen of the SMC stars have periods longer than 450^d and are therefore younger than 1 Gyr; eight have $P < 255^d$ and must be older than 5 Gyr. Periods have been determined for five IRAS sources; two of them have a period of respectively $\approx 800^d$ and 517^d [202] and the first has subsequently been shown to be C-rich. The old SMC cluster NGC 121 contains an O-rich Mira (V1) with a period of 140^d [176], which is comparable to Miras in Galactic globular clusters with similar periods [55]. In a field of $170 \times 170\,pc^2$ around the SMC cluster NGC 330 four LPVs with I-band amplitudes larger than 0.5^m have been found with periods of 186, 280, 365, and 400^d [160]. There are thus old (> 5 Gyr) and intermediate-age (1 to 3 Gyr) AGB stars in the SMC.

Data obtained in the monitoring programs to detect microlensing (EROS, MACHO, OGLE) are potential gold mines in the search for variables. In an LMC field of 0.25 square degrees, 1431 red variables have been found [205], but the amplitude distribution is not given, and it is not clear which (presumably) small fraction of these are periodic and with an amplitude $\Delta I > 0.5^m$. In another field in the LMC, of 0.5 square degree, about 750 stars show variability; there are 540 AGB stars among those. Forty-three are Miras, and 200 semiregulars [32]. Access to the very valuable EROS and MACHO databases is still limited; hopefully, they will soon become accessible to all.

The LPVs in the LMC have been divided into different age groups on the basis of their periods, and their distribution over the face of the LMC has been studied [90]. The oldest stars (> 5 Gyr) are spread uniformly over most of the LMC, but they do show a moderate concentration toward the bar. Radial velocities have been measured [91] and provide evidence for a kinematic spheroidal population. The youngest stars (< 1 Gyr) are projected against

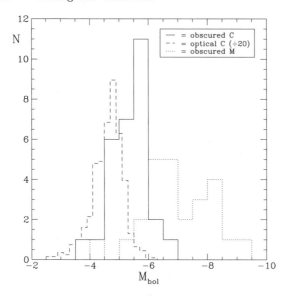

Fig. 8.18. The distribution of the bolometric magnitudes of C- and M-type stars in the LMC. From [188]

the bar and in the well-known star-forming region "Shapley Constellation III."

8.3.3 Carbon Stars

C-stars are relatively easy to recognize on object prism plates as long as they have thin CSEs; we will call them "optical C-stars." Several surveys have been carried out. In the LMC a total of about 11 000 C-stars are expected, of which so far less than 10% have been detected [13]. In striking contrast, 1700 of the 3100 C-stars expected in the SMC have already been identified. Existing near-IR survey data (DENIS and 2MASS) may be used to discover numerous new C-stars via photometric criteria (see the discussion of Figure 8.14 in [31]). C-stars with thick CSEs, or "obscured C-stars" [CW Leo (IRC +10216) remains the best-known example from the MWG] may escape detection even at near-IR wavelengths. In the LMC the ratio between the number of C-stars and of (AGB) M-stars is systematically higher in a ring around the edge of the AGB star distribution and this seems to imply that the abundances are systematically lower in the outer parts of the LMC [34].

Figure 8.18 shows the distribution of bolometric magnitude, M_{bol}, for samples of stars in the LMC. The "optical C-stars" populate the AGB mostly between $M_{bol} = -4$ and -5, while the "obscured C-stars" peak at a brighter M_{bol}, between -5 and -6. The third histogram represents a sample of O-rich stars with thick dust CSEs, including AGB stars and supergiants. On average, the AGB O-rich stars have higher luminosities than the C-stars, but there is some overlap. The LMC AGB stars have a range of ages and metallicities, and their luminosity function is plausibly explained [117].

Unlike the MWG, the MCs contain clusters of intermediate age. Three obscured C-rich LPVs have been found in clusters, two in the LMC and one in the SMC. The clusters have practically the same age (1.6, 1.7, and 2.0 Gyr), and the stars have similar periods (526, 450, and 491 days) and M_{bol} (−4.9, −5.0, and −5.0) [138]. The most luminous C-star in the LMC is an LPV with a mean luminosity, $M_{bol} \approx -6.8$, very close to the AGB-limit. This star has another peculiarity: It is surrounded by silicate dust, debris of its existence before it turned into a C star. This may well be a star in which hot bottom burning has just ended, allowing it to become C-rich at high luminosity [180].

In the MWG, CH stars are halo objects that are probably the result of binary mergers. A group of objects with somewhat similar spectra was found in the LMC and shown to have a high overall velocity dispersion, apparently identifying them as old objects. Surprisingly, their bolometric magnitudes are among the brightest found for LMC C-stars, with $-4 > M_{bol} > -6.2$ [58, 174]. If they are truly old objects, then they must have originated in binary mergers. However, it is possible that a combinations of complex kinematics and selection bias may have led to the large velocity dispersion and that these stars are actually among the youngest ($\lesssim 100$ Myr) and most massive of the LMC C-stars [174].

Intrinsic C-stars occur more frequently in the SMC than in the LMC: The ratio of C-stars to late M-giants (M5–M10) varies greatly from galaxy to galaxy, and even from one part of a galaxy to another; compare, e.g., the MWG bulge, where there are no C-stars, with the outer Galaxy, where there are many. There is also a general correlation between the C/M ratio and M_V, the absolute magnitude of the galaxy (see Figure 6.21 in [183]). These facts can be understood by the combined effects of metallicity and stellar mass [147]. The correlation between this C/M-ratio and the metal abundance is such that more C-stars are found at lower "metallicity," where the dredge-up of even small quantities of carbon will change the surface composition so that C/O > 1 (see Chapter 2); there will, of course, also be fewer M stars at low metallicity. In the Galactic bulge the stellar masses may be too low for dredge-up to occur at all. For intermediate-mass stars the (metallicity-dependent) action of hot bottom burning, which inhibits the production of C-stars, is also critical. It is clear from, e.g., Figures 6 and 7 of [117], that the fraction of C-stars in a galaxy will depend on the proportion of intermediate-mass stars with $M_i \lesssim 3.5\,M_\odot$, as well as on the metallicity.

Measurement of the radial velocities of AGB stars is useful in tracing the internal kinematics. This provides another way to determine the total mass (stars, dark matter, and gas). The LMC rotation curve has been measured from C-stars, and it follows that the total mass of the LMC is $(5.3 \pm 1.0) \times 10^9\,M_\odot$ [5]. The radial velocities of more than 500 C-stars lead to the conclusion that there are two populations of C-stars involved [74]. About 20% have a velocity dispersion of $8\,\mathrm{km\,s^{-1}}$ and belong to a young disk, while the bulk of the stars have a velocity dispersion of 22 km s^{-1} and could be part of an old disk population.

8.3.4 Lithium AGB-Stars

High lithium abundance has been detected in 29 S- and in 6 C-type stars in the LMC and SMC (out of 112 studied [169]), most of which have high luminosities, $-7.2 \lesssim M_{bol} \lesssim -6.0$; in fact, almost all such luminous (red) stars in the MCs have a high lithium abundance. Lithium enhancements are understood to occur as a consequence of hot bottom burning; see Chapter 2.

8.3.5 AGB Stars with High Mass-Loss Rates

For an unbiased survey for obscured AGB stars one needs to observe at mid-IR wavelengths. IRAS fulfills this requirement (see [158, 159]), but its sensitivity limit allowed only the detection of obscured stars in the MCs with luminosities $> 10^4$ L$_\odot$ ($-5.3 > M_{bol}$), whereas in the MWG obscured AGB stars are known with luminosities as low as 3000 L$_\odot$. In the LMC some 50 candidate obscured-AGB stars have been detected in various ways ([179] and earlier papers). The distribution of M_{bol} for these obscured stars, illustrated in Figure 8.18, shows that the luminosities of obscured C-stars are significantly higher than those of optical C-stars.

Near-IR monitoring of 12 obscured AGB stars in the LMC has shown that nine of these are large-amplitude variables with periods in the range of 530 to 1300^d and luminosities below an extrapolation of the logP/K relation [204]; the analysis of more extensive $JHKL$ photometry for seven stars shows good agreement for the periods, but higher luminosities for five of the seven sources ([198] and Whitelock et al. in preparation), bringing them into agreement with the PL relation.

1612 MHz OH masers have been detected in the LMC in four AGB stars and in one supergiant, all with colors $(K - L') \gtrsim 2$ [208]. The AGB stars have periods between 1260 and 1390^d, mean bolometric magnitudes in the range -6.2 to -6.9, and K band amplitudes of about 2^m.

In the SMC little work has been done so far on stars with thick CSEs. Near-IR photometry and optical spectroscopy of 30 IRAS sources leads to the conclusion that eight are O-rich and nine are C-stars; the type of the others has not been determined [77]; bolometric magnitudes have not been derived. Observations of a few stars are reported in [189, 209].

AGB stars in the LMC have ages varying from young (0.1 Gyr) to old ($\gtrsim$ 10 Gyr); this is true in the disk as well as in the bar. The distribution of the ages, or the star-formation history, remains uncertain. Whether a starburst took place in the last few Gyr is unclear; the present data neither rule out nor confirm such a burst.

8.4 M 31, M 32, NGC 205, and M 33

8.4.1 M 31, the Andromeda Galaxy

We quote some general properties from [183]. M 31 is a spiral galaxy of type SbI-II, at a distance of 760 kpc ($m-M = 24.4$), and thus 1 arcsec equals 3.7 pc. M 31 is viewed at a rather high inclination angle of 77°; its angular dimensions are 92× 197 arcmin. The mass is $(2-4) \times 10^{11}$ M$_\odot$, and this is marginally more than that of the MWG. M 31 has a nucleus that may contain a black hole. The bulge has an effective radius of (2.2 ± 0.2) kpc and is responsible for about half of the flux measured in the U, B, V, and R bands [191]. The disk, as traced by the OB associations, extends to 20 kpc from the center. It is surrounded by a halo that may have been populated by cannibalizing nearby dwarf galaxies; a stream of stars of solar metallicity has recently been detected in the M 31 halo. We also remind the reader that it was the detection in 1944 of individual red giants in M 31 that led Baade to propose the coexistence of two different stellar "populations."

Difficulties in making detailed observations of individual stars in M 31 are to be expected from the large distance (which makes the tRGB as faint as $K = 17.8$), the crowding (is a bright star really a star or an accidental conglomerate of several stars?), and interstellar extinction. It is therefore no surprise that the old- and intermediate-age populations have been studied less than the young population. The detection of over 300 PNe shows that AGB stars are present as well [61].

JHK-images have been obtained with the HST of five 90×90 pc^2 fields at distances of 0.57, 0.71, 1.35, 4.5, and 34 kpc from the center of M 31 [172]. Stars have been identified as faint as $K = 22$. The field stars (each field is centered on a globular cluster) may be assigned to, successively, the bulge, the disk, and the halo. The histogram of the M_K values of the field stars, i.e., their luminosity function, cuts off at $M_K = -7.5$, i.e., about one magnitude above the tRGB, and at about the same luminosity as the tip of the AGB in the most metal-rich globular clusters in the MWG. These M 31 AGB stars are thus probably as old as the globular clusters. From repeated observations one LPV has been found, serendipitously. An important conclusion of [172] is that "the fields surrounding the clusters have luminosity functions that cannot be distinguished from those in the NGC 6522 window."

An analysis of V and I photometry in fields in the bulge at distances of 0.71, 0.81, and 1.35 kpc from the M 31 center, shows that the brightest field stars have the same I magnitude as the brightest cluster stars, $I = 21.5$ or $M_I = -2.9$, [97] and most probably they are RGB stars. This leads to the interesting conclusion that there is no statistical difference between stars in globular clusters and in the field. Either the globular clusters and the field stars coincidentally had an identical history or, a more likely scenario, the field stars were born in now-dissolved globular clusters.

V and I band photometry down to $I = 22$ in five fields each 1.55×1.55 kpc^2 at distances varying from 4 to 32 kpc from the center shows that there are many red stars above the tRGB, which must be AGB stars [24, 25]. Narrow-band photometry enables these AGB stars to be separated into C- and M-types; spectra identify 48 C-stars and one S-star. Seven of the C-stars have enhanced ^{13}C (i.e., they are J-stars), three have enhanced lithium, and two have strong Hα emission, suggesting they are Mira variables. Most of the lithium-rich C-stars have larger bolometric magnitudes ($-3.6 \gtrsim M_{bol} \gtrsim -4.6$) (and lower luminosities) than similar stars in the MCs ($-4.6 \gtrsim M_{bol} \gtrsim -5.8$), if the bolometric luminosities, deduced from I_C and $V - I_C$, are correct. The luminosities of Galactic lithium-rich C-stars are somewhat uncertain, but they seem so low that the stars do not experience hot bottom burning and consequently do not produce lithium enhancements by that process. The S-star is probably one of the most luminous stars in the sample. The ratio of C- to M-stars increases with galactocentric radius, implying a metallicity gradient. The search is continued with the same technique in more fields [139].

Deep $VJHK$ images of a field of 440×440 pc^2 at 7 kpc from the center of M 31 show evidence for numerous large-amplitude variables [103], some with the colors of Miras. Ten very red ($H - K \gtrsim 0.9$) objects may be young clusters or AGB-stars with thick CSEs. However, the colors are not convincingly those of extreme AGB stars, and image crowding would seem a more plausible explanation. Population studies of the color-magnitude diagram in each field have been attempted, using theoretical isochrones, but are incomplete because these calculations were terminated at the beginning of the thermally pulsing regime. The deductions about the AGB population in M 31 are thus dubious.

8.4.2 M 31's Close Companions: M 32, NGC 205

The compact and close (less than 15 kpc from the center of M 31) elliptical companions *M 32* and *NGC 205*, represent important resolved examples of more distant systems, although the relationship of such small galaxies to the giant ellipticals remains controversial.

M 32: The red color and relatively blue main-sequence turnoff indicate an intermediate-age population, as does the detection of AGB stars that are considerably more luminous than those found in globular clusters. Some of the bright stars reported in early studies may have been the result of image crowding. Nevertheless, stars as bright as $M_K = -8.8$ have been detected throughout M 32, although it is unclear whether these are younger than 1 Gyr or just metal-rich. The spatial distribution of bright AGB stars follows the integrated light profile, showing no evidence of an age gradient [43]. It is suggested that the AGB stars either formed as part of a coherent galaxy-wide episode of star formation or that they originated in a separate system that merged with M 32. The integrated $(J-K)$-color and the CO index change with radius within a few tens of parsecs of the center, indicating that the population at the very center does differ from the surrounding galaxy. Until there are reliable models for the evolution of metal-rich AGB stars and good measurements of the luminosities and periods of the bright AGB stars in M 31 and M 32, the ages of these populations will remain very uncertain.

NGC 205 is considered, like NGC 185 (see below), to be a peculiar elliptical galaxy, becaus in addition to the normal population of old stars, it has bright early-type stars and dust clouds, indicating that limited star formation is still going on in its central regions. Luminous C-stars have been found in NGC 205, and thus there is an intermediate-age AGB population. Bright, up to $M_{bol} \approx -5.7$, AGB stars have been identified [108] that spread beyond the central regions where the more massive stars are concentrated. The detailed results of these studies may have been influenced by image crowding.

8.4.3 M 33 (NGC 598)

M 33 is a spiral galaxy (Sc II-III). By mass as well as by luminosity it is the third galaxy in the Local Group, after M 31 and the MWG. Its optical radius is 42 arcmin (10 kpc) and its OB-associations extend out to 24 arcmin (5.8 kpc). We view M 33 at an inclination angle of 56°. Its mass is estimated at $(0.8\text{–}1.4) \times 10^{10}\,M_\odot$, which is more than a factor 10 below that of M 31, while the optical diameter is half that of M 31. The galaxy has a semistellar nucleus, an exponential disk, and a halo, and it *may* have a bulge.

M 33 contains stars in all evolutionary stages, including AGB stars. Deep images in JHK and through two narrow filters show red stars with magnitudes down to $M_{bol} = -5.25$; these are luminous AGB stars [44, 123]. The existence of such stars suggests a star burst 1 to 2 Gyr ago. M 33 also contains more than 50 PNe [183].

8.5 The Remaining Galaxies of the Local Group (LG)

After having discussed the major galaxies of the LG, one by one we now consider the remaining galaxies. Van den Bergh [183] lists 35 LG members. Two galaxies, M 31 and MWG, stand out as the most massive by far; they form the double core of the LG. Inspection of the distances of each galaxy to the MWG and to M 31 leads to a natural division of the 35 galaxies into three groups. Table 8.1 contains 12 galaxies within 250 kpc from the MWG, Table 8.2 contains 14 galaxies within 300 kpc from M 31, and in Table 8.3 the remaining 9 may be found. Because the masses of the MWG and M 31 are very similar, it follows that in the first table the gravitational attraction of MWG is an order of magnitude larger than that of M 31. In the second table the attraction by M 31 dominates by an order of magnitude, and in the third table the gravitational attraction by M 31 is comparable to that by MWG.

The three tables give basic data on all LG galaxies; the contents of columns 1 to 4 has been taken from [183]. Column 1 contains the name; column 2, the morphological type; column 3, M_V, the absolute magnitude in the V band; column 4, d_{MWG}, the distance to the Sun (for the MWG the distance given is that of the Sun from the Galactic center); column 5, d_{M31}, the distance from M 31; column 6, the magnitude of the tRGB, i.e., the magnitude of the faintest AGB star that we will consider (i.e., $M_K < -6.6$). In column 7 we show the presence or absence of AGB stars as found in a search in refereed journals. It would have been a significant improvement if the simple "Y/N" could have been replaced by a quantity, e.g., the ratio of the number of luminous AGB stars as compared to the number of RGB stars. This, however, seems not yet possible.

AGB stars have now been detected in almost all of the more luminous dwarf galaxies, and have been looked for in the fainter galaxies to the same depth, but without success.

The C-stars are undoubtedly the best-studied AGB stars in these galaxies; for a review see [76]. Figure 8.19 shows a color-magnitude diagram with carbon star luminosities from various spheroidals. Between the dwarf spheroidals the total number of C-stars normalized to the total luminosity of the parent varies by a factor of ten. One should note, however, that the statistics are not good, in that there are very few C-stars in the low-luminosity dwarfs, e.g., *Ursa Minor* had only one (plus one CH star), and *Sextans* had none. Furthermore, many of these C-stars are not on the upper-AGB, and some have luminosities as low as $M_{bol} \approx -1.2$ and might thus be extrinsic C-stars. One should be aware of the likelihood that the searches for C-stars so far have been incomplete.

8.5.1 Galaxies Within 250 kpc of the MWG

The three most luminous galaxies, MWG, LMC, and SMC, are so close together that they have demonstrable gravitational interaction and may be considered as a 3-body system surrounded by a halo of dwarf spheroidals. The most luminous of those dwarfs, SDSG, is being torn apart by the MWG. There is a gap in absolute V luminosity of a factor of 10 between MWG and LMC and a similar gap (13:1) between SMC and SDSG. AGB stars have been found only in the most luminous of the spheroidals. Apart from the LMC and the SMC, Table 8.1 does not contain irregular galaxies. This must be real: Any galaxy such as IC 10 or NGC 6822 and that is within 250 kpc from the MWG should have been discovered if it existed.

The Sagittarius Dwarf Spheroidal Galaxy (SDSG). It has been appreciated for some time that merging dwarf galaxies have significantly influenced

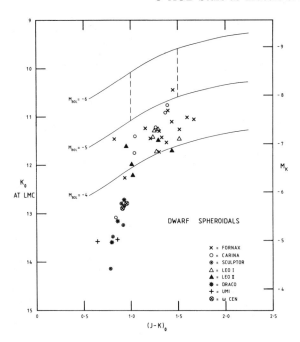

Fig. 8.19. An IR color-magnitude diagram for C-stars in dwarf spheroidals. The K_0 magnitudes illustrated assume an LMC distance modulus of 18.55, whereas in this chapter we assume 18.50. From [58]

the MWG. Nevertheless, the relatively recent discovery [95] of a hitherto unknown dwarf galaxy that is merging with the MWG took the astronomical community by surprise. We will identify this galaxy by the acronym SDSG.[3] Although SDSG has been classified as a dwarf *spheroidal*, morphologically it is a long thin ribbon, with tidal tails spreading around its orbit. It is being disrupted, and some of its stars are already considered to be in the MWG halo (see above). Several dynamical models have been proposed, e.g., [86]. Eight to ten Gyr ago the star formation may have been high, but it went into decline and has been low ever since [17]. SDSG is the nearest (by far) and the most luminous of the 9 dwarf spheroidals associated with the Milky Way (see Table 8.1). Its mean metallicity is $[Fe/H] \approx -1$, but there is evidence for a large range of values. The SDSG brought with it a number of globular clusters into the MWG; these clusters are quite old [107].

Although AGB stars are found in many dwarf spheroidals, SDSG is the only one with well studied Mira variables. All of these variables are C-rich and members of an intermediate-age population. There are, to date, 26 C-stars that are spectroscopically confirmed members of the SDSG, and the total population of such stars in the SDSG is estimated at around 100. Figure 8.20 shows that the majority fall clearly on an extended AGB. It would

---

[3] The reader is warned that another dwarf galaxy is known to be in the same constellation of Sagittarius; see Table 8.3. We will conform to the usage of calling this other galaxy "SagDIG."

Table 8.1. Galaxies within 250 kpc of the MWG

Name	Type	M_V	d_{MWG} [kpc]	d_{M31} [kpc]	m_K tRGB	AGB stars?
MWG	S(B)bcI-II	−20.9	8.0	760	8.0	Y
LMC	IrIII-IV	−18.5	50	790	11.9	Y
SMC	IrIV-V	−17.1	59	790	12.2	Y
SDSG	dSph(t)	−13.8	24	770	10.5	Y
Fornax	dSph	−13.1	138	750	14.2	Y
Leo I	dSph	−11.9	250	900	15.5	Y
Leo II	dSph	−10.1	210	860	15.1	N
Sculptor	dSph	−9.8	87	740	13.2	N
Sextans	dSph	−9.5	86	820	13.2	N
Carina	dSph	−9.4	100	820	13.5	N
Draco	dSph	−8.6	79	730	13.0	N
Ursa Minor	dSph	−8.5	63	740	12.5	N

seem natural to assume that they are associated with the most metal-rich component of the SDSG, but that is difficult to prove.

Of the 26 C-stars, eight have large-amplitude variations, $\Delta K > 0.4^m$, and are presumably Miras, and seven are nonvariable or have low amplitudes, $\Delta K < 0.3^m$. The remainder have not been sufficiently well studied, but are bluer and fainter than the others and therefore unlikely to have large amplitudes. It is curious that, although the very reddest stars have large amplitudes, the Miras and non-Miras are otherwise mixed in color and luminosity. In fact, the most luminous star shows only low amplitude variability, in contrast to the situation in globular clusters (see above). It is also difficult to distinguish between semiregular and Mira C-rich variables on the basis of their colors in the LMC or MWG. Approximate periods have been determined for five of the C-rich Miras; they range from 230 to 360^d. Using the logP/K relation for Miras in the LMC a distance modulus of $m-M = 17.4\pm0.2$ (30 kpc) is derived, in reasonable agreement with the value of 17.2±0.2 from RR Lyrae variables [201].

In the earlier section on the MWG halo we discussed dust-enshrouded C-stars in the halo, but with an orbit that associates them with the SDSG. We expect more such objects to show up in deep infrared surveys, e.g., 2MASS.

The Fornax Dwarf galaxy has the highest absolute visual magnitude, $M_V = -13.1$, of all the spheroidals associated with the MWG, apart from the curious case of the merging galaxy SDSG. Fornax also has the best-defined upper-AGB of any of the dwarf spheroidals. The star-formation history [154] is somewhat complex: Star formation was continuous from an early epoch $\gtrsim 12$ Gyr ago, until ≈ 3 or 4 Gyr ago and then it dropped off, although some star formation seems to have continued to less than 1 Gyr ago. The reddest

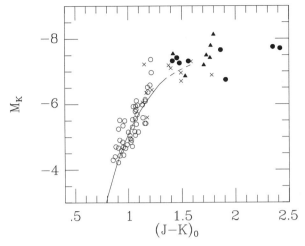

Fig. 8.20. An IR color-magnitude diagram for the SDSG. The open symbols represent O-rich stars, while the others represent C-rich stars as follows: closed circles, Miras; triangles, semiregulars, crosses have unknown variability characteristics. From [201]

and most luminous giants, many of which are C-stars, are centrally concentrated in the same way as the younger stars. The older populations are much more widely dispersed. Two of the globular clusters contain C-stars; although they are the most luminous stars in the clusters and probably on the AGB, they are below the tRGB.

Over 100 C-stars have been identified as well as a number of S, MS, and SC stars [13]. The reddest C-stars listed by [173] have colors $1.9 \gtrsim J - K_s \gtrsim 1.5$ and $-7.4 \gtrsim M_K \gtrsim -8.4$. They may therefore be Mira variables, with a maximum luminosity around $M_{bol} \approx -5.3$. Two other stars with photometry in 2MASS are very probably C-Miras with thick CSEs. Fornax also contains a PN.

In *Leo I* (also called the *"Regulus"* system) a color-magnitude diagram based on photometry in the V and I bands as deep as $I = 24$ shows a small number (50) of stars on the red giant branch and at I magnitudes above the tRGB [109]. The galaxy also has C-stars with luminosities above that of the tRGB; 33 are listed in [50]. Their I magnitudes range from the tRGB ($I = 18.0$) to 1 magnitude above that: These are AGB stars of an intermediate population. From V and I photometry it follows that 15 C-stars have bolometric magnitudes between -3.4 and -4.6, so that all but one are brighter than the tRGB. Recently, five obscured LPV C-stars have been reported (J. Menzies, private information) with luminosities up to $M_{bol} \approx -5.1$ and very red colors; see also [50]. There is thus ample evidence for the existence of AGB stars up to 1 magnitude above tRGB. Leo I also contains an old population of RR Lyrae stars [84].

Leo II had star formation at an enhanced rate between 6 and 11 Gyr ago, but little thereafter [128]. Six C-stars in Leo II are fainter and bluer than those of Leo I, with luminosities $-2.3 > M_{bol} > -4.3$; two are fainter than the tRGB. The detection of RR Lyrae stars proves the existence of a population older than 7 to 10 Gyr [167].

Sculptor contains C-stars but at a magnitude below the tRGB. A VI color-magnitude diagram based on HST measurements shows no stars brighter than $I = 18$, whereas the tRGB is at $I = 15.7$ [134]. The dominant population is old, and there has been little star-formation activity of the last few Gyr. Interestingly, Sculptor has two LPVs that could be AGB stars.

Sextans[4] is a very low surface brightness, dwarf spheroidal galaxy. Thirty-six RR Lyrae stars show the presence of a population older than ≈ 10 Gyr. A large number of blue stragglers are thought to represent an intermediate-age (around 2 to 4 Gyr) population [18], but an accompanying AGB component has not been identified. A solitary LPV may be on the AGB, but there is also much contamination [183].

Carina contains RR Lyrae stars indicative of an old population (> 10 Gyr) [183]. There has been a starburst between 7 and 5 Gyr ago and a more recent one, about 3 Gyr ago [92], but a VI color-magnitude diagram does not show stars brighter then the tRGB [127]. There are also two C-stars with luminosities above that of the tRGB [183].

Draco contains at most a few intermediate-age AGB stars [6]. There are about 260 RR Lyraes known [183]. A reconstruction of the star-formation history [6] shows that most stars formed 10 Gyr or more ago with a small burst about 2 Gyr ago. Three carbon stars [1] have luminosities below the tRGB, and most probably they are the product of mass exchange in a narrow binary (see Chapter 9).

Ursa Minor has only old, > 11 Gyr, stars and no AGB stars [183], except for four low-luminosity C-stars. color-magnitude diagrams based on $BVRI$ photometry show no stars, or perhaps a handful, above the tRGB, and the conclusion is that the U Mi system consists mainly of old stars. There are, however, some bright blue stars, and this would suggest ongoing star formation, but since there are no other signs that this happens, it is more attractive to explain these blue stars as blue stragglers [27].

8.5.2 Galaxies Within 300 kpc of M 31

This subset of the LG contains 14 galaxies, Table 8.2. The group has a large spiral galaxy at its center together with two smaller, but still relatively lu-

---
[4] This galaxy is to be distinguished from the galaxies Sextans A and Sextans B, which probably are outside of the LG [183]

Table 8.2. Galaxies within 300 kpc of M 31

Name	Type	M_V	d_{MWG} [kpc]	d_{M31} [kpc]	m_K tRGB	AGB stars?
M 31	Sb I-II	−21.2	760	–	17.9	Y
M 33	Sc II-III	−18.9	795	210	18.0	Y
M 32	E2	−16.5	760	–	17.9	Y
NGC 205	Sph	−16.4	760	–	17.9	Y
IC 10	Ir IV	−16.3	660	250	17.6	Y
NGC 185	Sph	−15.6	660	130	17.6	Y
NGC 147	Sph	−15.1	660	140	17.6	Y
And I	dSph	−11.8	810	70	18.1	N
And II	dSph	−11.8	700	140	17.7	N
And VI	dSph	−10.6	830	280	18.0	N
Pisces	dIr/dSph	−10.4	810	280	18.1	N
And V	dSph	−10.2	810	120	18.1	N
And III	dSph	−10.2	760	66	17.9	N
And VII	dSph	−9.5	690	220	17.7	N

minous, satellite galaxies. This core of one large and two smaller galaxies is surrounded by a halo of dwarf spheroidals. This description is remarkably similar to that of the group surrounding the MWG, if we ignore the fact that the two nearby satellites (LMC, SMC) are irregular in the case of the MWG and elliptic/spheroidal (M 32, NGC 205) in the case of M 31. The core of the M 31 group shows clear signs of gravitational interactions (e.g., in the deformation of an M 31 spiral arm near M 32), as is the case for the core MWG/LMC/SMC. The M 31 group differs from the MWG group in that the halo of small galaxies contains another spiral (M 33 and a dwarf irregular, IC 10). Both are, however, quite far away from the center of the group (M 33 at 210 kpc; IC 10 at 250 kpc).

IC 10 is an irregular galaxy with up to 20% interstellar matter. WR-stars have been found, and this demonstrates that there has been a recent starburst. IC 10 is the only starburst galaxy in the LG, and it may be the nearest blue compact dwarf (BCD) [148, 183]. The galaxy is at galactic latitude $b = -3.34°$, and there is much contamination by foreground stars and considerable extinction in the MWG [$E(B-V) = 1.0$]. The tRGB is around $K = 18$ and depends somewhat on the assumed foreground extinction. JHK photometry for the central area of IC 10 suggests the presence of many red stars brighter than the tRGB limit [23], and they will thus be AGB stars, if the crowding problems have been solved correctly.

NGC 185 has stars with ages within a wide range: RR Lyrae stars (10 Gyr or older) [151] and a significant number of red stars brighter than the tRGB [110, 118, 119]. A remarkable feature are the many stars in the center of the galaxy younger than 100 Myr, which suggest a recent starburst. This is reminiscent of what passed in NGC 205. In NGC 185 the brightest images may be a conglomeration of several objects. The galaxy contains PNe [149] and a cloud of gas of about $10^5 M_\odot$. NGC 185 and NGC 147 may form a double system [183].

In *NGC 147* a small number of relatively faint, $M_{bol} \gtrsim -5$, AGB stars have been identified that may represent a small intermediate-age population, about 5 Gyr old, [82], although they could also be older [145]. They should be monitored for variability.

Six dwarf spheroidal companions to M 31 are known as *And I* through *And VII*. *And IV* is a background galaxy, seen through M 31 [60]. *And VI* is also known as *Pegasus II*, and *And VII* as the *Cassiopeia* system.

And I (and several others of the smaller M 31 companions) have fewer upper-AGB stars than the spheroidal companions of the MWG and thus a smaller intermediate-age population. A red horizontal branch (HB) has been found in *And I* suggesting that the bulk of the population is around 10 Gyr, but the presence of blue HB stars and RR Lyraes provides evidence of a small older population [39].

And II is the only one of M 31's dSph companions in which upper-AGB C-stars have been found to date. A small number of C-stars in *And II* have $M_{bol} \approx -4.1$, suggesting ages around 6 to 9 Gyr according to [40]. The existence of a blue HB and of RR Lyraes argues for an additional old (> 10 Gyr) population. The RGB suggests a considerable spread of metallicity with a mean $[m/H] \approx -1.5$. A two-component model can explain the abundance spread.

And VI[5]. Color-magnitude diagrams made from deep CCD images in *BVI* do not contain stars above the tRGB [87].

The *Pisces* system (also called *LGS 3*) is probably associated with either M 31 or M 33. It is a low-luminosity, low-surface-brightness, gas-rich galaxy, of a type intermediate between dwarf irregular and dwarf elliptical; this is reflected in the rather small amount of gas for an irregular galaxy and a high amount for a spheroidal: 2% for Pisces, compared to 20% or more in other (dwarf) irregulars and to $< 0.1\%$ in the spheroidals. The star-formation rate in Pisces is one to two orders of magnitude smaller than in other irregulars [9]. If galaxies evolve from star-forming to passive systems, then Pisces may

---

[5] This galaxy has also been named the *"Pegasus"* galaxy, but this name has also been used for DDO 216, which is a nearby but different galaxy.

Table 8.3. The remaining galaxies of the Local Group

Name	Type	M_V	d_{MWG} [kpc]	d_{M31} [kpc]	m_K tRGB	AGB stars?
NGC 6822	Ir IV-V	−16.0	500	900	17.0	Y
IC 1613	Ir IV	−15.3	725	500	16.8	Y
WLM	Ir IV-V	−14.4	925	820	18.4	Y
DDO 216	Ir V	−12.3	760	410	17.9	Y
Leo A	Ir V	−11.5	690	1100	17.7	Y
Aqarius	-	−11.3	1025	1130	18.6	Y
SagDIG	Ir V	−10.7	1300	1540	19.2	Y
Phoenix	dIr/dSph	−9.8	395	840	16.5	Y
Tucana	dSph	−9.6	870	1320	18.2	N

represent the link between the two states, rendering it particularly interesting.

And V was recently (1998) discovered. In a *VI* color-magnitude diagram there are no stars above the tRGB [183].

And III contains an old population (blue HB stars and RR Lyrae variables) and does not show signs of an intermediate population in an HST *BV* color magnitude diagram [38].

And VII is also called the *Cassiopeia* galaxy. Deep *VI* photometry shows RGB stars but no red stars above the tip of the RGB [75].

8.5.3 The Remaining Galaxies of the Local Group

Table 8.3 shows mainly irregular galaxies; the Tucana dwarf galaxy is the only clear dwarf spheroidal (the Cetus galaxy will be a second case). All contain luminous AGB stars except the faintest companion, Tucana. The absolute V magnitudes in column 3 have an average significantly above that of the dwarf spheroidals in Tables 8.1 and 8.2, but this may be due to a selection effect: The search for dwarf spheroidals farther away than 250 kpc ($m - M > 22$) is probably incomplete.

NGC 6822 is a bright dwarf irregular galaxy, with some properties similar to those of the SMC. More than 900 known C-stars trace out a huge, slightly elliptical, halo and do not coincide with the H I cloud that surrounds the galaxy [112]. One AGB S-star has been found. *VRI* color-magnitude diagrams show a densely populated AGB extending to red colors. The star formation probably began at a very early epoch, 12–15 Gyr ago, and may have declined during the last few Gyr [67]. The last reference contains an informative discussion on the difficulties of modeling the AGB.

IC 1613 is a well-studied dwarf irregular within the Local Group. One hundred ninety-five C-stars, identified in a photometric survey [2], are distributed over a very large area, considerably greater than that occupied by H I. color-magnitude diagrams from VI photometry show a densely populated AGB, providing evidence for a large intermediate-age population [35]. One Mira variable was found recently [104]. The density of AGB stars is highest along the bar and decreases with increasing distance from the main body of the galaxy. The density is not enhanced in areas of current star formation. From the smooth luminosity function of the AGB stars one concludes that star formation was continuous between 2 and 10 Gyr ago. There is some evidence that the intermediate-age AGB population decreases in the vicinity of the H II regions in the galaxy. There are also suggestions of an age difference between AGB stars in the main body of the galaxy and those near the H II regions in the north-east. The former span a range in ages between 1 and 10 Gyr, while the latter may be younger than 1 Gyr.

Wolf Lundmark Melotte (WLM) is an isolated dwarf galaxy near the edge of the LG. A halo around WLM has been seen in deep VI-images. The youngest halo stars, which are also the brightest, have been estimated to be 10 Gyr old on the basis of the AGB tip luminosity, which is 0.8 to 0.9^m brighter in I than the tRGB [130, 131].

DDO 216[6] is a faint dwarf irregular galaxy in the LG. This galaxy combines dIr characteristics with those of dwarf elliptical or spheroidal systems, leading to the suggestion that it is a transition object. About 40 C-stars with a median value of $M_I = -4.3$ have been identified; they spread into the halo [16]. Intermediate-mass AGB stars show up in VRI color-magnitude diagrams [7, 8]. A dense concentration of extended-AGB stars is found along the major axis, following the same pattern as other stars. Outside of this zone the distribution is clumpy and probably marks the positions of older star-forming regions.

The *Sagittarius Dwarf Irregular Galaxy* (SagDIG) is the most remote galaxy associated with the LG. C-stars have been detected with $-4.4 \geq M_I \geq -5.4$ (see [51]). $BVRI$ photometry suggests that the few AGB stars above the tRGB represent a small population formed a few Gyr ago [111].

Aquarius or *DDO 210* is a faint dwarf irregular in the LG. Three particularly bright C-stars ($M_I = -4.8$) have been identified on the basis of their color [16]. One is at about 3.2 arcmin (950 pc) from the center.

---

[6] This galaxy has sometimes been called *"Pegasus"*; see the previous footnote.

Phoenix is situated on the western edge of the Magellanic Stream and, like *Pisces*, belongs to a class intermediate between dIr and dSph. Luminous AGB stars have been identified. Two of these stars are C-rich with $M_{bol} \approx -3.7$ and estimated ages of 8 to 10 Gyr. AGB stars as bright as $M_{bol} \approx -4.5$, are also present with ages claimed to be around 3 Gyr (although this luminosity is more plausibly associated with an AGB star of around 10 Gyr). The AGB stars, like the young population, are centrally concentrated, indicating that star formation has occurred recently and preferentially near the center [85].

Tucana is an isolated LG dwarf spheroidal, which appears, like *Ursa Minor*, to contain only old stars. C-stars have been looked for in vain with techniques that have been successful in the DDO 216 galaxy which is at almost the same distance [16].

Recently, a new member of the LG was found, called the *Cetus* galaxy [203]. The VI color-magnitude diagram shows a clear tRGB at $I = 20.5$, from which a distance of 775 kpc is derived or $(m-M) = 24.5$. There is no significant number of brighter red giants, and AGB stars have not been detected [153].

8.5.4 Conclusions About the Local Group

The star-formation history of each galaxy in the LG is of importance for the question of how the LG was formed and how it developed. This star-formation history may be read off from color-magnitude diagrams; various techniques have been used, but they all come down to comparing color-magnitude diagrams that have been observed with color-magnitude diagrams that have been derived from calculations of stellar evolution; for an example see [52]. Deep images in the V- and in the I band are the basis of most (observed) color-magnitude diagrams. It is the opinion of the authors of this chapter that more K band imaging followed by monitoring potential LPVs will enhance significantly the information on populations with an age between 100 Myr and a few Gyr and will thus fill in an uncertain part of the star-formation history.

The LG has some well-known overall properties. M 31 and the MWG have the highest mass by far and act as two gravitational centers of the whole group. Each has two nearest neighbors with considerably lower masses than either MWG or M 31, but still much more than any of the "little ones." All these little ones are dwarf spheroidals; they contain very little gas ($<0.1\%$), and they certainly contain an old population. The more luminous of these "little" spheroidals have luminous AGB stars in spite of a low gas content ($<0.1\%$, if we use the data in [183]). Are new stars forming from the debris left by previous generations of AGB stars? Gas-rich dwarfs are all irregular and are found far away from both M 31 and the MWG; they contain luminous AGB stars, and they contain between 20 to 40% of their mass in gas, with exceptions for DDO 216 and Pisces: 2% in both. One thus expects a scenario in which all (small) galaxies have formed as gas-rich irregulars, but this gas has disappeared for the galaxies closest to the giant galaxies MWG and M 31. Alternatively, the gas may have been driven out of these low-mass galaxies by supernovae. Such scenarios have been proposed as well (e.g., [120, 121]).

8.6 AGB Stars in Galaxies Outside of the Local Group

To estimate our chances of detecting AGB stars in galaxies outside of the LG we look at the detection limits at I and K and take the tRGB as a yardstick: $I_{trgb} = -4.0$, $K_{trgb} = -6.5$. With present-day techniques AGB stars can be detected out to $(m - M) = 30$ ($D = 10$ Mpc) in I and to $(m - M) = 29$ ($D = 6.4$ Mpc) in K. The edge of the Local Group is about 1 Mpc away. The large number of unexplored but accessible galaxies beyond this limit explains why the study of AGB stars is such a rapidly developing field. We have not made a complete search for literature on AGB stars in galaxies outside of the Local Group; the sample would have become incomplete by the time this book will be published. We present only some results representative of studies underway.

Three isolated galaxies: *IC 5152* is a dwarf irregular just beyond the LG at a distance of 1.7 Mpc or $(m-M) = 26.15$; a number of very red C-star candidates have been seen [210], and the galaxy may contain a significant intermediate-age population. *DDO 187* is a dwarf irregular galaxy. In VI-color-magnitude diagrams a weak AGB extends beyond the tRGB [11]; the existence of AGB variables requires confirmation. *DDO 190* is a dwarf irregular galaxy at (2.9 ± 0.2) Mpc, $(m-M) = 27.3$. An intermediate-age population appears to be present [10].

The Antlia-Sextans group (distance 1.4 Mpc, $m-M = 25.7$): This small group with only 5 members is on the periphery of the Local Group. Its largest member is NGC 3109, a dwarf irregular. VI photometry down to $I = 24$ of some 17 000 stars led to the identification of large numbers of AGB-stars [132]. JHK photometry has been obtained down to $K = 18$, and leads to age estimates around 1 Gyr for the AGB stars [4].

The Sculptor group (distance 2.5 Mpc, $m-M = 27.0$): Three of the dominant spiral galaxies, NGC 55, NGC 300, and NGC 7793, contain AGB-stars with ages between 0.1 and 10 Gyr. There is evidence in NGC 7793 that the star-forming rate in the center has been low over the last 1 Gyr [42].

The Sculptor group contains a dwarf irregular galaxy sometimes called SDIG (to confuse the unwary). From JK photometry a large population of AGB stars is recognized that is about 6 Gyr old and thus older than a recent outburst of star formation [83]. The metallicity is very low and the authors suppose that to reconcile early star formation and low metallicity, much of the gas must have been lost.

The *IC 342/Maffei 1 group* (3.6 Mpc, $m - M = 27.8$): *Maffei 1* is strongly obscured by the MWG. Probably the best that can be said is that the galaxy contains bright red giants that are probably on the AGB [45]. *NGC 1569* is a dwarf irregular galaxy intermediate, in metallicity and total luminosity, between the two Magellanic Clouds. It is the closest example of a star-bursting galaxy outside the LG, and has near its center two super star clusters, with young supergiants, WR stars, and star-forming H II regions. A well-populated AGB is seen in HST images despite severe crowding effects. The AGB stars are spread over the entire galaxy and are not confined to regions of current star formation [3].

The *M 81 group* (4.0 Mpc, $m - M = 28.0$): Using HST 25 LPVs have been found in two fields in *M 81* and the Mira $\log P/K$ relation used to confirm the distance derived from Cepheids [89]. In *NGC 2403* a few AGB stars may have been identified.

Five *blue compact dwarf (BCD) galaxies* observed with the HST contain old and intermediate-age stars: NGC 6789, Mrk 178, VII Zw 403, I Zw 36, and UGC A 290 [36, 53, 155, 156, 157]. The star-forming cores all exhibit significant populations of luminous AGB stars, indicating vigorous star formation several hundred Myr ago.

We close this chapter with a record: the galaxy *NGC 7252* at a distance of 64 Mpc ($m - M = 34.0$) is a merging galaxy that contains young star clusters. C-stars may have been identified in the integrated spectrum of the globular cluster W3 in this galaxy [136]. If confirmed, these are the most distant C-stars known to date.

8.7 Conclusions and Outlook

AGB stars, in practice mostly luminous AGB stars, have been detected in a major part of our MWG, at least well beyond the Galactic center. They are now also detected in increasing numbers in other galaxies, even beyond the LG. AGB stars are useful tracers of star-formation histories for times between those of the oldest stars, e.g., the RR Lyrae variables, and the youngest, Cepheids and O and early B-type stars. Near- and mid-infrared surveys have been most productive in the detection of these stars: AGB stars are infrared objects first of all. The properties of AGB stars are not uniquely determined by measuring one or two photometric magnitudes. When more properties are measured, AGB stars become increasingly useful as tracers of stellar populations. Pulsation periods are important because they indicate the initial stellar mass, but also a separation into carbon-rich and oxygen-rich is of great importance. The full luminosity function of AGB stars contains information on the history of the star-formation rate whereas the metallicity of a population can be derived from the separate luminosity functions of C- and M-type AGB stars. Maser stars contain important radial-velocity information. They are difficult to detect beyond, say, twice the distance to the Galactic center. Their further exploration must perhaps wait for the construction of the "square-kilometer array," SKA, for centimeter and decimeter radio astronomy.

All this argues that the *detection* of AGB stars is only the first step in studies of stellar populations in our and other galaxies; establishing in detail the further properties of AGB stars must be the next goal.

References

1. Aaronson, M. *ApJ*, 266, L11, 1983.
2. Albert, L., Demers, S., and Kunkel, W. E. *AJ*, 119, 2780, 2000.
3. Aloisi, A., Clampin, M., Diolaiti, E., et al. *AJ*, 121, 1425, 2001.
4. Alonso, M. V., Minniti, D., Zijlstra, A. A., and Tolstoy, E. *A&A*, 346, 33, 1999.
5. Alves, D. R. and Nelson, C. A. *ApJ*, 542, 789, 2000.
6. Aparicio, A., Carrera, R., and Martínez-Delgado, D. *AJ*, 122, 2524, 2001.
7. Aparicio, A. and Gallart, C. *AJ*, 110, 2105, 1995.
8. Aparicio, A., Gallart, C., and Bertelli, G. *AJ*, 114, 669, 1997.
9. Aparicio, A., Gallart, C., and Bertelli, G. *AJ*, 114, 680, 1997.
10. Aparicio, A. and Tikhonov, N. *AJ*, 119, 2183, 2000.
11. Aparicio, A., Tikhonov, N., and Karachentsev, I. *AJ*, 119, 177, 2000.
12. Ashman, K. M. and Zepf, S. E. *ApJ*, 384, 50, 1992.
13. Azzopardi, M. In Wing, R., editor, *IAU symposium 177: The carbon star phenomenon*, page 51. Kluwer Academic Publishers: Dordrecht, 2000.
14. Baade, W. *ApJ*, 100, 137, 1944.
15. Baade, W. *Evolution of Stars and Galaxies*. Harvard University Press: Cambridge, 1963.
16. Battinelli, P. and Demers, S. *AJ*, 120, 1801, 2000.
17. Bellazzini, M., Ferraro, F. R., and Buonanno, R. *MNRAS*, 307, 619, 1999.
18. Bellazzini, M., Ferraro, F. R., and Pancino, E. *MNRAS*, 327, L15, 2001.
19. Blanco, V. Distribution and motios on late-type giants. In Blaauw, A. and Schmidt, M., editors, *Galactic Structure*, page 241. University of Chicago Press; Chicago, 1965.
20. Blommaert, J. A. D. L., van der Veen, W. E. C. J., and Habing, H. J. *A&A*, 267, 39, 1993.
21. Blommaert, J. A. D. L., van der Veen, W. E. C. J., van Langevelde, H. J., Habing, H. J., and Sjouwerman, L. O. *A&A*, 329, 991, 1998.
22. Blommaert, J. A. D. L., van Langevelde, H. J., and Michiels, W. F. P. *A&A*, 287, 479, 1994.

23. Borissova, J., Georgiev, L., Rosado, M., et al. *A&A*, 363, 130, 2000.
24. Brewer, J. P., Richer, H. B., and Crabtree, D. R. *AJ*, 109, 2480, 1995.
25. Brewer, J. P., Richer, H. B., and Crabtree, D. R. *AJ*, 112, 491, 1996.
26. Butcher, H. *ApJ*, 216, 372, 1977.
27. Carrera, R., Aparicio, A., Martínez-Delgado, D., and Alonso-García, J. *AJ*, 123, 3199, 2002.
28. Catchpole, R. M., Whitelock, P. A., and Glass, I. S. *MNRAS*, 247, 479, 1990.
29. Christlieb, N., Green, P. J., Wisotzki, L., and Reimers, D. *A&A*, 375, 366, 2001.
30. Chu, Y.-H., Suntzeff, N., Hesser, J., and Bohlender, D., editors. *IAU symp. 190: New Views of the Magellanic Clouds*. ASP: San Francisco, 1999.
31. Cioni, M.-R., Habing, H. J., and Israel, F. P. *A&A*, 358, L9, 2000.
32. Cioni, M.-R., Marquette, J.-B., Loup, C., et al. *A&A*, 377, 945, 2001.
33. Cioni, M.-R., van der Marel, R. P., Loup, C., and Habing, H. J. *A&A*, 359, 601, 2000.
34. Cioni, M.-R. L. and Habing, H. J. *A&A*, 402, 133, 2003.
35. Cole, A. A., Tolstoy, E., Gallagher, J. S., et al. *AJ*, 118, 1657, 1999.
36. Crone, M. M., Schulte-Ladbeck, R. E., Hopp, U., and Greggio, L. *ApJ*, 545, L31, 2000.
37. Da Costa, G. S. and Armandroff, T. E. *AJ*, 109, 2533, 1995.
38. Da Costa, G. S., Armandroff, T. E., and Caldwell, N. *AJ*, 124, 332, 2002.
39. Da Costa, G. S., Armandroff, T. E., Caldwell, N., and Seitzer, P. *AJ*, 112, 2576, 1996.
40. Da Costa, G. S., Armandroff, T. E., Caldwell, N., and Seitzer, P. *AJ*, 119, 705, 2000.
41. Dame, T. M., Hartmann, D., and Thaddeus, P. *ApJ*, 547, 792, 2001.
42. Davidge, T. J. *ApJ*, 497, 650, 1998.
43. Davidge, T. J. *PASP*, 112, 1177, 2000.
44. Davidge, T. J. *AJ*, 119, 748, 2000.
45. Davidge, T. J. and van den Bergh, S. *ApJ*, 553, L133, 2001.
46. de Vaucouleurs, G. In Kerr, F. and Rodgers, A., editors, *IAU Symp. 20: The Galaxy and the Magellanic Clouds*, page 195. Australian Academy of Science: Canberra, 1964.
47. Deguchi, S., Fujii, T., Izumiura, H., et al. *ApJS*, 130, 351, 2000.
48. Deguchi, S., Fujii, T., Izumiura, H., et al. *ApJS*, 128, 571, 2000.
49. Deguchi, S., Matsumoto, S., and Wood, P. R. *PASJ*, 50, 597, 1998.
50. Demers, S. and Battinelli, P. *AJ*, 123, 238, 2002.
51. Demers, S., Dallaire, M., and Battinelli, P. *AJ*, 123, 3428, 2002.
52. Dolphin, A. E. *MNRAS*, 332, 91, 2002.
53. Drozdovsky, I. O., Schulte-Ladbeck, R. E., Hopp, U., Crone, M. M., and Greggio, L. *ApJ*, 551, L135, 2001.
54. Dwek, E., Arendt, R. G., Hauser, M. G., et al. *ApJ*, 445, 716, 1995.
55. Feast, M., Whitelock, P., and Menzies, J. *MNRAS*, 329, L7, 2002.
56. Feast, M. and Whitelock, P. A. In Giovannelli, F. and Matteucci, F., editors, *The Evolution of the Milky Way: Stars vs. Clusters*, page 75. Kluwer Academic Publishers: Dordrecht, 2000.
57. Feast, M. W., Glass, I. S., Whitelock, P. A., and Catchpole, R. M. *MNRAS*, 241, 375, 1989.
58. Feast, M. W. and Whitelock, P. A. *MNRAS*, 259, 6, 1992.

59. Feast, M. W. and Whitelock, P. A. *MNRAS*, 317, 460, 2000.
60. Ferguson, A. M. N., Gallagher, J. S., and Wyse, R. F. G. *AJ*, 120, 821, 2000.
61. Ford, H. C. and Jacoby, G. H. *ApJS*, 38, 351, 1978.
62. Freudenreich, H. T. *ApJ*, 492, 495, 1998.
63. Frogel, J. A. and Elias, J. H. *ApJ*, 324, 823, 1988.
64. Frogel, J. A., Terndrup, D. M., Blanco, V. M., and Whitford, A. E. *ApJ*, 353, 494, 1990.
65. Frogel, J. A. and Whitford, A. E. *ApJ*, 320, 199, 1987.
66. Fux, R. *A&A*, 345, 787, 1999.
67. Gallart, C., Aparicio, A., Bertelli, G., and Chiosi, C. *AJ*, 112, 1950, 1996.
68. Gardiner, L. T. and Noguchi, M. *MNRAS*, 278, 191, 1996.
69. Glass, I. S., Ganesh, S., Alard, C., et al. *MNRAS*, 308, 127, 1999.
70. Glass, I. S. and Lloyd Evans, T. *Nature*, 291, 303, 1981.
71. Glass, I. S., Matsumoto, S., Carter, B. S., and Sekiguchi, K. *MNRAS*, 321, 77, 2001.
72. Glass, I. S. and Reid, N. *MNRAS*, 214, 405, 1985.
73. Glass, I. S., Whitelock, P. A., Catchpole, R. M., and Feast, M. W. *MNRAS*, 273, 383, 1995.
74. Graff, D. S., Gould, A. P., Suntzeff, N. B., Schommer, R. A., and Hardy, E. *ApJ*, 540, 211, 2000.
75. Grebel, E. K. and Guhathakurta, P. *ApJ*, 511, L101, 1999.
76. Groenewegen, M. A. T. In Le Bertre, T., Lebre, A., and Waelkens, C., editors, *IAU Symp. 191: Asymptotic Giant Branch Stars*, page 535. ASP: San Francisco, 1999.
77. Groenewegen, M. A. T. and Blommaert, J. A. D. L. *A&A*, 332, 25, 1998.
78. Groenewegen, M. A. T., Oudmaijer, R. D., and Ludwig, H. *MNRAS*, 292, 686, 1997.
79. Habing, H. J. *A&A*, 200, 40, 1988.
80. Habing, H. J., Olnon, F. M., Chester, T., Gillett, F., and Rowan-Robinson, M. *A&A*, 152, L1, 1985.
81. Habing, H. J., Olnon, F. M., Winnberg, A., Matthews, H. E., and Baud, B. *A&A*, 128, 230, 1983.
82. Han, M., Hoessel, J. G., Gallagher, J. S., Holtsman, J., and Stetson, P. B. *AJ*, 113, 1001, 1997.
83. Heisler, C. A., Hill, T. L., McCall, M. L., and Hunstead, R. W. *MNRAS*, 285, 374, 1997.
84. Held, E. V., Clementini, G., Rizzi, L., et al. *ApJ*, 562, L39, 2001.
85. Held, E. V., Saviane, I., and Momany, Y. *A&A*, 345, 747, 1999.
86. Helmi, A. and White, S. D. M. *MNRAS*, 323, 529, 2001.
87. Hopp, U., Schulte-Ladbeck, R. E., Greggio, L., and Mehlert, D. . *A&A*, 342, L9, 1999.
88. Hughes, S. M. G. *AJ*, 97, 1634, 1989.
89. Hughes, S. M. G. In Bradley, P. and Guzik, J., editors, *A Half Century of Stellar Pulsation Interpretation*, page 390. ASP: San Francisco, 1998.
90. Hughes, S. M. G. and Wood, P. R. *AJ*, 99, 784, 1990.
91. Hughes, S. M. G., Wood, P. R., and Reid, N. *AJ*, 101, 1304, 1991.
92. Hurley-Keller, D., Mateo, M., and Nemec, J. *AJ*, 115, 1840, 1998.
93. Ibata, R., Irwin, M., Lewis, G. F., and Stolte, A. *ApJ*, 547, L133, 2001.
94. Ibata, R., Lewis, G. F., Irwin, M., Totten, E., and Quinn, T. *ApJ*, 551, 294, 2001.

95. Ibata, R., Gilmore, G., and Irwin, M. *Nature*, 370, 194, 1994.
96. Izumiura, H., Deguchi, S., Fujii, T., et al. *ApJS*, 125, 257, 1999.
97. Jablonka, P., Courbin, F., Meylan, G., et al. *A&A*, 359, 131, 2000.
98. Jackson, T., Ivezić, Ž., and Knapp, G. R. *MNRAS*, 337, 749, 2002.
99. Jensen, J. B., Tonry, J. L., Thompson, R. I., et al. *ApJ*, 550, 503, 2001.
100. Jiang, B. W., Deguchi, S., and Ramesh, B. *PASJ*, 51, 95, 1999.
101. Kerschbaum, F. and Hron, J. *A&AS*, 106, 397, 1994.
102. Kim, S., Staveley-Smith, L., Dopita, M. A., et al. *ApJ*, 503, 674, 1998.
103. Kodaira, K., Vansevicius, V., Tamura, M., and Miyazaki, S. *ApJ*, 519, 153, 1999.
104. Kurtev, R., Georgiev, L., Borissova, J., et al. *A&A*, 378, 449, 2001.
105. López-Corredoira, M., Hammersley, P. L., Garzón, F., et al. *A&A*, 373, 139, 2001.
106. Lançon, A., Mouhcine, M., Fioc, M., and Silva, D. *A&A*, 344, L21, 1999.
107. Layden, A. C. and Sarajedini, A. *AJ*, 119, 1760, 2000.
108. Lee, M. G. *AJ*, 112, 1438, 1996.
109. Lee, M. G., Freedman, W., Mateo, M., et al. *AJ*, 106, 1420, 1993.
110. Lee, M. G., Freedman, W. L., and Madore, B. F. *AJ*, 106, 964, 1993.
111. Lee, M. G. and Kim, S. C. *AJ*, 119, 777, 2000.
112. Letarte, B., Demers, S., Battinelli, P., and Kunkel, W. E. *AJ*, 123, 832, 2002.
113. Liebert, J., Cutri, R. M., Nelson, B., et al. *PASP*, 112, 1315, 2000.
114. Lin, D. N. C., Jones, B. F., and Klemola, A. R. *ApJ*, 439, 652, 1995.
115. Lindqvist, M., Habing, H. J., and Winnberg, A. *A&A*, 259, 118, 1992.
116. Maraston, C. *MNRAS*, 300, 872, 1998.
117. Marigo, P., Girardi, L., and Bressan, A. *A&A*, 344, 123, 1999.
118. Martínez-Delgado, D., Aparicio, A., and Gallart, C. *AJ*, 118, 2229, 1999.
119. Martínez-Delgado, D., Gallart, C., and Aparicio, A. *AJ*, 118, 862, 1999.
120. Mayer, L., Governato, F., Colpi, M., et al. *ApJ*, 559, 754, 2001.
121. Mayer, L., Governato, F., Colpi, M., et al. *ApJ*, 547, L123, 2001.
122. McGinn, M. T., Sellgren, K., Becklin, E. E., and Hall, D. N. B. *ApJ*, 338, 824, 1989.
123. McLean, I. S. and Liu, T. *ApJ*, 456, 499, 1996.
124. McWilliam, A. and Rich, R. M. *ApJS*, 91, 749, 1994.
125. Menten, K. M., Reid, M. J., Eckart, A., and Genzel, R. *ApJ*, 475, L111, 1997.
126. Messineo, M., Habing, H. J., Sjouwerman, L. O., Omont, A., and Menten, K. M. *A&A*, 393, 115, 2002.
127. Mighell, K. J. *AJ*, 114, 1458, 1997.
128. Mighell, K. J. and Rich, R. M. *AJ*, 111, 777, 1996.
129. Minniti, D. *AJ*, 109, 1663, 1995.
130. Minniti, D. *ApJ*, 459, 579, 1996.
131. Minniti, D. and Zijlstra, A. A. *AJ*, 114, 147, 1997.
132. Minniti, D., Zijlstra, A. A., and Alonso, M. V. *AJ*, 117, 881, 1999.
133. Miyazaki, A., Deguchi, S., Tsuboi, M., Kasuga, T., and Takano, S. *PASJ*, 53, 501, 2001.
134. Monkiewicz, J., Mould, J. R., Gallagher, J. S., et al. *PASP*, 111, 1392, 1999.
135. Montegriffo, P., Ferraro, F. R., Origlia, L., and Fusi Pecci, F. *MNRAS*, 297, 872, 1998.
136. Mouhcine, M., Lançon, A., Leitherer, C., Silva, D., and Groenewegen, M. A. T. *A&A*, 393, 101, 2002.

137. Nikolaev, S. and Weinberg, M. D. *ApJ*, 542, 804, 2000.
138. Nishida, S., Tanabé, T., Nakada, Y., et al. *MNRAS*, 313, 136, 2000.
139. Nowotny, W., Kerschbaum, F., Schwarz, H. E., and Olofsson, H. *A&A*, 367, 557, 2001.
140. Olivier, E., Whitelock, P., and Marang, F. *MNRAS*, 326, 490, 2001.
141. Ortiz, R., Blommaert, J. A. D. L., Copet, E., et al. *A&A*, 388, 279, 2002.
142. Payne-Gaposchkin, C. H. *The Variable Stars of The Large Magellanic Cloud.* Smithsonian Institution Press, 1971.
143. Reid, I. N., Hughes, S. M. G., and Glass, I. S. *MNRAS*, 275, 331, 1995.
144. Reid, M. J. *ARA&A*, 31, 345, 1993.
145. Renzini, A. *AJ*, 115, 2459, 1998.
146. Renzini, A. and Greggio, L. In Jarvis, B. and Terndrup, D. M., editors, *Bulges of Galaxies*. ESO Conf. and Workshop Proc. No. 35, 1990.
147. Renzini, A. and Voli, M. *A&A*, 94, 175, 1981.
148. Richer, M. G., Bullejos, A., Borissova, J., et al. *A&A*, 370, 34, 2001.
149. Richer, M. G. and McCall, M. L. In *Revista Mexicana de Astronomia y Astrofisica Conference Series*, volume 12, page 173, 2002.
150. Rogers, A. E. E., Doeleman, S., Wright, M. C. H., et al. *ApJ*, 434, L59, 1994.
151. Saha, A. and Hoessel, J. G. *AJ*, 99, 97, 1990.
152. Salaris, M., Cassisi, S., and Weiss, A. *PASP*, 114, 375, 2002.
153. Sarajedini, A., Grebel, E. K., Dolphin, A. E., et al. *ApJ*, 567, 915, 2002.
154. Saviane, I., Held, E. V., and Bertelli, G. *A&A*, 355, 56, 2000.
155. Schulte-Ladbeck, R. E., Hopp, U., Crone, M. M., and Greggio, L. *ApJ*, 525, 709, 1999.
156. Schulte-Ladbeck, R. E., Hopp, U., Greggio, L., and Crone, M. M. *AJ*, 120, 1713, 2000.
157. Schulte-Ladbeck, R. E., Hopp, U., Greggio, L., Crone, M. M., and Drozdovsky, I. O. *AJ*, 121, 3007, 2001.
158. Schwering, P. B. W. *A&AS*, 79, 105, 1989.
159. Schwering, P. B. W. and Israel, F. P. *A&AS*, 79, 79, 1989.
160. Sebo, K. M. and Wood, P. R. *AJ*, 108, 932, 1994.
161. Sevenster, M., Saha, P., Valls-Gabaud, D., and Fux, R. *MNRAS*, 307, 584, 1999.
162. Sevenster, M. N. *MNRAS*, 310, 629, 1999.
163. Sevenster, M. N., Chapman, J. M., Habing, H. J., Killeen, N. E. B., and Lindqvist, M. *A&AS*, 122, 79, 1997.
164. Sevenster, M. N., Chapman, J. M., Habing, H. J., Killeen, N. E. B., and Lindqvist, M. *A&AS*, 124, 509, 1997.
165. Sevenster, M. N., Dejonghe, H., and Habing, H. J. *A&A*, 299, 689, 1995.
166. Sevenster, M. N., van Langevelde, H. J., Moody, R. A., et al. *A&A*, 366, 481, 2001.
167. Siegel, M. H. and Majewski, S. R. *AJ*, 120, 284, 2000.
168. Sjouwerman, L. O., van Langevelde, H. J., Winnberg, A., and Habing, H. J. *A&AS*, 128, 35, 1998.
169. Smith, V. V., Plez, B., Lambert, D. L., and Lubowich, D. A. *ApJ*, 441, 735, 1995.
170. Stanek, K. Z. *ApJ*, 460, L37, 1996.
171. Stanimirovic, S., Staveley-Smith, L., Dickey, J., Sault, R., and Snowden, S. *MNRAS*, 302, 417, 1999.

172. Stephens, A. W., Frogel, J. A., Freedman, W., et al. *AJ*, 121, 2597, 2001.
173. Stetson, P. B., Hesser, J. E., and Smecker-Hane, T. A. *PASP*, 110, 533, 1998.
174. Suntzeff, N. B., Phillips, M. M., Elias, J. H., et al. *PASP*, 105, 350, 1993.
175. Te Lintel Hekkert, P., Caswell, J. L., Habing, H. J., et al. *A&AS*, 90, 327, 1991.
176. Thackeray, A. D. *MNRAS*, 118, 117, 1958.
177. Tiede, G. P. and Terndrup, D. M. *AJ*, 118, 895, 1999.
178. Totten, E. J., Irwin, M. J., and Whitelock, P. A. *MNRAS*, 314, 630, 2000.
179. Trams, N. R., van Loon, J. T., Waters, L. B. F. M., et al. *A&A*, 346, 843, 1999.
180. Trams, N. R., van Loon, J. T., Zijlstra, A. A., et al. *A&A*, 344, L17, 1999.
181. van den Bergh, S. *ApJ*, 517, L97, 1999.
182. van den Bergh, S. *ApJ*, 530, 777, 2000.
183. van den Bergh, S. *The Galaxies of the Local Group*. Cambridge University Press: Cambridge, 2000.
184. van der Marel, R. P. *AJ*, 122, 1827, 2001.
185. van der Marel, R. P. and Cioni, M. L. *AJ*, 122, 1807, 2001.
186. van Langevelde, H. J., Brown, A. G. A., Lindqvist, M., Habing, H. J., and de Zeeuw, P. T. *A&A*, 261, L17, 1992.
187. van Leeuwen, F., Feast, M. W., Whitelock, P. A., and Yudin, B. *MNRAS*, 287, 955, 1997.
188. van Loon, J. T., Groenewegen, M. A. T., de Koter, A., et al. *A&A*, 351, 559, 1999.
189. van Loon, J. T., Zijlstra, A. A., Whitelock, P. A., et al. *A&A*, 329, 169, 1998.
190. Vassiliadis, E. and Wood, P. R. *ApJ*, 413, 641, 1993.
191. Walterbos, R. A. M. and Kennicutt, R. C. *A&A*, 198, 61, 1988.
192. Weinberg, M. D. *ApJ*, 392, L67, 1992.
193. Weinberg, M. D. *ApJ*, 384, 81, 1992.
194. West, M. *Characteristics of the variability of OH/IR stars and estimation of their distances*. Master's thesis, Potchefstroomse Universiteit, 1998.
195. Westerlund, B. E. *The Magellanic Clouds*. Cambridge University Press: Cambridge, 1997.
196. Whitelock, P. In Dejonghe, H. and Habing, H., editors, *IAU Symp. 153: Galactic Bulges*, page 39. Kluwer Academic Publishers: Dordrecht, 1993.
197. Whitelock, P. and Catchpole, R. In Blitz, L., editor, *The Center, Bulge, and Disk of the Milky Way*. Kluwer Academic Publishers: Dordrecht, 1992.
198. Whitelock, P. and Feast, M. *Mem. Soc. astron. Ital.*, 71, 601, 2000.
199. Whitelock, P., Feast, M., and Catchpole, R. *MNRAS*, 248, 276, 1991.
200. Whitelock, P., Menzies, J., Feast, M., et al. *MNRAS*, 267, 711, 1994.
201. Whitelock, P., Menzies, J., Irwin, M., and Feast, M. In Whitelock, P. and Cannon, R., editors, *IAU Symp. 192: The Stellar Content of Local Group Galaxies*, page 136. ASP: San Francisco, 1999.
202. Whitelock, P. A., Feast, M. W., Menzies, J. W., and Catchpole, R. M. *MNRAS*, 238, 769, 1989.
203. Whiting, A. B., Hau, G. K. T., and Irwin, M. *AJ*, 118, 2767, 1999.
204. Wood, P. R. *A&A*, 338, 592, 1998.
205. Wood, P. R., Alcock, C., Allsman, R. A., et al. In le Bertre, T., Lebre, A., and Waelkens, C., editors, *IAU Symp. 191: Asymptotic Giant Branch Stars*, page 151. ASP: San Francisco, 1999.

206. Wood, P. R., Bessell, M. S., and Fox, M. W. *ApJ*, 272, 99, 1983.
207. Wood, P. R., Habing, H. J., and McGregor, P. J. *A&A*, 336, 925, 1998.
208. Wood, P. R., Whiteoak, J. B., Hughes, S. M. G., et al. *ApJ*, 397, 552, 1992.
209. Zijlstra, A. A., Loup, C., Waters, L. B. F. M., et al. *MNRAS*, 279, 32, 1996.
210. Zijlstra, A. A. and Minniti, D. *AJ*, 117, 1743, 1999.
211. Zinn, R. *ApJ*, 293, 424, 1985.

9 AGB Stars in Binaries and Their Progeny

Alain Jorissen

Institut d'Astronomie et d'Astrophysique, Université Libre de Bruxelles

9.1 The Binary–AGB Connection

Although binarity and AGB evolution are in principle disconnected concepts (a star need not be member of a binary system to evolve along the AGB), a rich world flourishes at their contact. The interest in discussing binary stars with an AGB component goes far beyond the possibility of deriving their masses. First, binarity is expected to alter the intrinsic properties of AGB stars (pulsation, mass loss, dust formation, etc.). Second, the importance of the rich progeny of binary stars involving AGB stars has been recognized recently. Various classes of stars with peculiar abundances (such as barium stars, CH stars, Tc-poor S stars, etc.) have indeed been identified as the progeny of binary systems formerly involving an AGB star. Mass transfer from the AGB component is held responsible for the chemical anomalies now observed at the surface of the companion star, which dominates the light of the system, since the AGB star has evolved into a dim white dwarf (WD). Therefore, two broad classes of binary systems must be distinguished in relation to AGB stars: (A) The primary star[1] is currently an AGB star, while the secondary is still an unevolved dwarf; (B) The secondary is a WD, while the primary is at any evolutionary stage (except WD) and has been polluted by the secondary when the latter was an AGB star.[2] The present chapter is organized as follows. Sections 9.2 and 9.3 deal with case A. They present, respectively, a brief review of the (poor) statistics of binaries among AGB stars, and of the impact of binarity on various properties of AGB stars. Section 9.4 deals with case B. Section 9.4.1.1 addresses the question of dis-

---

[1] Defined as the star dominating the visible spectrum.
[2] The case in which both stars are simultaneously on the AGB is expected to be rare, given the short time scale of the AGB evolution with respect to the total stellar lifetime. Such systems require that the two stars had initial masses differing by a very small amount. Recent discussions of the distribution of initial mass ratios q in binary systems (see the contributions by Mayor et al., by Tokovinin, and by Larson in IAU Symposium 200 [155] *The Formation of Binary Stars*) reveal that the case $q = 1$ is not especially frequent among binaries with periods larger than about 100 d, which are those relevant to the present discussion. Systems consisting of two AGB stars need not therefore be further discussed here.

tinguishing systems belonging to cases A and B when the primary is a giant. Section 9.4.2 then enumerates the various families of peculiar stars belonging to case B. Section 9.4.3.1 reviews their orbital properties, while Section 9.4.4 shows how these orbital properties make it possible to set constraints on the mass-transfer mode that took place in those systems.

9.2 AGB Stars in Binary Systems

9.2.1 Constraints on the Properties of Binary Systems Involving AGB Stars

Before discussing the methods available for identifying binary systems involving AGB stars (Section 9.2.2), it is interesting first to evaluate how the very large size of AGB stars sets a lower bound on the orbital separation—and hence on the orbital period—through the concept of the critical Roche radius.

9.2.1.1 Roche Radius

In the Roche description of a binary system (following the French physicist E. Roche 1820–1883), it is assumed that the gravitational fields of the two stars can be approximated by those of two point masses, that the orbits are circular, and that the two stars are in synchronous rotation with the orbital motion (so-called corotation). Under these conditions, the binary system appears to be at rest in a frame corotating with the orbital motion. In this frame, it is possible to define effective equipotential surfaces corresponding to the gravitational potential corrected for centrifugal effects due to the rotation of the frame (for a detailed description of the Roche model, see [68]). Equipotential surfaces are surfaces of constant Φ, where

$$\Phi = -\frac{\tilde{\mu}}{r_1} - \frac{1-\tilde{\mu}}{r_2} - \frac{x^2+y^2}{2}, \qquad (9.1)$$

with $\tilde{\mu} = M_1/(M_1+M_2)$, and $r_1 = [(x+1-\tilde{\mu})^2+y^2+z^2]^{1/2}$ is the distance to the star of mass M_1 lying at $(\tilde{\mu}-1, 0, 0)$, $r_2 = [(x-\tilde{\mu})^2+y^2+z^2]^{1/2}$ the distance to the star of mass M_2 lying at $(\tilde{\mu}, 0, 0)$, so that the center of mass corresponds to the origin of the coordinate system, and the orbital plane is the xy–plane. In (9.1), distances are expressed in units of the orbital separation, time is expressed in units of the orbital period, and masses in units of $(M_1 + M_2)$. The most important equipotential surface, from the point of view of binary-star evolution, is the critical surface whose intersection with the orbital plane forms a figure eight (Figure 9.1). The two cusped volumes that are enclosed by this critical surface are called the Roche lobes of the respective stars. They have a single contact point, along the line joining the two stars, called the inner Lagrangian point, usually denoted by L_1 and corresponding to the point located between the two stars where $d\Phi/dx = 0$. A system in which both components are within their Roche lobes is called *detached*. If one component fills its Roche lobe, the system is called *semidetached*, and when both components overflow their Roche lobes, the system is called a *contact binary* (Figure 9.1). Mass transfer occurs from the Roche-filling star to its companion in semidetached systems, and this phenomenon is usually referred to as *Roche lobe overflow* (RLOF). When involving an AGB star more massive than its companion, RLOF is usually considered to be dynamically unstable, and to lead to dramatic orbital shrinkage through a common envelope (CE) phase. Although this general statement will be somewhat attenuated in the detailed discussion presented in Section 9.4.4, it is nevertheless clear that the size of the Roche lobe constitutes a physical limit to the size that may be reached by an AGB star in a binary system. Conversely, knowing the size of an AGB star, it is possible to set a lower limit on the orbital separation of the binary system that is able to host it.

A useful approximation of the Roche radius, corresponding to the radius of the sphere having the same volume as the Roche lobe surrounding star 2, is given by (e.g., [68])

$$R_{R,2}/A = \begin{cases} 0.38 + 0.2 \log q, & 0.5 \leq q \leq 20, \\ 0.462 \left(\frac{q}{1+q}\right)^{1/3}, & 0 < q < 0.5, \end{cases} \quad (9.2)$$

where $q = M_2/M_1$, and A is the orbital separation. An alternative expression, valid for all q, is

$$R_{R,2}/A = \frac{0.49\, q^{2/3}}{0.6\, q^{2/3} + \ln(1 + q^{1/3})}. \quad (9.3)$$

9.2.1.2 Limitations of the Roche Lobe Concept in the Case of AGB Stars

The validity of the Roche approximation for mass-losing AGB stars has to be questioned in several respects. The Roche model described in Section 9.2.1.1 assumes that the two stars are point masses (a good approximation for centrally condensed objects like AGB stars), and that the only forces acting upon a test mass corotating with the binary system are the centrifugal pseudo-force and the gravitational forces of the two point-like stars. In particular, the radiation pressure and the shock waves imparting momentum on the escaping matter (from a reservoir of internal energy), as is the case for AGB winds, are not included in the Roche model. Therefore, the very concept of Roche lobe may be irrelevant for systems involving a mass-losing star where the wind-driving force may substantially reduce the effective gravity of the mass-losing star. Since both the radiation-pressure force (dP_r/dr) on the wind and the gravitational attraction decrease as r_2^{-2}, where r_2 is the distance to the AGB

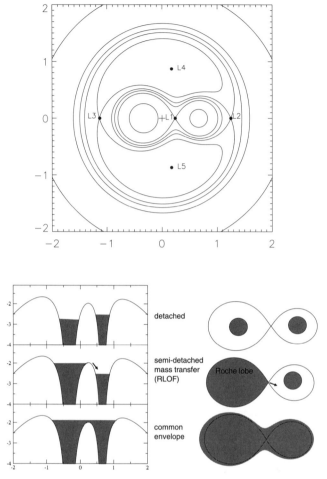

Fig. 9.1. Top panel: Sections in the orbital plane of Roche equipotentials, for a binary system with mass ratio $q = M_2/M_1 = 0.5$. The position of the Lagrangian point L_1 (black dot) is indicated, as well as that of the center of mass (cross) of the system.
Bottom panel: A schematic representation of the equipotentials as a function of distance along the line joining the two stars: **(top panel)** detached binary; **(middle panel)** semidetached binary; **(bottom panel)** contact binary

star, the net effect on the wind particle may be seen as a reduction of the effective gravity of the AGB star by the factor $(1 - f)$, where f is the ratio of the radiation pressure force per unit mass to the gravitational attraction (per unit mass) of the AGB star at any given point:

$$f = \frac{-\mathrm{d}P_r}{\rho\,\mathrm{d}r}\left(\frac{GM_2}{r_2^2}\right)^{-1}, \tag{9.4}$$

or

$$f \equiv \frac{1}{4\pi c G M_2} \int_0^\infty \kappa_\nu \, L_{\nu,2} \, d\nu, \quad (9.5)$$

since the radiation pressure force can be written

$$\frac{dP_r}{dr} = -\frac{1}{c} \int_0^\infty \kappa_\nu \, \rho \, \frac{L_{\nu,2}}{4\pi r_2^2} \, d\nu, \quad (9.6)$$

where κ_ν is the absorption coefficient per unit mass at frequency ν, c the speed of light, and $L_{\nu,2}$ the luminosity of star 2 in the frequency range $(\nu, \nu+d\nu)$. The existence of an extra force driving the mass loss will thus modify the shape of the equipotential surfaces, which are now defined as

$$\Phi = -\frac{\tilde{\mu}}{r_1} - \frac{(1-\tilde{\mu})(1-f)}{r_2} - \frac{x^2+y^2}{2} = \text{constant}, \quad (9.7)$$

where all symbols are the same as in (9.1) and it is assumed that star 2 is exerting radiation pressure on its wind. If f is assumed to be the same at all points, the equipotential surfaces may be easily derived from (9.7). Figure 9.2 shows that when f becomes larger than some critical value f_c (depending upon $\tilde{\mu}$, as given by curve b in Figure 9.3), there is no longer a critical Roche equipotential surrounding the two stars. In particular, when $f = f_c(\tilde{\mu})$, the critical Roche equipotential degenerates into a surface including *both* the inner Lagrangian point and the outer Lagrangian point opposite the mass-losing star, as shown in Figure 9.2 (to be compared with Figure 9.1). Matter flowing through L_1 will thus not necessarily be confined to the lobe surrounding the accreting component, but may escape from the system through the outer Lagrangian point, or may form a circumbinary disk. *This possibility is especially appealing in view of the frequent occurrence of such disks around post-AGB stars* (see Chapter 10). It should, however, be noted that for these modifications to the usual Roche geometry to become effective (they have been reported as well for X-ray binaries involving mass-losing supergiants [10, 62]), it is necessary that the wind remain optically thin; otherwise, it will be shielded from the irradiating star, and the extra force will disappear.

9.2.1.3 Radii of AGB Stars

The radius R of an AGB Mira star of mass M may be related to its pulsation period P using predictions from the pulsation theory fitted to linear diameters of Miras (as derived in [139] from angular size measurements and Hipparcos parallaxes):

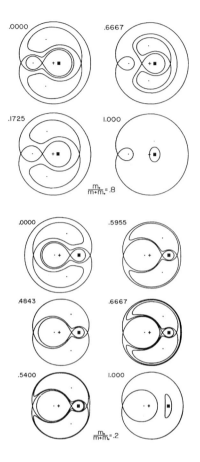

Fig. 9.2. Modification of the Roche equipotentials in the presence of an extra force driving the mass loss from star 2 (represented by a filled square symbol). The **upper four** correspond to $1 - \tilde{\mu} = 0.8$ and the six **lower panels** to 0.2, respectively. The various figures correspond to different f values, as labeled (with $f_c = 0.1725$ in the four **upper panels** and 0.5955 in the six **lower panels**). Compare with Figure 9.1. From [120]

$$\log R_{\rm Mira} = (\log P + 0.9 \log M + 2.07)/1.9 \text{ fundamental mode,}$$
(9.8)
$$\log R_{\rm Mira} = (\log P + 0.5 \log M + 1.40)/1.5 \text{ first overtone mode,}$$

where P is expressed in days, and R and M in solar units. The above relation also applies to those among the semiregular variables that fall along the same $P - L$ relation as the Mira variables. The question of whether Mira stars are pulsating in the fundamental or first overtone mode has been much debated (e.g., [149, 152, 153]), although recent evidence points to the fundamental mode (Chapter 2). The two possibilities will nevertheless be considered in the following discussion. Since AGB stars have very extended atmospheres, their radius is an ill-defined quantity. It is therefore necessary to specify that the above relations provide the so-called Rosseland radius (see [36] for a detailed discussion), which links the effective temperature to the bolometric luminosity. For non-Mira M stars, an empirical relationship between radii

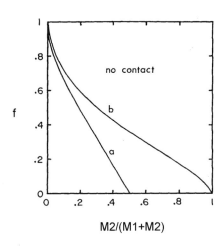

Fig. 9.3. The curve labeled b corresponds to $f = f_c(\tilde{\mu})$; i.e., the inner Lagrangian point and the external Lagrangian point opposite the mass-losing star share the same equipotential. Curve a corresponds to the situation where the two external Lagrangian points share the same equipotential (only possible when $1 - \tilde{\mu} < 0.5$). Systems with f lying above the curve labeled b (i.e., $f > f_c$) have no figure-eight equipotential surrounding the two stars. Systems to the left of curve a are topologically similar to the case $f = 0$ in the bottom panel of Figure 9.2. Systems between curves a and b are topologically similar to the case $f = 0$ in the upper panel of Figure 9.2. From [120]

and spectral types has been derived [25], with radii ranging from a median value of 50 R$_\odot$ at M0III to 170 R$_\odot$ at M7/8III. These values are somewhat smaller than the Mira radii (250 R$_\odot$ for $P = 300$ d and $M = 1$ M$_\odot$ in the fundamental mode and 385 R$_\odot$ in the first overtone), as obtained from (9.8).

9.2.1.4 AGB Stars in Detached Binary Systems

To hold within a binary system, the radius R of the AGB star (supposed to be a Mira variable) derived from (9.8) must be smaller than the critical Roche radius R_R expressed by (9.2) (a brief discussion of the consequences of the situation $R > R_R$ will be presented in Section 9.4.4). The orbital period for which a Mira of a given pulsation period (in either the fundamental or first overtone mode) fills its Roche lobe is displayed in Figure 9.4. It is derived from the above formulae (9.7) and (9.8), and from Kepler's third law (assuming typical masses for the AGB star and its companion, as indicated on the figure). Note that the results presented in Figure 9.4 assume that there is no radiation pressure acting on a wind (i.e., $f = 0$ in (9.7)). Accounting for such a radiation pressure may severely alter the results, as further discussed in Section 9.2.1.2.

These orbital periods are quite long (more than 1000 d), and are always much longer than the pulsation periods. The corresponding radial-velocity semiamplitudes K (Figure 9.4) may be computed from the relation [123]

$$K = 212.8\, f(M_{\text{AGB}}, M_{\text{comp}})^{1/3}\, P^{1/3}\, (1 - e^2)^{-1/2}, \qquad (9.9)$$

where K is expressed in km s^{-1}, $f(M_{\text{AGB}}, M_{\text{comp}}) = M_{\text{comp}}^3 \sin^3 i / (M_{\text{AGB}} + M_{\text{comp}})^2$ is the mass function expressed in M$_\odot$, e is the orbital eccentric-

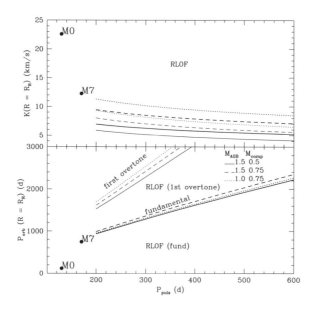

Fig. 9.4. Bottom panel: Orbital periods versus pulsation periods for Miras in semidetached systems, derived from the period–radius relations (9.8) for the fundamental and first overtone modes. Detached systems are located above the corresponding lines. The filled circles labeled M0 and M7 indicate the critical periods corresponding to the median radii of non-Mira M0 and M7 giants, respectively (adopting masses of 1.0 and 0.75 $M_\odot$ for the giant and its companion, respectively). **Top panel**: Orbital radial-velocity semiamplitude versus pulsation periods (thick line: fundamental mode; thin line: first overtone) for Miras in circular semidetached systems with $\sin i = 1$ (where i is the orbital inclination). Binary systems with larger semiamplitudes would involve Roche-lobe-filling AGB stars

ity, i the orbital inclination, and P the orbital period expressed in days. This quantity will prove to be important for comparison with the intrinsic radial-velocity variations of AGB stars (due to pulsations) to be discussed in Section 9.2.2.1.

9.2.2 Binary Systems Involving AGB Stars

Before discussing the methods available to identify binary systems involving AGB stars, it is necessary to define first observational criteria that may be used to identify an AGB star. The following three criteria have been used in this section: The star exhibits spectral lines from the unstable element technetium (Section 9.4.1.1), or is of spectral type C (although this may not be a sufficient condition given the intrinsic/extrinsic paradigm discussed in Section 9.4.1.2), or is a Mira variable. It must not be excluded, however, that some AGB stars might not bear these distinctive properties. Such "dormant" AGB stars, indistinguishable from RGB stars of similar luminosity or from massive supergiant stars, are thus missing from the statistics of binary AGB stars presented in this section. This remark applies especially to the class of VV Cep and ζ Aur binaries (late-type supergiants that are either genuine massive stars, or very luminous AGB stars, with an A-BV companion eclipsed by the extended atmosphere of the supergiant [14]), and to the many binary (often, but not always, symbiotic) systems involving a non-Mira M giant (RGB or AGB star? see especially [40] for a list of binary systems involving semiregular variable stars).

The methods available to find binary systems involving AGB stars are the following:

- Radial-velocity variations
- Presence of a visual companion
- Composite nature of the spectrum and/or symbiotic activity
- X-ray emission
- Shallow light curve of a Mira variable, with wide and flat minima
- Light variations associated with orbital motion (eclipses, Algol)
- Lunar occultations
- Imaging of CSE asymmetries

Binaries discovered by these methods, which are commented in the remainder of this section, are listed in Table 9.1. This table is, to the author's best knowledge, supposed to list all binaries involving an AGB star known as of mid-2001.

9.2.2.1 Radial-Velocity Variations: Intrinsic Versus Orbital

No convincing case of a spectroscopic binary involving an AGB star (defined according to the "working" criteria given at the beginning of Section 9.2.2) is currently known,[3] basically because orbital radial-velocity variations are confused by the intrinsic variations associated with the envelope pulsation (Figure 9.5; see also [40]). However, since spectroscopic orbits give (partial) access to the stellar masses, radial-velocity surveys are worth pursuing despite the confusion introduced by the pulsation. With a good understanding of these intrinsic radial-velocity variations, it may be hoped that orbital and intrinsic variations may in the end be disentangled [4, 40].

The semiamplitude of the radial-velocity variations associated with the pulsations of Mira and semiregular variable stars is displayed in Figure 9.6. For both the Mira and the semiregular variables, there is a good correlation between the radial-velocity semiamplitude and the visual amplitude. In contrast, Miras exhibit little correlation between their periods and radial-velocity

---

[3] The fact that all Tc-poor S stars are spectroscopic binaries (Table 9.7) does not contradict this statement, since these stars are not AGB stars.

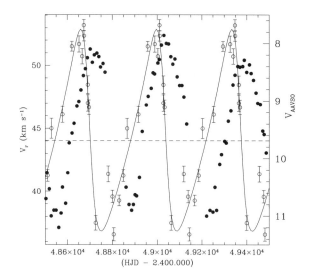

Fig. 9.5. AAVSO light curve (filled dots) for the CS Mira variable R CMi during three successive cycles, compared to its radial-velocity curve (solid line and open dots, from [135]). Radial-velocity measurements were not necessarily obtained at the indicated dates, as they were all folded onto the radial-velocity solution

semiamplitudes, which lie in the range 10–15 km s^{-1}. Semiregular variables have radial-velocity semiamplitudes increasing with the period, from a few km s^{-1} up to the values typical of Miras.

The above considerations thus allow one to draw a fairly accurate picture of the intrinsic variations exhibited by Mira stars, an essential prerequisite to search for binaries among AGB stars. From the comparison of Figures 9.4 and 9.6, it appears that the properties of the Mira intrinsic radial-velocity variations locate them in the forbidden (labeled RLOF in the upper panel of Figure 9.4) region of the period–semiamplitude diagram of *binary* systems. There is therefore in principle no overlap between the parameter ranges characterizing orbital and intrinsic radial-velocity variations of Miras. This property may in fact be of some help in correctly identifying the nature of the observed radial-velocity variations. Nevertheless, orbital variations remain difficult to detect, since for the reasonable choice of masses displayed in Figure 9.4, orbital variations of Miras in detached binary systems are smaller than the semiamplitudes of the *intrinsic* radial-velocity variations (Figure 9.6). *It is therefore clear that the detection of binary systems involving Mira stars with radial-velocity techniques is a difficult operation, because (i) the time scales involved are quite long (several years) and (ii) the orbital radial-velocity variations are smaller than the intrinsic variations. Binary systems will therefore*

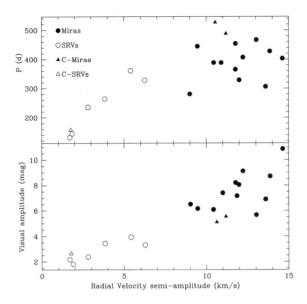

Fig. 9.6. Top panel: Correlation between the radial-velocity semiamplitude (as derived from the 1.5–2.5 μm CO lines) and the period of the visual lightcurve for stars from various classes of long-period variables.
Bottom panel: Same as top panel for the visual amplitude. Adapted from Figure 10 of [41]

basically show up as a long-term trend superimposed on intrinsic variations characterized by a shorter time scale.

Radial-velocity monitoring surveys of AGB stars have been scarce so far. A monitoring of 9 multiperiodic semiregular variable stars has yielded one possible binary [40] (AF Cyg, not listed in Table 9.1 because semiregular variables do not fulfill the strict AGB criteria outlined at the beginning of Section 9.2.2). Among the 81 long-period variables monitored in [4], binarity could be efficiently checked in 9 stars only, with one possible binary candidate emerging (S Cam). Among the 13 Mira S stars, 7 SC/CS stars, and 3 C stars monitored in [135], only one (S UMa) has revealed a possible signature of binary motion, but this signal might well in the end turn out to be a 1-y alias of the Mira pulsation period.

9.2.2.2 Visual Binaries

A list of Mira and semiregular variables in visual binary systems is provided in [109]. Only those systems with a separation smaller than 5″ have been listed in Table 9.1. The *Third Catalogue of Interferometric Measurements of*

Binary Stars[4] has been scanned for Mira variables and carbon stars. Among the 38 carbon and Mira stars with interferometric measurements found in the database, only X Oph is a firm detection, all the other cases being either ambiguous detections or nondetections. Several carbon stars are listed as double stars by the *Hipparcos Catalogue*, but subsequent speckle observations did not confirm their binary nature [76, 105]. Speckle observations of 33 semiregular and Mira variables aimed at detecting binary stars among late-type stars resulted in three positive detections, all listed in Table 9.1 [108].

The orbits available for X Oph and Mira Ceti [108] yield system masses of 20.9 and 4.4 $M_\odot$, respectively. Such a large value for X Oph appears very unlikely, and simply reflects the poor quality of the available parallax and of the orbital elements for such a long-period system. The situation appears somewhat better for Mira Ceti. Adopting a WD mass of 0.6 $M_\odot$, this preliminary orbit thus yields a mass of about 3.8 $M_\odot$ for Mira itself, but since its orbit is not yet fully constrained by the available observations, this mass still bears considerable uncertainties.

9.2.2.3 Composite Spectrum and Symbiotic Activity

The composite (or "symbiotic") nature of the spectrum has delivered several binary systems involving AGB stars (Table 9.1). At this point, it should be mentioned that the low discovery rate of binaries among non-symbiotic AGB stars may simply result from the fact that AGB stars in binary systems necessarily develop a strong symbiotic activity triggered by the interaction of their strong wind with their companion, as will be discussed below. Hence, most, if not all, binaries involving AGB stars may in fact appear as symbiotic stars. In this respect, it is interesting to note that the *Eighth Catalogue of the Orbital Elements of Spectroscopic Binary Stars* [8] contains 20 systems with M giant or supergiant primaries, among which 14 are in fact symbiotic systems. Moreover, it is worth noting that Mira Ceti itself is member of a binary system, which, despite its large orbital separation (Table 9.1), nevertheless exhibits some symbiotic activity [112]. Symbiotic stars are defined[5] as having peculiar spectra merging low-excitation absorption features (TiO bands, neutral metal lines) with high-excitation emission lines (like He II $\lambda 4686$; Figure 9.7). Nebular features like forbidden lines[6] requiring low densities, are frequently present as well. Moreover, far-UV spectroscopy reveals the presence of a hot

---

[4] Hartkopf et al., CHARA Contribution No. 4, 1999, available at http://www.chara.gsu.edu/DoubleStars/Speckle/intro.html

[5] See *Symbiotic stars probing stellar evolution* (edited by R. Corradi & J. Mikołajewska, Astron. Soc. Pacific Conf. Ser., 2003) for the latest developments on symbiotic stars.

[6] Forbidden lines correspond to radiative transitions not allowed by the first-order quantum-mechanical selection rules. These lines are not observed in high-density environments, where the collisional deexcitation of the upper level is much faster than its radiative deexcitation.

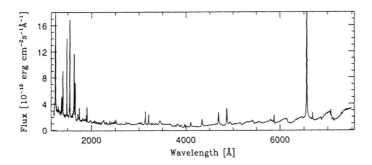

Fig. 9.7. Optical/ultraviolet spectrum of the symbiotic star AG Peg showing the hot UV continuum, the emission lines, the Balmer continuum in emission, the cool-star continuum and its absorption bands. From [92]

blackbody continuum (characterized by temperatures sometimes in excess of 10^5 K and luminosities in the range 10–1000 $L_\odot$) that makes it clear that symbiotic stars are in fact binary systems involving a cool giant and a hot compact star (either on the main sequence or, more generally, a WD) accreting matter from its companion. The hot continuum is emitted by the compact companion, which is heated by the accretion of the wind from the red giant star.

Symbiotic activity involves **eruptions** either caused by thermonuclear flashes at the surface of white dwarfs accreting at a slow rate ($< 10^{-9}$ $M_\odot$ yr^{-1}, as in the case AG Peg), or by instabilities in the accretion disk (for main-sequence stars accreting from a Roche lobe-filling giant like CI Cyg), **dust obscuration episodes** (in symbiotic Miras like R Aqr or BI Cru; see Section 9.3.2) and **reflection effect** caused by the heating of the cool component atmosphere facing the hot component.

Symbiotic stars form a heterogeneous family (see [9] for the most recent catalogue, comprising 188 symbiotic stars). A first classification involves the spectrum of the cool component: *Yellow* symbiotics comprise G or K giants, whereas *red* symbiotics comprise M giants. A second classification refers to their IR properties: *d-type* symbiotics ("dusty") have IR excesses indicating the presence of circumstellar dust and generally host a Mira variable (*d'-type* applies to dusty *yellow* symbiotics), whereas *s-type* ("stellar") symbiotics (not to be confused with stars of spectral type S) have infrared colors consistent with their photospheric temperature. Symbiotic Miras with an entry in the *General Catalogue of Variable Stars* (GCVS) are BI Cru (280 d), KM Vel (370 d), RX Pup (580 d), V704 Cen (?), HM Sge (540 d), V1016 Cyg (450 d), RR Tel (387 d), and R Aqr (387 d). The sample of symbiotic Miras has an average period of 424 d (with a range extending from 280 to 580 d), which is larger than the average for field Miras from the GCVS [147]. The reason why symbiotic stars tend to host long-period Miras is currently unclear (selection effect against long-period, obscured Miras in the GCVS, or rather stronger mass loss associated with long-period, high-luminosity Miras necessary to trigger the symbiotic activity?).

9.2.2.4 X-ray Emission

X-rays are normally not observed in single M giants lying to the right of the so-called dividing line, an almost vertical boundary in the HR diagram separating stars with hot coronae emitting X-rays (to the left) from stars with high mass loss (to the right) [44]. In a binary system, several processes may generate X-rays. First, X-rays may be produced at the shocks resulting from the collision of streams in the complex flow pattern associated with wind accretion in a detached binary system involving an AGB star (Section 9.3.1). Second, X-rays are generated when the gravitational energy of the AGB wind falling in the potential well of the companion star (of mass M_{comp} and radius R_{comp}) is converted into radiative energy when impacting the stellar surface. The accretion luminosity

$$L = G \, \dot{M}_{acc} \, M_{comp}/R_{comp}, \qquad (9.10)$$

where G is the gravitational constant and $\dot{M}_{acc}$ is the accretion rate, will either be radiated away in the form of hard X-rays if the infalling matter is optically thin, or if that matter is optically thick, half will be converted into thermal energy and half will be radiated away in the form of blackbody radiation (according to the virial theorem).

A search for X (or far-UV) sources among AGB stars is therefore an efficient way of finding binary systems involving AGB stars. In fact, this method is a corollary of the search for symbiotic activity, since symbiotics are generally strong X-ray emitters. The ROSAT all-sky survey of X-ray sources detected only 11 out of 482 M giants of luminosity classes I to III from the *Bright Star Catalogue* [44]. Of those 11 sources, one is the well-known symbiotic star R Aqr; another is 4 Dra, a triple system with a cataclysmic binary as a secondary; two have close G-type secondaries (η Gem and α Her), which are the likely sources for the observed coronal X-ray emission; and 3 sources probably result from spurious associations. None among the 4 stars remaining as candidate symbiotic systems exhibits the typical emission lines expected in the optical.

9.2.2.5 Light Curves

Light curves of long-period variables that are either shallow or have wide, flat minima may be indicative of the presence of a companion dominating the system light at minimum [91, 141]. These expectations are not borne out by spectroscopic inspection [38], which revealed composite spectra for only a few among these candidates. Among 23 shallow light curves for S stars, only W Aql, WY Cas, and T Sgr turn out to have a composite spectrum. Among the 55 candidate M stars, only R Aqr and possibly RU Her have composite spectra. Visual companions have also been found for Z Tau and X Oph.

The analysis of the MACHO light curves [152] raised the intriguing possibility that 25% of all AGB stars in the LMC might be semidetached binaries (see Chapter 2), but recent observations have challenged this claim [40].

9.2.2.6 Lunar Occultations

The binary nature of AGB stars may also be revealed by lunar-occultation observations, provided that the magnitude difference between the AGB star and its companion be not too large (< 3.5 mag [113]). Four candidate binary AGB stars have been reported so far [114].

9.2.2.7 Imaging of Asymmetries in CSEs

The copious amount of mass lost by AGB stars through winds feeds their CSEs. Despite the fact that all CSEs deviate to some extent from homogeneity and spherical symmetry (Chapter 7), stronger and very typical deviations may be expected in binary systems. Their detection by direct imaging may serve as an indicator of the binary nature of the underlying star. Specific and readily identifiable flow patterns may be expected in a binary system, as discussed in Section 9.3.1. Table 9.2 lists those stars where the CSE has been imaged and reveals complex structures most probably due to the binary nature of the underlying star.

A related diagnostic involves the detection of a circumbinary disk (Section 9.3.1) through the associated very narrow (FWHM $< 5\,\mathrm{km\,s^{-1}}$) CO radio line emission [55]. The detection of such features around the J-type carbon stars EU And and BM Gem [55] thus leads to the intriguing suggestion that the class of J stars might contain a large proportion of binary stars (further evidence in this respect is presented in Section 9.4.2.1).

9.3 Impact of Binarity on Intrinsic Properties of AGB Stars

9.3.1 Mass-Loss Geometry

The presence of a companion to the AGB star is expected to alter the geometry of the wind blowing off the AGB star. The geometry of mass flows in detached systems involving an AGB star is governed by the fact that the wind velocity v_∞ (typically 15 $\mathrm{km\,s^{-1}}$; Chapters 6 and 7) is of the same order of magnitude as the relative orbital velocity v_{orb} [on the order of

Table 9.1. A list of AGB stars (identified by their Tc-rich, C-rich or Mira nature) in binary systems. In column Var., M stands for Mira, SR for semiregular. The column Sep. provides either the angular semimajor axis of the relative orbit (X Oph only) or the angular separation on the sky. The column Comp. provides the spectral type of the companion when known (WD stands for white dwarf). Among symbiotic systems, only carbon symbiotics and Mira symbiotics with an estimated $P_{\rm orb}$ are listed (see [9] for an exhaustive catalogue)

Name	Sp.	Tc	Var.	$P_{\rm phot}$ [d]	$P_{\rm orb}$ [d]	Sep. ["]	Comp.
Composite spectra							
HD 172481	MIII	?	M	312			F2Ia (p-AGB)
RU Her	M6e-M9	y?	M	485			?[a]
IRC −20197[e]	M9	?	M	636		≈1	A-FV
W Aql	S6/6e	?	M	490			F5-8V
WY Cas	S6/6e	?	M	476			GV
T Sgr	S5/6e	y	M	394			F3IV
SZ Sgr	C7,3	y?	SRb	100:		1.9	A7V
TU Tau	C5,4	?	SR:	190			A2V:
DM−26°2983	C4,4	?	?	?			A5V
Symbiotic carbon stars							
UV Aur	C8,1Je	y	SRb	393.4		3.4	
SS 38	C9	?	M?	?			
AS 210	C	?	M?	?			
HD 59643	C6,2	?	SR?	70			
V335 Vul	Ce	?	SR	?			
Symbiotic Mira stars with estimated $P_{\rm orb}$[d]							
Mira Ceti	M7IIIe	y	M	331.6	$>3.6\times10^4$	0.6	Be/WD[b]
R Aqr	M7	y?	M	387	1.6×10^4?	0.055	?
V1016 Cyg	> M4	?	M	450	3×10^4?		?
Visual binaries							
HDE 308122 (= CpD -62° 1837)	M6-8e	?	M	320		4	K0III
RV Cam	M4II-III	?	SRb	101		0.07	
RY Cam	M3III	?	SRb	136		0.06	
RU Cap	M9e	?	M	346.9		1.6	
R Car	M4-8e	?	M	308.7		2	
T Dor	M5IIe	?	M	168		4.8	
π^1 Gru	S5,7	y	SRb	150		2.7	G0V
U Men	M3	?	SRa	407		0.6	K0III-IV
X Oph	M5/9IIIe	y?	M	334	1.8×10^5	0.34	K1III
SV Peg	M7	?	SRb	145		0.44	
SU Sgr	M7III	?	SRb	88		1.1	
Z Tau	S7.5,1e:	?	M	494		4	

Table 9.1. Continued

Name	Sp.	Tc	Var.	P_{phot} [d]	P_{orb} [d]	Sep. ["]	Comp.
Lunar occultations							
EI Tau	S	?	SRa	364		0.007	
V1203 Oph		?	M			0.050	
CGCS 3964	C	?				0.020	
ZZ Gem[c]	Ne	?	M	317		0.088	

[a] The composite spectrum is doubtful.
[b] The Be spectrum of the companion actually corresponds to an accreting WD.
[c] Binary nature not confirmed by speckle observations (see [75]).
[d] see [147] for a complete list.
[e] Also an OH maser.

Table 9.2. A list of AGB stars with a nonspherical CSE shaped by the binary nature of the underlying star. The various shapes encountered are discussed in Section 9.3.1

Name	Type	Shape	λ [µm]	Ref.
π^1 Gru	S(Tc) SRb	disk + fast polar jets	CO(J=2-1)	[61, 117]
V Hya	carbon Mira	fast bipolar wind, eclipses by CSE dust orbiting a comp. (17-yr period)	CO(J=2-1)	[60]
o Ceti	Mira	hook towards comp.	0.346	[57]
		quadrupolar struct.	1.65	[22]
		clump around comp.	11	[71]
EU And BM Gem	J(no-Tc)	circumbinary disk	CO(J=2-1)	[2, 55]
HD 44179 (="Red rectangle")	post-AGB	circumbinary disk	0.656	[100]
Various symbiotics		bipolar and/or spiral jet	0.658	[17]
M 2-9	PN hot WD+AGB	bipolar nebula, fast jet rotating with a $\approx$ 120 yr period	0.55	[69]

$K_{\text{AGB}} + K_{\text{comp}} = K_{\text{AGB}}(1 + M_{\text{AGB}}/M_{\text{comp}})$, where K is the semiamplitude of the velocity variations as given by Figure 9.4 for the AGB star]. Coriolis effects are therefore important, and lead to intricate flow patterns. In the classical description of the flow past a compact object [13], the flow is assumed to be plane-parallel at infinity and without any transverse density or velocity gradients (this situation holds, for example, in X-ray binaries involving fast winds from hot stars, where $v_\infty \gg v_{\rm orb}$). The gravitational field of the compact star bends the flow and makes the streams passing on each side of the compact star collide with each other. This collision results in the formation of an accretion column (tilted by an angle $\alpha = \arctan v_{\rm orb}/v_\infty$ with respect to the line joining the two stars), with matter falling onto the compact star from its rear side (as viewed from the mass-losing star) at a rate [130]

$$\dot{M}_{\rm acc, B\text{-}H} = -\alpha \mu^2 \frac{k^4}{[1 + k^2 + (c/v_\infty)^2]^{3/2}} \dot{M}_{\rm AGB}, \qquad (9.11)$$

where the subscript B-H stands for "Bondi–Hoyle," α is a parameter of order unity, $\mu \equiv M_{\rm comp}/(M_{\rm comp} + M_{\rm AGB})$, $k \equiv v_{\rm orb}/v_\infty$, $c_{\rm s}$ is the sound velocity, and $\dot{M}_{\rm AGB}$ is the wind mass-loss rate of the companion AGB star. No accretion disk is predicted from accretion in the Bondi–Hoyle regime, since the colliding flows are perfectly symmetric and there is thus no accretion of spin momentum either.

The situation relevant to detached binary systems involving an AGB star and characterized by $v_\infty \approx v_{\rm orb}$ is much more complicated than the Bondi–Hoyle regime, and must be investigated numerically. Figure 9.8 presents the flow structure from one of the very few 3D hydrodynamical simulations currently available [27, 78, 130].

The parameter space is quite vast (orbital separation, component masses, mass loss rate, AGB star corotating with the orbit or not, ratio wind/orbital velocities, presence of a wind blowing off the companion, etc.), and remains largely unexplored. General flow features emerging from the explored cases are as follows (Figure 9.8):

(i) Funneling Through L1

Even in detached binary systems, the wind from the mass-losing star is funneled through the inner Lagrangian point L_1 (Figures 9.1 and 9.8). This funneling results from the very geometry of the Roche potentials, and causes a substantial density enhancement and heating of the gas stream around L_1.[7] It is visible in Figure 9.8 as a curved stream between the two stars. Such a gas stream has been detected [122] in the 596 d detached binary HR 1105 (Tc-poor S star + WD) through the orbital modulation of the profile and intensity of the high-excitation line He I $\lambda 1083$ nm. This stream may also perhaps correspond to the hook-like appendage to Mira Ceti A observed in the UV [57] and pointing toward Mira B (Table 9.2).

(ii) Accretion Disk

---

[7] The wind leaving the AGB star from the hemisphere facing its companion is expected to be denser [28], and this effect, not taken into account in the simulation displayed in Figure 9.8, would even reinforce the density enhancement around L_1.

9 AGB Stars in Binaries and Their Progeny   479

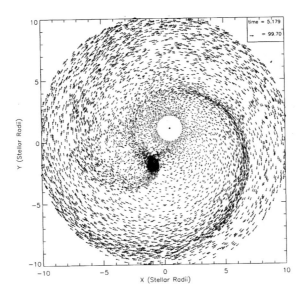

Fig. 9.8. Velocity field of the wind and accretion wake in the orbital plane, in a corotating frame (i.e., where the stars are at rest). The orbital period is 4.5 yr, $M_{\rm AGB} = 1.5\,{\rm M}_\odot$, $M_{\rm comp} = 1\,{\rm M}_\odot$, $v_\infty = 9\,{\rm km\,s}^{-1}$. The AGB star is in synchronous rotation with the orbit, and its radius is one-third of the orbital separation (or 0.8 R_R). From Figure 12b of [77]

The flow funneled through L_1 is curved by the Coriolis force, which introduces the asymmetry responsible for the formation of an accretion disk around the companion. This result contrasts with the prediction that no accretion disk forms in the Bondi–Hoyle wind-accretion regime. The formation of a disk has been observed in all the 3D numerical simulations currently available [78, 130]. It is stable (no flip-flop instability as observed in 2D simulations), thin, and tilted with respect to the orbital plane, and thus probably precessing. A precessing disk coupled with intermittent polar jets (although those are not obtained in the above simulations) might play a role in the formation of bipolar or even quadrupolar structures observed in several symbiotic, Mira and post-AGB binary stars (see Section 9.2.2.7). The prediction of an accretion disk forming during wind accretion is important, because it offers the possibility of accounting for the very commonly observed bipolar structures without invoking the much more transient and dramatic RLOF for producing the required disk, as in [93]. Observational evidence for accretion disks in detached binary systems has been found for o Ceti, 56 Peg (K0IIBa + WD), and HD 35155 (Tc-poor S + WD, $P = 640.5$ d).

(iii) Spiral Arm

The flow pattern of Figure 9.8 exhibits a very clear double spiral arm, a feature obtained in all the other 3D-simulations of wind accretion in detached binary systems. They may be seen as forming the boundaries of the accretion wake, strongly curved by the action of the Coriolis force. These spiral arms correspond to the shocks caused by the collision of the wind coming directly from the AGB star with the accretion wake.

(iv) Circumbinary Disk

The formation of a circumbinary disk seems common in binary systems with a mass-losing star (especially those involving post-AGB stars; see Chapter 10), as revealed by direct imaging, the clearest example being the "Red Rectangle" (HD 44179) [100]. Further, albeit indirect, support for the presence of circumbinary disks is provided by the orbital modulation observed in the light curve of several post-AGB systems (see Chapter 10), which is best interpreted by variable obscuration from dust trapped in such a disk. Finally, the very narrow CO circumstellar lines (FWHM < 5 km s^{-1}) exhibited by several post-AGB systems [55] also point to gas orbiting the system in a circumbinary disk, rather than flowing out with velocities of order 10–20 km s^{-1}, typical of AGB winds (Chapters 4 and 6).

A circumbinary disk does not, though, clearly emerge from the currently available 3D simulations of wind accretion in detached binaries. It is possible, however, that the extended spiral arm, winding around the system, mimicks a circumbinary disk (Figure 9.8). Alternatively, a circumbinary disk may form during RLOF if matter is expelled not only through L_1 but also through L_3 (Figure 9.1), as discussed in Section 9.2.1.2.

9.3.2 Mass Loss and Dust Formation

The flow pattern described in Section 9.3.1 includes several regions where the wind is compressed. These regions of high density will favor dust formation, provided that the corresponding temperature does not exceed the dust condensation temperature (about 1500 K, depending on its chemical composition; Chapter 5). Empirical evidence seems indeed to support the idea that dust formation may be easier in binary systems. First, dust obscuration episodes are frequent in symbiotic Miras and much rarer in single M-type Miras [94, 147, 148]. Similar behavior is observed in the K0III Ba3 + WD system HD 46407 ($P = 457$ d) [50]. This system involves a K giant with a low mass-loss rate where dust formation is not at all expected. Nevertheless, dust-obscuration episodes seem to be the best explanation for the observed light variations. Finally, dust seems to form in the wind of binary WC stars [21]. Although the latter two systems do not involve AGB stars, they serve to illustrate that dust may form in binary systems, despite rather hostile environments.

The dust formed in a binary system may be stored in the long-lived circumbinary disks observed in many instances, especially in systems involving post-AGB stars and silicate J-type carbon stars (Section 9.3.1 and Chapter 10).

9.3.3 Mass-Loss Rate

The possibility that stars in binary systems lose mass at a faster rate than their nonbinary counterparts must be considered seriously, since the presence of the companion will reduce the effective gravity of the mass-losing star, thus enhancing the mass loss [28].

This enhancement has been considered necessary to stabilize RLOF in those binary evolution channels leading to observed systems like Algols and short-period barium stars that would otherwise end up as cataclysmic variables (see Section 9.4.4). The key argument here is that the *companion-reinforced attrition process* (CRAP) [134] will reverse the mass ratio before the onset of RLOF, thus making it dynamically stable (see the discussion after (9.25)). As will be discussed in Section 9.4.4, CRAP is not, however, the only way to account for the short-period barium systems (see the discussion in relation to Figure 9.11). Besides these indirect theoretical arguments, the very large CO mass-loss rates observed for the binary AGB stars W Aql, WY Cas, and S Lyr [52] may provide some support to CRAP, although binary AGB stars like T Sgr and π^1 Gru have mass-loss rates close to the average for stars of their kind. Since orbital separations are not available for these systems, it is difficult to draw conclusive statements on the impact of binarity on the mass-loss rate. Nonetheless, detailed theoretical considerations indicate that mass loss from red giants in binary systems is enhanced by a factor $(R/R_R)^3$, where R is the radius of the mass-losing giant [28].

9.3.4 Masers

The maser activity of AGB stars and their binarity seem to be quite clearly anticorrelated. Direct evidence thereof is provided by the almost total lack of SiO masers among symbiotic stars containing a Mira variable [121], with the exception of the very wide binary R Aqr (Table 9.1). On the other hand, very wide systems like o Ceti, R Aqr, and X Oph (Table 9.1) lack the OH maser, which involves layers several 10^5 $R_\odot$ away from the star, in contrast to the SiO maser, which forms close to the photosphere (Chapter 7). A survey of IRAS sources [63] reveals that the region occupied by OH/IR sources in the IRAS color–color diagram also contains many stars with no OH masing activity. Furthermore, detected and undetected OH maser sources have identical galactic latitude distributions. These two facts find a natural explanation if binarity is the distinctive property between detections and nondetections. The role of binarity in preventing the maser activity was already suspected on theoretical grounds [39], since the presence of a companion periodically disturbs the layer where maser activity should develop. There should actually be a critical orbital separation for every kind of maser below which the companion sweeps through the corresponding masing layer [121], thus preventing its development ($A < 10$ AU: no masing activity; $10 \leq A(\mathrm{AU}) < 50$: SiO and H_2O masers possible but no OH maser; $A > 50$ AU: all masers can operate).

9.3.5 Coupling Between Pulsation and Orbital Motion

In systems that are close enough for tides to develop on the AGB star, the possibility exists for a resonance between the dynamic tide and a free oscillation mode, as suggested in the seminal paper [20]. This possibility has been investigated theoretically for close binaries involving upper main sequence stars (e.g., [151]), but not, to our knowledge, for AGB stars in binary systems. A possible coupling between the orbital motion and a (probably nonradial) pulsation has been reported for the symbiotic star AG Dra [29].

9.3.6 Abundances

The impact of binarity on abundances is clearly established for those post-AGB stars that are very metal-deficient, as a result of a fractionation process occurring in relation with the circumbinary disk (see Chapter 10 for a more detailed discussion). Another obvious link between abundances and binarity is the accretion by the companion of the AGB wind enriched in carbon and heavy elements. This mass-transfer process accounts for the zoo of chemically peculiar late-type stars (see Section 9.4).

Carbon stars belonging to the R and J types may also somehow bear their peculiar abundances from binarity, although the chain of events linking the two properties has not yet been uncovered. Binary-induced mixing rather than mass transfer may be at play here (Section 9.4.2.1).

9.4 The Progeny of AGB Stars in Binary Systems

Section 9.2 dealt with systems where the primary star is currently an AGB star. At this stage, the mass lost by the AGB star interacts with its companion and imprints several observable signatures (fast rotation, chemical anomalies, etc.). The remainder of this chapter is devoted to the description of the rich progeny of AGB stars in binaries, namely, binary systems with a WD component (i.e., the end product of the former AGB star) and a primary star polluted by the mass transferred from the former AGB star.

The evolution followed by a system consisting initially of two low- or intermediate-mass main-sequence stars is sketched in Figure 9.9. The systems where the primary star is an AGB star (according to the "working" definition given at the beginning of Section 9.2.2) correspond to cases 3 to 5 as well as 13 in Figure 9.9 (Note that cases 3–5 and 13 differ only in the nature of the companion: a main sequence star in the former case and a WD in the latter case). The present section deals with the evolution after the mass-transfer episode occurring at phase 5. That evolution will, in fact, depend on the mass-transfer mode itself (either wind accretion or RLOF), as discussed in Section 9.4.4. The mass-transfer scenario was originally invoked to account for the chemical anomalies exhibited by barium stars (corresponding to phase

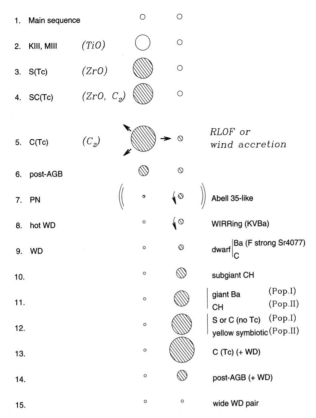

Fig. 9.9. The evolution of a system consisting initially of two low- or intermediate-mass main-sequence stars. The left column corresponds to the normal evolutionary sequence of single stars, while the right column represents the various classes of stars with chemical peculiarities specifically produced by mass transfer across the binary system. Hatched circles denote stars with atmospheres enriched in carbon or heavy elements

11 in Figure 9.9). These stars are G and K giants of too low a luminosity to be on the AGB, and whose overabundances of s-process elements can thus not be accounted for by internal nucleosynthesis (see also Section 9.4.2). Their origin remained a mystery until it was suggested [82] that the envelopes of barium stars may have been polluted by matter accreted from a former AGB companion (being now a dim WD [6]). Another important asset of the mass-transfer scenario is its ability to account for Tc-poor S stars. The observation in the spectra of *some* S stars, of lines from technetium, an element with no stable isotopes (Section 9.4.1.1), is a clear indication that these stars are TP-AGB stars where the s-process of nucleosynthesis is operating. It was unclear, however, why not all S stars exhibited Tc lines, until the mass-transfer scenario led to the formulation of the extrinsic/intrinsic paradigm;

namely, two different kinds of S stars should exist: (i) *intrinsic*, genuine TP-AGB S stars with Tc lines (phase 4 in Figure 9.9), and (ii) *extrinsic*, Tc-poor S stars whose chemical anomalies result from a mass-transfer episode as for barium stars (phase 12 in Figure 9.9). Since the time elapsed between phases 4 and 12 is in general much longer than the Tc half-life, Tc has ample time to decay, and is therefore absent in extrinsic, binary S stars belonging to phase 12. The extrinsic/intrinsic paradigm of S stars will be discussed in more detail in Section 9.4.1.2. In fact, the mass-transfer scenario makes several predictions that are open to verification, namely, (i) the Tc-rich/Tc-poor dichotomy of S stars (phases 4 and 12 in Figure 9.9); (ii) all barium stars and Tc-poor S stars should be binaries; (iii) the existence of many other kinds of extrinsic stars (as listed in the right column of Figure 9.9, from phases 7 to 12); (iv) the companion of extrinsic stars should always be WDs; (v) the possibility of transiently forming rapidly rotating stars from the accretion not only of mass but also of spin angular momentum (phases 7 and 8); (vi) the evolutionary link between dwarf Ba or C stars, giant barium stars, and Tc-poor S stars (which should thus have similar orbital elements). All these predictions will be confronted with the observations in the following sections.

9.4.1 Technetium and the Extrinsic/Intrinsic Paradigm of S Stars

9.4.1.1 Tc in Stars

Among elements lighter than bismuth, technetium (Tc, atomic number 43) is (along with promethium; Pm, atomic number 61) the only chemical element with no stable isotopes. Technetium (after the Greek word $\tau\epsilon\chi\nu\iota\kappa o$ meaning "artificial") was discovered in 1937 by Segré and Perrier in a piece of molybdenum that had been irradiated with deuterons. It is also a fission product of uranium. Its longest-lived isotope (^{98}Tc) has a half-life of 4.2×10^6 yr. The isotope ^{99}Tc (laboratory half-life of 2.1×10^5 yr) is the only one produced by the s-process (see Chapter 2). Lines of Tc have been found in the S star R And [89, 90] (see Chapters 1 and 2 for a description of the principal spectral features of S stars), thereby giving strong support to the operation of the s-process followed by dredge-ups in thermally pulsing AGB stars.[8] In fact, *the presence of Tc in a star may be considered as the most reliable signature of its AGB nature.* A comparison of modern high-resolution spectra of Tc-rich S stars and Tc-poor S and M stars is presented in Figure 9.10. High-resolution spectra ($\lambda/\Delta\lambda \geq 30\,000$) are required to detect the Tc I lines, since the strong resonance lines of Tc I lie in the violet, where blending is very severe in the

---

[8] Alternative scenarios for the production of Tc—either through spallation reactions at the stellar surface triggered by cosmic-ray protons, or through the photofission of the actinides originally present in the star, triggered by γ-rays from the H-burning shell—have appeared in the literature, but none of these is entirely satisfactory in the context of AGB stars.

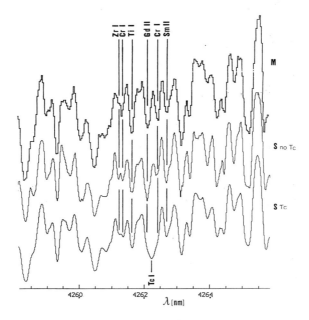

Fig. 9.10. Comparison of high-resolution spectra of S (lower two thin lines) and M stars (upper thick line) around the Tc I line at $\lambda 426.2$ nm. The spectra of the two S stars are almost identical, except for the absence of the Tc I line in the upper one. Note that lines from s-process elements are stronger in S stars than in M stars

spectrum of cool stars. For example, the Tc I $\lambda 426.227$ nm line is blended especially with Nd II $\lambda 426.223$ and to a lesser extent with Gd II $\lambda 426.209$ and Cr I $\lambda 426.237$ (Figure 9.10). Very cool stars and carbon stars emit very little flux in that spectral region, so that the use of the weaker intercombination line at $\lambda 592.447$ nm may be preferable.[9] In warmer stars ($T > 4000$ K), Tc is predominantly in the form Tc II, whose strong lines lie in the far UV at 264.70, 261.00, and 254.32 nm, accessible only through satellite observations. These UV lines have been used to look for Tc in barium stars and in the main-sequence A stars Sirius (α CMa A) and Vega (α Lyr), with negative results. This negative result bears some significance for barium stars and for Sirius, which exhibit heavy-element overabundances like S stars and are binaries with WD companions (Section 9.4.2). This result is precisely what is expected in the framework of the binary scenario sketched in Figure 9.9.

---

[9] Another intercombination line at $\lambda 608.523$ nm is strongly blended with a Ti I line, and its use is therefore not recommended [3]. This line is used in [7] to derive the frequency of Tc-rich stars among C stars, but this frequency is therefore questionable.

Table 9.3. Surveys of Tc in late-type giants (excluding supergiant stars). $N_{\rm tot}$ refers to the total number of stars in a given sample. The percentage of stars in each given Tc class (certain, probable and possible, doubtful, absent) is listed next. S and M stars have been subdivided into nonvariable or irregular variable stars, and long-period variable (LPV) stars

	Sample	$N_{\rm tot}$	Tc (%)			no Tc (%)			Ref.
			yes	prob+poss	tot	dbfl	no	tot	
MIII	non var.	17	0	0	0	76	24	100	[67]
	LPV	142	8	28	36	32	32	64	[67]
	All	159			33			67	[67]
S	non var.	16	38	12	50	19	31	50	[67]
	LPV	29	52	31	83	14	3	17	[67]
	All	45			71			29	[67]
	Henize	199	133		66±5	66		34±5	[136]

Since the pioneering work of Merrill, surveys searching for Tc in various kinds of stars (mostly M, S, and C stars) have been performed. Their results are summarized in Table 9.3 (see also footnote 9).

The important conclusions to be drawn from Table 9.3 are (i) that *Tc-rich stars are mostly LPVs* (with periods larger than 300 d) and (ii) *the existence of a substantial fraction of Tc-poor S stars and Tc-rich M stars* (Mira Ceti belongs to the latter). The latter group might appear puzzling at first sight, since Tc is not expected to be produced separately from the other s-process elements (hence Tc-rich stars should be of type S rather than M). Detailed s-process calculations in AGB stars [31] show, however, that in some cases Tc becomes detectable in the atmosphere before the M star has turned into an S star, thus making the existence of Tc-rich M stars possible. But conversely, all S stars should then be Tc-rich, making the existence of Tc-poor S stars puzzling in the framework of AGB evolution. To account for Tc-poor S stars, one might in principle imagine situations where successive dredge-ups are sufficiently far apart that Tc has time to decay below detectable levels before the next dredge-up occurs. S stars might therefore oscillate between the Tc and no-Tc states while evolving on the AGB. Tc-poor S stars exhibit, however, several other properties (like binarity, absence of IR excesses, low luminosities, high $T_{\rm eff}$, and large galactic scale height; see Section 9.4.1.2) that distinguish them clearly from Tc-rich S stars and that make an oscillation Tc-rich/Tc-poor on the AGB highly unlikely. It is the existence of such a correlation between the absence or presence of Tc and other stellar properties that provides the foundation for the extrinsic/intrinsic paradigm presented in Section 9.4.1.2.

Table 9.4. Respective properties of extrinsic and intrinsic S stars (compiled from [52, 54, 136, 138])

	intrinsic S	extrinsic S
Spectral definition	ZrO bands	
Spectral type	>S3	<S5
Tc	yes	no
Infrared excess at 12 μm	often	never
Mass loss rate ($M_\odot$ yr^{-1})	$>5\times10^{-8}$	$<5\times10^{-8}$
Luminosity	$> L_{1stTP}$	$< L_{1stTP}$
M_{bol}	-3 to -5.5	-2 to -4
Relative frequency:		
in magnitude-limited sample	$66 \pm 5\%$	$34 \pm 5\%$
in infinite column perpendicular		
to the galactic plane	$60^{+20}_{-50}\%$	$40^{-20}_{+50}\%$
Exponential scale height (pc)	200 ± 100	600 ± 100
Mass ($M_\odot$)	≈ 1.2 to 5	1.5 to 2
Evolutionary stage	TP-AGB	RGB or E-AGB

9.4.1.2 The Intrinsic/Extrinsic Paradigm for S Stars

The Tc-poor/Tc-rich dichotomy discovered among S stars [67] appears at first sight as quite an innocuous difference that reveals itself on high-resolution spectra only (Figure 9.10). It hides, however, a very profound difference in the evolutionary status of these two kinds of S stars, and is at the origin of the extrinsic/intrinsic paradigm whose scope goes far beyond the sole family of S stars. This paradigm states that Tc-rich (or "intrinsic") S stars are genuine TP-AGB stars, whereas Tc-poor (or "extrinsic") S stars acquired their chemical peculiarities (namely, carbon and s-process enrichment) through mass transfer across a binary system, at a time when the companion (now a cool WD) was a mass-losing AGB star (see Figure 9.9). Because the time elapsed since the mass-transfer episode is generally much longer than the half-life of ^{99}Tc, S stars that have been polluted by mass transfer should lack Tc lines. The binary nature of these Tc-poor S stars has indeed been clearly established (Section 9.4.3.1). But the difference between extrinsic and intrinsic S stars goes beyond the Tc dichotomy: A closer look reveals differences in their average luminosities, effective temperatures, infrared colors, and galactic scale heights, indicative of the different evolutionary stages characterizing the two kinds of S stars (Table 9.4). The respective properties of intrinsic and extrinsic S stars summarized in Table 9.4 clearly indicate that intrinsic S stars are low- or intermediate-mass TP-AGB stars, whereas extrinsic S stars are (mainly) low-mass stars either on the RGB or on the E-AGB.

The extrinsic/intrinsic dichotomy is also probably present in C stars, although no definite evidence in that sense has been obtained so far [1]. Besides the 3 Tc-poor C stars uncovered in [67], the only other hint at such a dichotomy among C stars possibly comes from the luminosity function of C stars in the SMC obtained by a deep survey [146]. This luminosity function is characterized by an extended wing of faint C stars, with M_{bol} going down to -1.5 (Chapter 3). In fact, this faint tail exactly overlaps the luminosity range covered by extrinsic S stars in the solar neighborhood (Table 9.4). Whether these faint C stars are actually extrinsic stars would need to be ascertained by specific observations. In any case, this particular example shows that it is important to uncover extrinsic stars in external systems before attempting to derive basic AGB properties like the minimum luminosity for the formation of carbon stars through the third dredge-up (as discussed in Chapters 3 and 8).

9.4.2 The Progeny of Binary AGB stars: The Zoo of Extrinsic, Peculiar Red Stars

This section identifies existing classes of peculiar red stars with the corresponding phase of the binary scenario sketched in Figure 9.9.

Phases 1 to 6 correspond to the normal evolutionary sequence of low- and intermediate-mass stars and do not require any further discussion here, except for phase 5, which is possibly associated with silicate J-type carbon stars.

9.4.2.1 Phase 5: Silicate J-Type Carbon Stars?

Several pieces of evidence [2, 154] currently point to the possible binary nature of a puzzling subtype of carbon stars, the J stars which combine a C-rich photosphere with O-rich dust in their circumstellar environment. Arguments suggesting their binary nature include the presence of a circumbinary disk, as inferred from very narrow CO(2-1) lines [55], where the O-rich silicate dust probably resides (a situation encountered in binary post-AGB stars as well; see Chapter 10).

Since J stars lack Tc as well as overabundances from other s-process elements, they do not bear the typical signature of mass transfer from a former AGB companion. Hence, if some causal link exists between the (suspected) binary nature of the J stars and their abundance peculiarities (C and Li overabundances, low $^{12}C/^{13}C$ ratios), it may reside in some kind of binary-induced mixing rather than in a mass-transfer episode.

R-type carbon stars are often considered as the progenitors of the more luminous J-type carbon stars, with which they share many abundance similarities. But this evolutionary link is difficult to reconcile with the *total absence* of binary stars among R stars [84]. In a sense, this very unusual circumstance may also possibly hint at a link between binarity and chemical peculiarities for this class of carbon stars (for instance, R stars could all be coalesced binary systems).

9.4.2.2 Phase 7: Binary Nuclei of PNe (Abell 35-like)

Among PNe with binary nuclei, there is a class consisting of the four objects Abell 35 (BD$-22°3467 =$ LW Hya), LoTr 1, LoTr 5 (HD 112313 = IN Com = 2RE J1255+255), and WeBo 1 (=PN G135.6+01.0). Their optical spectra are dominated by late-type (G-K) stars, but the UV spectra reveal the presence of extremely hot ($> 10^5$ K), hence young, WD companions. Related systems are HD 128220, consisting of an sdO and a G star in a binary system of period 872 d, and the yellow symbiotic/PN AS 201. In all these cases, the late-type star is chromospherically active and rapidly rotating. This rapid rotation is likely to result from some interaction between the binary components, either during a CE phase (see Section 9.4.4) or by accretion of spin angular momentum during wind accretion in a detached binary. The long orbital period, known with certainty only for HD 128220, and the similarity with the WIRRing systems described next, favor the latter hypothesis. These systems may thus fit in phase 7 of the binary paradigm, especially WeBo 1 and AS 201, whose late-type components have been shown to be barium stars [12, 102]. There is, however, no definite evidence for s-process overabundances in the other systems [30, 131].

9.4.2.3 Phase 8: WIRRing Systems (Rapidly Rotating KVBa + Hot WD)

The class of WIRRing (wind-induced rapidly rotating) stars was created [49] to describe a small group of rapidly rotating, magnetically active K dwarfs with hot WD companions discovered by the ROSAT Wide Field Camera and *Extreme Ultraviolet Explorer* (EUVE) surveys. From the absence of short-term radial velocity variations, it may be concluded that these systems have periods of a few months at least. Moreover, several arguments, based on proper motion, WD cooling time scale, and lack of photospheric Li, indicate that the rapid rotation of the K dwarf cannot be ascribed to youth. It has been suggested [49] that the K dwarfs in these wide systems were spun up by the accretion of the wind from their companion, when the latter was a mass-losing AGB star. The possibility of accreting a substantial amount of spin from the companion's wind has since been confirmed by *smooth-particle hydrodynamics* simulations of wind accretion in detached binary systems (see Section 9.3.1 and [130]). A clear signature that mass transfer has been operative in the WIRRing system 2RE J0357+283 is provided by the detection of an overabundance of barium [48]. Interestingly enough, the class of WIRRing stars is not restricted to dwarf primaries, since it probably also includes all yellow symbiotics with dust shells (the so-called d'-type) as well as HD 165141. Those systems share the properties of RS CVn systems and giant barium stars [53, 102]. The very long orbital period of HD 165141 ($P \approx 4800$ d) forbids the barium star from having been spun up by tidal effects, as is the case for RS CVn systems. This system, moreover, hosts the hottest WD

($T \approx 35\,000$ K) among barium systems. The cooling time scale of the WD is about 10 Myr, shorter than the magnetic braking time scale of the giant star (about 100 Myr) [130]. The mass transfer thus occurred recently enough so that the magnetic braking has not yet slowed down the giant star substantially. HD 165141, WeBo 1, and the yellow d' symbiotics are probably cases in which the mass transfer occurred when the accreting star was already a giant.[10] HD 128220 (G subgiant + sdO companion) may be another example of a subgiant star spun up by the accretion process [42].

9.4.2.4 Phases 9–10: Dwarf Ba/C Stars, Subgiant CH Stars

Except for those WIRRing stars discussed in the previous section, most extrinsic stars are expected to form as dwarfs, since (i) the stellar lifetime is longer on the main sequence than on the giant branch, and (ii) the cross section for wind accretion is independent of the stellar radius; wind accretion is therefore not less efficient for dwarf accretors than for giant ones (see (9.11)). About 30 dwarf C stars are known as of this writing (Table 9.5), and most of them have been uncovered very recently by the *Sloan Digital Sky Survey* [74] and by the *Two Micron All Sky Survey* [72]. Many more may be expected from these surveys in the coming years. Many dC stars appear to have halo kinematics. Their binary nature has not been demonstrated in all cases, but several exhibit the expected barium overabundances.

Dwarf Ba stars (some with C/O > 1) were identified [96] among the stars classified as "FV strong Sr $\lambda 4077$" in the Michigan spectral survey. The families of CH subgiants [11, 73] and blue metal-poor stars [106] also comprise stars near or slightly above the main sequence. As discussed in Section 9.4.3.1, these are heterogeneous classes of peculiar stars, merging post-mass-transfer binaries and single stars whose chemical peculiarities are of unknown origin.

9.4.2.5 Phase 11: Giant Barium Stars

As indicated in the introduction to Section 9.4, the mass-transfer scenario was originally introduced [82] to account for the overabundances of barium and other elements produced by the s-process of nucleosynthesis observed at the surface of giant barium stars. Giant barium stars represent about 1% of the giant G-K stars, and this frequency is consistent with the frequency of their "FV strong Sr $\lambda 4077$" progenitors among FV stars [97]. Early-type CH stars [58], with low $^{12}C/^{13}C$ ratios (of order 10), are the Pop. II analogues of barium stars. The metal-deficient barium stars [73] may also be Pop. II objects but with C/O ratios below unity, in contrast to the situation prevailing for CH stars.

---

[10] It is possible, though, that the giant appearance of the d' symbiotics is a transient phenomenon caused by the accreting dwarf star being driven out of thermal equilibrium by the rapid mass accretion (see Section 9.4.4.2).

Table 9.5. Currently known carbon dwarfs and subdwarfs. A + sign in the column labeled Ba indicates that Ba overabundances have been reported

Name	binary	Pop.	Ba	M_v	discovery paper
PG 0824+289	dCe + DA WD	disk	+	+10.4	[37]
CBS 311 (= SBS 1517+5017)	dC + DA WD	disk	?	+10.8	[66]
KA-2	?	disk	+	>+3.7	[110]
CLS 29?	?	disk	?	?	[133]
0041-295	?	disk?	?		[143]
0045-259	?	halo	?		[143]
G 77-61	yes	halo	+	+10.1	[23]
CLS 31	?	halo	+		[33]
CLS 50	?	halo	+		[32]
CLS 96	?	halo	+		[33]
LHS 1075	?	halo	?		[33]
GD 439	dC + WD	?	?	>+4	[65]
GH 7-21	?	?	?		[64, 145]
SA51-2C24	?	?	+		[16]
2048-348	?	?	?		[143]
LSR 2105+2514	?	?	?		[72]
LP 758-43	?	?	?		[72]
SDSS	17 candidates listed in [72]				[74]

9.4.2.6 Phase 12: Tc-Poor S or C in Pop. I, Yellow Symbiotics in Pop. II

The identification of Tc-poor S stars as mass-transfer binaries has been discussed at length in Section 9.4.1.2 and will not be repeated here. This section will instead focus on the relationships between Tc-poor S (or C) stars and various other families of peculiar red stars, namely, barium and symbiotic stars. The identification of Tc-poor S stars as descendants of the barium stars, first suggested in [47, 67], has now been definitely confirmed on the basis of the similarity of their orbital elements (see Figure 9.11 and Section 9.4.3.2).

A more controversial issue concerns the possible relationship between Tc-poor S stars and symbiotic stars, which is raised by the fact that both families consist of mass-losing giants and WD companions in systems with similar orbital periods [137]. This question is actually twofold: **(i) Do symbiotic systems exhibit overabundances of s-process elements? (ii) Do Tc-poor S stars exhibit some symbiotic activity?**

As far as question (i) is concerned, the answer seems negative for "red" symbiotics (i.e., with M giants as primary components), but is positive for "yellow" s-type symbiotics (i.e., with nondusty G or K giants as primary components). In fact, no S stars are known among the giant components of red symbiotics [95], but all yellow s-type symbiotics studied thus far are halo objects exhibiting overabundances of s-process elements [103, 118, 125, 126]. The fact that Pop. II symbiotics involve yellow giants rather than red giants is a likely consequence of the bluer location of the giant branch in low-metallicity populations. In that sense, s-type yellow symbiotics may be seen as the Pop. II analogues of the binary S stars, as both classes populate the upper end of the giant branch.[11] **It is in fact the absence of S stars among red symbiotics that is the real puzzle.** The difference between red symbiotics and Tc-poor S stars resides neither in their orbital parameters nor in their luminosities, as apparent from the similarity between the properties of the M+WD, noneruptive symbiotic systems like SY Mus or RW Hya, and the average Tc-poor S stars. A solution to that puzzle may come instead from the demonstration that red symbiotics actually belong to a more metal-rich population than the binary S stars. Peculiar red giants like carbon and barium stars are indeed known to be rarer in high-metallicity environments (see Chapter 8 and Section 9.4.3.1). This scarcity may be a consequence of the less-efficient operation of the s-process at high metallicities with $^{13}C(\alpha,n)^{16}O$ as the neutron source (Chapter 2). Conversely, carbon symbiotics are frequent in the low-metallicity environment of the Magellanic Clouds. The suggestion that red symbiotics in the Galaxy may belong to a rather metal-rich population is, however, not borne out by their subsolar carbon abundances [119], despite the claim that their infrared colors may be more typical of metal-rich, bulge M giants than of M giants in the solar neighborhood [150]. A detailed comparison of the kinematics and metallicities of red symbiotics and Tc-poor S stars is necessary to settle this question definitively.

As far as question (ii) is concerned, some Tc-poor S stars indeed exhibit symbiotic activity (Table 9.6; see also Section 9.3.1 and [137]), although rather weak in comparison to full-fledged symbiotics. The rarity of the symbiotic phenomenon among extrinsic S stars is fully apparent from the result of a high-resolution spectroscopic survey [136, 137] aiming at detecting Hα emission among 40 extrinsic S stars from the Henize sample (see Table 9.3). It uncovered only 2 stars with strong, broad (FWHM $\approx 70\,\mathrm{km\,s^{-1}}$) and double-peaked H$\alpha$ emission lines resembling the Hα profiles of symbiotic stars. Symbiotic activity in extrinsic S stars seems to be restricted, at least as far as Hα emission is concerned, to the period range 600–1000 d (Table 9.6), which is also where most red s-type symbiotic stars are found (excluding symbiotic novae and d-type symbiotics hosting Mira variables).

---

[11] Whether yellow symbiotics belong to phases 4–5 or 12 in Figure 9.9 cannot be answered at this stage, since the Tc content of these stars is currently unknown.

Table 9.6. S stars exhibiting symbiotic activity, according to various diagnostics: X-rays (X), UV continuum (in column UV cont., y means that excess UV flux is present, but does not match a clean WD spectrum), C IV λ155 nm and Mg II λ280 nm lines, Hα emission, and He I λ1083 nm. The stars have been ordered according to increasing orbital period. From [137]

Name	P [d]	X	UV cont.	C IV λ155	Mg II λ280	Hα em.	He I λ1083
Hen 108	197	?	?	?	?	no	?
Hen 147	335	?	?	?	?	no	?
HR 1105	596	?	y	y	y	?	var
Hen 137	636	?	?	?	?	moderate	?
HD 35155	642	var?	y	y	y	moderate?	var
Hen 121	764	?	?	y	?	strong	?
HD 191226	1210	?	WD	no	y	?	wk em.
Hen 119	1300	?	?	?	?	no	?
HD 49368	3000	?	y	y	y	no	absorption
HR 363	4590	var	y	no	?	?	wk em.
Hen 18	?	?	y	?	?	strong	?
Hen 134	?	?	?	?	?	moderate	?
ER Del	?	?	?	?	?	strong	?

9.4.2.7 Phase 14: Binary Post-AGB Stars

A detailed description of post-AGB stars is presented in Chapter 10. Of interest here are those post-AGB stars residing in binary systems and that in principle fit either in phase 6 (if the companion is a main sequence star) or in phase 14 (if the companion is a WD) of the binary scenario (Figure 9.9). Unfortunately, the mass functions of these SB1 systems do not always allow one to distinguish between a main-sequence and a WD companion. Most (though not all) binary post-AGB stars are very metal-deficient, and this correlation has been interpreted as the result of reaccretion, from a stable circumbinary disk, of gas depleted in refractory elements (see Chapter 10 and Section 9.3.1). The composition of the atmosphere of the binary post-AGB stars does not bear the s-process signature of the third dredge-up, either because the star left the AGB at an early stage, before these processes had a chance to operate, or as a result of this chemical fractionation (most s-process elements are refractory). Such a chemical fractionation does not seem to be observed in barium stars, despite the fact that these stars are supposed to be the descendants of the binary post-AGB stars in phase 6. At this point, it should be noted, however, that the possibility of such a fractionation has never really been investigated in detail for barium stars. In fact, one cannot totally exclude that the abundance pattern of barium or carbon stars is shaped both by the s-process nucleosynthesis *and* by the refractory properties of the elements (or even in some cases, of their oxides). In that respect, intriguing correlations seem to exist [70] between the abundance patterns observed in carbon stars and in SiC grains found in meteorites, which would clearly deserve further scrutiny. In any case, the apparent incompabitility between the abundances of barium and binary post-AGB stars needs to be properly understood. It is conceivable, for example, that barium stars are not at all the descendants of the binary metal-deficient post-AGB stars, if the latter left the AGB at a very early phase. This possibility is suggested by the similarity of the orbital elements of non-AGB M giants and post-AGB stars observed in Figure 9.11.

9.4.3 Orbital Elements of Extrinsic Stars

9.4.3.1 Binary Statistics of Extrinsic Stars

Field Stars

The results of radial-velocity monitoring aiming at deriving the frequencies of binaries among the various families of extrinsic stars is presented in Table 9.7. When account is made for various detection biases (due to pole-on or very long period orbits) [54], the fraction of binaries among (disk) subgiant CH stars, "strong" Ba dwarfs (but see below), mild barium stars, strong barium stars, Tc-poor S stars, and CH stars is compatible with 100%. Observed under similar conditions, normal giant stars in the field reveal about 10 to 30% of spectroscopic binaries [86]. This result is especially significant for barium stars with strong anomalies, since the monitored sample includes *all* such stars that are currently known. No similar conclusion can be drawn for dwarf C stars, since there has not been any systematic radial-velocity monitoring for this class of extrinsic stars.

The situation for dwarf barium stars is more confused. On the one hand, the binary frequency in a sample of "strong" Ba dwarfs (including "FV strong $\lambda 4077$" stars, blue metal-poor stars with Ba overabundances, disk subgiant CH stars, and Ba stars with spectral types earlier than G2 previously classified as giant stars, but which were shown to be dwarfs [98]) is compatible with 100%. But on the other hand, among the 6 barium dwarfs identified by the high-accuracy abundance survey of 200 F dwarfs in the solar neighborhood [26], only HR 4395 is a binary system (with a period of about 2000 d), whereas the other stars clearly have a constant radial-velocity after several years of monitoring. This situation is puzzling, since the Ba overabundances (in the range $0.23 \leq [Ba/Fe] \leq 0.68$) of these barium dwarfs cannot be ascribed to the normal galactic chemical evolution [26]. The situation is similarly puzzling for the *low-metallicity* CH subgiant stars cited in Table 9.7 that do not exhibit radial-velocity variations despite several years of monitoring.

In summary, binarity indeed appears to be a *necessary* condition to form the classes of peculiar stars listed in Table 9.7, with the exception of low-metallicity CH subgiants, mild Ba dwarfs, C dwarfs, and Tc-poor C-N stars,

Table 9.7. The fraction of binaries among the various families of extrinsic stars. The column labelled Refs. provides the references where the binary statistics and the orbital elements may be found

Family	Fraction of binaries	Refs.
dwarf Ba		
"FV str λ4077" and disk CH subgiants	28/30	[85, 98, 99]
blue metal-poor with Ba overabundances	100%?	[106, 127]
Ba dwarfs among FV from [26][a]	1/6	[98, 99]
low-metallicity CH subgiants	0/3	[107]
dwarf C	$\geq 4/31$	see Table 9.5
mild Ba	37/40	[54]
strong Ba	35/37	[54]
Tc-poor S	24/28	[54]
Tc-poor C-N	1/3?	[54]
J	100%?	[154]
CH	>10/12	[87]

[a] The Ba dwarfs are HR 107, HR 2906, HR 4285, HR 4395, HR 5338, HD 6434.

for which the situation remains unclear. Whether binarity is a *sufficient* condition to form the barium syndrome has not yet been conclusively answered. There seem indeed to exist giant stars that are *not* barium (or S) stars, although they are components of binary systems with orbital parameters like those of the barium systems. All red symbiotic stars are such cases. Another example is HD 160538 (=DR Dra). A detailed abundance analysis reveals no s-process enhancements whatsoever, while the giant belongs to a binary system with a hot WD companion and $P = 904$ d, $e = 0.07$, typical of barium systems. However, the mass of the WD should be derived to ascertain that its AGB progenitor was massive enough to evolve along the TP-AGB phase and to activate the s-process and the third dredge-up. Other examples that might appear at first sight puzzling are HD 128220 (a rapidly rotating G subgiant and a 0.55 ± 0.01 M$_\odot$ sdO star, in a system with $P = 871$ d and $e = 0.21$), HD 185510 (an RS CVn system consisting of a K0III and an sdO star) and HR 1608 (= K0IV + WD with $T_{\text{eff}} = 26\,000 \pm 1500$ K) in a system of $P = 903 \pm 5$ d and $e = 0.30 \pm 0.06$. However, the giant star in the latter system need not be a barium star, because the inferred WD mass ($0.40^{+0.2}_{-0.15}$ M$_\odot$) is too small for its progenitor to evolve along the TP-AGB (compare this mass with the core masses at the onset of the TP-AGB given in Chapter 3). The same holds for the sdO stars in HD 128220 and HD 185510 (about sdO stars, see also Chapter 10). Conversely, HD 33959C (F4V + DA), HD 74389 (A2V + DA), and β Crt (A2III-IV + DA) are binary systems with a WD companion discovered recently where the possible barium nature of the primary star would need to be checked.

Globular Clusters

The finding of CH stars in low-concentration globular clusters (like ω Cen, M2, M14, M22, and M55), and not in more concentrated ones [18, 83], provided the first hint that the overabundances of s-process elements observed in chemically peculiar red giants like CH stars were in fact related to binarity [81]. Somehow, binary systems hosting CH stars cannot survive in highly concentrated clusters. The dominant dynamical processes affecting binaries in globular clusters are the disruption of "soft" binaries through stellar encounters and the shrinking of "hard" binaries via energy exchanges with intruder stars. It seems most likely [18] that it is the latter process that accounts for the lack of CH stars in highly concentrated clusters.

In those, the shrinking of "hard" binaries seems efficient enough [18] to drive these binaries to separations (≤ 1 AU) that are too small to accommodate the necessary AGB progenitor of CH systems. This suspicion has been only partially confirmed, since the classical CH stars ROA 55 and ROA 70 in ω Cen have indeed been shown to be binaries [80], but *globular clusters seem to host nonbinary CH, barium, and S stars as well* (M22: [19], ω Cen: [80]). These stars are fainter than the luminosity threshold of the TP-AGB and thus neither internal nucleosynthesis nor mass transfer in a binary system can account for their chemical peculiarities! A scenario specific to globular clusters seems thus at work (see the discussion in [132, 140]).

9.4.3.2 The $(e, \log P)$ Diagram: Insights into the Mass-Transfer Mode

The $(e, \log P)$ diagram (where e is the eccentricity of the orbit and P its period) allows one to learn much about the evolution followed by a given class of binary systems. First, it shows at a glance the period range spanned by systems of a given family. Second, similarities or dissimilarities between $(e, \log P)$ diagrams of different families allow one to evaluate the likelihood of an evolutionary connection between them. Third, the distribution of eccentricities provides clues to the nature of the mass-transfer process (namely, RLOF versus wind accretion; see also Section 9.4.4).

Figure 9.11 compares the $(e, \log P)$ diagrams for the families of binaries described in Section 9.4.2 along with normal G-KIII stars in open clusters and field M giants used as comparison samples of (mostly) pre-mass-transfer binaries. The $(e, \log P)$ diagram of extrinsic systems differs from those of the pre-mass-transfer samples in two clear ways: (i) Many pre-mass-transfer giants are found with $P < 200$ d, whereas such short-period systems are quite rare among all extrinsic classes; (ii) the average eccentricity at any given period is much smaller for the extrinsic systems, without being zero, though. These differences between pre-mass-transfer giants and extrinsic systems are a clear indication that the latter have undergone mass transfer and possibly tidal interaction as well. It will be shown in Section 9.4.4 that the nonzero eccentricities of most of the extrinsic systems argue against Roche lobe overflow

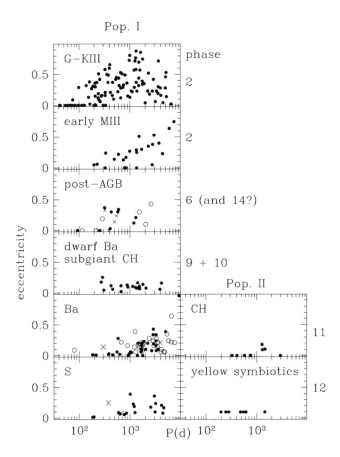

Fig. 9.11. Comparison of the $(e, \log P)$ diagrams for the families described in Section 9.4.2, referred to by the phase label of Figure 9.9 (written to the right of each panel). **Left** and **right** columns correspond to Pop. I and Pop. II systems, respectively. The comparison samples at phase 2 correspond to normal (hence in pre-mass-transfer binaries) G-K giants in open clusters, and to early M giants in the field. Post-AGB systems depleted in refractory elements are represented by black dots; systems not exhibiting the depletion pattern are represented by open dots; crosses denote systems with no abundance data. The crosses in the Ba panel correspond to the two pairs of the triple system BD+38°118. Ba1,2,3 ("mild" Ba) and Ba4,5 ("strong" Ba) stars are represented by open and filled dots, respectively. In the S panel, crosses correspond to two systems with unusually large mass functions. Since the eccentricities of symbiotic systems are generally poorly determined, they have all been set to 0.1. Orbital elements are taken from the references listed in Table 9.7, from Chapter 10 for post-AGB stars, from [88] for normal G-K giants, from [51] for early M giants, and from the references listed by [125] for the yellow symbiotic systems TX CVn, BD−21°3873, LT Del (=He2-467), and AG Dra

(RLOF; Section 9.2.1) as the mass-transfer mode, since this mode requires the mass-losing star to fill its Roche lobe, a situation where tidal interactions will very rapidly circularize the orbit [142]. Wind accretion is therefore preferred. In that mode, the system remains always detached, and the pollution of the extrinsic star is achieved by the accretion of a fraction of the wind ejected by the companion AGB star (Section 9.3.1). The requirement that the system have remained always detached seems, however, impossible to satisfy in systems with observed periods as short as 300 d (see Figure 9.4). RLOF seems thus unavoidable for the short-period systems. That conclusion is troublesome, however, because (i) several short-period systems *are not* in circular orbits (especially among post-AGB and dwarf Ba systems), and (ii) RLOF from an AGB star with a deep convective stellar envelope is generally expected to lead to a CE phase (at least when the mass-losing AGB star is the more massive component), and to a dramatic orbital shrinkage (Section 9.4.4). The end product of such an evolution is a cataclysmic variable with an orbital period of a few hours, much shorter than the shortest period observed among the extrinsic systems. The very existence of extrinsic systems with periods too short to be accounted for by the wind accretion process indicates that these systems must have avoided the dramatic fate outlined above. Several ways out of this channel (which sometimes does operate, though, since cataclysmic variables *do* exist!) will be sketched in Section 9.4.4. At this point, it will simply be mentioned that the most detailed simulations of mass transfer available to date for extrinsic systems [56, 104] include prescriptions for wind accretion, RLOF, tidal interaction, and for possibly enhanced mass loss from the AGB star due to the presence of the companion (CRAP, Section 9.3.3). This last ingredient actually turns out to be an essential one: If a strong mass loss from the AGB star reverses the mass ratio before the onset of RLOF, RLOF will no longer lead to a CE phase nor to dramatic orbital shrinkage (see Figure 9.14 and Section 9.4.4). These simulations face, however, two major difficulties (Figure 9.12): (i) They cannot reproduce the short-period high-eccentricity systems present among extrinsic stars, and (ii) they predict too many extrinsic systems with long periods ($> 10^4$ d). In that respect, it should be stressed that no observational bias alters the period range ($80 < P(\mathrm{d}) \lesssim 10^4$) covered by extrinsic systems (Figure 9.12), especially for strong barium stars, since the *complete* sample of strong Ba stars has been monitored. Figure 9.12 therefore reveals that there is both a short- and a long-period cutoff to produce extrinsic systems: If the period is too long, the system is too wide for wind accretion to be efficient and to lead to significant pollution (one example is Mira Ceti itself; Table 9.1). On the other hand, if the period is too short, another evolutionary channel (leading to cataclysmic variables or to binary sdB stars?) must have been followed [24, 79]. In that respect, it might be interesting to search the primary components of cataclysmic variables for chemical peculiarities similar to those exhibited by extrinsic systems.

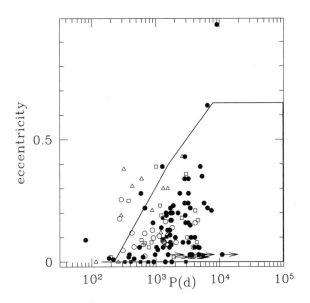

Fig. 9.12. Comparison of the $(e, \log P)$ diagrams for observed extrinsic systems (filled circles: barium stars; open circles: Ba dwarfs and subgiant CH stars; open triangles: post-AGB stars; open squares: S stars; filled squares: CH stars; arrows mark systems with only a lower limit on their orbital period) and for a synthetic binary population (falling within the thick lines) computed with detailed prescriptions for wind accretion, stable RLOF, and tidal interaction [56]. The displayed synthetic population is that representing the best fit to barium systems, although the model overproduces long-period systems and is unable to account for short-period, high-eccentricity systems observed among post-AGB and dwarf Ba systems

Finally, Figure 9.11 fully confirms the hypothesis of an evolutionary link between barium and Tc-poor S stars, which are distributed identically in the $(e, \log P)$ diagram but differ only in terms of their effective temperatures. The same conclusion is reached from the similarity of their mass-function distributions (Section 9.4.3.3). In contrast, there are dissimilarities between the $(e, \log P)$ diagrams of post-AGB systems, dwarf Ba, and barium stars. If an evolutionary path indeed links these families, physical processes altering the eccentricity and operating specifically in some of these families must be invoked to account for these dissimilarities. For example, tidal circularization of the closest systems ($P < 300$ d) may occur when the Ba and S giants ascend the RGB [142]. This process may account for the fact that among the short-period systems, the average eccentricity of giant Ba stars is considerably smaller than that of dwarf Ba systems (Figure 9.11). Similarly, post-AGB systems with short periods have large eccentricities despite the fact that tidal circularization must have operated during the AGB evolution. The similarity between the $(e, \log P)$ diagrams of post-AGB systems and early M giants might indicate instead that the post-AGB systems did not evolve very far up the AGB. Otherwise, processes driving the eccentricity upwards in post-AGB systems must be invoked, like (i) an exchange of angular momentum between the orbit and the circumbinary disk [5] that frequently occurs in those systems (see Section 9.3.1 and Chapter 10), or (ii) isotropic mass loss occurring preferentially during periastron passages [129].

9.4.3.3 Comparison of Mass Functions: Insights into the Galactic Population and into the Nature of the Companion

The mass functions of extrinsic systems convey information on the masses of the two components, and hence on the galactic population to which they belong as well as on the nature of the companion (i.e., WD versus main sequence). They may also be used to confirm the evolutionary link between the various families already discussed in Section 9.4.3.2 in relation to the $(e, \log P)$ diagram. In spectroscopic binaries with only one observable spectrum, as most of the extrinsic systems are, the individual masses are not directly accessible; only the mass function $f(M) = M_2^3 \sin^3 i / (M_1 + M_2)^2$ (where M_1 is the mass of the observable component, M_2 that of the unseen component, and i is the inclination of the orbital plane on the plane of the sky) can be derived directly from the observations. To extract even the mass ratio $Q \equiv M_2^3/(M_1 + M_2)^2$ requires resorting to statistical methods to eliminate the inclination i of the orbital planes, which may be assumed to have their poles distributed randomly on a sphere. Classical inversion methods, like Lucy–Richardson, may be used for that purpose [54] and provide the distribution of Q. The Q distributions of extrinsic systems are typical in that they are quite peaked, contrary to the situation prevailing for normal G-K giants in binary systems (Figure 9.13).

The narrow Q distribution typical of extrinsic systems may be seen as the signature of the WD nature of the companions, as required by the binary paradigm sketched in Figure 9.9. Since Q depends much more sensitively upon M_2 than upon M_1, the narrow range spanned by the WD masses, as observed for field WDs [111], results in a peaked Q distribution for extrinsic systems. In contrast, since the companions of normal giants in binary systems need not be WDs, their masses may span a much wider range (the only proviso being $M_2 \leq M_1$).

Direct detection of the WDs in extrinsic systems by their UV radiation has been achieved in very few cases. Among 15 giant barium stars ever looked at, WD companions have been detected with certainty ([6] and references therein) for ζ Cap, ζ Cyg, 56 Peg, HR 774, HR 4474, HR 5058, HD 104979, and HD 165141 (a tentative detection has been reported for ν^2 Cas, whereas firm or tentative WD detections for HD 65699 and ξ^1 Cet are irrelevant, since these stars are no longer considered to be barium stars). This relatively low success rate may be attributed to the fact that the WD companions

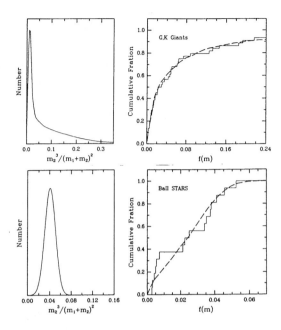

Fig. 9.13. Comparison of the mass-function (**right panels**) and Q ($\equiv M_2^3/(M_1 + M_2)^2$; **left panels**) distributions of barium stars (**lower panels**) and normal G-K giants (**upper panels**). From [87]

are generally very cool ($T_{\rm eff} < 10^4$ K) and these systems hosting luminous giant stars are generally far away. Among 9 Tc-poor S stars that were observed with the *International Ultraviolet Explorer* satellite, only HD 191226 revealed clear signatures of a WD companion (o^1 Ori is another, special, case, since the S star is in fact Tc-rich), whereas a UV continuum has been detected for HR 363, HR 1105, HD 35155, and HD 49368 (Table 9.6 and references therein). The somewhat larger detection rates may in this case be attributed to the fact that the larger mass-loss rates of S stars as compared to barium giants trigger some kind of symbiotic activity and an associated UV continuum (Section 9.4.2.6). At this stage, the two puzzling Tc-poor S stars HD 191589 and HDE 332077 should be mentioned. Instead of the WD expected in the framework of the binary paradigm, ultraviolet observations revealed A-type companions, consistent with their outlying mass functions of 0.395 and 1.25 $M_\odot$, respectively. The evolutionary status of these Tc-poor S stars is currently unknown, especially since the current radial-velocity data provide no hint that these systems might be triple.

Direct detection of the WD companion has also been achieved for a few dwarf C stars (see Table 9.5). A very interesting possibility offered by the Ba dwarfs is a statistical estimate of the mass of their companion (and not just

Table 9.8. Average masses $\overline{M}_1$ of the extrinsic star in the different classes as derived from their $f(M)$ distribution, for two different values of the companion average mass $\overline{M}_2$. N is the number of available orbits

Family	N	$\overline{Q}$	$\overline{M}_1$		Ref.
			$\overline{M}_2 = 0.60$	$\overline{M}_2 = 0.67$	
		[M$_\odot$]	[M$_\odot$]	[M$_\odot$]	
CH	8	0.095	0.9	1.1	[87]
dBa + sgCH	14	0.080	–	1.25±0.25	[99]
Barium (strong)	36	0.049	1.5	1.9	[54]
Barium (mild)	27	0.035	1.9	2.3	[54]
Barium (total)	63	0.043	1.65	2.0	[54]
S (Tc-poor)	17^a	0.041	1.6	2.0	[54]

^{a} The two peculiar S stars HD 191589 and HDE 332077 were not included (see text).

of Q as for the other extrinsic systems). The mass of the Ba dwarf may be derived from a spectroscopic determination of the gravity, effective temperature, and metallicity combined with a comparison to theoretical evolutionary tracks for given masses. The distribution of the companion masses is then derived by inverse methods from the observed $f(M)$ and inferred primary-mass distributions. The best fit is obtained with a Gaussian distribution for the WD masses centered at 0.67 M$_\odot$ with a dispersion of 0.09 M$_\odot$ [99]. This is the first direct estimate of the average mass of the companion of extrinsic stars. It is in fact close to the average mass of field WDs [111].

Table 9.8 provides the average Q values and the corresponding masses for various families of extrinsic stars. Masses are derived for two different choices of the average WD mass, namely, $M_2 = 0.60$ and 0.67 M$_\odot$, the latter corresponding to the value directly derived from the analysis of dwarf Ba stars. Barium and Tc-poor S stars have very similar average masses, confirming once again that their different spectroscopic appearance simply results from a difference in their effective temperatures (see also Section 9.4.3.2), but they belong, in fact, to the same galactic population.[12] As expected, CH stars have an average mass (on the order of 1.1 M$_\odot$) that is significantly smaller than that inferred for Ba and S stars (about 1.9–2.3 M$_\odot$, adopting the same average WD mass of 0.67 M$_\odot$ for the three families). This conclusion based on their orbital properties is consistent with the halo kinematics of CH stars. Dwarf barium and subgiant CH stars appear to be intermediate between CH and giant Ba stars. Interestingly enough, these population differences seem to be reflected as well by differences in the corresponding $(e, \log P)$ diagrams

---

[12] A closer look at the $f(M)$ distribution of barium stars [54] reveals that mild barium stars have a slightly larger average mass than strong barium stars (Table 9.8), a conclusion supported by slightly different kinematic properties [15].

(Figure 9.11), although the interplay between galactic population and orbital dynamics is far from being obvious and still needs to be clarified.

The $f(M)$ distribution of post-AGB binaries is very different from that of the other extrinsic systems. Although some post-AGB systems have mass functions similar to those of the other extrinsic families, several have $f(M)$ in the range 0.14–0.97 $M_\odot$. This dichotomy confirms that post-AGB binaries are a mixture of systems with main-sequence (phase 6 of Figure 9.9) and WD (phase 14) companions.

9.4.4 The Mass-Transfer Mode, or How to Avoid Orbital Shrinkage Due to Unstable RLOF in Extrinsic Systems

This final section is devoted to a discussion of an important problem that arises from the period distribution of all extrinsic systems, as noted in Section 9.4.3.2 in relation to Figures 9.11 and 9.12: These systems that host WD companions have periods in a range that seems incompatible with the standard prescriptions for mass transfer from an AGB star. The observed periods seem too large for being the outcome of dynamically unstable RLOF (resulting in strong orbital decay triggered by a CE phase, and leading to cataclysmic variables) and, at the same time, too short to have been able to host an AGB star in a detached system. The main issue addressed in this section is to identify which one among the various possible mass-transfer modes operated in extrinsic systems: (i) wind accretion in a detached binary; (ii) RLOF in a semidetached system occurring on a dynamical (ii-a) or thermal (ii-b) time scale; (iii) RLOF in a semidetached system occurring on much longer nuclear or orbital time scales. The mass-transfer mode controls the outcome of the process and shapes the post-mass-transfer binary system in several respects:

(a) The efficiency of the *accretion* by the companion star (labeled 1 in the following) of the matter lost by the AGB star (labeled 2) will be high in RLOF modes, which are close to being conservative, i.e., $dM_1 = -\eta\, dM_2$ with $\eta \approx 1$. In contrast, in the wind accretion mode, η is on the order of a few percent only [130].

(b) The *reaction of the accreting star* to the matter pouring down upon it will depend on the value of the accretion time scale $M_1/\dot{M}_1$ (which is in turn fixed by the mass-transfer rate, and hence by the mass-transfer mode; Section 9.4.4.1), as compared to the dynamical and Kelvin–Helmholtz time scales. The catastrophic fate generally associated with RLOF results from the fact that the fast RLOF mass-transfer rates are responsible for short accretion time scales that make the accreting star swell, thus possibly embedding the system in a CE (Section 9.4.4.2). In turn, this CE exerts a drag on the orbiting stars that may result in a substantial orbital shrinkage.

(c) The RLOF mode is necessarily preceded by a phase of *tidal circularization of the orbit* (Section 9.4.4.3). Therefore, post-RLOF systems have circular orbits, as opposed to wind-accretion systems where circularization is incomplete.

9.4.4.1 The Various Mass-Transfer Modes and Associated Stability Criteria

In the framework of the Roche model (Section 9.2.1), the operation of wind accretion (mode i) as opposed to RLOF (modes ii or iii) is controlled by the ratio $R_2/R_{R,2}$ being smaller than unity, where R_2 is the radius of the mass-losing star and $R_{R,2}$ is the Roche radius (9.2) around it (in the following, subscripts 2 may be omitted when there is no confusion possible). The RLOF criterion $R_2/R_{R,2} \geq 1$ is in fact too severe for AGB stars, since the scale height of the density stratification in their atmosphere is a significant fraction of their photospheric radius (Section 9.2.1.3). Therefore, the mass flow through the inner Lagrangian point L_1 becomes nonnegligible long before the photosphere reaches the critical Roche equipotential [101].

Which one among modes (ii-a), (ii-b), or (iii) will operate in a given situation depends on the stability of RLOF mass transfer. The linear stability analysis amounts to a comparison of the first-order variations of the stellar and Roche radii under the effect of mass loss and subsequent orbital evolution:

$$R_2(t_0 + \Delta t) = R_2(t_0) + \frac{\mathrm{d}R_2}{\mathrm{d}M_2}\frac{\mathrm{d}M_2}{\mathrm{d}t}\Delta t, \tag{9.12}$$

$$R_R(t_0 + \Delta t) = R_R(t_0) + \frac{\mathrm{d}R_R}{\mathrm{d}M_2}\frac{\mathrm{d}M_2}{\mathrm{d}t}\Delta t. \tag{9.13}$$

Stability requires that after mass loss, $\Delta M_2 = \mathrm{d}M_2/\mathrm{d}t\,\Delta t\,(<0)$, the star be still contained by its Roche lobe. Assuming that $R_2(t_0) = R_R(t_0)$ is satisfied initially, the stability condition then becomes $R_2(t_0 + \Delta t) \leq R_R(t_0 + \Delta t)$, or $\zeta \geq \zeta_R$, where

$$\zeta_R \equiv \frac{\mathrm{d}\ln R_R}{\mathrm{d}\ln M_2} \tag{9.14}$$

and

$$\zeta \equiv \frac{\mathrm{d}\ln R_2}{\mathrm{d}\ln M_2}. \tag{9.15}$$

The radius–mass exponent ζ expresses how the radius of the mass-losing star adapts to a loss of mass. The stability condition simply expresses that if the stellar radius shrinks in response to mass loss ($\zeta > 0$), it must shrink faster than the Roche lobe (hence $\zeta \geq \zeta_R$). A star responds to mass loss on two time scales. The immediate response is (usually[13]) adiabatic. Hydrostatic equilibrium is almost immediately restored, but negligible heat transport occurs on this dynamical time scale τ_{dyn}:

---

[13] very extended AGB stars, $R \geq 500\,R_\odot$, with thin envelopes have Kelvin–Helmholtz time scales that become smaller than the dynamical time scales [101].

$$\tau_{\text{dyn}} = \left(\frac{R_2^3}{GM_2}\right)^{1/2}, \qquad (9.16)$$

with G the constant of gravitation. The radius–mass exponent characterizing this adiabatic readjustment is denoted by ζ_{ad}. The star then regains thermal equilibrium by redistributing its internal energy to account for its new distribution of energy sources and mass. This occurs over the Kelvin–Helmholtz time scale τ_{KH}:

$$\tau_{\text{KH}} = \frac{G\,M_2\,M_{\text{env}}}{R_2\,L_2}, \qquad (9.17)$$

where M_{env} is the envelope mass of the AGB star. The corresponding change in radius is characterized by the exponent ζ_{th}.

The stability condition thus amounts to comparing these responses to the change in the dimension of the star's Roche lobe due to the orbital evolution and the changing mass ratio, and thus can be written $\zeta_R < (\zeta_{\text{th}}, \zeta_{\text{ad}})$. If this is not satisfied, then mass transfer runs to the fastest unstable time scale.

Adiabatic Radius–Mass Exponent ζ_{ad} of AGB Stars

The adiabatic radius–mass exponent for (static) AGB stars with a large convective stellar envelope is best represented by that of a condensed polytrope [128], where the central point mass represents the degenerate core (with mass fraction $m = M_{c,2}/M_2$), and the convective envelope is represented by a polytrope of index n ($= \frac{3}{2}$ for a perfect gas, in the absence of radiation pressure). The function $\zeta_{\text{ad}} = \zeta(n = \frac{3}{2}; m)$ can be reasonably well fitted (to better than a percent) by the function [128]:

$$\zeta_{\text{ad}} = \frac{2m}{3(1-m)} - \frac{1-m}{3(1+2m)} - 0.03\,m + \frac{0.2m}{1 + (1-m)^{-6}}. \qquad (9.18)$$

For $m = 0$, the above relation yields $\zeta_{\text{ad}} = -\frac{1}{3}$, as expected for a polytrope of index $n = \frac{3}{2}$ corresponding to a fully convective star. Thus, *the adiabatic response to mass loss of AGB stars with small core masses ($m < 0.2$) is an increase of their radius ($\zeta_{\text{ad}} < 0$).* The adiabatic radius–mass exponent ζ_{ad} is positive for $m > 0.2$, and goes to infinity as $m \to 1$ (Figure 9.14). Thus, *AGB stars with large cores ($m > 0.2$) contract as an adiabatic response to mass loss.*

The Thermal Radius–Mass Exponent ζ_{th} of AGB Stars

The thermal radius–mass exponent ζ_{th} for AGB stars may be derived from the Stefan–Boltzmann law (expressing that the star is in thermal equilibrium) $L = 4\pi R^2 \sigma_R T_{\text{eff}}^4$ (where σ_R is the Stefan–Boltzmann constant), the position of the evolutionary track $L = L(T_{\text{eff}}, M)$ in the HR diagram, and the core-mass luminosity relation implying $\partial L/\partial M = 0$. This yields [116]

$$\zeta_{\text{th}} \approx -0.3. \qquad (9.19)$$

At this point, it should be stressed that ζ should actually be derived from models including the dynamics of the AGB stellar envelope (Chapter 4). A first step in that direction may be found in [115] (equation 12.48), which provides an expression for ζ that includes some properties of dynamical envelopes like the gravitational luminosity and the mass-loss timescale.

The Roche Radius–Mass Exponent ζ_R

The Roche radius–mass exponent ζ_R must be computed by considering how, as a consequence of mass loss dM_2, R_R varies due to variations of the mass ratio and of the orbital separation (or, alternatively, of the orbital angular momentum). Thus, setting $R_{R,2} = A f_2(q)$, where $q = M_2/M_1$,

$$\zeta_R = \frac{\partial \ln R_R}{\partial \ln M_2} = \frac{d \ln A}{d \ln M_2} + \frac{d \ln f_2(q)}{d \ln q} \frac{d \ln q}{d \ln M_2}, \tag{9.20}$$

where $d \ln q / d \ln M_2 = 1 + q\eta$, with $\eta \equiv -dM_1/dM_2 \geq 0$ being the accretion rate by the companion star. So-called conservative mass transfer corresponds to $\eta = 1$. The term $d \ln f_2(q)/d \ln q$ depends upon the geometry of the Roche lobe. It should be computed for the generalized equipotentials (9.7) taking into account the extra force driving the mass loss. In the absence of a compact formula for $f_2(q)$ in this general case, (9.3) will be used instead, but the reader should keep in mind that this treatment thus provides only approximate results. Finally, the term $d \ln A/d \ln M_2$ may be rewritten in terms of the changes in the orbital angular momentum J associated with the mass loss/transfer processes. Since

$$J = \left(\frac{G M_1^2 M_2^2 A}{M_1 + M_2}\right)^{1/2} (1 - e^2)^{1/2}, \tag{9.21}$$

it follows that[14]

$$\frac{d \ln A}{d \ln M_2} = 2 \frac{d \ln J}{d \ln M_2} + \frac{q}{1+q}(1 - \eta) + 2(\eta q - 1). \tag{9.22}$$

The term $d \ln J/d \ln M_2$ accounts for all possible causes of variations of the orbital angular momentum associated with the mass loss/transfer processes.

Of relevance to the present purpose are the situations where the AGB star loses mass through an isotropic wind, of which a fraction α escapes to infinity, a fraction η is accreted by the companion, and a fraction δ forms a circumbinary ring of radius $A_r = \gamma^2 A$ from which it then escapes to infinity (with $\alpha + \eta + \delta \equiv 1$). Assuming that the AGB star is a point source (thus neglecting magnetic braking), the losses in orbital angular momentum are

---

[14] Only circular orbits will be considered in the following, since the orbit is supposed to have been circularized by tidal spin-orbit interaction prior to RLOF (Section 9.4.4.3).

easily computed by realizing that the escaping matter carries away the specific orbital angular momentum (i.e., per unit mass) h_i at the corresponding location i in the system [128]:

$$\partial J = h_2 \, \alpha \, \partial M_2 + h_{\text{ring}} \, \delta \, \partial M_2, \tag{9.23}$$

where $h_2 \equiv J_2/M_2 = \frac{1}{1+q}\frac{J}{M_2}$, $h_{\text{ring}} = \gamma J/\mu$, and $\mu = M_1 M_2/(M_1 + M_2)$ is the reduced mass of the system (in deriving h_{ring}, $\gamma \gg 1$ was assumed). The change in angular momentum thus can be written

$$\partial \ln J = \left(\frac{\alpha}{1+q} + \gamma \, \delta \, (1+q) - \lambda \right) \partial \ln M_2, \tag{9.24}$$

where the last term accounts for the drag exerted by the escaping wind on the orbital motion (with $\lambda \approx 0.05$ according to the hydrodynamical simulations of [130]). Inserting (9.22) and (9.24) into (9.20) finally yields

$$\zeta_R = 2\left(-\lambda + \frac{1-\eta-\delta}{1+q} + \gamma\delta(1+q)\right) + \frac{q}{1+q}(1-\eta)$$
$$+ 2\eta q - 2 + \frac{\mathrm{d} \ln f_2(q)}{\mathrm{d} \ln q}(1+\eta q). \tag{9.25}$$

The Roche radius–mass exponent derived from (9.25) is compared to the adiabatic and thermal radius–mass exponents in Figure 9.14. Remembering that stable mass transfer corresponds to $\zeta > \zeta_R$, we see that several situations may arise:

(i) *isotropic mass loss from highly non-conservative RLOF ("quasi-wind" accretion).* This case corresponds to the very special circumstance where RLOF is highly nonconservative (i.e., $\eta \ll 1$) and yet leads to isotropic mass loss from the system. Such a situation may perhaps be encountered when the geometry of the Roche equipotential is strongly modified by the extra force driving the mass loss (see Figure 9.2; remember, however, that the analytical developments presented here are not fully consistent in such situations, since the appropriate $\mathrm{d}\ln f_2(q)/\mathrm{d}\ln q$ term has not been used in (9.20)). The mass transfer is then dynamically stable, except when $\eta > 0.1$ for massive AGB stars ($M_2 > 3.5\,\mathrm{M}_\odot$). The mass transfer may become thermally unstable for low mass ratios q.

(ii) *quasi-conservative mass transfer.* This situation is either dynamically ($q > 1$) or thermally ($1 > q > 0.7$) unstable, except for small mass ratios $q < 0.7$. A strong wind prior to RLOF may thus stabilize RLOF if it drives q below 0.7.

(iii) *mass loss from a circumbinary ring.* The mass lost from a circumbinary ring carries away a very large amount of orbital angular momentum that makes the Roche radius shrink rapidly, and consequently strongly destabilizes mass transfer.

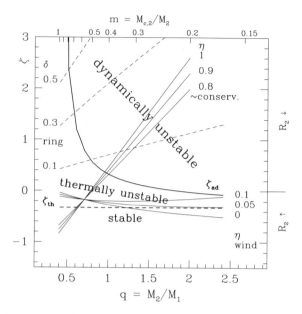

Fig. 9.14. Comparison of the adiabatic exponent ζ_{ad} (thick solid line, (9.18)), thermal exponent ζ_{th} (thick dashed line; (9.19)) and Roche exponents ζ_R (thin lines) as a function of the mass ratio M_2/M_1, where M_2 is the mass of the AGB star of relative core mass $m = M_{c,2}/M_2$, and M_1 is the mass of the companion. To fix the ideas, $M_1 = 1.5$ M$_\odot$ and $M_{c,2} = 0.6$ M$_\odot$ have been adopted. The Roche exponent ζ_R has been computed for different mass-transfer modes: isotropic wind with a fraction η ($\ll 1$) accreted by the companion, quasi-conservative evolution ($\eta \approx 1$), and mass loss through a circumbinary ring of radius $\gamma = A_r/A = 2$ for a fraction δ of the mass ejected in the wind. The labels along the right-hand vertical axis indicate whether the radius R_2 of the mass-losing AGB star increases or decreases as a result of mass loss

In summary, according to the above considerations, **RLOF involving an AGB star will be stable in two situations**: (i) The radiation pressure driving the wind opens up the critical Roche equipotential so as to make RLOF highly nonconservative (see the discussion in Section 9.2.1.2), without storing too much mass in a circumbinary disk, or (ii) RLOF is quasi-conservative, but a former wind episode has reduced the mass ratio below the critical $q = 0.7$ value at the onset of RLOF. However, it should be realized that even though the mass-transfer process does not lead to a runaway, the normal AGB wind mass loss may reach very large values that are comparable to the thermal mass-transfer rates anyway (see Chapter 6), so that the reaction of the accreting star is of much concern, even for stable RLOF, as discussed in the next section.

9.4.4.2 Reaction of the Accreting Star

If RLOF is unstable or if the wind mass-transfer rate is high, the outcome of the mass-transfer process will depend sensitively upon the reaction of the accreting star to the matter being dumped onto it. The worst situation is the one in which a contact system is formed because the accreting star has swollen so much as a reaction to accretion that it has filled its own Roche lobe. The reaction of the accreting star follows the same principles as those describing the reaction to mass loss, except that ΔM_1 is now positive [144]. If the accretion time scale $M_1/\dot{M}_1$ is faster than the Kelvin–Helmholtz time scale of the accreting star, the star cannot regain thermal equilibrium, and its radius evolution will obey the adiabatic exponent $\zeta_{\mathrm{ad},1}$. From considerations based on the run of the specific entropy in the outer layers of stars, it may be concluded [115, 144] that stars with a deep convective stellar envelope have $\zeta_{\mathrm{ad}} < 0$ (see Figure 9.14); hence they contract as a reaction to accretion, whereas stars with radiative envelopes have $\zeta_{\mathrm{ad}} > 0$; hence they expand as a result of rapid accretion. Therefore, the risk of forming a contact system does exist only for upper main-sequence companions having radiative envelopes (i.e., $M_1 > 1.2\,\mathrm{M}_\odot$, as is the case for most of the extrinsic systems but the CH stars; see Table 9.8). In this case, the instability criterion can be written

$$M_1/\dot{M}_1 < \tau_{\mathrm{HK},1}, \tag{9.26}$$

or, assuming to fix the ideas that the AGB star loses mass on a thermal time scale and that a fraction η is accreted by the companion,

$$-\frac{M_1}{\eta \dot{M}_2} = \frac{M_1}{M_2}\frac{\tau_{HK,2}}{\eta} < \tau_{\mathrm{KH},1}, \tag{9.27}$$

or

$$\frac{M_{\mathrm{env},2}}{\eta M_1}\frac{R_1}{R_2}\frac{L_1}{L_2} < 1. \tag{9.28}$$

This instability criterion is generally largely satisfied (since the mass-losing star 2 is an AGB star, and the accreting star 1 is supposed to be an upper main-sequence star), so that the accreting star will swell. Calculations predict a 10-fold (respectively, 100-fold) increase of the radius of a $2\,\mathrm{M}_\odot$ star for an accretion rate of $5\times10^{-5}\,\mathrm{M}_\odot\,\mathrm{yr}^{-1}$ (respectively, $2\times10^{-4}\,\mathrm{M}_\odot\,\mathrm{yr}^{-1}$) [59]. These factors are reached already after a few $0.1\,\mathrm{M}_\odot$ have been accreted. A CE will develop in systems close enough that this radius increase will make the accreting star fill its Roche lobe. This condition has been made more explicit in Table 9.9, which provides the orbital period below which a common envelope will form for a system consisting of a $3\,\mathrm{M}_\odot$ AGB star transferring mass onto a $2\,\mathrm{M}_\odot$ main-sequence companion (in that derivation, an initial radius of $1.7\,\mathrm{R}_\odot$ is assumed for the main sequence companion, and (9.2) is adopted for the Roche radius). Since systems involving AGB stars necessarily have periods of a few hundred days, Table 9.9 reveals that

Table 9.9. Critical orbital periods below which a CE forms when a $2\,M_\odot$ main-sequence star accretes mass from a $3\,M_\odot$ AGB star at the rate $\dot{M}_{\text{acc}}$. The relation between the mass accretion rate and the radius increase R_{acc}/R_0 is taken from [59]

$\dot{M}_{\text{acc}}$ [$M_\odot\,\text{yr}^{-1}$]	R_{acc}/R_0	$P_{\text{max}}(\text{CE})$ [d]
5×10^{-5}	10	18
2×10^{-4}	100	562

accretion rates larger than about $10^{-4}\,M_\odot\,\text{yr}^{-1}$ are needed for a CE to form. The final outcome is, however, uncertain, since such large accretion rates may in fact trigger a strong wind from the accreting star that would then impede accretion, thus preventing the formation of a CE [34]!

At the stage where the two components of the binary system are engulfed by a common gaseous envelope, the CE is generally not corotating with the binary system. As a consequence of the velocity difference between the binary components and the CE, they experience a drag force that works to reduce the binary separation. A large part of the energy deposited into the CE will be in the form of heat, though some fraction will be used to eject the envelope. Although the detailed processes by which the orbital energy is transferred into the CE are complex ones (see [46] and references therein), their global outcome may be encapsulated into the parameter

$$\alpha_{\text{CE}} = \frac{\Delta E_{\text{bind}}}{\Delta E_{\text{orb}}}, \tag{9.29}$$

describing the efficiency with which orbital energy is used to eject the CE. In that relation, ΔE_{bind} is the effective binding energy (gravitational minus thermal) of the CE, and ΔE_{orb} is the variation in the orbital energy of the binary, between the beginning and the end of the process. Assuming that the initial configuration consists of two stars surrounded by a CE of diameter $\approx 2A_0$ (where A_0 is the initial separation), the CE must be ejected from the gravitational field of both components. Under these circumstances, (9.29) can be written approximately as

$$\frac{(M_1+M_2)(M_2-M_{2,\text{f}})}{2A_0} - E_{\text{th}} \approx \alpha_{\text{CE}} \left(\frac{M_1 M_{2,\text{f}}}{2A_\text{f}} - \frac{M_1 M_2}{2A_0} \right), \tag{9.30}$$

where the mass of the AGB star varies from M_2 initially to $M_{2,\text{f}}$ at the end of the process, and E_{th} is the thermal energy stored in the CE. In spite of the uncertainties involved, the knowledge of α_{CE} coupled with the knowledge of the binding energy of the CE allows one to estimate the final separation of the emerging binary. For AGB stars, several processes are likely to reduce the effective binding energy of the stellar envelope [46] (such as the recombination energy in the hydrogen and helium ionization zones, excitation of nonradial pulsation modes, shock-heating, dust-driven winds, etc.). With those extra energy sources reducing the effective binding energy of the stellar envelope, not so much energy ought to be extracted from the orbit to expel it, thus reducing (and even perhaps suppressing) the orbital decay. As shown in [35], this reduction of the effective binding energy of the AGB stellar envelope by its sole thermal energy is already sufficient to limit the orbital shrinkage to an extent that barium systems like HD 77247 with periods not shorter than 80 d are allowed to form.

In summary, the previous discussion has shown that there are in fact many ways to avoid the dramatic fate usually thought to be associated with RLOF involving AGB stars.

9.4.4.3 Tidal Interaction and Orbital Circularization

Another important consequence of *any mass-transfer mode involving RLOF* is the tidal circularization of the orbit and the synchronization of the AGB spin with the orbital motion. These two properties result from the fact that as the AGB star comes close to filling its Roche lobe, tides will set in. In general, the stellar envelope cannot adjust instantaneously to the tides because of various sources of viscosity. The tides will therefore not point exactly toward the companion star, which will thus exert a torque on the AGB star. These torques act to synchronize the AGB spin with the orbital motion. Synchronization thus requires the conversion of some spin angular momentum into orbital angular momentum. On the other hand, tides are necessarily accompanied by differential mass motions that will slowly dissipate orbital energy into heat. The gradual lowering of the mechanical energy of the system through this energy dissipation will drive the system closer to its minimum-energy state. For a given mass distribution and fixed total angular momentum, a system has the least mechanical energy when it rotates like a rigid body (e.g., [124]). For a detached binary, this minimum-energy state corresponds to a circular orbit with synchronized spins.

When the orbital angular momentum is much larger than the spin angular momentum, then $\tau_{\mathrm{sync}} \propto \left(\frac{A}{R}\right)^6$, whereas $\tau_{\mathrm{circ}} \propto \left(\frac{A}{R}\right)^8$, so that synchronization proceeds faster than circularization [45]. Nevertheless, as the AGB star nears its Roche lobe, (R/A) becomes of order unity, and circularization is achieved rapidly. Once a circular orbit has been achieved, it will almost certainly remain circular.[15] Therefore, a post-mass-transfer system with a noncircular orbit has almost certainly not gone through an RLOF phase.

The outcome of the eccentricity in the framework of wind accretion is much more difficult to predict, since it depends on the details of the flow

---

[15] A process that may turn a circular into an elliptic orbit is a sudden (i.e., occurring on a time scale much smaller than the orbital period) isotropic loss of mass as in supernova explosions, which is probably not relevant here.

pattern. If the wind is not too much disturbed by the accreting star, and if it is expelled in a spherically symmetric fashion by the synchronously rotating mass-losing star (the so-called Jeans mass-loss mode), the angular momentum in the relative frame attached to the mass-losing star will not change, and so the eccentricity will not change [43]. In the case that there is some exchange of momentum between the escaping wind and the accreting star, the eccentricity will change, and its evolution will depend on the flow pattern. Several expressions describing the evolution of the orbital eccentricity in the wind-accretion mode have been proposed in the literature [43, 56, 130]. They correspond to different (inequivalent) parameterizations of the angular momentum transfer. In all cases, the relative eccentricity variation $\Delta e/e$ is found to be proportional to $-\Delta M_1/M_1$; i.e., the eccentricity will decrease in the same proportion as the relative mass increase of the accreting star. This generally does not exceed 10% for wind accretion in a relatively wide binary. Thus, eccentricity variations due to angular momentum transfer during wind accretion remain small. The eccentricity distribution of the barium stars (Figure 9.11) is satisfactorily reproduced in the framework of the wind accretion scenario, by combining the transfers of angular momentum by wind and tides [56], although those models fail to reproduce the period distribution, as discussed previously in relation to Figure 9.12.

9.5 Conclusions and Outlook

This last section provides a list of open questions relating to AGB stars in binary systems:

- The binary paradigm for extrinsic stars is generally well confirmed by current observational data, except for the following questions:
  (i) What is the origin of the seemingly nonbinary Ba dwarfs identified among F dwarfs in the solar neighborhood [26]? The same question applies to the family of low-metallicity subgiant CH stars reported in [107].
  (ii) What is the evolutionary status of the Tc-poor S stars HD 191589 and HDE 332077, which host A-type companions instead of the expected WD?
  (iii) The existence and frequency of extrinsic stars among giant C stars remain to be established.
  (iv) Is binarity a sufficient condition to form, e.g., barium systems, or in other words, are there *normal* (non-barium) giants in binary systems with a WD companion massive enough (i.e., ≥ 0.6 M$_\odot$) for its progenitor to have gone through the TP-AGB phase? If so, what is the other parameter besides binarity needed to form barium stars (e.g., low metallicity)? In a sense, this is an indirect way to identify the conditions required for the s-process and the third dredge-up to operate in AGB stars.

- The puzzling suggestion arising from the analysis of MACHO light curves [152] that 25% of all AGB stars in the LMC belong to semidetached binaries should be further explored.
- More hydrodynamic simulations exploring the parameter space characterizing wind mass-loss from AGB stars in binary systems should be performed to better characterize the different regimes leading to the various possible flow structures (spiral arm, bipolar nebula, circumbinary disk, inner Lagrangian point stream) developing in those binary systems.
- The link between symbiotic stars and extrinsic stars remains to be clarified, and in particular, why red symbiotic stars never exhibit overabundances of s-process elements. A possible hint could come from a kinematic or metallicity analysis of red symbiotic stars to assess whether they belong to a more metal-rich population than Tc-poor S stars do. Another open question relating to symbiotic stars is to identify the parameter(s) controlling their symbiotic activity, not shared by Tc-poor S stars, and explaining why the latter stars exhibit only weak symbiotic activity despite orbital elements identical to those of full-fledged symbiotics.
- Why does chemical fractionation operate in binary post-AGB stars, but leaves (apparently) no signature in their progeny (the barium and Tc-poor S stars)?
- If some cataclysmic variables indeed result from the same evolutionary channels as those producing extrinsic stars (except that RLOF is dynamically unstable in the former case), a search for heavy-element overabundances in the primaries of CVs might prove successful.
- If the $(e, \log P)$ diagrams of binary systems belonging to different galactic populations are indeed significantly different (typically, binaries involving older, or, equivalently, less massive, stars seem to have lower eccentricities on average at any given period), the origin of this interplay between galactic population and orbital dynamics should be clarified. In other words, synthesis of binary populations as in [56] should include metallicity as a parameter impacting on, e.g., the AGB radius and the mass loss rate. These improved models should also aim at properly reproducing the period distribution of barium stars. This requires a careful assessment of the RLOF stability, in order to avoid dynamical RLOF and dramatic orbital shrinkage. This assessment in turn requires the evaluation of the adiabatic and thermal radius–mass exponents in a fully dynamical AGB model (including pulsation and mass loss). The possibly stabilizing effect of a wind from the accreting star, and of the modified Roche geometry in the presence of a radiation-driven wind from the mass-losing star, should also be properly evaluated.
- An evaluation of the consequences of a possible coupling between the AGB pulsation and the dynamical tides raised by the companion would be of interest.

References

1. Abia, C., Domínguez, I., Gallino, R., et al. *ApJ*, 579, 817, 2002.
2. Abia, C. and Isern, J. *ApJ*, 536, 438, 2000.
3. Abia, C. and Wallerstein, G. *MNRAS*, 293, 89, 1998.
4. Alvarez, R., Jorissen, A., Plez, B., et al. *A&A*, 379, 305, 2001.
5. Artymowicz, P., Clarke, C. J., Lubow, S. H., and Pringle, J. E. *ApJ*, 370, L35, 1991.
6. Böhm-Vitense, E., Carpenter, K., Robinson, R., Ake, T., and Brown, J. *ApJ*, 533, 969, 2000.
7. Barnbaum, C. and Morris, M. *BAAS*, 182, 876, 1993.
8. Batten, A. H., Fletcher, J. M., and McCarthy, D. G. *Publ. DAO*, 17, 1, 1989.
9. Belczyński, K., Mikołajewska, J., Munari, U., Ivison, R. J., and Friedjung, M. *PASP*, 146, 407, 2000.
10. Bolton, C. T. *ApJ*, 200, 269, 1975.
11. Bond, H. E. *ApJ*, 194, 95, 1974.
12. Bond, H. E., Pollacco, D. L., and Webbink, R. F. *AJ*, 125, 260, 2003.
13. Bondi, H. and Hoyle, F. *MNRAS*, 104, 273, 1944.
14. Carpenter, K. G. In Kondo, Y., Sisteró, R. F., and Polidan, R. S., editors, *IAU Symp. 151: Evolutionary Processes in Interacting Binary Stars*, page 51. Kluwer Academic Publishers: Dordrecht, 1992.
15. Catchpole, R. M., Robertson, B. S. C., and Warren, P. R. *MNRAS*, 181, 391, 1977.
16. Chiu, L.-T. G. and Kron, R. G. *Nature*, 299, 702, 1982.
17. Corradi, R. L. M., Ferrer, O. E., Schwarz, H. E., Brandi, E., and García, L. *A&A*, 348, 978, 1999.
18. Côté, P., Hanes, D. A., McLaughlin, D. E., et al. *ApJ*, 476, L15, 1997.
19. Côté, P., Pryor, C., McClure, R. D., Fletcher, J. M., and Hesser, J. E. *AJ*, 112, 574, 1996.
20. Cowling, T. G. *MNRAS*, 101, 367, 1941.
21. Crowther, P. A. *MNRAS*, 290, L59, 1997.
22. Cruzalebes, P., Lopez, B., Bester, M., Gendron, E., and Sams, B. *A&A*, 338, 132, 1998.
23. Dahn, C. C., Liebert, J., Kron, R. G., Spinrad, H., and Hintzen, P. M. *ApJ*, 216, 757, 1977.
24. de Kool, M. and Ritter, H. *A&A*, 267, 397, 1993.
25. Dumm, T. and Schild, H. *New Astr.*, 3, 137, 1998.
26. Edvardsson, B., Andersen, J., Gustafsson, B., et al. *A&A*, 275, 101, 1993.
27. Folini, D. and Walder, R. *Ap&SS*, 274, 189, 2000.
28. Frankowski, A. and Tylenda, R. *A&A*, 367, 513, 2001.
29. Gális, R., Hric, L., Friedjung, M., and Petrík, K. *A&A*, 348, 533, 1999.
30. Gatti, A. A., Drew, J. E., Lumsden, S., et al. In Habing, H. J. and Lamers, H. J. G. L. M., editors, *IAU Symp. 180: Planetary Nebulae*, page 105. Kluwer Academic Publishers: Dordrecht, 1997.
31. Goriely, S. and Mowlavi, N. *A&A*, 362, 599, 2000.
32. Green, P. J., Margon, B., Anderson, S. F., and MacConnell, D. J. *ApJ*, 400, 659, 1992.
33. Green, P. J., Margon, B., and MacConnell, D. J. *ApJ*, 380, L31, 1991.
34. Hachisu, I., Kato, M., and Nomoto, K. *ApJ*, 470, L97, 1996.

35. Han, Z., Eggleton, P. P., Podsiadlowski, P., and Tout, C. A. *MNRAS*, 277, 1443, 1995.
36. Haniff, C. A., Scholz, M., and Tuthill, P. G. *MNRAS*, 276, 640, 1995.
37. Heber, U., Bade, N., Jordan, S., and Voges, W. *A&A*, 267, L31, 1993.
38. Herbig, G. H. *Kleine Veröff. Remeis-Sternwarte*, 4, No 40, 164, 1965.
39. Herman, J. and Habing, H. J. *Phys. Rep.*, 124, 255, 1985.
40. Hinkle, K., Lebzelter, T., Joyce, R., and Fekel, F. *ApJ*, 123, 1002, 2002.
41. Hinkle, K. H., Lebzelter, T., and Scharlach, W. W. G. *AJ*, 114, 2686, 1997.
42. Howarth, I. D. and Heber, U. *PASP*, 102, 912, 1990.
43. Huang, S. S. *AJ*, 61, 49, 1956.
44. Hünsch, M., Schmitt, J. H. M. M., Schröder, K., and Zickgraf, F. *A&A*, 330, 225, 1998.
45. Hut, P. *A&A*, 99, 126, 1981.
46. Iben, Jr, I. and Livio, M. *PASP*, 105, 1373, 1993.
47. Iben, Jr, I. and Renzini, A. *ARA&A*, 21, 271, 1983.
48. Jeffries, R. D. and Smalley, B. *A&A*, 315, L19, 1996.
49. Jeffries, R. D. and Stevens, I. R. *MNRAS*, 279, 180, 1996.
50. Jorissen, A. In Sterken, C. and de Groot, M., editors, *The Impact of Long-Term Monitoring on Variable Star Research*, page 143. Kluwer Academic Publishers: Dordrecht, 1994.
51. Jorissen, A., Famaey, B., Dedecker, M., et al. *Rev. Mex. Astron. Astrophys.*, in press, 2003.
52. Jorissen, A. and Knapp, G. R. *A&A*, 129, 363, 1998.
53. Jorissen, A., Schmitt, J. H. M. M., Carquillat, J. M., Ginestet, N., and Bickert, K. F. *A&A*, 306, 467, 1996.
54. Jorissen, A., Van Eck, S., Mayor, M., and Udry, S. *A&A*, 332, 877, 1998.
55. Jura, M. and Kahane, C. *ApJ*, 521, 302, 1999.
56. Karakas, A. I., Tout, C. A., and Lattanzio, J. C. *MNRAS*, 316, 689, 2000.
57. Karovska, M., Hack, W., Raymond, J., and Guinan, E. *ApJ*, 482, L175, 1997.
58. Keenan, P. C. *PASP*, 105, 905, 1993.
59. Kippenhahn, R. and Meyer-Hofmeister, E. *A&A*, 54, 539, 1977.
60. Knapp, G. R., Dobrovolsky, S. I., Ivezić, Z., et al. *A&A*, 351, 97, 1999.
61. Knapp, G. R., Young, K., and Crosas, M. *A&A*, 346, 175, 1999.
62. Kondo, Y., McCluskey, G. E. J., and Gulden, S. L. In Boldt, E. and Kondo, Y., editors, *X-Ray Binaries (NASA SP-389)*, page 499. NASA: Washington D.C., 1976.
63. Lewis, B. M., Eder, J., and Terzian, Y. *AJ*, 94, 1025, 1987.
64. Liebert, J. *ApJ*, 204, L93, 1976.
65. Liebert, J., Dahn, C. C., Gresham, M., and Strittmatter, P. A. *ApJ*, 233, 226, 1979.
66. Liebert, J., Schmidt, G. D., Lesser, M., et al. *ApJ*, 421, 733, 1994.
67. Little, S. J., Little-Marenin, I. R., and Bauer, W. H. *AJ*, 94, 981, 1987.
68. Livio, M. In Shore, S. N., Livio, M., and van den Heuvel, E. P. J., editors, *Saas-Fee Advanced Course 22: Interacting Binaries*, page 135. Springer-Verlag: Berlin, 1994.
69. Livio, M. and Soker, N. *ApJ*, 552, 685, 2001.
70. Lodders, K. and Fegley, B. J. *ApJ*, 484, L71, 1997.
71. Lopez, B., Danchi, W. C., Bester, M., et al. *ApJ*, 488, 807, 1997.
72. Lowrance, P. J., Kirkpatrick, J. D., Reid, I. N., Cruz, K. L., and Liebert, J. *ApJ*, 584, L95, 2003.

73. Luck, R. E. and Bond, H. E. *ApJS*, 77, 515, 1991.
74. Margon, B., Anderson, S. F., Harris, H. C., et al. *AJ*, 124, 1651, 2002.
75. Mason, B. D. *AJ*, 112, 2260, 1996.
76. Mason, B. D., Martin, C., Hartkopf, W. I., et al. *AJ*, 117, 1890, 1999.
77. Mastrodemos, N. and Morris, M. *ApJ*, 497, 303, 1998.
78. Mastrodemos, N. and Morris, M. *ApJ*, 523, 357, 1999.
79. Maxted, P. F. L., Heber, U., Marsh, T. R., and North, R. C. *MNRAS*, 326, 1391, 2001.
80. Mayor, M., Meylan, G., Udry, S., et al. *AJ*, 114, 1087, 1997.
81. McClure, R. D. *Mem. Soc. Astron. Italiana*, 50, 15, 1979.
82. McClure, R. D. *PASP*, 96, 117, 1984.
83. McClure, R. D. *ApJ*, 280, L31, 1984.
84. McClure, R. D. *PASP*, 109, 256, 1997.
85. McClure, R. D. *PASP*, 109, 536, 1997.
86. McClure, R. D., Fletcher, J. M., and Nemec, J. M. *ApJ*, 238, L35, 1980.
87. McClure, R. D. and Woodsworth, A. W. *ApJ*, 352, 709, 1990.
88. Mermilliod, J.-C. In Milone, E. and Mermilliod, J.-C., editors, *The Origins, Evolution, and Destinies of Binary Stars in Clusters*, page 95. ASP: San Francisco, 1996.
89. Merrill, P. W. *ApJ*, 116, 21, 1952.
90. Merrill, P. W. *Science*, 115, 484, 1952.
91. Merrill, P. W. *PASP*, 68, 162, 1956.
92. Mikołajewska, J. In Mikołajewska, J., editor, *Physical Processes in Symbiotic Binaries and Related Systems*, page 3. Copernicus Foundation for Polish Astronomy: Warsaw, 1997.
93. Morris, M. *PASP*, 99, 1115, 1987.
94. Munari, U. and Whitelock, P. A. *MNRAS*, 237, 45P, 1989.
95. Mürset, U. and Schmid, H. M. *A&AS*, 137, 473, 1999.
96. North, P., Berthet, S., and Lanz, T. *A&A*, 281, 775, 1994.
97. North, P. and Duquennoy, A. *A&A*, 244, 335, 1991.
98. North, P. and Duquennoy, A. In Duquennoy, A. and Mayor, M., editors, *Binaries as Tracers of Stellar Formation*, page 202. Cambridge University Press: Cambridge, 1992.
99. North, P., Jorissen, A., and Mayor, M. In Wing, R. F., editor, *IAU Symp.177: The Carbon Star Phenomenon*, page 269. Kluwer Academic Publishers: Dordrecht, 2000.
100. Osterbart, R., Langer, N., and Weigelt, G. *A&A*, 325, 609, 1997.
101. Pastetter, L. and Ritter, H. *A&A*, 214, 186, 1989.
102. Pereira, C. B., Cunha, K., and Smith, V. V. In Corradi, R., Mikołajewska, J., and Mahoney, T. J., editors, *Symbiotic Stars Probing Stellar Evolution*, in press. ASP: San Francisco, 2003.
103. Pereira, C. B., Smith, V. V., and Cunha, K. *AJ*, 116, 1977, 1998.
104. Pols, O., Karakas, A. I., Lattanzio, J. C., and Tout, C. A. In Corradi, R., Mikołajewska, J., and Mahoney, T. J., editors, *Symbiotic Stars Probing Stellar Evolution*, in press. ASP: San Francisco, 2003.
105. Pourbaix, D., Platais, I., Detournay, S., et al. *A&A*, 399, 1167, 2003.
106. Preston, G. W. and Sneden, C. *AJ*, 120, 1014, 2000.
107. Preston, G. W. and Sneden, C. *AJ*, 122, 1545, 2001.
108. Prieur, J. L., Aristidi, E., Lopez, B., et al. *ApJS*, 139, 249, 2002.

109. Proust, D., Ochsenbein, F., and Pettersen, B. R. *A&AS*, 44, 179, 1981.
110. Ratnatunga, J. PhD thesis, Australian National Observatory, 1983.
111. Reid, I. N. *AJ*, 111, 2000, 1996.
112. Reimers, D. and Cassatella, A. *ApJ*, 297, 275, 1985.
113. Richichi, A., Baffa, C., Calamai, G., and Lisi, F. *AJ*, 112, 2786, 1996.
114. Richichi, A., Calamai, G., and Stecklum, B. *A&A*, 382, 178, 2002.
115. Ritter, H. In Wijers, R. A. M. J., Davies, M. B., and Tout, C. A., editors, *Evolutionary Processes in Binary Stars*, page 223. Kluwer Academic Publishers: Dordrecht, 1996.
116. Ritter, H. *MNRAS*, 309, 360, 1999.
117. Sahai, R. *A&A*, 253, L33, 1992.
118. Schmid, H. M. *A&A*, 284, 156, 1994.
119. Schmidt, M. and Mikołajewska, J. In Corradi, R., Mikołajewska, J., and Mahoney, T., editors, *Symbiotic Stars Probing Stellar Evolution*, in press. ASP: San Francisco, 2003.
120. Schuerman, D. W. *Ap&SS*, 19, 351, 1972.
121. Schwarz, H. E., Nyman, L.-A., Seaquist, E. R., and Ivison, R. J. *A&A*, 303, 833, 1995.
122. Shcherbakov, A. G. and Tuominen, I. *A&A*, 255, 215, 1992.
123. Shore, S. N. In Shore, S. N., Livio, M., and van den Heuvel, E., editors, *Saas-Fee Advanced Course 22: Interacting Binaries*, page 1. Springer-Verlag: Berlin, 1994.
124. Shu, F. H. and Lubow, S. H. *ARA&A*, 19, 277, 1981.
125. Smith, V. V., Cunha, K., Jorissen, A., and Boffin, H. M. J. *A&A*, 315, 179, 1996.
126. Smith, V. V., Cunha, K., Jorissen, A., and Boffin, H. M. J. *A&A*, 324, 97, 1997.
127. Sneden, C., Preston, G. W., and Cowan, J. J. *AJ*, in press (astro-ph/0304064), 2003.
128. Soberman, G. E., Phinney, E. S., and van den Heuvel, E. P. J. *A&A*, 327, 620, 1997.
129. Soker, N. *A&A*, 357, 557, 2000.
130. Theuns, T., Boffin, H. M. J., and Jorissen, A. *MNRAS*, 280, 1264, 1996.
131. Thevenin, F. and Jasniewicz, G. *A&A*, 320, 913, 1997.
132. Thoul, A., Jorissen, A., Goriely, S., et al. *A&A*, 383, 491, 2002.
133. Totten, E. J., Irwin, M. J., and Whitelock, P. A. *MNRAS*, 314, 630, 2000.
134. Tout, C. A. and Eggleton, P. P. *MNRAS*, 231, 823, 1988.
135. Udry, S., Jorissen, A., Mayor, M., and Van Eck, S. *A&AS*, 131, 25, 1998.
136. Van Eck, S. and Jorissen, A. *A&A*, 360, 196, 2000.
137. Van Eck, S. and Jorissen, A. *A&A*, 396, 599, 2002.
138. Van Eck, S., Jorissen, A., Udry, S., Mayor, M., and Pernier, B. *A&A*, 329, 971, 1998.
139. Van Leeuwen, F., Feast, M. W., Whitelock, P. A., and Yudin, B. *MNRAS*, 287, 955, 1997.
140. Vanture, A. D., Wallerstein, G., and Suntzeff, N. B. *ApJ*, 569, 984, 2002.
141. Vardya, M. S. *A&AS*, 73, 181, 1988.
142. Verbunt, F. and Phinney, E. S. *A&A*, 296, 709, 1995.
143. Warren, S. J., Irwin, M. J., Evans, D. W., et al. *MNRAS*, 261, 185, 1993.
144. Webbink, R. F. In Pringle, J. E. and Wade, R. A., editors, *Interacting Binary Stars*, page 39. Cambridge University Press: Cambridge, 1985.

145. Wegner, G. *MNRAS*, 171, 529, 1975.
146. Westerlund, B. E., Azzopardi, M., Breysacher, J., and Rebeirot, E. *A&A*, 303, 107, 1995.
147. Whitelock, P. A. *PASP*, 99, 573, 1987.
148. Whitelock, P. A. In Tateuki, M. and Sasselov, D., editors, *Pulsating Stars: Recent Developments in Theory and Observation*, page 31. Academic Press, 1998.
149. Whitelock, P. A. and Feast, M. *MNRAS*, 319, 759, 2000.
150. Whitelock, P. A. and Munari, U. *A&A*, 255, 171, 1992.
151. Willems, B. and Aerts, C. *A&A*, 384, 441, 2002.
152. Wood, P. R. *Publ. Astron. Soc. Australia*, 17, 18, 2000.
153. Wood, P. R. and Sebo, K. M. *MNRAS*, 282, 958, 1996.
154. Yamamura, I., Dominik, C., de Jong, T., Waters, L. B. F. M., and Molster, F. J. *A&A*, 363, 629, 2000.
155. Zinnecker, H. and Mathieu, R., editors. *IAU Symp. 200: The Formation of Binary Stars*. ASP: San Francisco, 2001.

10 Post-AGB Stars

Christoffel Waelkens[1] and Rens B.F.M. Waters[2]

[1] Instituut voor Sterrenkunde, Katholieke Universiteit Leuven
[2] Astronomical Institute 'Anton Pannekoek', University of Amsterdam

10.1 Introduction

The final phase of the evolution of low- and intermediate-mass stars is characterized by rapid changes in the structure and properties of the star and its circumstellar envelope (CSE); see Chapter 2. While it ascends the Asymptotic Giant Branch (AGB), its luminosity rises to values several thousand times that during core-hydrogen burning. Thermal pulses result in substantial variations in luminosity that lead to drastic changes in the chemical composition of the envelope. Pulsations and radiation pressure on dust cause a dense stellar wind of variable strength, obscuring the object from view at short wavelengths; see Chapters 4 and 6. At the tip of the AGB, mass-loss rates have risen to such high values that this essentially removes the hydrogen-rich stellar envelope on timescales of about 10^4 years, thus effectively terminating the AGB phase; see Chapter 7. The central star subsequently evolves to higher temperatures on a timescale that depends on mass (and may be as short as several 100 years), while layers close to its core become exposed. At the same time, the ejected CSE slowly expands and cools, revealing the central star. The interplay between the AGB CSE-remnant, the ensuing fast, tenuous stellar wind from the central star, and its hardening radiation field result in a fascinating spectacle of shaping of the ejected material and of processing of its composition: A planetary nebula (PN) is born.

Post-AGB stars (here after PAGBs) therefore form an interesting interface between physical and chemical processes that occur at low temperatures (e.g., dust nucleation) and those that require high energies, generated by shocks and/or ultraviolet radiation. In addition, studies of the properties of the central star give clues concerning its chemical evolution that are impossible to obtain while the star is still on the AGB. Post-AGB stars represent a crucial phase in the evolution of the matter ejected into the interstellar medium (ISM). For these reasons, PAGBs have been the focal point of an increasing number of detailed studies in the past 15 years. This chapter focuses on the observed properties of PAGBs and their relation to AGB stars and PNe.

10.2 Observational Definition of a Post-AGB Star

It is useful first to consider what objects can be classified as PAGBs. This definition is complicated by the fact that in many cases the central star is not directly observable, and only information about its CSE is available. When the star has just left the AGB, this CSE will naturally still closely resemble that of an AGB star, making a distinction very difficult. This is especially true for high-luminosity (-mass) PAGBs, since the rate at which these objects increase their surface temperature is very short compared to the expansion timescale of the CSE. It is therefore important to consider both the stellar properties and those of the circumstellar environment. Several studies have used stellar and circumstellar properties to find PAGBs.

A practical definition of a PAGB star using the central star properties is by means of spectral type and luminosity class, see, e.g., [125]. The location of PAGBs in the HR diagram as derived from evolutionary calculations suggests spectral types in the range K to B, with luminosity class I to III as a reasonable delimiter. The inclusion of M stars can result in confusion with red giants and early AGB stars. Stars with spectral type O are hot enough to ionize the AGB ejecta, and these are thus classified as central stars of PNe; they are discussed in the context of PNe. Luminosity classes below III are not in the proper range of luminosities expected for PAGBs. An exception to this point are the so-called "AGB-manqué" stars, which evolve to higher temperatures at much lower luminosities, and are often classified as sdO. These objects are not strictly speaking PAGBs, since they never ascended the AGB. A secondary classification criterion is the chemical composition of the photosphere, which usually shows substantial deviations from solar. The galactic latitude in combination with the spectral type and luminosity class can also be considered as a secondary classification criterion: Population I stars of luminosity class I to III tend to be found close to the galactic plane, and the presence of a supergiant at high galactic latitude is therefore not expected. However, low-mass stars are abundant at high galactic latitude, and when these stars evolve from AGB to PN they have a low surface gravity and are classified as supergiants, making them easily detectable against the background stellar population.

The circumstellar environment is the second obvious selection criterion to define a PAGB star. In fact, many known or candidate PAGB stars have been found on the basis of the infrared spectral energy distribution (SED), which is characterized by thermal emission from cool dust, with typical temperatures of 100 to 200 K. This dust was ejected by the star when it was still on the AGB, and has had no time yet to cool significantly. In many cases, the early expansion of the CSE results in a substantial reduction of the line-of-sight optical depth toward the central star on very short timescales, resulting in the detection of an optical counterpart with spectral type in the range given above; this strengthens the identification as PAGB star, and gives the spectral energy distribution its characteristic double-peaked appearance; Figure 10.1.

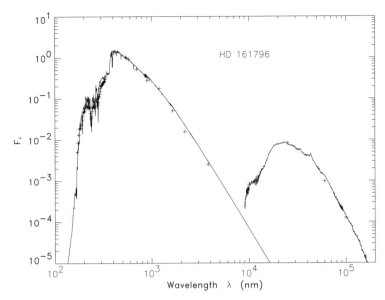

Fig. 10.1. Spectral energy distribution of the optically bright PAGB star HD161796. Plus signs represent data taken with the IUE satellite for the UV, ground-based telescopes for the optical and near-IR, and the IRAS satellite for the mid-IR. The continuous lines represent a Kurucz model (with E(B-V) = 0.25) to the stellar photosphere and the ISO SWS-LWS spectrum. This object is a typical low-mass PAGB star; higher-mass objects tend to be very obscured in the optical

Stars may cease to pulsate when their stellar envelope mass drops below a certain critical value, and this then marks the beginning of the PAGB phase. In that case the absence of variations in the infrared luminosity and/or in the OH maser strengths may be used to identify dust-enshrouded O-rich PAGB stars [58]. This event also leads to an increase of stellar temperature and the start of the PAGB evolution of the star. Such PAGB objects are therefore distinguishable from AGB stars by their lack of variation in both IR luminosity and OH maser line strength. An example of such an object is OH17.7–2.0 [22, 61].

There are two classes of objects that, on the basis of their photospheric abundances, pulsation properties, and/or circumstellar material have been classified as PAGB stars: the RV Tau stars and the R CrB stars. Both classes of objects are known to have some, or substantial amounts of, circumstellar dust, giving support to their evolved nature. The presence of R CrB stars in the Magellanic Clouds allowed the determination of luminosities [2], confirming the hypothesis that these stars are evolved low- or intermediate-mass stars. The extreme H-deficiency of the R CrB stars suggests that their evolution was not "standard," and some mechanism efficiently removed the entire

H-rich stellar envelope. This may be related to binary evolution, e.g., [73], or to a thermal pulse after the star left the AGB [28].

PAGBs may not always have substantial circumstellar matter. This depends on the timescale at which the central star increases its temperature, compared to the timescale at which the AGB CSE disperses into the ISM [147]. Low-luminosity stars may take up to 10^5 years to reach spectral type O, and by that time, no material near the star remains to produce a visible PN.

There are several other, different, types of objects that have properties similar to those of PAGBs. The most obvious group of stars is that of post-red-supergiants, which are on a blueward loop in the HR-diagram. As is the case for PAGBs, these objects are characterized by cool central stars surrounded by a large amount of circumstellar gas and dust. If one does not know their distance, they are easily mistaken for PAGBs: Estimates of the mass in the CSE, expected to be higher in post-red supergiants, depend on distance. In some cases, the expansion velocity of the detached CSE can be used to classify an object: Massive red supergiants tend to have wind expansion velocities in the range 20–40 $\mathrm{km\,s^{-1}}$, while AGB stars have expansion velocities in the range 10–20 $\mathrm{km\,s^{-1}}$. Examples of post-red supergiants are IRC+10 420 [76] and AFGL 4106 [42, 112]. A second class of objects that may lead to confusion with PAGBs is that of Herbig Ae/Be stars, young intermediate-mass pre-main-sequence stars; for reviews see, e.g., [189].

It is important to realize that severe selection effects probably affect the current census of PAGB candidates. A criterion based on the properties of the central star favors the detection of less-obscured, and hence probably low-mass, objects. A criterion based on the circumstellar properties, on the other hand, is likely to lead to a sample containing spectacular objects (e.g., the Red Rectangle) with huge amounts of circumstellar material, which may have resulted from mass transfer in a binary system.

10.3 Observed Properties of PAGB Stars: The Central Star

10.3.1 Spectral Types and Luminosity Classes

Bidelman [10] was the first to point out the occurrence of stars with the spectra of A- and F-type supergiants at high galactic latitudes, where massive supergiants are not expected to exist. His original sample, consisting of the objects HR 6144 (A7 Ib), 89 Herculis (F2 Ia), and HD 161796 (F3 Ib), has since been extended to some tens of objects, a selection of which is listed in Table 10.1. The view that (most of) these stars are low-mass objects caught in a late phase of evolution broke through in the 1980s because of new evidence: the detection of substantial infrared excesses, interpreted in terms of dusty CSE ejected during a previous AGB phase; the direct detection of

Table 10.1. Some optically bright PAGB stars

Object	HD	m_V	MK type	IR excess	remarks	Ref.
Red Rectangle	44179	8.8	B9 II-III	Y	pec. binary	
	46703	9.2	G I	Y	pec. binary	[101]
	52961	7.4	F I *	Y	pec. binary	[179]
	56126	8.2	F5 I	Y		[91]
U Mon	59693		F8 Ibe	Y	RV Tau	[78]
HR 4049	89353	5.5	pec	Y	pec. binary	[94]
RU Cen	105578	9.0	G2 Ie	Y	RV Tau	[78]
SX Cen	107439	9.6	G3 Ie	Y	RV Tau	[78]
HR 4912	112374	6.8	F3 Ia	N		[92]
SAO 173329		10.6	F2 I	Y	binary	[124]
SAO 239853		9.3	A9 Iab	Y		[124]
SAO 240664	114855	8.4	F5 I	Y		[123]
EN TrA	131356			Y	RV Tau	[124]
SAO 225457	133656	7.5	A1 I	Y		[123]
SAO 243756		9.8	B8 Ia	Y		[123]
HR 6144	148743	6.4	A7 Ib	N		[10]
	161796	7.0	F3 Ib	Y		[10]
89 Her	163506	5.4	F2 Ia	Y		[10]
AC Her	170756		F2 Ibpe	Y	RV Tau	[78]
	172324	8.2	A0 Iab			[155]
	172481	9.0	F2 Ia	Y		[135]
	187885	8.6	A I	Y		[161]
HR 7671	190390	6.4	F1 III	N		[102]
	213985	8.8	A2 I	Y		[181]
BD+39.4926		9.2	A7-F0 I	N	pec. binary	[181]

circumstellar structures by means of imaging; and the discovery that several of these objects are low-metallicity stars.

Quite probably, Table 10.1 contains at most a few massive supergiants, but it remains to be seen how every individual object fits into the PAGB scenario. In terms of current understanding, our favorite genuine (carbon rich) PAGB star is HD 56126, for which spectral type, luminosity class, IR excess, circumstellar morphology, and chemical composition all concur with theoretical expectations. However, for several objects, such as HR 4912, for which also the low overall metallicity is indicative of a low mass, no infrared excess has yet been recorded. Moreover, as will be discussed further below, the evolution of several objects in Table 10.1 has been affected by mass transfer in a binary system.

The selection of PAGB stars from optical spectra holds a bias toward the lowest-mass objects, which are not so much fainter intrinsically than more massive PAGB stars, but are longer-lived and less obscured because their

Table 10.2. Some IRAS-selected faint PAGB stars

IRAS number	m_V	MK type	Ref.
02229+6208	12.1	G8-K0 Ia	[66]
07340+1115	12.6	G5 Ia	[66]
10215−5916	8.7	G5 I	[69]
12175−5338	9.3	A9 Iab	[69]
17119−5926	var	B3e	[129]
18062+2410	11.5	B1 IIIpe	[128]
18095+2704	10	F3 Ib	[68]
19244+1115	11.2	F8 Ia	[69]
20004+2955	8.9	G7 Ia	[69]

CSEs are much thinner. Optical and near-infrared follow-up studies of IRAS objects with colors typical of transition objects between the AGB and PNe have yielded a large number of fainter candidates, which fulfill the criteria for more massive PAGB stars, see e.g., Table 10.2. One hundred and ten PAGB stars have been identified out of 187 IRAS objects by multiwavelength observations [43, 44, 104]; these PAGB stars cover the spectral range from B to M in a rather homogeneous way. A second search in the IRAS database yielded a list of 48 early-B-type PAGB candidates, some of which, such as Hen 1357 and IRAS 18062+2410 are currently undergoing rapid evolution [128, 129, 130]. Other studies of objects with IRAS colors typical for PNe have yielded several objects for which the absence of a radio continuum suggests that the central stars have not yet reached the temperatures needed to ionize the nebula [158, 160]. A preliminary analysis of the spectra of the faint central stars shows that they have late-B or early-A spectral types [159]. Finally, several more PAGB candidates with F- and G-supergiant spectra have been identified by Kwok [87]. A list of 220 PAGB objects, among which B-, F-, and G-spectral types dominate, has been compiled from data in the literature [153]. Not unexpectedly, mostly the hotter objects cluster in the region of the IRAS color-color diagram occupied by the PNe.

On the cool side of the PAGB evolutionary tracks, candidates have mostly been selected from displaying the OH maser emission typical for terminal AGB stars but having apparently ceased the pulsation that is thought to drive AGB mass loss. The thick dust CSEs associated with the maser phenomenon most often preclude an optical spectral classification of the central star, but the detection of CO absorption lines in the infrared spectra of at least a few objects [162] suggests that the coolest PAGB stars may be found in such samples.

A group of PAGB candidates for which infrared excesses have never been observed consists of high-galactic-latitude B-type stars, which again are not easily understood as massive stars. Though evidence is strong that some such objects may be indeed genuine Population I stars, e.g., PHL 346 [178], a significant subsample displays a metal deficiency pointing to a Population II nature, e.g., [107]. Since at the relevant effective temperatures PAGB tracks intersect with the main sequence, the spectral classification of these objects is not suitable for attesting their PAGB nature.

It is clear from the above that selection criteria still may heavily bias the present census of PAGB stars, so that a confrontation of number counts of stars of given spectral type with lifetimes on theoretical evolutionary tracks is difficult. In addition, the exact significance of some spectacular nebulae (with a poorly known central star) remains to be settled. Since several of these bright nebulae were among the first objects to be discussed in terms of PAGB evolution, they still often act as prototypes, but may in fact turn out to be rather peculiar objects. Among these we note AFGL 618, the center of which suffers some 100 mag of extinction [97], and AFGL 2688 (the Egg Nebula; also named CRL 2688), in which an F supergiant may reside [36]. The former may be a reliable example of a very massive progenitor, and so of a very rapidly evolving star, but the peculiar geometry and kinematics the latter displays do not appear to be exemplary for PAGB stars as a class, see Figure 10.6.

10.3.2 Luminosities

Accurate parallaxes are not available for any of the known PAGB stars. Some probable PAGB stars have been found among the UV-bright stars in globular clusters. In a study of the location of the UV-bright stars in ω Cen, the two brightest such objects, V1 and ROA 24, were shown to be PAGB stars with luminosities close to $\log(L/L_\odot) = 3.25$, corresponding to an absolute magnitude $M_V = -3.4$ [51]. The two Post-AGB A-F Stars in the globular cluster NGC 5986 also have $M_V = -3.4$ [18], confirming an earlier suggestion [17] that PAGB stars may even serve as useful extragalactic candles. In fact, a preliminary test based on the observations of seventeen objects with $0 \leq B-V \leq 0.5$ in the halo of M31 yielded the right distance modulus for the Andromeda galaxy [17].

More luminous PAGB stars must certainly exist, since [163] found thermally pulsing AGB stars in the LMC up to $M_{bol} = -7.5$, which is close to the limit expected from theory. The short evolutionary time scales for such objects once they leave the AGB renders their detection fairly unlikely, however.

The literature on galactic PAGB stars is somewhat polluted by luminosity estimates that are based on spectroscopic criteria, such as the equivalent width of the OI-line at 778 nm. It is obvious, however, that the low-mass nature of PAGB stars leads to an overestimate of the luminosity when criteria calibrated for Population I supergiants are applied. It appears convenient to adopt a value of -3.4 for the absolute visual magnitude of the lowest-mass PAGB stars, and a somewhat smaller value for objects for which a higher-mass nature is likely.

Finally, it should be noted that [195] were able to estimate the luminosity of the PAGB candidate HD172481 by applying the period-luminosity relation for Mira variables to the companion, which is a Mira with a period of 312 days, and derived a luminosity some 10^4 $L_\odot$.

10.3.3 Photospheric Abundances

The study of the photospheric abundances of PAGB candidates provides a natural test of the nature of these objects, since, as compared to massive supergiants, they should reflect their old age in a subsolar average metallicity and their PAGB nature in dredge-up-induced anomalies. It deserves to be noted that PAGB stars present a precious advantage over AGB stars concerning the detailed study of abundances, since their spectral types tend to cluster in a range where model atmosphere analysis in the optical range is the most reliable and spectra are not crowded; moreover, their effective temperatures are such that the energy distribution peaks in the spectral region that is most easily observable.

Abundance analyses of PAGB stars have initially been focused on the brightest high-latitude supergiants, and when it was realized that this selection induced a bias against the higher-mass objects, several studies on more embedded F-type PAGB supergiants were carried out. An independent path has been the analysis of the spectra of high-latitude B-type stars. During the study of high-latitude supergiants it was realized that a class emerged of extreme objects, which will be discussed in Section 10.5.

The hypothesis that the different selection criteria for PAGB stars indeed lead to a sample of evolved low-mass stars has received confirmation by the fact that subsolar iron abundances are consistently found for such stars. For the brightest high-latitude supergiants, iron deficiencies range from moderately low ([Fe/H] $\approx$ –0.3) for Bidelman's original objects HR 6144, HD 161796, and 89 Her – though [172] concludes from a chemical analysis that HR 6144 is a Population I star – to more than an order of magnitude lower for objects such as HD 56126, HR 4912, and HR 7671. A series of papers by the Belfast group also led to the identification of several high-latitude B-type stars with severe iron deficiencies; see for example [138].

The situation, however, becomes much more confusing when it comes to compare the results with the (uncertain) expectations from the third dredge-up that PAGB stars must have undergone, i.e., a global enhancement with respect to Fe of CNO abundances, a high C/O ratio, and a substantial enhancement of s-process elements . It turns out that so far, among PAGB candidates in the field, only the so-called 21 µm sources, which display a distinctive solid-state feature at this wavelength (Section 10.4), conform to theoretical expectation. In fact, essentially only for these field sources have the predicted s-process enhancements been detected (Figure 10.2), some of them being among the most s-process-enriched objects known till now. It has to be underlined, however, that the globular cluster ω Centauri contains

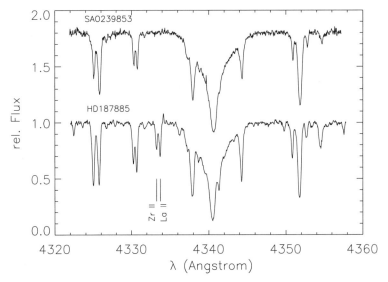

Fig. 10.2. Section of the optical spectrum of two evolved supergiants with similar spectral energy distributions and photospheric properties. While the O-rich object SAO 239853 does not show evidence of s-process enhancements, the C-rich object HD 187885 displays strong such enhancements. The latter object also is a 21 μm source. From [166]

three UV-bright stars, which display a large Fe-deficiency and mild s-process enhancements [54]; these objects also show a significant enhancement of CNO elements with respect to Fe, but the lower enhancement for carbon than for nitrogen and oxygen does require a non standard dredge-up scenario.

In Table 10.3 we summarize the abundance results for different known 21 μm sources, as they were homogenized with respect to atomic line lists and atmospheric modeling by [166]. For the object IRAS04296+3429, [83] list [Fe/H] = 0.2. With this value and the others listed in the table, it turns out that all 21 μm sources have C/O values in excess of one, i.e., they descend from carbon stars.

The s-process enhancements listed in Table 10.3 are averages over the lower- and larger-mass s-nuclei. When a distinction is made between both, a clear relation emerges between their relative values and the total s-process enhancement, in the sense that more enhanced objects tend to be mostly so in the heavier elements ([166]). A higher [hs/ls]-index implies a stronger integrated neutron irradiation, and thus the dredge-up efficiency is strongly linked with the neutron production in the intershell region, and thus the proton engulfment; see also Chapter 2.

The only other putative PAGB star in the field for which an s-process enhancement has been found is HR 7671 [102], with [s/Fe] = 0.6. This object, which does not display an infrared excess, is also peculiar in the sense that its

Table 10.3. Photospheric abundances of some 21 μm PAGB stars

Object	[Fe/H]	[C/Fe]	[N/Fe]	[O/Fe]	[s/Fe]
IRAS 04296+3429	-0.6	0.8	0.4		1.6
IRAS 05341+0852	-0.8	1.0	0.7	0.6	2.1
HD 56126	-1.0	1.1	0.9	0.8	1.6
HD 187885	-0.6	1.0	1.0	0.7	1.2
IRAS 22223+4327	-0.3	0.3	0.2	-0.1	1.0
IRAS 23304+6147	-0.8	0.9	0.5	0.2	1.6

total CNO abundances are not enhanced, and that carbon is even depleted, besides showing a rather high lithium abundance of [Li/H] = 1.4. On the other hand, O-rich objects such as HD 161796 and SAO 239853, which satisfy most criteria to be considered as genuine PAGB stars, tend to be solar or even depleted as far as their [s/Fe] ratio is concerned. Apparently, these stars did not experience enough dredge-up to become C-rich and s-process enhanced. AGB evolution models predict that only stars in a certain mass range are likely to become C-rich; low-mass stars tend not to, and high-mass stars may undergo hot bottom burning, preventing them from becoming C-rich PAGB stars. It is then likely that objects such as HD 161796 and HR 7671 are of lower initial mass than the 21 μm sources. It should be noted also that the limit in mass at which the transition to a C star can occur depends on initial metallicity, low-mass metal-poor stars being more easily turned into C-stars.

In general, for the non-21 μm sources, no clear pattern emerges concerning the CNO abundances [100, 164], and it is fair to say that the chemical composition of these objects is not properly understood. The typical CNO enhancement is small; in some objects nitrogen is definitely enhanced, suggestive of mixing of CNO-cycle products; however, the expected accompanying carbon deficiency is rarely observed.

Also, the chemical composition of the high-latitude B-type PAGB candidates cannot easily be explained in terms of PAGB evolution, since for these objects carbon is severely underabundant [33]. Around one of these objects (LS IV –12°111), a nebula has been observed, the composition of which is similar to that of the halo PN BD+33°2642 [116]. It does not seem likely that these metal-poor high-latitude B-type stars evolve from the high-latitude F–G supergiants discussed here. A possible progenitor for such objects may be HD 107369, which was detected as a low-gravity metal-deficient A-type halo star by [21] and studied in detail by [164], who derived [Fe/H] = −1.1 and found it to be carbon-deficient. Also, this object lacks a detectable IR excess. It is then not certain that the high-latitude B stars have undergone standard AGB evolution between the horizontal branch and their current stage. Considering also the similarity of their chemical composition with that of some halo PNe, the term "proto-PN" may be more appropriate for them than "PAGB star".

10.3.4 Pulsations of Post-AGB Stars

Post-AGB evolutionary tracks cross the upper Cepheid instability strip; hence pulsations are expected to occur in PAGB objects and may contain useful diagnostics on their internal structure. A rather vast literature on the F-G high-latitude supergiants listed in Table 10.1 shows that most are photometric and spectroscopic variables on time scales on the order of a month.

The main overall characteristic of the variability may be its irregularity: Amplitudes vary between 0.1 and 0.3 mag, and cycle-to-cycle variations in the "period" occur. It appears that the analysis of long data strings, covering several subsequent years, may finally yield rather well defined periods. Most of the photometric variations of 89 Her during the preceding 30 years of 89 Her have been fitted with a period of 62.65 d [39], a period of 63.3 d from another analysis of data for 89 Her, and of 41 d for HD 161796 [40], of 36.8 d for HD 56126 [9] and a period of 28 d using Geneva photometry of HR 7671 [176]. Linear nonadiabatic analyses of pulsations of luminous low-mass stars are able to account for the rather irregular behavior that is observed [1, 9], thus providing additional evidence for the low-mass nature of these objects.

A distinct class of more classical pulsating variables, which appears to be directly relevant to the issue of PAGB stars, is constituted by the RV Tauri stars [183, 98]. These stars are defined as luminous (I-II) mid-F to K stars, which show alternating deep and shallow minima in their light curves, and have periods (defined with respect to successive deep minima) between 50 and 150 days. Since the pioneering theoretical work by Christy [27], the alternance of deep and shallow minima is explained as a nonlinear phenomenon, be it a resonance between the fundamental radial pulsation period and an overtone [114] or chaotic behavior [23]. After Gehrz [46] and Lloyd-Evans [38] pointed out that many RV Tauri stars present infrared excesses, Jura [78] proposed to consider them all as PAGB objects. In our Table 10.1, four classical RV Tauri stars (SX Cen, RU Cen, AC Her, U Mon) are listed, and also HD 52961 [179] and EN TrA [38] present the characteristic RV Tauri star variability. We will consider these stars further in Section 5, since it appears that binarity is an important ingredient for their understanding in the context of PAGB stars.

10.3.5 Present-Day Mass Loss

By definition, PAGB stars are objects for which the "superwind" phase has ceased, and for which the central star is not yet hot enough to initiate the fast ionizing wind which will further shape the PN. Nevertheless, some mass loss from PAGB stars appears to occur, since Hα emission is observed for all (pulsating) PAGB stars. Clearly, as shown by [156], PAGB mass loss can shorten the transition time scale significantly.

Observations of the morphologies of CSEs are obviously of high importance in tracing the striking evolution from the often fairly spherically symmetric outflows observed for OH/IR stars, see Chapter 7, to the varied aspects of PNe, which are often elliptical, bipolar, or point-symmetric. Observational results on the morphology of the CSEs of PAGB stars are discussed in Section 10.4.

Following [89], the shaping of PNe requires the interaction of the fast wind driven by radiation from the hot central star and the more slowly expanding CSE ejected on the AGB. The observational data discussed in Section 10.4 contain plenty of evidence of shock-induced enhancements in PAGB CSEs, even suggesting that radiation-driven winds are not the only cause of the faster winds in these objects. Based on high-resolution ^{13}CO maps of "Minkowski's Footprint" (M1–92), which has a central star with $T_{\text{eff}} \approx 20000\,\text{K}$, [24] suggest that it may derive its energy to drive the wind from reaccretion of part of the CSE. For He 3–1475, an object that is still an OH maser, [14] detect outflow velocities as high as $500\,\text{km\,s}^{-1}$. A fairly fast PAGB wind, detected through molecular lines and that strongly affects the circumstellar morphology, is also present in objects such as AFGL 618, AFGL 2688, and OH231.8+4.2 [80, 199]. For some of these objects, there is evidence that the outflows may originate from accretion energy in a binary system, but it is not unlikely that also single PAGB stars may have a substantial wind.

10.4 Observed Properties of Post-AGB Stars: The Circumstellar Envelope

In most known PAGBs, the detached CSE dominates the spectral energy distribution at infrared and submillimeter wavelengths. It contains the mass most recently ejected by the star when it was on the tip of the AGB. It can therefore give important information about the physical and chemical conditions in the AGB wind in the short period before it terminated the high mass loss. Comparison of the composition and geometry of PAGB shells with those of the winds of high mass loss AGB stars can lead to a better understanding about the mechanisms that are responsible for the "superwind" phase that effectively ends the AGB.

10.4.1 Thermal Emission from Dust

Almost all information about the composition of PAGB dust CSEs is obtained from infrared spectroscopy. This is because the emission of most PAGB dust CSEs peaks in this wavelength range, and the most important vibrational resonances are in the mid-infrared. These resonances are important diagnostics of the chemical composition of the dust as well as the size and shape of the dust particles.

Depending on the C/O ratio in the atmosphere of the star, either an O-based or a C-based chemistry is present; see Chapters 4 and 5. The dust that condenses in the stellar wind reflects this chemistry. Examples of dust CSEs with O-rich and C-rich chemistry are given in Figures 10.3 and 10.4. In O-rich stars, the dust is mainly composed of amorphous silicates, with strong, broad resonances near 10 and 18 µm. In addition, crystalline silicates with bands near 23.5, 27.5, and 33.5 µm have been found with the Infrared Space Observatory (ISO; e.g., [112, 187]). In exceptional cases the crystalline dust can be a major fraction of the total dust mass. Kuiper Airborne Observatory (KAO) spectra of the Frosty Leo revealed the presence of strong crystalline H_2O ice bands near 43 and 60 µm [41]. Crystalline H_2O ice has now been found in several O-rich PAGB stars and is also observed in OH/IR stars.

In a C-rich PAGB star, the dust consists of amorphous carbon, SiC (which has a resonance near 11.2 µm), and a strong, broad peak near 26 µm, often referred to as the "30 µm feature" [113]. This band is also detected in C-stars with a high mass loss rate and has been attributed to MgS, e.g., [50, 118]. It turns out that the 30 µm band is actually at 26 µm in AGB stars, and shifts to about 35 µm in the PN NGC 7027 [64]. Such a large shift is difficult to explain with grain size and/or shape effects, and may point to additional components (contributing either as a separate species or as a coating).

PAGB stars have prominent emission in the famous UIR bands at 3.3, 6.2, 7.7, 8.6, and 11.3 µm, usually attributed to polycyclic aromatic hydrocarbons (PAHs) [3] . PAHs are not found in C-rich AGB stars, and appear only when the star begins to increase its temperature to values above about 5000–6000 K. This suggests a substantial modification of the dust composition after it has been ejected by the AGB star. In the peculiar PAGB star HR 4049, prominent bands at 3.43 and 3.52 µm are present [45, 111]. These bands have recently been identified as due to nanodiamonds [57].

Some cool C-rich PAGB stars show a strong band near 21 µm [91], which is actually near 20.3 µm [174], and is seen only in cool PAGB stars. This band has been identified with TiC [175]. In order to form TiC, very high densities are required, which leads to mass-loss-rate estimates in excess of $10^{-3} M_\odot$ yr^{-1}. The mass observed in PAGB shells of stars with the TiC band is 0.1–0.5 $M_\odot$ [108], suggesting a lifetime of this high-mass-loss phase of only a few hundred years. The short lifetime of this phase may explain why no AGB star with TiC 20.3 µm band emission is observed.

The chemical evolution of AGB stars from C/O < 1 to > 1 predicts that the dust surrounding PAGB stars should either be O-rich or C-rich. The short evolutionary time-scales involved imply that the central star has a chemical composition similar to that of the dust. However, the actual situation is more complex, and we discuss some examples below.

In PNe with a cool [WC] central star, there is strong emission from PAHs in the 3–15 µm wavelength range, e.g., [32, 185], in line with the C-rich nature of the central star. However, the bulk of the cool dust in these stars turns out to be O-rich [31, 185], pointing to a very rapid chemical evolution from O-rich and H-rich to C-rich and H-poor. A thermal pulse at the very end of

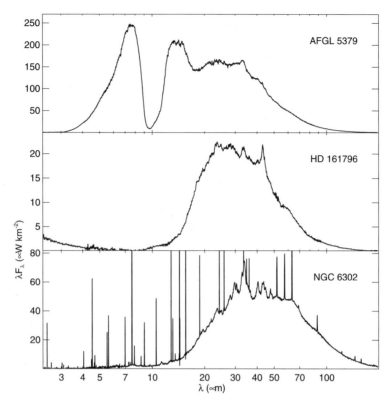

Fig. 10.3. Spectral evolution of O-rich AGB and PAGB stars. Top: heavily enshrouded OH/IR star, with deep absorptions at 9.7 and 18 μm due to amorphous silicates. At long wavelengths, narrow emission bands from crystalline silicates are seen. Middle panel: young PAGB star, showing thermal emission from the CSE. Note the prominent crystalline H_2O ice bands at 43 and 60 μm. At short wavelengths the Rayleigh–Jeans tail of the energy distribution of the central star is visible. Bottom panel: PN spectrum, dominated by strong emission from fine-structure transitions of ions in the ionized part of the AGB CSE-remnant. Also visible are crystalline silicate bands at 24 and 34 μm

the AGB, or when the star has just left the AGB and the stellar envelope mass is low, may be causing this rapid chemical evolution [63].

While in cool [WC] stars the different dust components are believed to be radially separated and reflect the chemical evolution of the central star, there are several objects with massive circumstellar dust disks whose dust composition is believed to be different from the dust in the polar regions. A well-studied example is the Red Rectangle [30], which hosts a wide binary system [169]. The X-shaped nebula is C-rich, as indicated by the appearance of PAHs in the nebula, e.g., [55], but the ISO spectrum longwards of 20 μm is dominated by crystalline silicates [186]. It is suggested that this O-rich

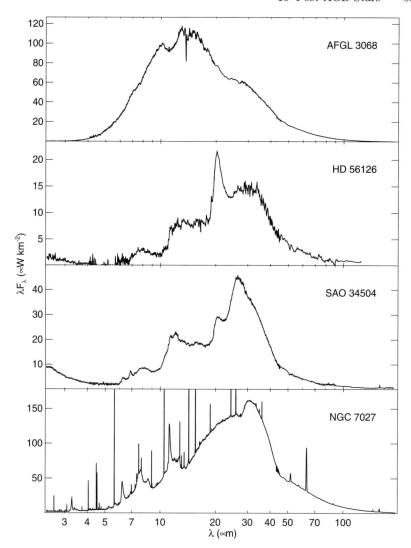

Fig. 10.4. Spectral evolution of C-rich AGB and PAGB stars. Top: heavily enshrouded so-called infared carbon star, with SiC absorption near 11.3 μm and a prominent emission band near 26 μm, attributed to MgS. Middle panels: young PAGB stars, showing thermal emission from the detached envelope. Note the band at 20.3 μm (often referred to as the 21 μm band), which is very strong for HD 56126, and is attributed to TiC. At short wavelengths the Rayleigh–Jeans tail of the energy distribution of the central star is visible. Bottom panel: PN spectrum, dominated by strong emission from fine-structure transitions of ions in the ionized part of the AGB CSE-remnant. The 30 μm band has shifted to about 35 μm

dust is located in the dense equatorial disk, which was created when the more evolved star in the system was still O-rich. Due to interaction with the companion, this mass could not leave the system and was stored in a circumbinary disk. Subsequently, the central star evolved to a C-rich chemistry and produced the X-shaped, C-rich nebula. We will come back to this point when we discuss binary PAGB stars.

10.4.2 The Molecular Circumstellar Envelope

PAGB stars are characterized by a rich variety of molecular species, almost exclusively found in the detached AGB CSE-remnant. The central star is too warm to contain molecules, except perhaps in the coolest G-type central stars. In the case of binaries, the companion may be evolved and have a molecule-rich stellar atmosphere (an example is OH231.8+4.2). Most molecules have been detected in emission at (sub)millimeter and radio wavelengths, but also at infrared and optical/UV wavelengths; in the latter case, these detections usually are in absorption against the stellar photospheric continuum radiation, e.g., [7, 65]. The molecules found reflect the chemistry of the AGB CSE-remnant, and, depending on the C/O ratio, are dominated by an O-rich or a C-rich chemistry.

Compared to their AGB progenitors, PAGB star CSEs show roughly similar molecular properties, however, with some notable trends that mark the rapid increase in temperature of the central star and changes in the nature of its stellar wind; see also Chapter 7. The increasing intensity of the UV radiation of the central star will create a photodissociation region in the innermost regions of the envelope, where previously chemical equilibrium prevailed. This can be inferred from the increased abundance from AGB to PN of photodissociation and ion chemistry products (such as CN, HCO^+), as well as from the detection of typical PDR finestructure lines [26]. The molecular emission of different isotopes provides important clues as to the nature of the progenitors of the PAGB stars, but beam size effects as well as optical depth effects often hamper a reliable derivation of the $^{12}C/^{13}C$ ratio from CO. Limits on isotopic ratios in PAGB stars do not conflict with values derived for AGB stars or planetary nebulae; see, e.g., [6].

O-rich AGB stars have a relatively sober chemistry, with CO, H_2O, CO_2, OH, SO, SO_2, and SiO as some of the most abundant molecules. Maser emission from SiO, H_2O, and OH characterizes AGB stars of a range of mass-loss rates. Interferometric maps of this maser emission shows that the H_2O and SiO masers originate in regions close to the star, while the OH maser is located in the outer photodissociation region, where it is pumped by mid-IR radiation from circumstellar dust. Most maser emission disappears quickly after the AGB, typically in less than 10^3 years. Only few PPN with SiO maser emission are found [119]. The OH maser, however, seems to persist for a longer time scale, presumably because it originates in the outer regions of the CSE. For the CSE of the OH maser source Roberts 22, a dynamical time scale of ≤ 440 yr is found [145]. An interesting case is IRAS07027–7925, which has an OH maser, while the central star has already evolved to a cool [WC] type [200]; the transition between O-rich AGB star and C-rich, H-poor PAGB star may have taken about 500 years. The bipolar nebula IRAS19312+1950 shows maser emission from H_2O and SiO [115], while a group of bipolar PAGB stars studied by [199] shows high-velocity ($\approx 50\,\mathrm{km\,s^{-1}}$) OH maser emission. It has been suggested [199] that the bipolar, disk-like nature of these sources may enhance the PAGB lifetime to $\approx 10^4$ yr by reaccretion of some previously ejected material.

Carbon stars, and their descendants the C-rich PAGB stars, are characterized by a very rich chemistry; some of the most abundant molecules are CO, C_2H_2, HCN, CS, C_2, MgS, and SiC. The best-studied C-rich PAGB star in this respect is AFGL 2688, in which to date about 20 molecular species have been found [122]. As mentioned above, many PAGB stars show emission in the UIR bands, often identified with PAHs. These large molecules are excited by the increasingly intense optical and near-UV radiation field of the central star. It may very well be that PAHs play a key role in the formation of C-rich dust; so far, however, they have not been detected in C-rich AGB stars, and it cannot be excluded that they are produced in the early phases of the evolution of the gas/dust CSE after the AGB.

The most abundant molecule H_2, while not detected in AGB stars, is prominently seen in the near-IR spectra of many PAGB stars through emission in the S(1), $v=1\rightarrow 0$, line at $2.12\,\mu m$. This line is indicative of the presence of shocked H_2, presumably resulting from the interaction between a faster PAGB wind and the AGB CSE-remnant. The presence of this H_2 emission is closely correlated with the bipolar nature of the (planetary) nebula, e.g., [193, 201]; the H_2 emission is concentrated in the two rings that are usually found in bipolar PNe [81]. However, some bipolar PAGB nebulae with M-type central stars show no H_2 emission [192]. This suggests that the onset of H_2 emission is not related to the formation of the bipolar structure but occurs later in the evolution of the PAGB star.

10.4.3 Imaging

The geometry of PAGB CSEs has been determined from imaging at optical, infrared, and radio wavelengths. At short (optical, near-IR) wavelengths, light from the central star is scattered by circumstellar dust, while at long (IR, millimeter) wavelengths the thermal emission from the dust particles heated by the central star dominates. In addition, molecular emission from, e.g., H_2 and CO has been imaged at high angular resolution. It has become evident from these images that matter is often distributed in a highly axisymmetric manner. It is likely that this axisymmetry is the cause of the bipolar shape of virtually all PNe. High-resolution imaging has been instrumental in revealing the complex structure of PAGB CSEs. The Hubble Space Telescope (HST)

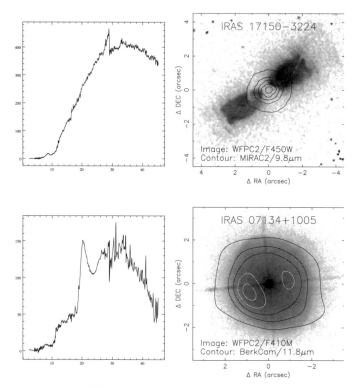

Fig. 10.5. left panels: ISO-SWS spectra of two PAGB stars with moderate (lower panels, IRAS17150) and extreme (top panels, HD 56126 = IRAS07134) polar to equatorial density contrast. The images show HST/WFPC2 observations of both stars with mid-IR (wavelength about 10 µm) contours superimposed. The images are taken from [157]. Note that in the O-rich source IRAS17150 the 10 µm amorphous silicate band is in absorption, indicating a substantial dust optical depth, in agreement with the optical images. The C-rich source HD 56126 shows a plateau starting near 12 µm; all dust bands are seen in emission, consistent with its much less extreme polar/equatorial density contrast

and other high-resolution telescope images have provided evidence for disks, jets, lobes, rings, and halos.

Meixner and collaborators [108, 109] have imaged the thermal dust emission of PAGB CSEs at 10 and 20 µm using mid-IR array cameras. A detailed analysis of C-rich PAGB stars with the 21 µm dust feature shows that these objects have an axisymmetric dust distribution with density contrast between pole and equator between 18 and 90; it is likely that even more extreme ratios occur in the highly bipolar objects.

At optical wavelengths, ground-based imaging of objects such as the Red Rectangle [30, 146], the Egg Nebula (AFGL2688) [96, 117], and the Frosty Leo [136, 137] showed the presence of a dark absorbing dust lane and bright polar

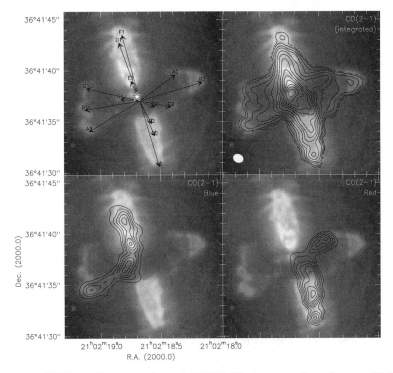

Fig. 10.6. CO(J=2–1) observations of AFGL2688 superposed on the near-IR HST image from [141]. The HST image shows reflected light from the AGB CSE-remnant as well as shocked molecular gas. The upper right and two lower panels show, respectively, the integrated CO(2-1) intensity, and the blue-shifted (–80 to –60 km s^{-1}) and red-shifted (–22 to –2 km s^{-1}) CO emission. The arrows in the upper left panel indicate multiple bipolar outflows. From [34]

lobes. This is strongly suggestive of a dense, dusty equatorial disk seen edge-on, with optically thin scattering lobes. The spectacular images provided by the HST have revealed an incredible richness of structure in the ejecta surrounding PAGB stars (e.g., [139] and references therein, [67, 88]). Ueta et al. [157] divide PAGB nebulae into two categories: star-dominated emission with a faint scattering nebula, and dust-scattering dominated, with a faint or completely obscured central star. These two classes of objects probably reflect a difference in density contrast between pole and equator; Figure 10.5. These differences may be caused by a difference in progenitor mass, the more massive progenitors being more strongly bipolar and concentrated toward the galactic plane. Several objects show a complicated, multipolar geometry, e.g., [34, 140, 142, 143], which suggests that several epochs of collimated outflow must have occurred with a changing plane of symmetry; Figure 10.6.

A remarkable pattern of concentric rings has been detected, e.g., [90, 144] . These rings are also seen in the AGB star IRC+10216 (see, e.g., [35, 106]), where they extend up to several tens of arcsec from the star, and in several PNe, such as NGC 7027 and NGC 6543 [8]. These observations show that the rings are created while the star is on the AGB and remain visible throughout the PN phase. Their roughly round morphology indicates that the large-scale structure of the wind during this episode is roughly spherical. The rings are due to light scattered from dust particles in the circumstellar CSE. In IRC+10216 the density contrast between the ring and interring material is of order a factor 3 [106]. The latter authors find evidence for an increase of the ring spacing with distance from the star. If the rings are due to density fluctuations in the stellar outflow, time scales range between 150 and 1000 years. Such time scales are incompatible with known time scales of variability in AGB stars: They are much longer than typical stellar pulsation time scales, but much shorter than thermal pulse time scales.

It is unclear at present what mechanism could be responsible for the creation of the rings. A possibility [151] is that the rings are due to instabilities in the gas–dust interaction in the accelerating outflow of the AGB star. From calculations of the geometry of AGB winds in detached binaries [105], a hydrodynamical instability triggered by the orbital motion of the companion was found to result in a spiral density pattern that may reproduce the observed structures. In this model, objects with rings should be (very wide) binaries. Also, a magnetic wind model has been proposed to explain the rings [152].

The large-scale geometry of the winds of AGB stars is not yet well established; see Chapter 7. Speckle observations and interferometry at near-IR and mid-IR wavelengths suggest that AGB winds are inhomogeneous, non-spherical, and of course time-variable. However, the pronounced axisymmetry observed in many PAGB stars does not seem evident in most AGB stars. For instance, the detached shells seen around C-rich AGB stars are fairly round (both in dust and in CO gas; e.g., [75, 120, 121]). The question arises what causes the winds of AGB stars to develop this axisymmetry, on time scales that are short compared to their AGB lifetime. Several mechanisms have been proposed in the literature. Even a small amount of rotation influences the structure of the loosely bound atmosphere of an AGB star [133], and can cause the dust–forming layers to move inward in the equatorial regions, resulting in a denser outflow near the equator (in dust–driven winds the wind density is closely related to the locus of dust formation). The rotation may result in a confinement of the outflow to the equatorial regions [74]. Soker proposed that the interaction with a companion, which may be substellar in mass, can produce a highly flattened geometry of the wind. Indeed, there are a number of known binaries in wide systems known among PAGB stars with a highly axisymmetric distribution of gas and dust, presumably in a circumbinary disk (e.g., the Red Rectangle and HR 4049). This underpins the importance of a companion on the final outcome of AGB evolution, both in terms of the evolution of the AGB star and of its CSE.

Table 10.4. Orbital periods, eccentricities, and mass functions for binary PAGB stars

Object	Period [days]	e	f(M)	Ref.
SAO 173329	116	0.0	0.026	[170]
HD 213985	259	0.0	0.97	[169]
89 Her	288	0.19	0.00083	[190]
HD 44179	318	0.38	0.049	[169]
HR 4049	429	0.31	0.143	[169]
HD 46703	600	0.34	0.28	Hrivnak, priv. comm.
AC Her	1194	0.12	0.25	[170]
HD 52961	1305	0.3	0.46	[167]
EN TrA	1534	0.3	0.72	[167]
U Mon	2597	0.43	0.92	[131]

10.5 Binary Post-AGB Stars

For a general introduction on binarity in AGB stars the reader is referred to the preceding chapter by A. Jorissen.

10.5.1 Basic Data

As stated earlier, the bipolar appearance of many PAGB CSEs and PNe can probably be properly explained by the evolution of an equatorial enhancement of the superwind of late AGB stars. However, in several cases it could have its origin in a binary nature of the central object [105]. While the binary hypothesis may be plausible for transition objects such as AFGL 2688, the large extinction of the central parts of this nebula so far have prevented an unambiguous detection of a central binary.

On the other hand, once spectroscopic programs were initiated to analyze the chemical composition of optically bright PAGB candidates, it was rapidly realized that a substantial fraction of these candidates presented radial-velocity variations indicative of orbital motion. The orbital periods, excentricities, and mass functions, which were derived for PAGB candidates, are listed in Table 10.4, where the binaries are ranked according to increasing orbital period. To this table, also BD+39°4926 should be added, but the real orbital period is probably shorter than the value of 775 days claimed by [84] on the basis of a limited number of observations (Van Winckel, priv. comm.).

Photometric variability is associated with the orbital motion of some of these stars, in different ways. The shortest-period object in the sample, SAO 173329, displays ellipsoidal variations [15]. For HR 4049 [177] and HD 213985 [182] variable circumstellar extinction with the orbital period is observed, and was interpreted in terms of a model with an inclined disk with a maximum obscuration during inferior conjunction of the primary. On the other hand, the brightness of HD 52961 is at its maximum at inferior conjunction, suggesting that in this wider binary the dust, which causes the long-term brightness variations, is located around the secondary [167]. U Mon is an RV Tauri star of the RVb-type, which means that a long-term cycle occurs in the mean brightness, which may also be ascribed to obscuration by dust surrounding the companion. Finally, for HD 44179, the central star of the Red Rectangle, which is seen only indirectly through scattered light emerging from an edge-on disk, the observed brightness variations appear to be due to the variation of the scattering angle during the orbit [180].

In all but three cases, the observed mass functions are consistent with a companion of lower initial mass than the present primary, i.e., a companion that still is a relatively inconspicuous main-sequence star. The three exceptions are HD 213985, EN TrA, and U Mon, for which the high mass functions suggest that appreciable mass transfer must have occurred. By all means, for almost all binaries in the sample a specific binary evolution scenario is needed to explain their present location in the HR-diagram, since the orbital periods are too short to accommodate an AGB star; see Chapter 9.

10.5.2 Binary induced depletion of the photosphere

Several objects in Table 10.4, in particular HD 44179, HD 52961, and HR 4049, but also BD+39°4926, present severe photospheric composition anomalies, being characterized by an extreme Fe, Mg, and Si deficiency ([Fe/H] being -4.8 for HR 4049 and HD 52961, and about -3.0 for the two other stars), but fairly normal abundances of carbon, nitrogen, oxygen, and sulphur. The abundance anomalies of HR 4049 may be due to the depletion into dust of elements with high condensation temperatures [16, 173]. This hypothesis received strong confirmation when [165] found for HD 52961 a normal abundance of Zn, after CNO and S the most common element, with the fifth-highest condensation temperature.

Since the photospheres of these stars are much too hot for dust to form in them, [188] suggested that the dust depletion occurred in the long-lived disks occurring in these systems, and thus that the photospheres consist of reaccreted gas. In HD 52961, the "clean" gas layer apparently is able to survive despite the RV Tauri type pulsation the star displays! The reaccretion scenario is confirmed by the fact that all these stars show a C/O ratio of order unity, i.e., essentially the C and O atoms that were locked into CO molecules before the gas has reaccreted.

Similar, but milder, depletions are also observed in HD 46703 [19], which is an irregularly pulsating star. In a series of papers Giridhar, Gonzalez and coworkers showed that the chemical composition of several RV Tauri stars is also determined by depletion rather than dredge-up processes ([47, 48, 49, 52, 53, 170]). Besides underscoring that depletion onto circumstellar dust and subsequent reaccretion of gas may be a widespread phenomenon (though

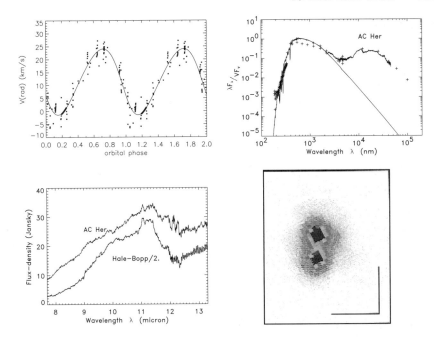

Fig. 10.7. Properties of the RV Tauri star AC Her. Top left panel: radial velocity variations showing the binary nature of AC Her [167]; Top right panel: optical to far-IR spectral energy distribution, showing a large IR excess due to circumstellar dust. Lower right diagram: Spatial distribution of the 18 μm emission, consistent with a limb-brightened edge-on torus [79]. Lower left diagram: comparison of the 10 μm silicate band in AC Her with that of the solar system comet Hale–Bopp; the peak at 11.3 μm of the silicate band in both objects is due to the prominent presence of crystalline silicates

[48] suggest that it is inefficient around stars with a low initial composition), these findings led [167] to the hypothesis that all dusty RV Tauri stars may be binaries; Figure 10.7. Since the RV Tauri stars in Table 10.4 happen to be the four binaries with the longest periods, it may unfortunately still take some time before this hypothesis has been thoroughly tested.

Among the stars in Table 10.4, only two objects do not show any abundance anomalies suggesting that dust depletion has occurred, i.e., the interacting binary SAO 173329 and 89 Her. It is not clear what prevented the process to occur in the latter object, though the clear oxygen richness of its dust, which contrasts with the carbonaceous features observed around HR 4049 and HD 44179, may suggest that the depletion scenario is more efficient when C-rich dust is present. We also note that 89 Her has by far the lowest mass function of the objects in the sample.

10.5.3 Impact of Binarity on (Post)-AGB Evolution

The occurrence of such a large fraction of binaries among proposed PAGB candidates, and certainly the short orbital periods involved, clearly raises the question whether these objects are at all relevant in the framework of PAGB evolution; see Chapter 9. Indeed, no terminal AGB star can fit inside an orbit with a period less than some 600 days. On the other hand, the C-rich dust surrounding several of these objects suggests that the central stars have experienced the thermal pulses necessary to reverse the C/O ratio in their outer layers.

The existence of Ba stars with periods of order a year observationally proves that binary AGB stars can find ways to avoid the rapid Roche-lobe overflow and common-envelope ejection that lead to central stars of PNe with short periods and, ultimately, cataclysmic variables. An important theoretical ingredient may be the tidal enhancement of the mass-loss rate, which may cause the mass ratio to evolve toward a value where stable Roche-lobe overflow is possible [154]. This scenario may account for the shortest-period Ba stars, the orbits of which tend to be circularized as Roche-lobe overflow scenarios predict.

However, in an $(e,\log P)$-diagram, eccentric orbits still occur for PAGB stars below the periods at which Ba stars are circularized, and also for dwarf Ba and subgiant CH systems [77]. It is possible that these systems have acquired a substantial eccentricity due to tidal interaction with their circumsystem disks, as was proposed for pre-main-sequence binaries by [99]. Another possibility, which needs further investigation, is that the short-period binary PAGB candidates were forced to leave the red AGB early because of tidally induced mass loss, but retained a stellar envelope mass too high to evolve rapidly toward the PN phase. In such a picture, objects such as HR 4049 and the central star of the Red Rectangle are not truly PAGB stars, but rather are blue AGB stars. Such a hypothesis implies that their lifetime in their present phase of evolution may be substantial; this would agree with their fairly numerous occurrence, and would also be consistent with the huge extent of the Red Rectangle nebula. Moreover, the oxygen richness observed for the circumbinary disks of both the Red Rectangle [186] and HR 4049 [25] shows that mass transfer occurred when the stellar envelope was still O-rich, while the C-rich dust resides in the outflows generated after the mass-transfer episode; it is not unreasonable that a sequence of thermal pulses occurring in a stellar envelope that has lost much of its mass because of mass transfer may rapidly lead to a turnover of the C/O ratio.

The case of the RV Tauri stars remains most surprising. Their long orbital periods suggest that they are the best candidates to have escaped mass transfer; yet, the mass functions impose secondary masses that are uncomfortably close to the putative initial masses of the primaries [168], and the depletion pattern of the abundances suggests substantial circumstellar processing. Moreover, it is striking that these objects have been classified initially as a group on the basis of the alternating minima in their *pulsational* light curves. How could binary evolution affect the pulsational behavior of these objects? A first explanation certainly is that mass transfer is responsible for evolving the stars towards the zone in the HR-diagram where Cepheid-like pulsations are driven. Moreover, the nonlinear behavior accounting for the RV Tauri phenomenon appears to be linked with a high luminosity-to-mass ratio [23], and mass transfer may indeed contribute to increase this ratio.

10.6 Confrontation of Observations with Theory

10.6.1 Post-AGB Evolutionary Tracks

Theoretical PAGB evolutionary tracks for single stars were initially thought to depend only on the remnant mass when the star left the AGB [126, 127]. A first complication of this simple picture arises when it is taken into account that mass loss also occurs during the PAGB phase. In fact, PAGB mass-loss rates and the mechanisms that cause them are poorly known, and ad hoc assumptions are needed in order to account for the effect of mass loss in stellar-evolution calculations. These assumptions often rely on ad hoc extrapolations of the mass-loss rates that occur at the end of the AGB phase (early PAGB evolution) and during the PN phase (late PAGB evolution).

Initial attempts to parameterize the early PAGB mass loss [147, 198] assumed a continuation of the superwind, approximated by a Reimers-type rate (see Chapter 3), until a certain temperature of the central star was reached. More recent developments invoke a recipe for a more gradual decrease of the mass loss; e.g., it is assumed that the superwind continues after the AGB until the effective temperature of the central star has doubled, and for the subsequent evolution mass-loss rate is adopted that is an extrapolation of the radiation-driven wind of central stars of PNe [171]. For the later phases, [12] follows a similar line of reasoning, but with a different parameterization.

Since the mass contained in the stellar envelope of a PAGB star is small, it is clear that the disappearance time scale of this envelope, and hence the time spent on a PAGB track, depends crucially on the external mass loss, as it does also on mass decrease of the envelope due to continuing nuclear burning. As far as the latter is concerned, a distinction must be made according to whether the central star left the AGB as a H-burning or a He-burning star. Considering the typical time scales computed for the thermal-pulse cycles [70], it is expected that some 15% of PAGB stars are He-burning objects, some 70% are H-burning objects, and for the remaining 15% both nuclear energy sources contribute in comparable fractions [11]; see Chapters 2 and 3 for a discussion of thermal pulses.

It is natural that models with high mass-loss rates until an effective temperature of some 5000 K is reached will lead to a clustering of PAGB stars between late A and late F spectral types, which is consistent with observations. However, it should be stressed again that observational selection effects favor the detection of PAGB stars around precisely these spectral types, so that it still appears premature to determine early PAGB mass loss-rates from number counts.

Better constraints on the validity of PAGB evolutionary tracks are obtained from their implications on later stages of evolution, i.e. the PN, pre-WD, and WD stages of evolution. The shorter transition times found by [12] as compared with [171] for the largest-mass central stars appear to be more consistent with the kinematical ages derived for the youngest PNe [149]. For the lower-mass central stars, the agreement between the different models is better, and the lifetimes found are consistent with observations of PNe. It should be pointed out, however, that the transition time scales for objects with core masses less than $0.65\,M_\odot$ (initial masses less than $2\,M_\odot$) are so long that the CSE of such objects is already dispersed before the central star acquires the temperature needed to ionize a nebula, so that it is likely that most observed transition objects may never become true PNe.

10.6.2 The Occurrence of Thermal Pulses During Post-AGB Evolution

Since the PAGB lifetimes (a few times 10^4 yr) are not so much shorter than the thermal-pulse cycle (10^4–10^5 yr), the circumstance can occur that a star experiences a final thermal pulse during its PAGB phase. It is customary to distinguish between a "late" thermal pulse, which occurs on the horizontal part of the track, and a "very late" pulse, when the object is already evolving on the WD cooling track. Such flashes lead to a rapid inflation of the remaining stellar envelope, and thus to a rapid evolution back toward the red-giant domain, so that the object becomes a so-called born again AGB star [70]. Following the pulse, the remaining hydrogen is mixed into high-temperature layers and burns very fast, after which the star continues as a He-burning AGB star. A "very late pulse" scenario may apply directly to the three rapidly evolving objects V605 Aql, FG Sge, and V4334 Sgr (Sakurai's object), and is invoked to explain the R CrB stars [28], that are C-rich objects which undergo episodic dust formation events, while "very late" or "late" thermal pulses may explain the peculiar abundances of some pre-WDs as well as the WC-type central stars of PNe.

While standard PAGB evolution calculations predict that appreciable amounts of hydrogen should remain present in the stellar envelopes, observations show that many hot PAGB stars are severely depleted in hydrogen, which in some cases accounts for less than 1% in mass. Extreme abundances are observed for the PG 1159 stars, which are hot pre-WDs. Typical surface abundances for such objects are (He, C, O) = (33, 50, 17) by mass [37, 194]. Among the central stars of PNe, some 20% are hydrogen deficient [110], while the composition of the others is solar-like. An important subgroup of the H-rich central stars is that of the Wolf–Rayet (WC) central stars, which typically contain 40% of He by mass, 50% of C, and 10% shared essentially by O and, in some cases, Ne [85, 86].

A variety of scenarios is possible to explain these stars, involving "late" or "very late" pulses, or even a final pulse at the very end of the AGB itself [11]. A critical observation to identify the specific history of an object is the H-abundance [62]: It should be appreciable when the pulse occurred when the star left the AGB; for a "late" pulse it should amount to a few percent, due to convective mixing during the second AGB stage; for a "very late" pulse, which occurs in stars where most hydrogen has been burned, essentially no photospheric hydrogen would survive the "born again" phase. The explanation of the appreciable O-abundances in these stars requires a fair amount of mixing, since O occurs only at the bottom of the H-rich intershell. Iben and McDonald [71] performed calculations involving pulse-induced mixing and showed that such a scenario would indeed lead to modest surface O-enhancements. A quantitative agreement with the observations seems within reach if the effects of convective overshoot are taken into account for the AGB star from which the objects evolve [63]: Indeed, AGB models with overshoot provide intershell abundances that are close to those observed for WC central stars and PG 1159 stars.

The likelihood that a thermal pulse at the very end of the AGB phase or early in the PAGB phase may be relevant for the explanation of WC central stars is strengthened by the ISO observations of the dust in the PNe surrounding WC stars [31, 185]. The $\lambda < 15\,\mu m$ region of this dust shows prominent emission from C-rich dust, but at longer wavelengths the dust emission is dominated by (crystalline) silicates. The O-rich dust then traces the AGB outflow before the last thermal pulse, which inverted the C/O-ratio, occurred, while the C-rich dust has been formed in the outflow consequent to the final pulse. The prominence of PAH features in the C-rich dust then suggest that the final pulse did not remove all hydrogen, as would occur in "very late" pulses and also to some extent in "late" pulses [62].

With hydrogen abundances ranging from a lower limit of 10^{-8} to 10^{-1} times solar [93], the R CrB stars may fit in this scenario in terms of a late or a very late pulse. R CrB owes its name as a variable star partly to the radial pulsations it displays, a property that it shares with similar objects such as RY Sgr and V854 Cen. Its most spectacular variability, which also occurs in nonpulsating C-rich objects, is, however, the occurrence of deep brightness minima, which are due to the episodic formation of and subsequent obscuration by carbon dust. The photospheric C-abundances amount to 25% in some objects, leaving little doubt that R CrB stars are evolved low-mass stars. As far as we are aware, s-process enhancements have only be reported for the R CrB star U Aqr [20]. An explanation of the R CrB stars in terms of the merging of two degenerate objects has been proposed [72, 191]. The late-pulse scenario was first proposed by [134], and was further developed by [148]. While the double-degenerate scenario may apply as an explanation for several specimens in the zoo of faint hot stars, the late-pulse scenario may apply for at least a fraction of the R CrB stars. The most convincing evidence is formed by the striking similarities in the behavior of the objects V605 Aql, FG Sge, and V4334 Sgr, which were caught during their rapid evolution toward R CrB status, and indeed hint at a recent final pulse [29].

V605 Aql brightened by some five magnitudes between 1917 and 1919, and was initially classified as a slow nova [197]. From a spectrum taken in 1921, Lundmark [103] noted the similarity with a cool R CrB star. During subsequent years it showed episodes of fading and brightening typical of an R CrB star [59, 150]. FG Sge started to brighten in 1894, and continued to do so slowly until about 1970, with its spectral type evolving from B4 when it was first recorded in 1955 to G2 in 1975 [196], and since then it is slowly fading, again with a spectral type and obscuration events typical of R CrB stars. The behavior of V4334 Sgr is more similar to that of V605 Aql, since less than 600 days after its outburst in 1996 its effective temperature fell below 6000 K and it became another new R CrB star. All three objects occur in PNe that predate the outbursts.

Several authors [56, 132, 150] found that the inner part of Abell 58, the PN in which V605 Aql resides, is depleted in hydrogen, and a H-deficiency of two orders of magnitude is found for the early spectra of V4334 Sgr [4]. These observations suggest that a very late thermal pulse is responsible for the outburst of these objects, and the evolution of the R CrB phase has been very fast. On the other hand, strong H-lines are present in the nebula surrounding FG Sge [82], and also, the photospheric spectrum of this object does not point to severe H-deficiency [60]; [13] computed several models for PAGB stars undergoing a He-flash during horizontal evolution, and found best agreement for FG Sge with a model with a mass of $0.61\,M_\odot$.

Of these three objects, V4334 Sgr is the one that attained maximum brightness ($m_V = 10.5$) at an epoch when sensitive high-resolution spectrographs were available and enabled detailed chemical analysis during the rapid phase of evolution. Between days 140 and 550 after the outburst, the H-abundance decreased by another order of magnitude, the Li-abundance increased from 3.6 to 4.2, the s-process elements Rb, Sr, Y, and Zr increased in logarithm by 0.9 (lower limit), 0.7, 1.0, and 0.5, respectively, and it was possible to explain these variations as a result of mixing and nucleosynthesis following the final He-shell flash [5]. For most other elements, no important changes were observed. The (initial) metallicity of the object is clearly subsolar, and the high abundances of the CNO elements and of the s-process elements indicate that at outburst the object was a C-rich PAGB star. In FG Sge important s-process enhancements occurred at the end of its rapid phase [95], reaching their maximum and current level in the early 1980s [184]. In the H-burning late-flash scenario proposed by [13], these surface s-process enhancements are due not to nucleosynthesis during the flash, but rather to the mixing with the deeper layers that occurs when the stellar envelope convection attains its maximal depth, when the star reaches its highest luminosity and coolest temperature.

10.7 Conclusions and Outlook

The selection of objects and topics in this chapter on PAGB stars was intended to review the whole subject, but may nevertheless reflect to some extent the personal bias of the authors. This seems to be inevitable for some time to come still, since the objects claimed to belong to the class are most diverse as far as their characteristics are concerned. It is not unlikely that some putative PAGB stars have in fact not terminated standard AGB evolution; this may be true in particular for binaries, the blueward evolution of which may be due to mass transfer rather than to a genuine end of the AGB phase. Renaming, as is often done, the class "proto-planetary nebulae" does not solve all problems in this respect, since observational selection favors the detection of low-mass objects, many of which may evolve too slowly ever to develop a PN.

By all means, the study of transition objects appears to be rich in diagnostics about the late evolution of low- and intermediate-mass stars. Their warm spectra and high visual luminosities render them most suitable for abundance analysis and tests on nucleosynthesis; the study of the morphology of their circumstellar media and of the dynamics of the molecular material contains important clues to an understanding of how outflows develop from nearly symmetrical around AGB stars to the diverse appearance of PNe; the rich solid-state spectra of the thermal dust emission add important knowledge to the rapidly developing field of circumstellar mineralogy. On the other hand, our relatively poor knowledge of accurate masses and luminosities of these objects will continue to hamper their usefulness as probes of theoretical evolutionary tracks.

The prospects of increasing our knowledge of PAGB evolution, and accordingly the risk that this chapter may be outdated in the not-so-distant future, are excellent, however. The current revolution in ground-based observations at high angular resolution will certainly profoundly affect the field, as well at optical–infrared wavelengths as in the submm and mm domain. The rich harvest of the ISO mission allows us to anticipate that first SIRTF, and later ESA's Herschel mission and the mid-infrared spectrometer for JWST will add important knowledge on the PAGB outflows, while the astrometric satellite GAIA should finally settle the ambiguities concerning the fundamental properties of the central stars.

References

1. Aikawa, T. and Sreenivasan, S. In Bradley, P. and Guzik, J., editors, *A Half*

Century of Stellar Pulsation Interpretation: A Tribute to Arthur N. Cox, page 234. ASP: San Francisco, 1998.

2. Alcock, C., Allsman, R. A., Alves, D. R., et al. *ApJ*, 554, 298, 2001.
3. Allamandola, L. J., Tielens, G. G. M., and Barker, J. R. *ApJS*, 71, 733, 1989.
4. Asplund, M., Gustafsson, B., Lambert, D. L., and Kameswara Rao, N. *A&A*, 321, L17, 1997.
5. Asplund, M., Lambert, D. L., Kipper, T., Pollacco, D., and Shetrone, M. D. *A&A*, 343, 507, 1999.
6. Bachiller, R., Forveille, T., Huggins, P. J., and Cox, P. *A&A*, 324, 1123, 1997.
7. Bakker, E. J., van Dishoeck, E. F., Waters, L. B. F. M., and Schoenmaker, T. *A&A*, 323, 469, 1997.
8. Balick, B., Wilson, J., and Hajian, A. R. *AJ*, 121, 354, 2001.
9. Barthès, D., Lèbre, A., Gillet, D., and Mauron, N. *A&A*, 359, 168, 2000.
10. Bidelman, W. P. *ApJ*, 113, 304, 1951.
11. Blöcker, T. *Ap&SS*, 275, 1, 2001.
12. Bloecker, T. *A&A*, 299, 755, 1995.
13. Bloecker, T. and Schoenberner, D. *A&A*, 324, 991, 1997.
14. Bobrowsky, M., Zijlstra, A. A., Grebel, E. K., et al. *ApJ*, 446, L89, 1995.
15. Bogaert, E. *Multispectrale studie van de aard en de veranderlijkheid van optisch-heldere sterren met circumstellaire stofschillen*. PhD thesis, Katholieke Universiteit van Leuven, 1994.
16. Bond, H. E. In Michaud, G. and Tutukov, A., editors, *IAU Symp. 145: Evolution of Stars: the Photospheric Abundance Connection*, page 341. Kluwer Academic Publishers: Dordrecht, 1991.
17. Bond, H. E. In Habing, H. and Lamers, H., editors, *IAU Symp. 180: Planetary Nebulae*, page 460. Kluwer Academic Publishers: Dordrecht, 1997.
18. Bond, H. E., Alves, D. R., and Onken, C. In *American Astronomical Society Meeting*, volume 196, page 4110, 2000.
19. Bond, H. E. and Luck, R. E. *ApJ*, 312, 203, 1987.
20. Bond, H. E., Luck, R. E., and Newman, M. J. *ApJ*, 233, 205, 1979.
21. Bond, H. E. and Philip, A. G. D. *PASP*, 85, 332, 1973.
22. Bowers, P. F. *A&AS*, 31, 127, 1978.
23. Buchler, J. R. and Kovacs, G. *ApJ*, 320, L57, 1987.
24. Bujarrabal, V., Alcolea, J., and Neri, R. *ApJ*, 504, 915, 1998.
25. Cami, J. and Yamamura, I. *A&A*, 367, L1, 2001.
26. Castro-Carrizo, A., Bujarrabal, V. ., Fong, D., et al. In Szczerba, R. and Górny, S. K., editors, *Post-AGB Objects as a Phase of Stellar Evolution*, page 409. Kluwer Academic Publishers: Dordrecht, 2001.
27. Christy, R. F. *ApJ*, 145, 337, 1966.
28. Clayton, G. C. *PASP*, 108, 225, 1996.
29. Clayton, G. C. *Ap&SS*, 275, 143, 2001.
30. Cohen, M., Anderson, C. M., Cowley, A., et al. *ApJ*, 196, 179, 1975.
31. Cohen, M., Barlow, M. J., Sylvester, R. J., et al. *ApJ*, 513, L135, 1999.
32. Cohen, M., Tielens, A. G. G. M., and Bregman, J. D. *ApJ*, 344, L13, 1989.
33. Conlon, E. S. In Sasselov, D., editor, *Luminous High-Latitude Stars*, page 33. ASP: San Francisco, 1993.
34. Cox, P., Lucas, R., Huggins, P. J., et al. *A&A*, 353, L25, 2000.
35. Crabtree, D. R., McLaren, R. A., and Christian, C. A. In Kwok, S. and Pottasch, S., editors, *Late Stages of Stellar Evolution*, page 145. D. Reidel Publishing Company: Dordrecht, 1987.

36. Crampton, D., Cowley, A. P., and Humphreys, R. M. *ApJ*, 198, L135, 1975.
37. Dreizler, S. and Heber, U. *A&A*, 334, 618, 1998.
38. Evans, T. L. *MNRAS*, 217, 493, 1985.
39. Fernie, J. D. *ApJ*, 306, 642, 1986.
40. Fernie, J. D. and Seager, S. *PASP*, 107, 853, 1995.
41. Forveille, T., Morris, M., Omont, A., and Likkel, L. *A&A*, 176, L13, 1987.
42. Garcia-Lario, P., Manchado, A., Parthasarathy, M., and Pottasch, S. R. *A&A*, 285, 179, 1994.
43. Garcia-Lario, P., Manchado, A., Pych, W., and Pottasch, S. R. *A&AS*, 126, 479, 1997.
44. Garcia-Lario, P., Manchado, A., Suso, S. R., Pottasch, S. R., and Olling, R. *A&AS*, 82, 497, 1990.
45. Geballe, T. R., Noll, K. S., Whittet, D. C. B., and Waters, L. B. F. M. *ApJ*, 340, L29, 1989.
46. Gehrz, R. D. *ApJ*, 178, 715, 1972.
47. Giridhar, S., Lambert, D. L., and Gonzalez, G. *ApJ*, 509, 366, 1998.
48. Giridhar, S., Lambert, D. L., and Gonzalez, G. *ApJ*, 531, 521, 2000.
49. Giridhar, S., Rao, N. K., and Lambert, D. L. *ApJ*, 437, 476, 1994.
50. Goebel, J. H. and Moseley, S. H. *ApJ*, 290, L35, 1985.
51. Gonzalez, G. *AJ*, 108, 1312, 1994.
52. Gonzalez, G., Lambert, D. L., and Giridhar, S. *ApJ*, 481, 452, 1997.
53. Gonzalez, G., Lambert, D. L., and Giridhar, S. *ApJ*, 479, 427, 1997.
54. Gonzalez, G. and Wallerstein, G. *AJ*, 108, 1325, 1994.
55. Grasdalen, G. L., Sloan, G. C., and Levan, P. D. *ApJ*, 384, L25, 1992.
56. Guerrero, M. A. and Manchado, A. *ApJ*, 472, 711, 1996.
57. Guillois, O., Ledoux, G., and Reynaud, C. *ApJ*, 521, L133, 1999.
58. Habing, H. J., van der Veen, W., and Geballe, T. In Kwok, S. and Pottasch, S., editors, *Late stages of Stellar Evolution*, page 91. D. Reidel Publishing Company: Dordrecht, 1987.
59. Harrison, T. E. *PASP*, 108, 1112, 1996.
60. Herbig, G. H. and Boyarchuk, A. A. *ApJ*, 153, 397, 1968.
61. Herman, J. and Habing, H. J. *A&AS*, 59, 523, 1985.
62. Herwig, F. *Ap&SS*, 275, 15, 2001.
63. Herwig, F., Blöcker, T., Langer, N., and Driebe, T. *A&A*, 349, L5, 1999.
64. Hony, S., Waters, L. B. F. M., and Tielens, A. G. G. M. *A&A*, 390, 533, 2002.
65. Hrivnak, B. J. *ApJ*, 438, 341, 1995.
66. Hrivnak, B. J. and Kwok, S. *ApJ*, 513, 869, 1999.
67. Hrivnak, B. J., Kwok, S., and Su, K. Y. L. *ApJ*, 524, 849, 1999.
68. Hrivnak, B. J., Kwok, S., and Volk, K. M. *ApJ*, 331, 832, 1988.
69. Hrivnak, B. J., Kwok, S., and Volk, K. M. *ApJ*, 346, 265, 1989.
70. Iben, I. *ApJ*, 277, 333, 1984.
71. Iben, I. and MacDonald, J. In Koester, D. and Werner, K., editors, *White Dwarfs*, page 48. Springer-Verlag: Berlin, Heidelberg, New York, 1995.
72. Iben, I. and Tutukov, A. V. *ApJS*, 54, 335, 1984.
73. Iben, I. J., Tutukov, A. V., and Yungelson, L. R. *ApJ*, 456, 750, 1996.
74. Ignace, R., Cassinelli, J. P., and Bjorkman, J. E. *ApJ*, 459, 671, 1996.
75. Izumiura, H., Waters, L. B. F. M., de Jong, T., et al. *A&A*, 323, 449, 1997.
76. Jones, T. J., Humphreys, R. M., Gehrz, R. D., et al. *ApJ*, 411, 323, 1993.
77. Jorissen, A. In Le Bertre, T., Lebre, A., and Waelkens, C., editors, *IAU Symp. 191: Asymptotic Giant Branch Stars*, page 437. ASP: San Francisco, 1999.

78. Jura, M. *ApJ*, 309, 732, 1986.
79. Jura, M., Chen, C., and Werner, M. W. *ApJ*, 541, 264, 2000.
80. Kastner, J. H., Weintraub, D. A., Gatley, I., and Henn, L. *ApJ*, 546, 279, 2001.
81. Kastner, J. H., Weintraub, D. A., Gatley, I., Merrill, K. M., and Probst, R. G. *ApJ*, 462, 777, 1996.
82. Kipper, T. In Jeffery, C. and Heber, U., editors, *Hydrogen Deficient Stars*, page 329. ASP: San Franscisco, 1996.
83. Klochkova, V. G., Szczerba, R., Panchuk, V. E., and Volk, K. *A&A*, 345, 905, 1999.
84. Kodaira, K., Greenstein, J. L., and Oke, J. B. *ApJ*, 159, 485, 1970.
85. Koesterke, L. *Ap&SS*, 275, 41, 2001.
86. Koesterke, L. and Hamann, W.-R. In Habing, H. J. and Lamers, H. J. G. L. M., editors, *IAU Symp. 180: Planetary Nebulae*, volume 180, page 114. Kluwer Academic Publishers: Dordrecht, 1997.
87. Kwok, S. *ARA&A*, 31, 63, 1993.
88. Kwok, S., Hrivnak, B. J., and Su, K. Y. L. *ApJ*, 544, L149, 2000.
89. Kwok, S., Purton, C. R., and Fitzgerald, P. M. *ApJ*, 219, L125, 1978.
90. Kwok, S., Su, K. Y. L., and Hrivnak, B. J. *ApJ*, 501, L117, 1998.
91. Kwok, S., Volk, K. M., and Hrivnak, B. J. *ApJ*, 345, L51, 1989.
92. Lambert, D. L., Luck, R. E., and Bond, H. E. *PASP*, 95, 413, 1983.
93. Lambert, D. L. and Rao, N. K. *Journal of Astrophysics and Astronomy*, 15, 47, 1994.
94. Lamers, H. J. G. L. M., Waters, L. B. F. M., Garmany, C. D., Perez, M. R., and Waelkens, C. *A&A*, 154, L20, 1986.
95. Langer, G. E., Kraft, R. P., and Anderson, K. S. *ApJ*, 189, 509, 1974.
96. Latter, W. B., Hora, J. L., Kelly, D. M., Deutsch, L. K., and Maloney, P. R. *AJ*, 106, 260, 1993.
97. Lequeux, J. and Jourdain de Muizon, M. *A&A*, 240, L19, 1990.
98. Lloyd Evans, T. In Le Bertre, T., Lebre, A., and Waelkens, C., editors, *IAU Symp. 191: Asymptotic Giant Branch Stars*, volume 191, page 453. ASP: San Francisco, 1999.
99. Lubow, S. H. and Artymowicz, P. In Duquennoy, A. and Mayor, M., editors, *Binaries as Tracers of Stellar Formation*, page 145. Cambridge University Press: Cambridge, 1992.
100. Luck, R. E. In Sasselov, D., editor, *Luminous High-Latitude Stars*, page 87. ASP: San Francisco, 1993.
101. Luck, R. E. and Bond, H. E. *ApJ*, 279, 729, 1984.
102. Luck, R. E., Bond, H. E., and Lambert, D. L. *ApJ*, 357, 188, 1990.
103. Lundmark, K. *PASP*, 33, 314, 1921.
104. Manchado, A., Garcia-Lario, P., Esteban, C., Mampaso, A., and Pottasch, S. R. *A&A*, 214, 139, 1989.
105. Mastrodemos, N. and Morris, M. *ApJ*, 523, 357, 1999.
106. Mauron, N. and Huggins, P. J. *A&A*, 349, 203, 1999.
107. McCausland, R. J. H., Conlon, E. S., Dufton, P. L., and Keenan, F. P. *ApJ*, 394, 298, 1992.
108. Meixner, M., Skinner, C. J., Graham, J. R., et al. *ApJ*, 482, 897, 1997.
109. Meixner, M., Ueta, T., Dayal, A., et al. *ApJS*, 122, 221, 1999.

110. Mendez, R. H. In Michaud, G. and Tutukov, A., editors, *IAU Symp. 145: Evolution of Stars: the Photospheric Abundance Connection*, volume 145, page 375. Kluwer Academic Publishers: Dordrecht, 1991.
111. Molster, F. J., van den Ancker, M. E., Tielens, A. G. G. M., et al. *A&A*, 315, L373, 1996.
112. Molster, F. J., Waters, L. B. F. M., Trams, N. R., et al. *A&A*, 350, 163, 1999.
113. Moseley, H. *ApJ*, 238, 892, 1980.
114. Moskalik, P. and Buchler, J. R. *ApJ*, 366, 300, 1991.
115. Nakashima, J. and Deguchi, S. *PASJ*, 52, L43, 2000.
116. Napiwotzki, R., Heber, U., and Koeppen, J. *A&A*, 292, 239, 1994.
117. Ney, E. P., Merrill, K. M., Becklin, E. E., Neugebauer, G., and Wynn-Williams, C. G. *ApJ*, 198, L129, 1975.
118. Nuth, J. A., Moseley, S. H., Silverberg, R. F., Goebel, J. H., and Moore, W. J. *ApJ*, 290, L41, 1985.
119. Nyman, L., Hall, P. J., and Olofsson, H. *A&AS*, 127, 185, 1998.
120. Olofsson, H., Bergman, P., Lucas, R., et al. *A&A*, 353, 583, 2000.
121. Olofsson, H., Carlstrom, U., Eriksson, K., and Gustafsson, B. *A&A*, 253, L17, 1992.
122. Omont, A. In Szczerba, R. and Górny, S. K., editors, *Post-AGB Objects as a Phase of Stellar Evolution*, page 357. Kluwer Academic Publishers: Dordrecht, 2001.
123. Oudmaijer, R. D. *A&A*, 306, 823, 1996.
124. Oudmaijer, R. D., van der Veen, W. E. C. J., Waters, L. B. F. M., et al. *A&AS*, 96, 625, 1992.
125. Oudmaijer, R. D., Waters, L. B. F. M., van der Veen, W. E. C. J., and Geballe, T. R. *A&A*, 299, 69, 1995.
126. Paczynski, B. *Acta Astronomica*, 20, 47, 1970.
127. Paczynski, B. *Acta Astronomica*, 21, 417, 1971.
128. Parthasarathy, M., García-Lario, P., Sivarani, T., Manchado, A., and Sanz Fernández de Córdoba, L. *A&A*, 357, 241, 2000.
129. Parthasarathy, M., Garcia-Lario, P., de Martino, D., et al. *A&A*, 300, L25, 1995.
130. Parthasarathy, M., Gauba, G., Fujii, T., and Nakada, Y. In Szczerba, R. and Górny, S., editors, *Post-AGB Objects as a Phase of Stellar Evolution*, page 29. Kluwer Academic Publishers: Dordrecht, 2001.
131. Pollard, K. H. and Cottrell, P. L. In Stobie, R. and Whitelock, P., editors, *IAU Colloq. 155: Astrophysical Applications of Stellar Pulsation*, page 409. ASP: San Francisco, 1995.
132. Pottasch, S. R., Mampaso, A., Manchado, A., and Menzies, J. In *IAU Colloq. 87: Hydrogen Deficient Stars and Related Objects*, page 359. D. Reidel Publishing Co.: Dordrecht, 1986.
133. Reimers, C., Dorfi, E. A., and Höfner, S. *A&A*, 354, 573, 2000.
134. Renzini, A. In Westerlund, B., editor, *Stars and Star Systems*, page 155. D. Reidel Publishing Co.: Dordrecht, 1979.
135. Reyniers, M. and Van Winckel, H. *A&A*, 365, 465, 2001.
136. Roddier, F., Roddier, C., Graves, J. E., and Northcott, M. J. *ApJ*, 443, 249, 1995.
137. Rouan, D., Lacombe, F., Omont, A., and Forveille, T. *A&A*, 189, L3, 1988.
138. Ryans, R. S. I., Dufton, P. L., Mooney, C. J., et al. *A&A*, 401, 1119, 2003.

139. Sahai, R. In Szczerba, R. and Górny, S. K., editors, *Post-AGB Objects as a Phase of Stellar Evolution*, page 53. Kluwer Academic Publishers: Dordrecht, 2001.
140. Sahai, R., Bujarrabal, V., Castro-Carrizo, A., and Zijlstra, A. *A&A*, 360, L9, 2000.
141. Sahai, R., Hines, D. C., Kastner, J. H., et al. *ApJ*, 492, L163, 1998.
142. Sahai, R., Nyman, L., and Wootten, A. *ApJ*, 543, 880, 2000.
143. Sahai, R. and Trauger, J. T. *AJ*, 116, 1357, 1998.
144. Sahai, R., Trauger, J. T., Watson, A. M., et al. *ApJ*, 493, 301, 1998.
145. Sahai, R., Zijlstra, A., Bujarrabal, V., and Te Lintel Hekkert, P. *AJ*, 117, 1408, 1999.
146. Schmidt, G. D. and Witt, A. N. *ApJ*, 383, 698, 1991.
147. Schoenberner, D. *ApJ*, 272, 708, 1983.
148. Schoenberner, D. In *IAU Colloq. 87: Hydrogen Deficient Stars and Related Objects*, page 471. D. Reidel Publishing Co.: Dordrecht, 1986.
149. Schoenberner, D. In Habing, H. and Lamers, H., editors, *IAU Symp. 180: Planetary Nebulae*, volume 180, page 379. Kluwer Academic Publishers: Dordrecht, 1997.
150. Seitter, W. C. *Mitteilungen der Astronomischen Gesellschaft Hamburg*, 63, 181, 1985.
151. Simis, Y. J. W., Icke, V., and Dominik, C. *A&A*, 371, 205, 2001.
152. Soker, N. *ApJ*, 540, 436, 2000.
153. Szczerba, R., Górny, S. K., and Zalfresso–Jundziłło, M. In Szczerba, R. and Górny, S. K., editors, *Post-AGB Objects as a Phase of Stellar Evolution*, page 13. Kluwer Academic Publishers: Dordrecht, 2001.
154. Tout, C. A. and Eggleton, P. P. *MNRAS*, 231, 823, 1988.
155. Trams, N. R., Waters, L. B. F. M., Lamers, H. J. G. L. M., et al. *A&AS*, 87, 361, 1991.
156. Trams, N. R., Waters, L. B. F. M., Waelkens, C., Lamers, H. J. G. L. M., and van der Veen, W. E. C. J. *A&A*, 218, L1, 1989.
157. Ueta, T., Meixner, M., and Bobrowsky, M. *ApJ*, 528, 861, 2000.
158. van de Steene, G. C. and Pottasch, S. R. *A&A*, 299, 238, 1995.
159. Van de Steene, G. C., van Hoof, P. A. M., and Wood, P. R. *A&A*, 362, 984, 2000.
160. van de Steene, G. C. M. and Pottasch, S. R. *A&A*, 274, 895, 1993.
161. van der Veen, W. E. C. J. and Habing, H. J. *A&A*, 194, 125, 1988.
162. van der Veen, W. E. C. J., Habing, H. J., and Geballe, T. R. *A&A*, 226, 108, 1989.
163. van Loon, J. T., Groenewegen, M. A. T., de Koter, A., et al. *A&A*, 351, 559, 1999.
164. Van Winckel, H. *A&A*, 319, 561, 1997.
165. Van Winckel, H., Mathis, J. S., and Waelkens, C. *Nature*, 356, 500, 1992.
166. Van Winckel, H. and Reyniers, M. *A&A*, 354, 135, 2000.
167. Van Winckel, H., Waelkens, C., Fernie, J. D., and Waters, L. B. F. M. *A&A*, 343, 202, 1999.
168. Van Winckel, H., Waelkens, C., Fernie, J. D., and Waters, L. B. F. M. *A&A*, 343, 202, 1999.
169. Van Winckel, H., Waelkens, C., and Waters, L. B. F. M. *A&A*, 293, L25, 1995.
170. Van Winckel, H., Waelkens, C., Waters, L. B. F. M., et al. *A&A*, 336, L17, 1998.

171. Vassiliadis, E. and Wood, P. R. *ApJS*, 92, 125, 1994.
172. Venn, K. A. *ApJ*, 449, 839, 1995.
173. Venn, K. A. and Lambert, D. L. *ApJ*, 363, 234, 1990.
174. Volk, K., Kwok, S., and Hrivnak, B. J. *ApJ*, 516, L99, 1999.
175. von Helden, G., Tielens, A. G. G. M., Hrivnak, B. J., et al. *Science*, 288, 313, 2000.
176. Waelkens, C. and Burnet, M. *Informational Bulletin on Variable Stars*, 2808, 1, 1985.
177. Waelkens, C., Lamers, H. J. G. L. M., Waters, L. B. F. M., et al. *A&A*, 242, 433, 1991.
178. Waelkens, C. and Rufener, F. *A&A*, 201, L5, 1988.
179. Waelkens, C., Van Winckel, H., Bogaert, E., and Trams, N. R. *A&A*, 251, 495, 1991.
180. Waelkens, C., Van Winckel, H., Waters, L. B. F. M., and Bakker, E. J. *A&A*, 314, L17, 1996.
181. Waelkens, C., Waters, L. B. F. M., Cassatella, A., Le Bertre, T., and Lamers, H. J. G. L. M. *A&A*, 181, L5, 1987.
182. Waelkens, C., Waters, L. B. F. M., Van Winckel, H., and Daems, K. *Ap&SS*, 224, 357, 1995.
183. Wahlgren, G. M. In Sasselov, D., editor, *Luminous High-Latitude Stars*, page 270. ASP: San Francisco, 1993.
184. Wallerstein, G. *ApJS*, 74, 755, 1990.
185. Waters, L. B. F. M., Beintema, D. A., Zijlstra, A. A., et al. *A&A*, 331, L61, 1998.
186. Waters, L. B. F. M., Cami, J., de Jong, T., et al. *Nature*, 391, 868, 1998.
187. Waters, L. B. F. M., Molster, F. J., de Jong, T., et al. *A&A*, 315, L361, 1996.
188. Waters, L. B. F. M., Trams, N. R., and Waelkens, C. *A&A*, 262, L37, 1992.
189. Waters, L. B. F. M. and Waelkens, C. *ARA&A*, 36, 233, 1998.
190. Waters, L. B. F. M., Waelkens, C., Mayor, M., and Trams, N. R. *A&A*, 269, 242, 1993.
191. Webbink, R. F. *ApJ*, 277, 355, 1984.
192. Weintraub, D. A., Huard, T., Kastner, J. H., and Gatley, I. *ApJ*, 509, 728, 1998.
193. Weintraub, D. A., Kastner, J. H., and Gatley, I. In Szczerba, R. and Górny, S. K., editors, *Post-AGB Objects as a Phase of Stellar Evolution*, page 377. Kluwer Academic Publishers: Dordrecht, 2001.
194. Werner, K. *Ap&SS*, 275, 27, 2001.
195. Whitelock, P. and Marang, F. *MNRAS*, 323, L13, 2001.
196. Whitney, C. A. *ApJ*, 220, 245, 1978.
197. Wolf, M. *Astronomische Nachrichten*, 212, 167, 1920.
198. Wood, P. R. and Faulkner, D. J. *ApJ*, 307, 659, 1986.
199. Zijlstra, A. A., Chapman, J. M., te Lintel Hekkert, P., et al. *MNRAS*, 322, 280, 2001.
200. Zijlstra, A. A., Gaylard, M. J., Te Lintel Hekkert, P., et al. *A&A*, 243, L9, 1991.
201. Zuckerman, B. and Gatley, I. *ApJ*, 324, 501, 1988.

Index

absorption band
  TiO, 24
  ZrO, 24
accretion, 475, 503
  Bondi–Hoyle, 477
  disk, 478
  wind, 478
acetylene, 261
AFGL catalog
  significance, 6
AGB star
  final luminosity, 18, 26
  initial mass, 26
  initial–final mass relation, 132
  lifetime, 18, 34
  overall structure, 12
  spectra, 15, 23, 152
AGB-manqué star, 520
ammonia, 265
angular diameter, 97, 156, 161, 163, 232
  first measurements, 7
atmosphere
  continuous absorption and scattering, 184
  convection, 204
  definition of, 149
  heating and cooling, 192
  levitation, 207
  pulsating models, 214
  radiation field, 182
  radiative equilibrium, 191
  shock waves, 207
  static model, 189
  transition zone, 220
  turbulent pressure, 202
  velocity field, 157, 165, 229

Baade's windows, 426

barium star, 489, 490, 494, 496, 500
binaries
  spectroscopic, 469
  symbiotic, 472, 491
  visual, 471
binary stars
  close, 414
  common envelope, 414, 496, 509

C/O ratio, 175, 359, 371
Cameron–Fowler mechanism, 68
carbon star, 23, 175
  dwarf, 490, 494
  first detection, 2
  J-type, 480, 488
  luminosity function, 136
  R-type, 488
CH
  giant, 494, 496
  subgiant, 490, 494, 500
chemistry
  circumstellar, C-rich, 270, 535
  circumstellar, O-rich, 279, 534
  equilibrium, 173, 251, 303
  ion-neutral reactions, 286
  negative ions, 273
  neutral–neutral reactions, 286
  shock, 255
chromosphere, 158, 269
circumstellar atoms/molecules
  abundance estimates, 345, 357
  excitation, 332
  species observed, 348, 349
circumstellar envelope, 220
  atom/molecule composition, 349
  brightness distribution
    dust, 362
    lines, 335, 343

observed, 356, 369, 371
chemical composition, 247
column density
  beam-averaged, 334
  incremental, 335
  radial, 334
  source-averaged, 334
dust composition, 367
extinction, 248, 329
geometry, 371
kinematics, 377
small-scale structure, 381
wind acceleration, 380
convection, 204
  3D models, 206
  criterion, 28
  semi-, 28
convective stellar envelope, 65
core
  carbon–oxygen, 31
  convective, 28
  helium flash, 28, 108
  hydrogen burning, 27
  mass–interpulse relation, 116
  mass–luminosity relation, 37, 110
corundum, 249, 261, 263
cyanopolyynes
  abundance estimates, 357
  chemistry, 272

diopside, 263
disk
  accretion, 478
  circumbinary, 478, 488, 496, 538
dust
  composition, 367
  formation, 177, 212, 480
  grain size, 261, 370
  growth rate of grains, 179, 180
  nucleation, 304
  opacity, 188
  pre-solar grains, 249
  properties, 369
  radiation, 362
  spectral signatures, 365, 367
dust-driven wind, 293
  back warming, 315
  drag force, 299, 301
  drift instability, 312

drift velocity, 299, 328
dynamical time scale, 301
effect of binarity, 309
effect of magnetic field, 309
effect of stellar rotation, 309
instability, 298
light pressure on dust, 297, 327
  first calculations, 9
momentum coupling, 299
position coupling, 299
single-fluid versus two-fluid description, 299
two-fluid approach, 312
velocity law, 327
dust-enshrouded stars, 438
dust-induced shock wave, 310
dust-to-gas mass ratio, 370

early-AGB, 28, 109
eccentricity, 496, 511
Eddington approximation, 308
effective temperature, 116, 139, 160, 161
elemental abundances
  determination, 230, 235
  in AGB stars, 234
  in PNe, 140
energy balance
  atmospheric gas, 192
  circumstellar dust, 361
  circumstellar gas, 331
Euler equation, 294

finite-difference scheme, 301
first dredge-up, 28, 43, 109
fractional abundance
  circumstellar estimates, 357
  definition, 329
  theory, 345

gas-dust interaction, 264, 312
gas/dust return, 399
Gibbs free energy, 308
graphite, 249

H_2 density distribution, 266, 329
hibonite, 249
high-galactic-latitude B-type stars, 524
hot bottom burning, 65, 113

HR diagram and AGB stars, 23
hydrocarbons
    abundance estimates, 357
    chemistry, 257, 270

ice mantle, 265
infrared flux method, 233
instability
    dynamic, 291
    Kelvin–Helmholtz, 291
    radiative, 291
    Rayleigh–Taylor, 291
    Richtmyer–Meshkov, 291
intershell convective zone (ISCZ), 46
ionization
    by cosmic rays, 268
    by radioactivity, 268
IRAS
    significance, 6
IRC catalog
    significance, 6
irregular variable, 80
isotopic ratios, 249, 359

light curves, 80, 474
line absorption, 185
line profiles
    absorption, 344
    emission, 335, 343
    masers, 342
    observed, 339, 344
    scattering, 343
lithium giants, 65, 438
LMC, 431
Local Group, 441
local thermodynamic equilibrium, 150, 247
luminosity function
    AGB stars, 433
    carbon stars, 436

M star, 23, 175
M31, 439
    companions, 440, 446
magnetite, 261
maser, 342, 351
maser stars, 413, 481, 488
    first detection, 7
mass continuity equation, 294, 328

mass functions, 500
mass loss
    asymmetric, 282
    binary systems, 481
    early papers, 8
    significance for stellar evolution, 37
mass-loss rate
    dependences, 391, 397
    distributions, 392
    Reimers law, 118
mass-loss-rate estimators
    CO radio lines, 385
    dust features, 390
    IR colors, 389
    monochromatic fluxes, 388
    OH masers, 386
    spectral energy distributions, 390
meteorite
    aluminium, 269
    Bishnapur, 249
    isotopic ratios, 249
    Murchison, 250
    Orgueil, 249
methane, 265
Mie theory, 298
Milky Way Galaxy
    bar, 416, 425
    bulge, 416, 425
    companions, 442
    disk, 416
    Galactic center, 416, 420
    globular clusters, 430
    halo, 416, 430
    outer disk, 429
Mira Ceti, 472, 486
Mira variable, 80
    distribution in Milky Way Galaxy, 418
molecular density distribution, 329

nucleation
    classical theory, 307
    iron grain, 263
    silicate, 263
nucleosynthesis, 43
    r−process, 49
    s−process, 24, 46, 49
    s−process elements, 484, 491, 493
    s−process, main component, 50

558   Index

$s-$process, strong component, 50
$s-$process, weak component, 50
during hot bottom burning, 121
early studies, 5
hs–elements, 54
ls–elements, 54
production of ^{12}C, 46
production of ^{13}C, 48
production of Al, 63
production of CNO, 125
production of F, 56
production of Li, 68
production of Mg, 62
production of Na, 59
production of Ne, 59
production of Pb, 50
sources of neutrons, 46
stellar yields, 80, 125
surface abundances, 118

OH/IR star, 413
olivine, 263
opacity distribution function, 197
opacity sampling, 198
overluminosity, 113

Paczynski relation, 36
period–luminosity relation, 85
perovskite, 263
photochemical shells, 265
photodissociation, 266
　CO radius, 267, 330
　OH radius, 386
planetary nebulae, 489
　astrophysical explanation, 11
　role in AGB evolution, 41
polycyclic aromatic hydrocarbons, 250, 531, 535
post AGB star, 479, 493, 496
　$s-$process elements, 526
　binaries, 539
　carbon deficiency, 528
　central star, 522
　circumstellar envelope
　　concentric rings, 537
　　dust disk, 537
　　dust emission, 530
　　imaging, 535
　　light scattering, 537
　molecular envelope, 534
　depletion of abundances, 540
　hs/ls ratio, 527
　impact of stellar companion, 542
　luminosities, 525
　mass loss, 529
　observational definition, 520
　OH maser, 521
　photospheric abundances, 526
　pulsations, 529
　significance of thermal pulses, 544
　spectral energy distribution, 520
　subsolar Fe abundance, 526
　theoretical evolutionary tracks, 41, 543
post red supergiant, 522
Prasad–Tarafdar mechanism, 268
pre-AGB phase, 27
　evolution, 108
pulsating stars, 142
pulsation
　effect on atmosphere, 206
　first overtone, 466
　fundamental mode, 92, 466
　large-amplitude models, 96
　linear models, 90
　nonlinear models, 98
　overtones, 92

quasi-stationary layer, 260

R CrB star, 521
radial-velocity variations, 469
radius
　radius–period relationship, 465
　Roche, 462, 504
red giant branch star
　first models, 4
　tip of the red giant branch, 412, 433
Roche
　lobe overflow, 496
　radius, 462, 504
rotation-temperature diagram, 347
RV Tau star, 521

S star, 484, 491, 494, 496, 500
sdO star, 520
second dredge-up, 31, 43, 109
self-shielding, 266

semiregular variable, 80
shell hydrogen burning, 27
shock wave, 157, 247, 252
silicates, 367
   amorphous, 262
   crystalline, 262, 367
silicon chemistry, 276
SMC, 431
solar neighborhood, 418
spectral energy distribution, 362
spectroscopy, 227
   far-IR, 351
   mid-IR, 351
   near-IR, 152
   radio, 348
   visual, 152, 353
spinel, 250, 263
state of matter, 165
statistical equilibrium, 171
stellar models
   history, 3
stellar models, 27, 28, 31
stellar population, 411
   single, 411
   synthesis, 129
stellar wind, 221
   acoustic-wave-driven, 224, 292
   critical point, 294
   dust-driven, 222, 224
   inner region, 226
   light pressure on grains, 297
   MHD-wave-driven, 223, 292
   momentum equations, 297
   sonic point, 295
   subsonic region, 295
sulphur chemistry, 278
supersaturation, 261
superwind, 39, 118
   first use of the word, 10

surface gravity, 233
surface tension, 308
surveys
   infrared, 368
   radio lines, 355

technetium, Tc, 25, 484
temperature distribution
   circumstellar dust, 362
   circumstellar gas, 331
terminal velocity
   distributions, 377
   theory, 327
thermal pulse, 32, 109
   cycle, 36, 109
   discovery, 5
thermal runaway, 33
thermally-pulsing AGB, 32, 109
third dredge-up, 39, 119
   efficiency, 41, 119
21μm-sources, 526

variability of AGB stars, 16
variables
   large-amplitude, 26, 413
   long-period, 26, 413, 433
      first detection, 2
   mass of progenitor, 84
   short-period, 420
   short-period blue, 420
   short-period red, 420

white dwarf
   mass distribution, 395
   role in AGB evolution, 41
wind equation, 294

X-rays, 474

Zeldovich factor, 308

ASTRONOMY AND ASTROPHYSICS LIBRARY

Continued from page ii

Modern Astrometry
By J. Kovalevsky

Astrophysical Formulae 3rd Edition (2 volumes)
Volume 1: Radiation, Gas Processes, and
  High Energy Astrophysics
Volume 2: Space, Time, Matter, and Cosmology
By K.R. Lang

Observational Astrophysics 2nd Edition
By P. Lena, F. Lebrun, and F. Mignard

Astrophysics of Neutron Stars
Editors: V.M. Lipunov and G. Börner

Galaxy Formation
By M.S. Longair

Supernovae
Editor: A.G. Petschek

General Relativity, Astrophysics, and Cosmology
By A.K. Raychaudhuri, S. Banerji, and A. Banerjee

Tools of Radio Astronomy 3rd Edition
By K. Rohlfs and T.L. Wilson

Atoms in Strong Magnetic Fields
Quantum Mechanical Treatment and Applications in Astrophysics and Quantum Chaos
By H. Ruder, G. Wunner, H. Herold, and F. Geyer

The Stars
By E.L. Schatzman and F. Praderie

Physics of the Galaxy and Interstellar Matter
By H. Scheffler and H. Elsässer

Gravitational Lenses
By P. Schneider, J. Ehlers, and E.E. Falco

Relativity in Astrometry, Celestial Mechanics, and Geodesy
By M.H. Soffel

The Sun
An Introduction
By M. Stix

Galactic and Extragalactic Radio Astronomy 2nd Edition
Editors: G.L. Verschuur and K.I. Kellermann

Reflecting Telescope Optics (2 volumes)
Volume I: Basic Design Theory and Its Historical Development
Volume II: Manufacture, Testing, Alignment, Modern Techniques
By R.N. Wilson

Tools of Radio Astronomy
Problems and Solutions
By T.L. Wilson and S. Hüttemeister

1-MONTH